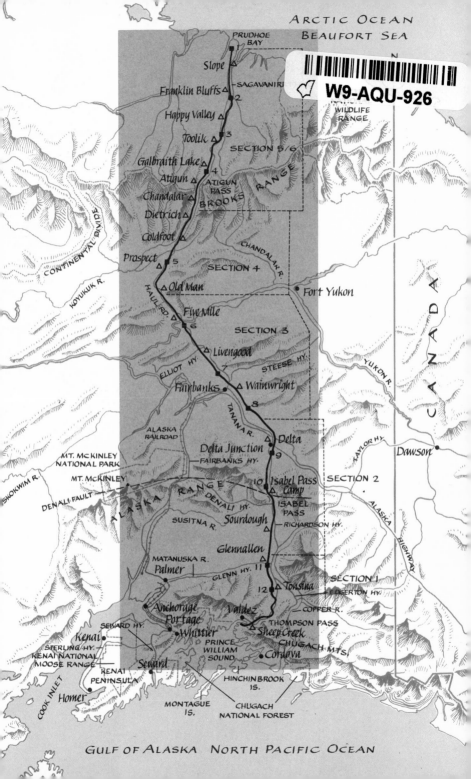

JOURNEYS DOWN THE LINE

JOURNEYS DOWN THE LINE

Building the Trans-Alaska Pipeline

ROBERT DOUGLAS MEAD

Doubleday & Company, Inc., Garden City, New York 1978

ISBN: 0-385-11578-4
Library of Congress Catalog Card Number 77-92226

To
My mother, Madge Parker Mead,
with a lifetime of gratitude

CONTENTS

Between the idea
And the reality
Between the motion
And the act
Falls the Shadow.

      —T. S. Eliot,
       The Hollow Men

JOURNEYS DOWN THE LINE

Chapter 1

THE LAND OF BURNING SNOW

The world, though made, is yet being made; this is still the
morning of creation.—John Muir

You breathe and move carefully, forewarned. The weather has
turned cold again, the temperature down from a normal ten or
twenty below to minus forty-five, which, with the light, dry, metallic wind that buffs the shallow, island-guarded bays and inlets
of this Arctic coast, translates to a chill factor of minus seventy
or seventy-five; bared to this cold and this light and constant
wind, your flesh can freeze white and hard in half a minute, the
freeze probing deeper till the flowing blood, which is life,
thickens and stops, the veins break, the flesh begins to die. But
that chill factor does not mean much: layered in wool or down,
face covered, or in the caribou and seal skins which to the Eskimos are the Old Way and sure—one layer of fur turned in to
warm and caress the skin, the other out, resisting the cold—you
embrace and hold in your body's pulsing heat and are secure.
Only, your hands and especially your feet are vulnerable, sweating, the moisture transmitting and dispersing the precious
warmth to the infinite cold universe of air. Half an hour or an
hour: hands and feet will warn you with aching before they
freeze and cease to feel, and if you are wise and have far to go to
the shelter of other men, you will stop now, dig a shallow
trench, roof it with slabs of the hard, wind-polished snow which
with a good knife cuts like plywood under a band saw, and

crawl in to wait for the cold to lift; and hope that it does not last long enough to kill. If you have heavy work to do or try to run—but you cannot run from this cold, it is the clear mind as much as your body's mindless warmth that keeps you alive—you feel the heavy ache of it deep in the cavity of your chest. Lungs also can freeze. You move with care.

The sun here observes the ancient druidical solstice of Christ's birth: on December 24th it goes below the rim of earth and does not show itself again till the 18th of January. Yet in that three weeks of the sun's absence the darkness is qualified (and not in itself depressing—that is another much-repeated lie about the Arctic): a twilit day swells toward noon in the familiar rhythm of the temperate zone, surrounded by the colors of dawn and dusk, while the sun struggles toward the horizon but does not cross it, as in the hard birthing of a first child. And when it does reappear—remote, small, refracted through the long angle of earth's air, with only the burnished-copper promise of warmth, not yet the reality—you rejoice like a druid. The sun lives: earth lives; man lives.

It is a month exactly since the sun's return. Now, past noon, the sun, due south, nears the end of its narrow arc and slips toward the rim, the four-hour day shading from sunrise into sunset, bounded at either end by the long twilight. The sky, cloudless, deepens to a pale blue; here and there, bronze light washes the edge of a wind furrow in the hard, gray-white snow and seems to float on the surface, transparent—you see the snow color through it, as through a lens. The wind lifts the dust of snow, whirls, drops and scatters it in its ceaseless and never-finished labor of scouring all things flat, detesting irregularity. In its brief spring and summer, maps tell you, this land is dotted with innumerable ponds and small lakes, like puddles of standing water in the spring-thawed furrows of a cornfield, and that is what they are, but hugely enlarged—the sun's diffracted light penetrates not much beyond a foot, and below that zone of thaw, that delicate, living skin, the earth is frozen to a depth of a thousand feet, two thousand feet. Permafrost: gravel, sand and boulders, alluvial muck, offscourings of vanished glaciers, remote mountain ranges that have been ground to dust by two hundred million years of wind and water. There is no sign now of these many ponds. The land stretches flat—none more so on earth—as far as eye can carry, an immense coastal plain fifty or a hundred miles wide, five hundred long, and it is as if, holding the eyes steady, you can see it all in one long glance, but there is no way of judging distance, scale. The landscape is almost featureless: or, precisely, every minute corrugation in the glaze of snow is magnified, and you with difficulty distinguish small and near from large and distant. Human, you crawl across this level landscape like a slug across the sea floor, and that, you realize, is what it has been, what it is: the sea above you is a sea of air. A few miles north, where the Arctic Slope slides into the Arctic Sea, only a shallow lip of ice, like the rim of a saucer, marks the transition, and you could cross it, hardly noticing, and leave the land behind, wander always northward across the winter ice until you reach the Pole, where all directions are south and there is no farther north to go. The sea now out there is frozen solid to its shallow depth, and its light tides and gently revolving currents do not disturb its shoreline: sea and land are united and at peace.

It is a landscape, then, such as men imagined on the moon, before they went there, walked it, saw it for themselves. But: not dead; not even wholly asleep. A little way off now a form materializes and moves across the land like a shadow, gray-white fur against the gray-white snow: an Arctic fox stepping his silent deliberate dogtrot over wind ridge and wind crease, incurious in this season and indifferent to your human presence, hunting; intent on the voles, lemmings, ground squirrels, marmots that are its food, burrowed under the frozen snow. Far out on the deep pack ice there will be white polar bears haunting the open leads and blowholes for seals, which in the spring will return to this coast to den their young, and with them, homing from their winter mountains to the south, there will come back to their calving grounds parts of the two great caribou herds centered to the east and west along the Arctic Slope—and the undersized gray-black wolves that follow the herds, preying on the infirm and young, the stragglers. And—a scattering of the great hump-shouldered golden grizzlies, myriads of shore-birds drawn across the continental flyways to breed here and nest and hatch their young; in the few weeks of summer when the sun opens the river mouths and breaks the shore ice and the Arctic winds, altering, carry it out and north, the rivers will swarm with returning fish and, offshore, the great blue-black whales and small white belugas, feeding. Indeed, for as long as these life forms have existed, this Arctic plain has been a focus of their comings and goings, and men with them, likewise migrant, though rarely to settle long and in numbers, coming in season to hunt, returning to more certain and abundant food sources east and west along the coast; passing through, between these settlements, for trade. Their beginnings, known or inferred, are reckoned in human thousands of years rather than the land's geological millions: the generations-long passage that drew the solitary hunters, their women and young, from Asia across a Bering Strait shallowed by an ice age and east along this northern coast until they reached Greenland, little more than a millennium ago, and they could go no farther east. These were the men archaeologists, rooting the floors of their huts and the stone tent circles of their camps, gathering the chippings from their tools and weapons, called proto-Eskimo, ancestors of the present northern men, who to themselves are Inuit —the people, mankind—or, more emphatically, Inupiat, the *real* people, distinct from all others who are not.

In the south, the sun has dropped below the edge but still shines pink on a narrow band of cloud closing down on the horizon like the lid of a box. To the north now the sky is changing from deep blue to slate, and with it the snowscape changes its color, only a little less smoothly uniform. Then against that darkness a mile or two off—but distance and scale remain equivocal —you discover a wonder: the circular glow, fuzzy at the edges, of a great light, all its brightness swallowed up in the immensity of the darkening sky. The light means men, brought here for some barely imaginable purpose. Beside it, you can see now, a tapered, openwork tower pricks at the sky, only a little higher than the light, like a pin stuck in a map. The wonder grows stranger. Then to one side of that distant tower a new light bursts upward in a slow explosion, growing; white, then yellow, darkening to orange and smoking, lifting and swelling in great rolling fireball puffs of flame. Unprepared, you would say the men have set fire to the snow, that it is the cold earth itself burning, but the colors are the colors of burning oil, and they signify discovery: beneath this Arctic plain there is enough oil pressured by the weight of two thousand feet of permafrost and thousands more of porous rock to fountain upward at the hypodermic prick of the drill bit suspended from the derrick—the "openwork tower" which to oil men is simply a rig. The men on the rig, for safety, have piped the oil off to a burning pit, dug in hope of this find, and lighted it. Later, men will claim to have seen this flame a hundred miles off, flying north from Fairbanks, over the mountains. Words here, like distance, are relative, but it is a big fire; big enough for a god to warm his hands at.

The date is one to remember: February 18, 1968, a few minutes past three in the early Arctic evening. Not many anywhere have yet heard of this Alaskan place, let alone what is being done here, but it will be called Prudhoe Bay, for the only distinguishing feature on this part of the coast, a rounded, shallow inlet two miles east. The name honors a highborn naval crony of the unfortunate British explorer Sir John Franklin, who passed this way a century and a half ago; yet another seeker after a Northwest Passage.

If you learned your first lessons in geography, as I did, in the mid-30s, then what you know of oil and coal is tangled with images of steaming ancient jungles, giant tree ferns, and great dinosaurs browsing with broad-footed deliberation across oozing tropical marshlands. The reality is different, as witnesses the oil found at Prudhoe Bay and, more and more, in the whole belt of northern lands that rims the polar sea. Or at any rate the current theory of the reality: if we knew precisely and with certainty how oil is made, we would, I suppose, make it, with less toil and anguish than we have in finding it.

The first reality is what the geologists call sediment. Air, water and ice flow, penetrating the pores of the underlying rock, breaking it off in slabs, boulders, carrying all along, grinding, pulverizing, until the dust can be ground no finer; until there is no further minimal slope down which to flow, or the mass for a time comes to rest against some more resistant geologic form. Mixed with this mass of flowing dust are the remains of whatever life stuff has been present, from single cells to the immense complexities of plants, trees, sentient animals, all reduced, by the common chemistry of air and water, the mechanics of heat and cold, to the few simple elements from which they were made; inter-mingled, recombined. All this is the action of runoff, bearing the sediments with their cargo of dissolved life from high ground to low, dropping them finally in patient layers along the seas' edge; which you can observe in miniature, in the span of a tide-fall, along any ocean beach. Patience: the layers of sediment, packed down, accumulate no more, perhaps, than one eighth of an inch per year, roughly the growth ring of a slow-growing tree in a dry year, but time is their true dimension, time that if not infinite in the meaning of the word, literally boundless, extends beyond any human comprehending. A million years of such layering—and a million years, a million of earth's revolutions, is the basic unit on a scale that reckons in hundreds of millions—would lie ten thousand feet thick on earth's brittle underlying crust of rock, where it began; or so it would thicken if the process that made it could simply cease. But the process does *not* stop, no more than earth itself grows tired and stops revolving.

So: worked on by the same planetary forces that carve and smooth the earth's face everywhere, the layered sediments accumulate, and more—together, they possess enormous mass, which becomes a new force in itself, a factor in the process. Under this mass, earth's crust bends, sinks, bowl-like, like the skin of a pudding in which a finger has been pressed—a form that to geologists is a syncline. Around the rim, responding in symmetry to the same downward-pressing force, the earth rebounds, ridging upward in sloping, layered folds, which are called anticline, a new source for runoff and sedimentation. Other effects of mass go on more or less simultaneously. Under its pressure, the loose granules of sediment pack together, solidify, become stone, but altered, held apart by jagged edges and therefore porous, per-meable, structured like the cells of a microscopic sponge. Heat, the product of this same intense pressure, transforms the liquid life stuff bound with the granules of rock, reforming the atoms of its two fundamental elements, hydrogen and carbon, in molecu-lar combinations that, again, by any human reckoning are infinite.

These effects of relentlessly accumulating mass are micro-scopic. Others are large. The layers, solidified, press together until abruptly they break and crack apart as if slashed by a knife, and a whole vertical segment of the mass slips downward, the layers now discontinuous; this slash is called a fault. Other forces, still larger and only partly understood, work on the sedimentary mass. It rests at the edge of a continental matrix, a segment of earth's crust floating on the molten magma beneath—a form which we call a tectonic plate, and, since floating, *moving* with the earth's motion. These plates, our continents and sea floors, grind together along their edges, press under one another, tilt and rise, break off, rebound. So, back from the edge, the continents echo and vibrate with the shock, earth shudders, and mountain ranges rear up, miles high, like the wrinkles in a blanket on a half-made bed. The earth pulses with the rhythm of a heart, only at intervals too long by a million years for humans to more than guess. *The earth is alive.* If we fail to perceive its life, it is our perception that is at fault, our definition too small.

The liquid we call oil is one product of this ceaseless interplay of forces, as incidental to earth's life as sweat to a man's. Whether it occurs or collects is subject to further conditions: an impermeable floor to the sedimentary basin where it gathers, and a dome on top, like the lid of a pot, also impermeable, lest the accumulating weight of sediments pressure the liquid out and disperse it; the occurrence of faulting—cracking and slippage of the layered, compacted sediments, thereby preventing the liquid from escaping at the sides; and the original life-bearing dust of rock, formed to a spongelike permeable solidity, such as sandstone, in which the broth of hydrogen and carbon atoms can pressure cook and separate, floating upward, from the sea water in which it began—when oil is found, it is always afloat on the salt water of a submerged ancient sea; inland lakes are filled and buried by the same process of sedimentation, but even the greatest of them has too transient a life, compared with the oceans, to live out the process that produces oil. Geologists, scratching and probing the surface of earth's skin, can guess at the shapes below that might have made it possible for oil to form and collect: noting the folding and layering of rock exposed by weather, inferring its age from embedded fossils, plotting syncline and anticline from minute variations in magnetism or the re-echoing of surface sound waves, as from a series of small explo-

sions. There is as much of art and luck in these investigations as
of science. Only the drilling of a well can show whether the geologist's guess was true—whether the basin-and-dome structure inferred by this means, hundreds or more likely many thousands
of feet below, actually contains oil; more often than not, it
doesn't. As a matter of probability, finding oil at the end of this
indirect, cumbersome, and expensive search is less sure than the
proverbial needle in the haystack: given the assurance the needle
was there in the first place, you could with patience and unlimited time sort through the stalks of hay till you in fact found it,
but that is an assurance the oil hunter does not have, and his
haystack is the earth itself; in the continental United States,
about one new well in ten finds oil; the chance of enough oil to
be worth producing is one in a hundred. Even actual oil oozing
on the surface, as it does at many points on the earth and along
the margins of its seas, is no proof of an actual pool of oil below;
it may mean only that the oil, once formed, has not been held in
place and is gone.

The particular oil found at Prudhoe Bay on that day in February had its source *to the north:* in sediments and life chemicals
draining south from an extension of the Canadian shield, the
skeletal underlying rock on which much of North America is
built, long since sunk under the Arctic Ocean. The North Slope
(or, properly, the Arctic Slope) was then a *south* slope and the
shallow edge of a·shallow sea covering much of what became
northern Alaska. The sediments then were covered, subsided, the
higher ground to the north sank, the old sea filled and rose. A
continental grinding together of tectonic plates (possibly a part
of the same event that sheered off that northern reach of the Canadian Shield) upfolded a wall of mountains to the south, the
Brooks Range, and a new reservoir of offpouring sediments. The
earlier runoff from the north and the newer from the mountains
to the south outpaced the wearing away of time and weather.
The oil-bearing deposit is now mantled ten thousand feet deep.
Reasoning down through all those layerings, pondering their fos-
sil clues, comparing the rock with types identified elsewhere, the
geologists locate the oil's origin at the intersection of two
geological periods, the Triassic and Permian (later, this ambiguity will be certified in a double name: Permo-Triassic). The

oil, that is, was laid down about 225 million years ago; an age that the human mind can reckon and reason on but not truly conceive. And now it has been found.

The rig, as you come nearer and begin to see it, is an Eiffel Tower reduced in scale, a hundred and fifty feet high from the snow plain to the crown block at the top. Function determines the rig's size and form: to lift and lower, with man-size blocks and a complex rigging of steel cable, the 90-foot stands of pipe, three 30-foot joints to a stand, which, threaded together, turn the drill bit at the bottom of the hole. The rig's four-legged base is hidden by a cluster of fuel tanks to power its diesel engines, by the long, steel-walled shed where the drilling mud is mixed for pumping down the hole (literally mud, light gray and thick as whipped cream, but of a special kind, based on a clay called bentonite; heavy enough to balance the pressure of any gas encountered in the hole; its particles swelling, when wet, so as to hold in suspension the fragments of rock broken by the bit and carry them up the hole to the surface). The drilling floor, where the actual work of drill pipe and bit is managed, is twenty feet off the ground, walled with canvas to keep off the wind, hold in a little heat. From its corners, big lights shine down, glowing green in the Arctic twilight. Ninety feet higher, where the rig narrows, a plywood-sheltered platform is set in for a man to maneuver and rack the stands of pipe as they are winched into or out of the hole. Around the base of the rig huddle a couple of small Quonset huts where the drilling crews eat and, at the end of their twelve-hour tours, try to sleep; some trucks and a bulldozer, engines throbbing, never turned off against the freezing of oil, grease, batteries; and long stacks of pipe and casing, laid out flat where they can be lifted into the rig as needed. By custom and state law, the drill site carries a series of identifications— lease owner and operating company, drilling permit number, a coded reference to a state map, signifying its exact location; and its name, Prudhoe Bay State No. 1.

You reach the drilling floor up a narrow, open ladderlike stairway clinging to one side of the rig. The steel treads slip under your feet, the stairway flexes and sways with your weight and the light gusting wind, and you take a steadying grip on the steel-pipe railings at either side. It is a climb to be made with deliberation, and only if you have business up there, known and acknowledged: the well being drilled here is a wildcat in the vocabulary of oil that goes back to the beginnings in northwest Pennsylvania when, maybe, the fierce small cats were still around you on the wooded hillsides and you could hear their screeching and caterwauling in the night—exploring new territory where no oil has yet been proved to exist by the only proof that counts. The well is also, from another standpoint, a tight hole, whose secrets—whether oil is found or not and, if it is, how much and of what kind, at what depth—will be fiercely guarded. Strangers are spies and unwelcome.

Up on the drilling floor things are quiet, in preparation for the test. A couple of torpedo-shaped heaters whine, pumping heat across the floor from corners, out of the way, but the heat spills upward to the unroofed cold. Down below somewhere the diesels that power the winches and the rotating drill still throb, idling. When the rig is making hole, there will be a team of roughnecks on the floor—lead tongman, backup man, pipe threader—fitting the stands together as the string drills down, ninety feet to a stand; or threading them apart again as the string comes up to change a dulled drill bit, a process that may eat up a whole twelve-hour tour and, apart from the time, costs money, a couple of thousand dollars per bit. Out of sight, but part of this team, are the derrickman, high up on his monkey board; the motorman, a couple of roustabouts, the semiskilled labor of oil; drivers for the trucks and forklifts and Cats. On the floor, quarterbacking at the controls, eyes on his crew and the battery of gauges that tell him what's happening down where the bit is, is the driller, supervised—coached, if you like—by a man who most likely has worked up through all these skills and earned the modest-sounding title of tool pusher. All these men by custom draw their pay from the drilling company (Loffland on this well) working under contract to the lease owner, supplying the rig and the crews to run it—twenty-four hours a day, every day, one crew per twelve-hour tour, until the bit hits oil or, more often, only more rock, and the dry hole is abandoned, the rig dismantled, packed up, shipped out. The men work six weeks of such days, then break for three and a paid flight to Fairbanks or Anchorage; it takes three crews, staggered on this schedule, to keep the rig running.

Now, with the test beginning, the crew has gone back to a Quonset to eat or sleep, except for the driller on duty at the controls, bulky in his mud-spattered, quilted coverall, with the shiny regulation aluminum safety hat on his head. A drilling engineer and a geologist have joined him, representing the lease operator, the recently formed Atlantic Richfield Company; and a test operator contracted from yet another company. It is the engineer and geologist who collect and analyze the rock cores brought up in the course of drilling, schedule the tests, and are finally responsible for the millions of dollars of hopeful oil money drilled down the hole; not a job that leaves much time for sleep.

A valve slips open at the bottom of the hole, 8,800 feet down. Gas rises through the column of mud, is caught and piped through a series of valves, meters, small-bore chokes, and is led out to the end of a pipe hanging over a pit dug a safe distance from the rig. The object of the test is to find out what is down there, to measure the pressure and quantity of its flow. There is nothing else to do with the gas that presently is blowing into the pit but burn it—and the oil that follows, forced upward by the pressure of the gas. Someone tosses a torch into the pit and the gas ignites from the end of the pipe, where it meets the air; this is the flaring light, turning smoky when the oil comes, that would have been visible across the frozen plain, to anyone watching. The engineer left the drilling floor and went over to the pit to watch the stuff burn, changing color, warming from its warmth.

Early the next morning, a messenger flew to Fairbanks to telephone the find to an Arco district manager in Anchorage—there were no telephones from Prudhoe Bay and the radio was too public for the secret. The gas was calculated at a flow of 22 million cubic feet per day, the oil at something over 2,200 barrels, figures so large it is not easy to make human connections with them. An average American oil well produces fifteen or eighteen barrels per day, and it mostly does not come out under pressure but has to be pumped. Gas-production figures, for comparison, are hard to come by: in the United States most gas is produced in some combination with oil, and the distinction, oil from gas, is less sharp than it might seem; because of that and because, for many years, the illusion of controlled low prices for gas has made it unattractive to produce, there are not many U.S. wells that produce nothing but gas. Nevertheless: the gas in this test flowed at a rate perhaps twenty times the national average for gas-producing wells; or fifty or sixty times that for all wells that produce some gas. But more important, there was oil, and in even grander—princely—quantities. To put the test quantities in homely terms, the 2,200 barrels of oil converted to gasoline (the barrel is a standard measure of 42 gallons) would run a car—an average car with an average driver—for nearly two centuries. The 22 million (cubic feet) of gas would fuel a family stove, cook all its meals, for about two thousand years.

The pay from that hole, in the oil vocabulary, looked to be big. But it had been a long time coming.

As early as the 1850s, Russian traders, their minds on the dwindling and erratic supply of furs that brought them there, had noticed seepages of oil along the west shore of Cook Inlet, in south-central Alaska. The local Indians, like the Indians everywhere in North America where oil collected in surface hollows, found modest uses for it—as a medicine, as a sealant easier come by than spruce gum. There was some talk among the Russians of commercial uses for the oil—there was a long-standing trade at Baku, from similar sources—but lethargically they did nothing; the oil had too little value for its distance from populations and markets, whether in Russia itself, in China, the terminus of the Russian fur trade, or in recently Americanized California. In the years before and after 1900, Americans—when not distracted by the succession of major gold strikes that eventually galvanized nearly every corner of the territory—prospected the same area and even drilled a few halfhearted wells, but without hitting enough oil or gas to repay production. About the same time however, to the east, near Katalla at the mouth of Prince William Sound, drilling in another area of ancient oil seeps was rewarded with the discovery of a producible oil field. Over the next thirty years, it was tapped by 44 shallow wells and, at the rate of a dozen barrels a day, fed a small refinery at Katalla, which in turn supplied diesel fuel and a little gasoline to the local fishermen—

until in 1933 the refinery burned and seemed not worth rebuilding.

Meanwhile, although Alaska remained as generally unmapped and unsurveyed as in the Russian time, prospectors and the honorable gentlemen of the U. S. Geological Survey, poking around north of the Brooks Range, had also noticed oil seeps at several locations—enough to suggest huge reservoirs of oil under the permafrost, if it ever became worth while hunting them. One of the things the Navy learned from World War I was that its ships, expensively converted from coal, had developed a powerful thirst for oil and were no use without it. Hence, it seemed prudent to set aside tracts of public land that showed prospects for supplying that thirst if, in an unlikely future, American commercial production ever ran short. That was one of the purposes that in 1923 led Warren Harding to Alaska, and his never-explained death: to proclaim a Naval petroleum reserve in an area lined out on the sketchy USGS map of Arctic Alaska. This amounted to 36,000 square miles, nearly half the area of the entire Arctic Slope north of the Brooks Range, but since it had no other apparent use or value, the gesture was less grand than it sounds: Congress had provided few means for acquiring land in Alaska, so that virtually all the Territory remained in the public domain —it did not much matter which departments of government were ostensibly in charge. The new tract was designated Naval Petroleum Reserve No. 4 and for want of a more descriptive name has been known ever since simply as Pet 4. (The three other and more accessible reserves, established a few years earlier, are Elk Hills and Buena Vista Hills in California, Teapot Dome in Wyoming.) As the people would presently discover, Harding tended to put his trust in officials with a knack for turning public wealth to private profit—such as the oil at Teapot Dome. Whether he (or those anyway who told him what to say and think) was motivated only by Naval security may be doubted, but it did not matter. The lines were drawn on the map, the President took ship for Seattle and died, the scandals became public; the land returned to its ancient repose.

During World War II, when Alaska revealed itself as a prize in the grand Pacific strategy—a sword pointed at Tokyo but one that could be reversed to strike the American heartland; a base from which half-trained Russian pilots ferried fleets of bombers to the defense of Moscow and Leningrad (to this day the wrecks are landmarked across the Alaskan mountainscape)—the Navy once more gave thought to its Arctic reserve. An exploration program begun in 1944 over the next eight years sank 37 test wells: one of these discovered a gas field south of Point Barrow and, with cheap heat and electricity produced from it, provided a motive for gathering the Inuit into Barrow, transforming the ancient village into something like a city; another well to the east along the coast found a little oil. The program defined one other gas field in the Brooks foothills, two other oil fields too limited for production, but hardly seemed to have repaid the $60 or $100 million it cost the Navy. It did, however, disclose a good deal about the geology of the Arctic Slope, in particular a subsurface formation that was named the Barrow Arch, following the coast from Barrow east—the kind of structure that might serve to wall in oil between the Brooks Range and the sea, if any had ever formed. This knowledge, purchased by the Navy, became public property. As such, it attracted the attention of two oil companies from opposite poles of the industry: Richfield, a small producer based in the American Southwest, with a knack for discovering oil and developing it with the help and money of bigger companies; and British Petroleum which, from a strike in Iran in 1908, had grown, with periodic doses of government investment, into one of the industry's largest corporations, with a sure hand on perhaps 20 or 30 per cent of the world's known oil reserves.

As a small, domestic company among international giants, the Richfield Oil Corporation was in the position of picking up the leavings among its big rivals' oil lands, surviving by wit, caution, and luck. In consequence, its capacity to refine and market oil increasingly outran its ability to produce: to keep going, it was forced to buy more and more of its oil from others, resulting in higher costs and declining profits. This predicament pointed the company toward Alaska, where it had been among those contracted by the Navy to manage the Pet 4 exploration—its geologists had had a part in the official USGS reports that resulted and in fact knew their contents from direct experience. In a list of prospects drawn up when the program ended, the coastal land east of Pet 4 and around Prudhoe Bay was Richfield's first choice, but this area, still federally owned, was not yet open to oil leasing. Second on the list was the Kenai Peninsula forty miles down the Cook Inlet from Anchorage, but that had been set aside in 1941 as a national moose range. With the help of a group of Alaskan businessmen bent on development, statehood, and profit for themselves, Richfield managed to persuade the Department of the Interior, which was in charge, that moose and oil were compatible. Leases in hand, it started drilling in 1956 and by July of the following year had hit a major field near the Swanson River. An oil rush followed, fresh discoveries on both sides of the inlet (and under it), and, on January 3, 1959, statehood, conferred by a Congress at last persuaded that the northern land might after all be self-supporting and not a permanent charge on the public.

British Petroleum (or BP, as it is commonly known) was drawn to Alaska by the same motive as Richfield, shared, indeed, with every other oil company—the necessity of feeding the machinery of its trade with oil; but in the context of a global strategy. In 1951, facing a dithering and in theory sympathetic British Labour government, a wild-eyed Iranian, Mohammed Mossadeq, had been chosen premier for the purpose of nationalizing the BP oil monopoly, in the turmoil that followed stopping production, driving out the British technicians and managers, and finally deposing the young Shah. Five years later, another nationalist, Egypt's self-elected president Gamal Abdel Nasser, undertook a nationalization even more threatening not merely to Great Britain but to Europe as a whole: the seizure of the Suez Canal, through which most of the vital Mideastern oil still flowed to Europe. Both crises were in time smoothed over. Mossadeq was overthrown, the Shah restored, BP brought back, though with its monopoly diluted by the participation of its American rivals. (The coup was said to have cost the CIA a modest $700,000.) In Egypt, the canal was reopened and began to function more or less normally under its new masters, and in the meantime had become obsolete—the Europeans had responded by sponsoring a new breed of tankers too big to pass through the shallow canal, big enough to make it economic to transport the oil the long way around from the Persian Gulf to the Baltic, 15,000 miles by way of the Cape of Good Hope. Yet disquiet remained. The company's Mideastern oil riches were not after all so safe as they had seemed, and after fifty years of orderly ease it would be necessary to scramble after new sources that were. BP geologists in London, studying the USGS reports on Pet 4, reached the same conclusion as Richfield's, and in a world-wide context: an attractive prospect. To the motive of corporate and indeed national survival was added another nearly as compelling: entrance into the American market, the world's largest and most profitable, where BP, alone among the world's great oil companies, had never yet been represented.

This proved a delicate enterprise, entering the American market through the back door of Alaska. The British government, enmeshed in the same financial troubles that have been continuous since World War II, discouraged the export of British capital to chase rainbows in the American Arctic. Hence, although rated eleventh among all the world's corporations, BP was compelled, like Richfield, to proceed with extreme caution and a tight budget, picking up cheaply, where it could, likely oil land that its better-heeled rivals had overlooked. It managed by buying up leases on an apparently worn-out California oil field, arranging for its revival, and using the $2 or $3 million a year thus generated to finance a more ambitious hunt; and by making a joint exploration agreement with the prosperous but oil-poor Sinclair Oil Corporation. By the time BP reached this point, however, the Cook Inlet boom was already on and land costs had climbed too high for the company's limited means. It turned of necessity to the Arctic Slope, where leasing was beginning to open up and the other companies, busy in the south, were showing little interest.

In 1958, the Department of the Interior opened Alaska north of the Yukon to oil leasing under a system designed to encourage exploration, known as simultaneous filing—the "simo." Those interested could apply for leases on tracts of about four square miles; if there were more than one application for a tract, the winner was drawn from a hat. The costs were nominal—a $10 filing fee plus 25¢ per acre and a 12½ per cent royalty on any oil actually produced—but even so there were not many takers. A few citizens in Fairbanks, encouraged by the play around Cook Inlet, put in for tracts here and there and in time made fortunes by reselling to oil companies at stiffly competitive prices. Only land already designated for some purpose was excluded, meaning chiefly the 23 million acres of Pet 4. The simo was repeated in 1964 with similar results. By then the acreage fee had doubled to 50¢ (it still cost $10 to file) and another 9 million acres of Arctic Slope had been withdrawn to form the Arctic National Wildlife Range, east of Prudhoe Bay along the Canadian border.

In the winter of 1963, using a Canadian rig barged down the Mackenzie River to the mouth of the Colville River and tractored eighty miles inland to the northern foothills of the Brooks Range, BP began serious exploration of its Arctic leases. Through the following year and a half, splitting costs and operations with an increasingly skeptical Sinclair, BP drilled eight wells along the eastern border of Pet 4, in the vicinity of the modest Navy discoveries. The results were like those of the earlier program: unprofitable but instructive. Apart from one new field too small to produce at this distance from markets, the drilling and the accompanying on-the-ground geological and seismic surveys had identified two promising structures, one near the Colville delta, the other around Prudhoe Bay.

The impetus for further exploration came with Alaskan statehood. Congress, nervous about Alaska's solvency, provided an unprecedented endowment: the new state would be allowed to pick from the public domain 103,350,000 acres of free land; in addition, it was entitled to another million acres under a 1956 law granting land that could be sold or leased to generate income for purposes of mental health. (Earlier, the territory had been given 100,000 acres to finance the establishment of a university.) The timetable for selection was a leisurely twenty-five years from statehood—and it would take at least that long to survey whatever lands the state picked. Except for 400,000 acres of national forest land, the state was excluded from the more than 80 million acres already designated as national parks, wildlife refuges, petroleum reserve, and so on; one way or other, when the selection process was completed, the federal government would still be landlord to well over half the state.

Although the process was inherently slow, the new state had a powerful incentive to claim its endowment and start turning it into cash. The first selections were for land that already had obvious worth—around Anchorage and Fairbanks; on the Kenai

Peninsula south of the Moose Range, where there might be more oil, and coal deposits had long been known but little used (at Homer, instead of driftwood, you fire your clambake with lumps of coal picked up along the beach); and along the Alaska Railroad and the state's few highways. Then, early in 1964, pondering the Pet 4 reports and the advice of its own geologists, some of whom had been rock-sampling in the north since the 1920s, the state put in for what remained of the Arctic Slope—chiefly the 125 miles of coast centered on Prudhoe Bay, between Pet 4 on the west and the Arctic National Wildlife Range on the east. When the application received tentative approval in October, the state, urged by BP, announced that the land would be opened to leasing.

This time the oil leases would be subject to competitive bidding. That is, the state first invited nominations from oil companies prepared to compete. Land actually to be offered was decided from these suggestions, and those who wanted in on the play then submitted sealed bids by a stated deadline—a cash bonus per acre, accompanied by a certified check for 20 per cent of the total, on top of the fixed annual rental of $1 per acre for ten years and the standard royalty on oil produced, if any. (In addition, state law imposed what was called a production, or severance, tax on top of the royalty, at this time about 4 per cent.) The state reserved the right to reject bids it deemed too low, but it was still possible for a lucky individual to pick up potential oil land cheap and sell it dear, though the risk was now a good deal stiffer than in the federal simos. As it turned out, there were not many takers. When the bids were opened, on December 8, 1964, BP Alaska, a corporation created for the purpose, but still cash poor, had teamed up with Sinclair to walk off with most of the chips: 125 tracts, or 318,000 acres, at an average bonus of $6 per acre. As it happened, the state had offered first the western part of its northern domain, near the Colville River border of Pet 4.

In July 1965, the state auctioned the other half of its Arctic oil dowry, the area around Prudhoe Bay. By now, Sinclair, discouraged by the energetic but fruitless drilling so far, declined to send any more good money after bad; the Canadian rig, worked north from the foothills to the coast, was stacked where it

started, at the mouth of the Colville, ready for shipment back up
the Mackenzie. Even so, and in the face of livening interest from
other oil companies, BP, by a canny placing of its bets, managed
to capture nearly a quarter of the 403,000 acres offered, this time
at about $17.50 per acre, half the price bid by its rivals on other
tracts.

The chief of these rivals was Richfield, bankrolled, in return
for a half share in any discoveries made on the leases, by the
Humble Oil and Refining Company, the American operating
subsidiary of Standard Oil of New Jersey. (Since then, in an
effort to assert a single identity, it has changed most of its corporate names to Exxon*—but then and now, by whatever name, it
is the biggest of all the world's big corporations.) By one of
those curious coincidences that abound in the oil business, a substantial part of Richfield was actually owned by the Pittsburgh-
based Sinclair Oil Company, the personal creation of the same
Harry Sinclair who in 1923 supplied the cash that produced the
Teapot Dome scandal. Following the sale, Richfield, eager for
the hunt, dismantled a rig and shipped it north to Fairbanks by
rail, with drill pipe, bunkhouses, supplies—3,000 tons; and for
the final lift over the mountains chartered a Lockheed C-130
Hercules, a plane capable of carrying twenty-five tons in a tail-
loading, boxcar-size hold 41 feet long. Known affectionately as
the Herc, the plane would be the vital link to the explorations
that followed.

At this juncture, with Alaska's future in effect stacked on the
runway at Fairbanks, a new player took a hand in the game,
with a sharp eye for distant goals and the straight-line will and
intelligence for pursuing them: a man named Robert Anderson,
who a few weeks before the July lease sale had been brought in
as chairman of a small and stodgy Philadelphia oil company
known as Atlantic Refining. As things turned out, the Arctic
Slope was among his goals, and Richfield proved to be the means
of getting there.

* The Exxon name became official on November 1, 1972. Oil company
names change so frequently that keeping track of their identities over a period of years is something of a problem. I shall use the common, gas-pump
name in most cases, but where the full corporate designation is required,
it will be the one extant at the time referred to.

Richfield was the successor to an earlier California company gone bankrupt in the early 30s and rescued by Sinclair and the Cities Service Company, which between them thereafter controlled its board of directors. Thirty years later, the Department of Justice concluded that this arrangement, too similar to the kind the antitrust laws were meant to stop, was in restraint of trade and in 1962 undertook a suit against all the parties concerned. As the suit dragged on and Richfield's Alaskan ventures showed signs of prospering, the company's chairman, an oversized archetypal Texan named Charles Jones, came to feel that it would end in his company's being sliced up and sold off piecemeal like the shopworn shirts and neckties in a bankrupt haberdashery; the only way out was merger with another company agreeable to the bloodhounds of the Department of Justice. Mr. Jones accordingly began shopping through a list of prospects that included most of the medium-to-big famous names of oil, with BP at the top; another curious coincidence. Eventually, he worked his way down to Atlantic and its new and youthful head, Robert Anderson. The two men had a serious talk, Mr. Anderson liked what he heard, and on September 16, 1965, an agreement was signed which became effective January 3rd with the approval of the two sets of stockholders. Atlantic, which had failed in the July 1965 auction through niggardly bids of $6 an acre, was now back in the game with the Richfield leases among its assets. Robert Anderson and his company, renamed Atlantic Richfield, presently shortened to Arco, had taken a first large step toward the Olympian heights occupied by the oil giants.

With the legal formalities disposed of, the chartered Herc began its shuttle over the mountains in February 1966, with Bob Anderson on board for the first run. By March, the gear was assembled, the camp set up, and the well spudded in; it was called Susie Unit No. 1, for a neighboring benchmark. The site chosen was seventy miles inland from the coast, in the same general zone in which BP had suffered its prolonged disappointment a short distance west; although the land by now belonged to the state, an older federal lease still in force would lapse unless drilling was begun at once, and the company was therefore obliged to make its effort in this unpromising location. Drilling, suspended in May when the Arctic spring thawed the ground surface to mush and the mosquitoes rose in their multitudes from eggs quick-frozen in the fall, was resumed with freeze-up, and, in January 1967, 13,517 feet and $9 million later, was abandoned; the hole was dry.

A decision now had to be made as to whether to go on drilling elsewhere or find some better use for the money, such as putting it in a comfortable savings account. But that is not how oil companies operate if they mean to stay in business, and, with the advice of the Richfield geologists and the consent of Humble, whose half interest was part of the Richfield inheritance, Anderson agreed on one more try. The rig and camp were dismantled, loaded on sleds, and skidded north to Prudhoe Bay by Cat train. This took time. It was late April before drilling could begin on the well catalogued as Prudhoe Bay State No. 1, and it had penetrated only 590 feet, still in permafrost, when the season ended and the work shut down.

The drilling went faster when it started up again on November 18th. Within a month, a test at 7,000 feet showed gas, and two weeks later, in a second test at 8,600 feet, the gas bubbled up with such force that it appeared the well was about to blow. In shutting down to prevent this, the crew lost the bottom of the string, it jammed, and all efforts to fish it out failed. It was necessary to pump cement down the hole to stop it up and angle the pipe around the plug before drilling could resume. All this took up three nervous and expensive weeks. Another month of drilling and a third test, at about 8,800 feet, on February 18th produced not only gas but oil—and the great rolling, smoky flare that lit the Arctic darkness like a bonfire of victory. At this distance in time and with a layman's hindsight, one would say the drillers had hit something big—several strata ballooning with gas and oil straining to burst forth, or else a single zone of extraordinary thickness. Arco, however, doggedly went on drilling and kept its thoughts to itself. There was a fourth test in March at about 9,800 feet, with similar results, and the well was finally completed at 12,005 feet. The stuff was down there, and in quantity. The next question was how far the zone extended—how much— and to find out, a new well was started seven miles to the southeast, near where the many-channeled Sagavanirktok River spills its burden of gravel and sediment into the Beaufort Sea, on the east side of Prudhoe Bay. The name, a whole sentence in Inupiaq, describes the strength and swiftness of the river's current. Since the name does not trip nimbly on an English tongue, the well was designated Sag River State No. 1, retaining, as it happens, the root verb that means "to flow."

The oil business goes about its affairs in a curious atmosphere of concealment and publicity: it recoils from revelation, but its secrets are about as private as a congressional investigation. From December on, watchful rivals would have known in general what Arco was up to at Prudhoe Bay, and anyone interested —and a good many were—could pick up the details from an off-duty roughneck in any Fairbanks saloon for the price of a drink. Given, therefore, that the company had to say *something* official about its findings, the secretive caution with which it proceeded is instructive—and characteristic. Interpreting the announcements is like chewing the fibers of truth from the veiled reportage of a Soviet newspaper; the language is esoteric and takes some learning.

Thus, the day following the third big test, the *Wall Street Journal* carried a small item reporting only that Arco had taken an "oil-saturated core sample . . . from Prudhoe Bay State No. 1"; nothing was said of the large quantities of both gas and oil that flowed in the test, until June 25th when Arco likewise admitted that oil and gas had been found in the Sag River well. (This time, also, the project was correctly identified as a joint venture with Humble.) Meanwhile, with the fourth test in March, the company had let out that it had run "an encouraging test of oil and gas," and in the company's annals this day, March 13th, became Discovery Day, though the honor might equally have gone to the February or January test days; later, in Senate testimony, Arco executives in fact mentioned January as the big day, but they did not, at any rate, begin to act on their discovery till March. But we anticipate.

The first well, Prudhoe Bay State No. 1, was designated the Discovery Well, while the second, Sag River State No. 1, is known as the Confirmation Well—it showed, that is, that Arco had found a field, an extensive layer of oil and gas rather than an isolated and anomalous pocket. Since the confirmation came at about the same depth as the discovery and anyone who took the trouble to look up the location in the state land records would learn that it was nearly seven miles off, it was evident that the field was a big one; the question now was how big—how much oil was down there. For an answer, Arco and Humble turned to the Houston oil engineering and consulting firm of DeGolyer and MacNaughton.

Estimating a new oil field is a mixture of scientific measurement and less objective kinds of judgment. For a start, from the drilling and test records, one learns the thickness of the oil-producing zone, or "pay." The same records, coupled with seismic studies of the region's geology, indicate the extent of the oil-bearing formation. Core samples show the volume of oil to a given volume of rock. From these and similar kinds of evidence, one knows roughly how much oil the formation contains, but how much can actually be gotten out is a separate and trickier question: the answer depends on such things as the porosity of oil-saturated rock (that is, how readily the liquid will filter through into the well holes); the pressure of gas above the oil, if any, and of water below; the character of the oil itself—oils differ greatly from one another, rather more, say, than vintage wines, and hence differ in what they are worth, what they will sell for. As important as any of these factors is the producer's judgment as to where the oil can be marketed and at what price, how much it will cost to get it there—factors that, in turn, inter-act with the estimates of quantity and quality. For example, the earlier explorations in Pet 4 had found as much as 100 million barrels of oil, a volume that in Texas or even around Cook Inlet would be regarded as a bonanza; but in the Arctic would cost more to produce and transport than it would ultimately sell for. Similarly, if the field, like most now producing in the United States, has little or no pressure of its own, you can get the oil out with pumps or by techniques known as secondary or tertiary recovery—drilling additional wells down which you can pump water, steam, chemicals, or gas under pressure to force the oil out. Since all these methods cost money, you can afford to use them only when the combination of market price and trans-portation cost will yield a profit at the end: otherwise you will have consumed, in effect, more energy than you have produced.

Balancing all these probabilities, the oil companies generally expect to get out no more than 30 or 40 per cent of the oil actually present. The figure that results, which in fact leaves considerable latitude for interpretation, judgment, and error, is known as a field's proved or recoverable reserve.

On July 16, 1968, their own calculations certified by DeGolyer and MacNaughton, Arco and Humble announced the Prudhoe Bay reserve at 5 to 10 billion barrels of oil. In November, another estimate, apparently by the same methods, came to 30 to 40 billion barrels, and there were at least rumors that the total was as high as 100 billion. After further drilling and for reasons that will become evident as we proceed, the two companies and others with leases in the area settled on an extremely conservative estimate: 9.6 billion barrels; and 26 trillion cubic feet of gas. Among other things, these figures are deliberately limited to the contents of the main Prudhoe Bay formation, known as Sadlerochit; two more have been identified (Lisburne and Kuparuk), and there are almost certainly others both in the area and offshore, including the shallow bay itself, but the companies have refrained from publishing estimates, and 9.6 billion has remained official ever since.

Whatever the reality it would finally signify, the DeGolyer and MacNaughton report had an effect that may properly be called electrifying. Oil companies that had been cool to the Slope or, like BP, suffered disappointment now went back to their lease records to refigure their chances and began marshaling rigs and crews for rush shipment to the Arctic. In New York, the Stock Exchange hastily stopped trading in Arco stock; when trade was reopened five days later, the price had risen by $34.50 per share. The boom was on.

We have a tendency in the United States to think of great public events, such as the Prudhoe Bay discovery, as *happening to someone else;* as abstract. It is a state of mind encouraged by the pampered demagoguery of our politicians and the day-by-day fragmentation and inconsequence of much of our press (not to mention the antic showmanship of television). The attitude is and, I believe, should be foreign to us; it has been our luck and talent to regard our public institutions, both governmental and corporate, not as abstract entities standing apart, on a different plane of being, but as personal and continuous with ourselves; having immediate and concrete effects on our actual lives; and which we can act on in turn, and finally subject, at least collectively, to our will. In no industry is this clearer than in the oil industry, despite the air of monolithic apartness and secrecy in which it likes to appear. Hence, to understand what had happened at Prudhoe Bay, it is necessary to go back behind the flurry of stock trading that the discovery set off and hunt its meaning in the national life and the ways in which, for more than a century, that has been fueled by oil and intertwined with the oil business. Indeed, it is high time we all knew far more of these sober matters than we do. The alternative is to linger in a world of children, where eggs grow in cartons, clothes sprout from racks in a store, and gasoline comes out of the nozzle of a pump.

It has been often said that the United States consumes a disproportionate share of the world's energy, and of its oil in particular, and that this disproportion results from Americans' blowing their fuel out the tailpipes of joy-riding cars and is therefore culpable; with the corollary that in so doing we are consuming a finite and irreplaceable resource. All three of these propositions veer between half-truth and lie, but they have been repeated so often as to have the ring of axioms. The reality is different and more complicated, both more threatening and, in a sense, more hopeful.

A nation's livelihood is measured by the value of the goods it produces and of the services its people perform. The total value of all these things is what economists call the gross national product—the GNP. Since the 1920s and despite a decline that lasted through the depressed 30s, the U.S. GNP has multiplied about thirteen times, while the population has less than doubled. In other words, the per capita product—the value, in effect, of what each of us produces—has risen about eight times, from $847 per capita in the boom year of 1929 to $6,594 in 1974. During the same period, our per capita consumption of all forms of energy has increased at about the same rate as the total population; with the result that by 1974 every barrel of oil consumed (or the equivalent in coal, natural gas, and so on) was producing about $107 in GNP—*five times* the rate at the end of the 1920s. Thus, while our consumption of energy has increased, our efficiency in using it has grown far more rapidly, and the difference represents the vaunted rise in the American standard of living—the growing abundance of goods and services for all of us. Again during the same thirty or forty years, U.S. oil consumption has increased both absolutely and in relation to other forms of energy—from about 25 per cent to 44 per cent of the total. Without belaboring the point, it seems clear that American prosperity, both nationally and individually, coincides with the supply of energy, and of oil above all. The reasons are fairly obvious: oil is more abundant than natural gas, more adaptable in its uses; more efficient than coal (one and a half times the heat value by weight); more easily produced, and, again, more adaptable. Oil, gas, coal: there have not been and are not yet any serious alternatives.

The surplus that equals national and personal prosperity does not mean that our energy resources have flowed into private pleasures and frivolities. Since World War II, the great era of economic growth, about one third of the energy supply has gone into manufacturing, while transportation and electricity have consumed about one fourth each, and direct household and commercial uses together have taken about one fifth; and these proportions have remained fairly constant. In each of these sectors, oil has supplied an increasing percentage of the power consumed, and with the disappearance of the coal-fired steam engine it accounts for about 95 per cent of the fuel used in transportation—meaning primarily the trucks that deliver your food to the supermarket rather than the car that takes you to work or off on a summer vacation; according to one recent estimate, automobiles account for only about 15 per cent of U.S. oil consumption.

Since World War II, world-wide consumption of oil and of all forms of energy has grown even faster than in the United States. Hence, while in 1965 American energy consumption was about one third that of the world as a whole, by 1975 it was down to about 28 per cent; the proportion of oil in the world energy supply grew at about the same rate as in the United States. In effect, in its patterns of using oil and other energy sources, the world followed the American lead.

It was natural for this pattern to derive from the United States. Although some use of oil from surface seepages is probably as old as human society, oil as an industry began with the drilling of Colonel Edwin Drake's famous well near Titusville in northwest Pennsylvania in 1859, and it was Americans, by and large, who took the lead in extracting the liquid, finding uses for it and methods of refining to enhance its usefulness—and in trading the product and their ideas throughout the world. From the time of the Civil War on, the United States remained the great producer and consumer of oil and of all of its products, and both production and demand followed a fairly constant upward curve. All through this era and until after World War II, production and demand remained essentially in balance: if the United States imported comparatively modest quantities of crude oil and some refined oil products (mostly from Venezuela, later from Canada), U.S. demand included a similar proportion of oil and its products that were exported for the same reasons—economies of transportation or supply (preference for one kind of crude over another in a particular market or for a particular refinery).

Somewhere in the mid-50s, however, the supply and demand equation began to get out of step. Production continued to rise, but consumption was rising a little faster, and the difference was made up partly by new techniques of extraction (secondary and tertiary recovery, mentioned earlier) or of utilizing the available supply (growing use of the heavier and wetter, oil-like, components of natural gas, once regarded as a useless nuisance); and by increased imports, more now of necessity than for convenience or economy. Another way of looking at this change is to compare the production and demand with the proved reserves— the amount of oil estimated to be in the ground that we can actually afford to extract. In 1950, U.S. reserves were about 14 times the annual production of oil (excluding natural gas liquids, which do not figure in the reserve calculations). In 1970, when production had nearly doubled and the cautious estimate of the Prudhoe Bay discovery had been added to the reserves total, the ratio was 13.4:1, while in 1974, despite a decline in both production and reserves the figure was still 13.1:1. In contrast, the ratio of reserves to demand declined drastically: from over ten to one in 1950 to just over seven to one in 1970 and under six to one in 1974. That is, again, although rising production was largely matched by new reserves, consumption rose at a considerably faster rate. The change looks—and is—alarming, given our very large dependence on oil for most of the things that as a society we take for granted, and given also the absence, so far, of any obvious substitute. At the same time, it is important to bear in mind that the reserves figures are not absolute. To say, that is, in any year that oil reserves amount to twelve or ten times annual production does not mean that in ten or twelve years the last American oil well will burp up its last barrel of oil and run dry. Within the system, oil flows in a continuous cycle from exploration to production to refining to distribution. Exploration and discovery are inherent in this process, and actual reserves at any point—the oil in the ground that is worth producing—are relative to economic factors: the potential profit that pays for and rewards the effort and cost of exploration and of the more expensive recovery techniques. There are old oil fields in Pennsylvania, Texas, and California that may have been taken out of and returned to the reserves pool more than once in the years since discovery. The difference has not been in the amount of oil actually present but in the incentives for getting it out. *That* is the absolute. Hence, while American reserves, and the world's, may indeed be finite in an abstract or philosophical sense—and perhaps in a geophysical sense as well—the practical meaning of that proposition is not so clear-cut. It is well to be cautious in drawing conclusions.

While this disquieting shift in the old balance between oil production and demand was occurring, two contradictory developments were taking place whose final effects are still not altogether clear. On the one hand, the oil companies followed every other U.S. industry, encouraged in part by favorable corporate tax laws, in the immense expansion of overseas trade and development which changed the so-called multinational corporation from an exotic rarity, as it had been in the 30s, to a commonplace economic reality; one signal result of this movement, as we noted earlier, was the entrance of several of the major U.S. oil companies into the previously forbidden territory of Iran. Generally speaking, and particularly in the Mideast, the new oil fields to which the United States thus gained access were bigger than any at home and their oil came out under its own pressure, obviating investment in energy or cash for special recovery techniques. Both these fundamental facts meant that the oil could be produced and sold for less than most domestic oil. Furthermore, the fields were so situated that their production could be taken to market by tanker, the cheapest of all methods, an advantage enhanced by the turn to big tankers that followed the 1956 closing of the Suez Canal—big tankers are inherently more efficient in the water than small ones, apart from the fact that, for their carrying capacity, they cost less to build and operate; and if they happen to be of foreign registry, they have yet another advantage in cost, in that long-standing U.S. policy has worked to price U.S. ships and crews out of international competition, virtually driving them from the seas.

In any case, the consequence was a sudden abundance of cheap imported oil at the moment when it was beginning to look as if the nation needed it. What this situation promised was a range of benefits to the society as a whole so large and pervasive as to be literally incalculable, and—incidentally, I should say—individual convenience, and enlarged profits for the oil companies concerned. One might therefore have expected it to be greeted with national rejoicing and dancing in the streets. Instead, there came a howl from the smaller domestic oil companies—the "independents"—whose production comes from the more costly American fields and who feared that cheap and abundant imported oil would put them out of business. A solicitous Congress, in 1958, responded with an ingenious law aimed at limiting imports to 12.5 per cent of U.S. production and pricing what did get in up to the level of the native product. There were reasons for this besides the protection of constituents' profits and their ability to make campaign contributions: limiting the national dependence on imported oil; encouraging exploration for new reserves; maintaining a vital national industry against the chance of an emergency that might make the imports impossible to deliver. Nevertheless, the law, since it ignored the various realities, was about as effective as King Canute's commanding the tide not to rise. Demand continued to outrun production. Ways were found of letting in the necessary imports. In 1960, imported oil already supplied nearly 19 per cent of U.S. consumption and by 1968, when the Prudhoe Bay discovery was made, the figure was about 22 per cent; and that was only a beginning—by 1977 it was closer to 50 per cent and still rising with no sign of letup.

All through this long dance of supply, demand, and money, there were signs, if anyone had cared to look at them squarely—but it does not seem that public functionaries, whether they sit in Congress or on corporate boards, are capable of looking at things squarely or very far ahead—that all might not be well; that U.S. dependence on imported oil, growing willy-nilly, might lay the groundwork for national disaster. As early as 1948, Venezuela had taken a step toward nationalizing its oil industry (brought into being, of course, by foreign investment, notably Humble's) by granting itself 50 per cent of the profits, though not of the costs, rather than the established and contractual 12½ per cent royalty. In 1951, Iran tried to go the whole way, and failed. In 1956, the Suez shutdown suggested that the unpredictable politics of the Middle East might cut off the oil supply altogether, but that scare was traumatic chiefly for Europe (and for the American oil companies supplying much of Europe's oil); the United States was still comparatively immune and in fact, by raising its own production, was able to make up much of the European deficit. Then in 1960, Standard of New Jersey, Humble's parent, proposed to knock a few pennies off the price it would pay for Middle East oil—and hence off its royalties. Five of the oil-producing countries responded by meeting in Baghdad for the purpose of resisting price cuts and, if possible, controlling the supply of oil and thereby its price and profit and their own royalties: Saudi Arabia, Iran, Iraq, Kuwait, and Venezuela, with the latter, which had already had practice in bringing oil companies to heel, supplying the fervor and the strategy. Thus began the Organization of Petroleum Exporting Countries—OPEC. The members' interests seemed at first too divergent for effective common action, but in June 1967, with a new war between Israel and Egypt and the Suez Canal once more shut down, the three Arab states made the experiment which has since borne such unwholesome fruit: the use of their oil for political as well as economic ends, by cutting off the supply to those importing nations which would not accede to their policies. The experiment did not work—Iran and Venezuela cared more about profit than politics —but, looked at in retrospect, it seems obvious that the exporters would in time learn from it and learn to hang together in the future to make it work; as they indeed have. It was, in short, a sign.

Meanwhile, the industrialized liberal nations—Western Europe, Japan—had taken a view of energy rather different from the United States. In the United States, the oil-import quota system protected the domestic producers' profits but was linked with a policy of keeping the consumers' energy costs low. (The arbitrarily low gas prices, controlled by the Federal Power Commission since 1956, are only an extreme case.) The Europeans and the Japanese, on the other hand, more dependent on imported oil than the United States or the world at large, used various means to restrict the consumption of energy, in particular high prices by way of stiff taxes. What they created thereby was a strong incentive for efficient use of all fuels. Thus, by 1974, U.S. productivity had climbed dramatically since the 1920s—to about $107 of GNP per capita for each barrel of oil or equivalent fuel consumed. The corresponding figure for West Germany, however (with per capita GNP nearly as high), was over $180; for Sweden (with nearly identical per capita GNP and an even greater dependence on imported oil), about $164; and so for all the rest, with even perennially ailing Great Britain slightly ahead in this measure of efficiency. To put it another way: each of these nations, per capita, was consuming half the fuel used in the United States while producing up to 80 or 90 per cent more of the goods and services that are the national livelihood. The comparison is, of course, obviously approximate: apart from great differences in population, the industrial areas in Europe and Japan are compact and concentrated, the distances smaller, and transportation necessarily represents a smaller part of the GNP. Even so, the comparison is suggestive: the United States had grown used to living beyond its means.

Among the liberal nations, only Canada, with a tenth of the U.S. population spread over even greater distances, used more fuel per capita and got less from it in GNP—about $77.

Comparison with the Soviet Union is instructive in other ways. While per capita GNP and fuel consumption were low compared with the other industrial nations, its fuel productivity—the value of its per capita GNP for each barrel of fuel consumed—was about half that in the United States: $53 against $107. Looking at Soviet oil-production figures, however, one would say that they could perhaps afford that kind of inefficiency. Indeed, all through the 1960s, when Russia, with a long series of oil discoveries and rapidly rising reserves, finally entered the ranks of the world's leading oil producers, the Soviet supply and demand situation was exactly the reverse of that in the United States. Thus, in 1960, the United States was producing about two and a half times as much oil as the Russians, and U.S. production continued to grow to what turned out to be the peak year, 1970, at an average rate of about 3.5 per cent per year. Russian production, however, was increasing nearly three times as fast all through the decade and by 1974 had actually passed that in the United States, making the Soviet Union the world's leading producer. Russian consumption, on the other hand, was growing at a slightly lower rate than production, with the result that about one quarter of the national production was constantly available for export, for economic or political gain—but the two ideas, in the Soviet view, are inseparable.

It is not necessary to imagine Communist bogeymen under one's bed to find these differences disturbing. To paraphrase Charles Dickens in another connection: spend sixpence more a year than you gain and the result is misery; sixpence less, and happiness. And in this case, a sixpence—half of one per cent between growth rates in demand and production, in both nations—is about all the difference there has been.

Personally, however, I have not seen anything in a lifetime of watching to persuade me that the Soviet Union has altered its long-run goal of reorganizing the world, or all it can reach, for its own benefit—though that long run may be calculated in generations or centuries. Whether the means to that end are to be economic or military, oil is the indispensable ingredient. Moreover, in that perspective, Soviet strategy in Africa and western Asia since World War II looks like an immense flanking movement enveloping the Middle East, with the region's oil reserves as the immediate prize—more than half the world's reserves, ten times the American—and an even grander dominion not far beyond. The Soviet concern for the various "movements of national liberation" is then the concern of the 'fox for the welfare of the hens.

All this, at least in outline, was the meaning of that flame that lighted up the February evening at Prudhoe Bay in 1968; and, finally, the national survival, whether you see that only in personal terms or through the lens of freedom that the nation supports—but neither are these two ideas separable. Whether anyone saw that meaning whole, or whether that was even possible, is doubtful. What the oil men saw was oil, and, as the estimates came in and were engraved in the industry's habitual caution, brilliant comparisons. For instance. The last great American oil discovery was the East Texas field, which in October 1930 exploded in its epic gush of waste and rejoicing and, still producing, remains the principal reason why there are millionaires in Texas. But the East Texas reserves were estimated at around five billion barrels when first discovered—those at Prudhoe Bay were *double!* In effect, with the opening of a valve, American oil reserves had leaped upward by one third, roughly from thirty to forty billion barrels, and the nation's lifeline of energy had been lengthened in proportion. Finally, in 1968, that nagging gap between supply and demand amounted to only about 2 million barrels a day that had to be filled out with imports, and at the usual ratio to reserves, the new field might be made to produce at that rate; with the additional reserves already hinted by drilling and geology (though guarded from the public view), it would be possible to survive—as a company, as a nation—even the worst foreseeable disruption in imports, self-sustaining as in the days of the Texas gushers.

Even that much was much to see, at the time. Except for oil men and stockbrokers, the flame was invisible, but lives—which would be caught up and whirled in the currents of heat and burning air—began to change.

Among these, Robert Anderson, the only one of the oil company heads who remained in place from discovery to production, had the luck to be lying in a New Mexico hospital, near his ranch at Roswell, when the first news came. He had broken his hip, skiing, the break held together with pins. Reports came daily from Prudhoe Bay, as from Arco's other sectors; years afterward, when I talked to him, he remembered the winter tests, what they had shown, their dates. By way of therapy, he was out of the hospital and at Palm Springs, sunning and swimming, when the results came in and the decision was made to move the rig over to the mouth of the Sag and try for confirmation.

The reports on which the decision was based came from men like Jim Savage, who later would be in charge of all of Arco's North American oil exploration and hence intimately connected with the Alaskan reality. He was still a fairly junior geologist, inherited from Richfield and assigned to the newly merged company's Los Angeles office. As such, he kept track of the tests and their results as they trickled from the wells at Prudhoe Bay, all that frozen spring of 1968, and entered the telephone lines from Fairbanks to Anchorage, Los Angeles, Dallas, Philadelphia. Years later, in his managerial office high above Dallas, where drill cores and samples of crude from Prudhoe Bay shared the walnut shelves with his books, he remembered those beginnings and talked about them in the careful language of a geologist and an oil man, giving me by the way, an old friend, my first lessons in geology, Alaska, and its oil.

"The first thing we heard," Jim reflected, "was that this initial wildcat had been drilled and that it was a very unusual geologic set of circumstances and had numerous oil pay. Of course that was pretty exciting, but nobody knew the limits of the field or exactly how big it might be—doesn't, oftentimes, mean too much to get one particular well. The magnitude of the field really became apparent when the second well was drilled; came in with the same producing reservoirs in it. It became apparent that we might really have a *monster* of a situation!"

A couple of thousand miles northeast of Los Angeles, a Welsh engineer named W. J. Darch—Bill, with American or Canadian informality—had been handed a problem at a place in Canada called Margaret Lake. BP, attempting to gain a foothold in Canada as in the United States, had found some oil at Margaret Lake, but it was difficult stuff—no pressure of its own to force it out of the ground, too heavy to pump. Bill Darch had been put in charge of the Margaret Lake project and told to find some way of extracting the oil. By the spring of 1968, he thought he'd found a method: "I'd put up a case to spend a million bucks on a pilot plant—generating steam to decrease the viscosity of the oil and provide some push to get it out of the ground. It looked pretty good that I would in fact be getting this money to start the next phase of the development. Then we came up with the Prudhoe Bay discoveries. And I was told, quite clearly, that BP *didn't want their money chasing rainbows.*" Bill stayed on till the end of the year but there was not much more to do. He went back to London and from there, over the next few years, his talents would take him to most of the world's other difficult places where oil occurs and finally, at a critical moment, to the driver's seat of the company charged with transporting the oil from Prudhoe Bay.

In Houston, another engineer, Frank Therrell, working for the Humble Pipeline Company, found he had a new assignment: design a pipeline across Alaska to carry the oil. Humble, whose interest in the Arctic Slope so far had been the avuncular one of banker to Arco's hopes and good intentions, had decided it was time to take up the role of Big Brother. It was mid-March, within days of the fourth big test of Prudhoe Bay State No. 1 and the decision to try a second well and find out if the drillers had truly hit the monster that they seemed to have. To get started on the job, Frank needed a map. He scouted the Houston map stores. There *were* no maps of Alaska. He settled finally for a big wall map of North America designed for a classroom— Alaska, Canada, the United States, Mexico, in pale geographers' colors—rolled it up and carried it back to his office under his arm, like an overlong French loaf. (A few years after, when I went to a Philadelphia store on the same errand, I was offered a choice of several Alaskan maps with a lot of detail—including the route of the pipeline Frank Therrell had helped to plot. The world and its mapmakers had begun to take note of Alaska.)

Ripples from the find at the edge of the Beaufort Sea spread outward, touching other shores, other lives. In mid-July, Brock Evans, the Seattle-based Northwest representative of the Sierra Club (Alaska, with its two hundred members, came with the territory), landed in Juneau by ship with the club's president, Ed

Wayburn. The two men had been cruising the islands of south-east Alaska, studying areas still more or less wild that they hoped to protect from logging and other forms of development; in the days at sea, with these immediate problems on their minds, neither had given thought to what might follow if the drilling at Prudhoe Bay succeeded—indeed, had been no more conscious, probably, than the rest of us, that there *was* any drilling. They checked into the Baranof Hotel (named for the manager of the Russian American Company who, at Sitka in the first twenty years of the 19th century, created such order in the Russian colonies as prevailed to the end); bought a copy of the local daily, the *Empire,* carried their bags up to their room; and received a shock. It was the day Arco let out the first wary estimate of the reserves it had found, and the report supplied the headline in the newspaper—GREAT OIL STRIKE—and the talk around the hotel and the bars of Juneau. They talked it over, taking it in, and, Brock Evans remembered, "We realized right then that the implications would be enormous for the values we were talking about in the North. We just hadn't anticipated the tremendous crunch." So, if the oil companies were rushing, with hardly a pause for breath, to move their oil from discovery to exploitation, another and quite different view was already being formulated, and an opposing company of troops, self-enlisted, was beginning to form ranks. Brock Evans had been drawn to his job chiefly by discovery of western Washington and its mountains. The strategies of the contest to come would take him first to Alaska—Anchorage, Fairbanks, Juneau—for weeks at a time, then permanently to the national capital where finally the conflict would be resolved. There would not be much time for mountains.

At Prudhoe Bay it is still that day in February, cold late day in a winter that had been hard even for the Arctic, but you have seen what you came for—the gas blowing through the end of its narrow pipe and then the oil, the flame—and there is not much left but to wait and count, watch gauges, and take samples. You climb down the unsteady, clattering stairway, holding the guard-rails, and move off north, imagining the land's end and the sea. Icecap: a limitless dome of ice covering and containing the axis earth spins on, which in imagination and until you can see it close is featureless as a hockey rink with all the northern lands for a curb. It would be good to see it now, as it is, but the cold, after sun's brief appearance, is down to fifty below and will go lower, and the actual sea is four miles north; risky, you think, to get to and come back. There is the bay, nearer, perhaps possible, with the rig to steer by and its light. There is no other landmark. You turn east.

The darkness is like other darkness but it lies like a fog on the gray shimmer of the frost-hard snow; the elements here are less distinct than in the soft lands where our cosmogonies were formulated. The sky, though, is pricked with sharp stars in not quite familiar constellations, but they anyway are real. The sound of the engines throbbing under the rig is behind you and gone, as if spun from this end of earth, nothing to stop or hold it. Your boots slip and stick on the snow furrows with a sound like splitting wood, and that also is real; and the warm breath hoarded in your mouth. A shadow starts, moves off, and you stop, eyes straining through the haze of darkness—the fox again, or another, caught something this time?—but it is gone. Small events here are magnified—so the Arctic people, Inuit, winter-gathered in their driftwood, sod-domed huts, made stories of the life outside, souls of animals and men and the bodiless spirits which are shadows, materializing from the darkness, vanishing; waking dreams. If the earth lives, it sleeps now and we are its dreams—the men back there watching in the pumping warmth of the drilling floor, in the cities, planning, deciding, acting, from what is learned here.

Pain first and then sleep: the bay in this cold is farther than you reckoned, and you had best turn back now, while you can. In the full dark the rig is gone, but the fire still pulses in the place where the rig and its crew must be, and you steer for it. At the turn of the winter day, at this fountainhead of all the earth's cold, a fire has been lighted that will not soon be put out. It is life.

Chapter 2

GREAT LAND

> Congress has deemed that things in Alaska are national, not
> state-wide: as if going through two or more states.
>
> —Al Mongin

I lay back in the enveloping, foam-rubber-padded seat of the
747 and tried to read by the single overhead light. Around me in
the half-darkness everyone else seemed to be plugged in to ear-
phones, eyes fixed on the screen let down at the front of the
cabin where Barbra Streisand and a truck were engaged in tech-
nicolor comedy; or tuned to calming music, like the Japanese beside me, one finger moving in tranquil arcs, keeping time—the
plane would touch down in Anchorage for refueling and a foray
into the duty-free shop for furs, whiskey, native jade, then resume its northern loop toward Tokyo. I brought my eyes back to
the page. John Muir, making his own discovery of Alaska in
1879, by the slow and not unhazardous ship from Seattle, north
through the Alexander Archipelago, named to honor a liberal
czar, each island bigger than a small Eastern state:

> Along the sides of the glacier we saw the mighty flood
> grinding against the granite walls with tremendous pressure,
> rounding outswelling bosses, and deepening the retreating
> hollows into the forms they are destined to have when, in the
> fullness of appointed time, the huge ice tool shall be with-
> drawn by the sun. Every feature glowed with intention,
> reflecting the plans of God. Back a few miles from the front,

the glacier is now probably but little more than a thousand
feet deep; but when we examine the records on the walls, the
rounded, grooved, striated, and polished features so surely
glacial, we learn that in the earlier days of the ice age they
were all over-swept, and that this glacier has flowed at a
height of from three to four thousand feet above its present
level, when it was at least a mile deep.

Standing here, with facts so fresh and telling and held up
so vividly before us, every seeing observer, not to say geologist, must readily apprehend the earth-sculpturing, land-
scape-making action of flowing ice. . . .

The movie ended, lights came on around me, and my neighbor
lifted the shade on his window for a look. I caught a glimpse of
enormous mountains that seemed to reach up like twisted hands,
gloved in fresh snow, to catch the plane in its flight, and ran to
an unoccupied porthole to watch.

Today, if you enter Alaska from the central or eastern United
States, you see the shaping and sculpturing of the earthforms as
Muir saw them, but all at once, successive ages of folded rock
and flowing ice crowded together as in a speeded-up film and
more than eye or mind can take in. From Edmonton the plane
angles northwest to cross the border at the natural frontier of the
St. Elias Mountains, a cluster of immense peaks, some rising to
nearly 20,000 feet; then, turning west along the Alaskan south
coast, it follows the folds of a dozen ice fields and glaciers
flowing southwest toward the Gulf of Alaska, comes out over
water where Prince William Sound claws at the coast and its
protecting armor of mountains, and, lifting across a final glacier,
slopes down Captain Cook's Turnagain Arm to the thrusting
breast of Anchorage, a flat peninsula of gravel, sand, and muck
dropped by a vanished sea of ice. That day in September, the
high peaks punched upward through a floor of cloud, red-brown
in the bright sunlight where the slopes were too steep for snow
or ice to hold—fountains, Muir called them, from which the glaciers flow. Where the cloud cleared, the mountain folds were
woven together in a web of knife-edge ridges, the valleys between brimmed with ice in alternating ribbons of white and the
black débris of mountains ground to glacial till: you can *see* the

ice flow, a new earth in process of formation—mountains lifting, ice grinding and scouring to make way for life.

We clustered at the porthole, men with cameras, binoculars, or simply staring with our eyes, awed, dazzled. We were past the mountains and the ice fields now. Prince William Sound spread below us like a map, exactly detailed in the brown and green of steeply wooded shores, blue-green of deep water stippled with a gold wash of sunlight: another of Cook's names, honoring a corpulent, ineffectual son of George III. Beside me a gaunt, tall man, sandy-mustached, pointed—a place down there somewhere, up a jagged inlet, the village where he was born and grew up; he was an Alaskan coming home after years abroad. That pipeline, someone else said (this was the fall of 1975 and the pipeline from Prudhoe Bay had started building, aimed at a terminus near Valdez, up a fjord at the head of the Sound)— that pipeline must be down there somewhere too. How long was it going to be, anyway? Since the pipeline was my reason for coming, as it was for a good many other travelers to Alaska just then, and I had been making an effort to learn such things, I ventured that the distance was still officially seven hundred eighty-nine miles; but as the route kept changing slightly in the course of construction, it looked as if it would come out to about eight hundred—Oh, no! the Alaskan said, as if I'd belittled his land. Why, he said, it's eight hundred miles just *from Anchorage to Fairbanks!* The distance from Prudhoe would have to be twice that. The question was answered and the others nodded, acknowledging the authority of one born to the land. I did not try to correct them.

The Alaskan's mistake was understandable—everything known from childhood is enlarged in memory—but characteristic; Alaskans, like Texans, tend to brag their state's admitted bigness to titanic size. One of the oldest Alaskan jokes is aimed at the Lone Star people—shut up about your state or we'll cut ourself in half and then you'll only be the *third* biggest. The very name *Alaska* derives from a much-altered Aleut word invariably translated as "the *great* land," though to the island-dwelling Aleuts it meant simply that fringe of their world which was *not* island and where they did not live; which we would call "mainland."

Popularizing books about Alaska make the point of bigness by means of a map in which the state is superimposed on the other forty-eight: the tip of the southeast archipelago then lies about on Charleston, South Carolina, while Attu, the westernmost of the Aleutians, reaches to Los Angeles, and Barrow, the most northern point in Alaska, touches International Falls, which as it happens is the most northern town in the contiguous states. Describing it requires superlatives—there are a good many others—but also qualification. Alaska is like the King crab, a crustacean often exhibited in the lobbies of Anchorage hotels and occasionally found on their menus: a bundle of spindly legs with the reach of a man's arms, five or six feet, and at the center a small, plated body the size of a moderate Chesapeake oyster; a candidate for the state animal, if there were one. The *King* in King crab is one of those words reverently capitalized in the Alaskan newspapers, as something peculiar to and definitive of the state, along with King salmon, the state fish; *Outside,* meaning any place that is not Alaska, particularly the rest of the United States; *Native,* an Indian, Eskimo, or Aleut, in any combination; and *Bush,* the nonurban parts of Alaska or, sometimes, simply the antithesis of Anchorage, Fairbanks, or Juneau.

Alaska is indeed extensive, but the substance is less than it looks on a map or in the tourist brochures put out by the Department of Commerce and Economic Development. Canada's Northwest Territories, for instance, have three times the area and, with one tenth the population, are even more thinly settled; and perhaps for that reason turn up less often in Alaskan conversation than Texas, California, or Oklahoma. Moreover, Alaska shrinks on acquaintance, particularly if you are the kind of person who lives in a town and lives from the kind of work that towns afford, as most of us are. There is a lot of land, but much of it is not accessible, or not without notable effort and expense; or not usable except as a temporary camp, to touch lightly and move on.

Alaska's 7,600 miles of highway (about half paved) amounts to one seventy-seventh (say 1.3 per cent) of the mileage in the United States as a whole, in proportion to the total land area. There are simply not many roads of any kind, and those that exist lie within the quadrant of Alaska bounded by the Yukon River and the south coast, between Mount McKinley and the Ca-

nadian border. In the panhandle of islands, glaciers, and high
mountains known as Southeast where through accidents of history and geography the state capital, Juneau, is still situated,
there are, practically speaking, no roads at all; the only way of
getting to Juneau from any place else in the state is in a plane,
by way of a mountain-walled airport that, even on the good days
when it is not sealed off by snow, rain, or fog, has one of the
scariest approaches in the world, attested by frequent accidents.
(One of the rewards of a flight from Anchorage to Juneau is,
rather often, a night in Seattle when the pilot decides not to risk
his neck on Juneau's weather and its airport, though the detour
is so common that the airline no longer picks up the hotel bill.)
Agitation to move the capital anywhere else is therefore at least
as old as statehood, and it seems likely, after a series of referenda
on the subject, that it is one of the benefits Alaskan oil will begin
to pay for before the 1970s are out.

If you take the absence of roads as signifying wilderness, and
wilderness as one of the values for which Alaska exists and is
worth preserving, then this disproportion is a virtue, but it has
some paradoxical consequences. Although most of the state is indeed thinly settled and more or less wild, that quarter that can
be reached by road is heavily used, hardly less so in summer
than Yosemite, Yellowstone, the Adirondacks, or the White
Mountains. For a visit to Mount McKinley, it is well to reserve a
campsite weeks or months in advance; both there and in the
state-run parks and campgrounds, the apparently pristine
streams where a camper would draw his water are posted with
warnings against various intestinal risks. Moreover, as a matter of highway engineering and topography, much of the wild
country actually visible from the roads is at a remove, like the
scenery in a travel film: roadbound, you are tantalized by the
sight of mountain ranges grander than the Alps, high tundra as
free and open as the English downs—but insulated by ten or
fifty or a hundred miles of sodden muskeg, ferocious rivers, glaciers it would be foolhardy to attempt. The U. S. Army, for the
sake of a telegraph line, began cutting a pack and dog trail from
Valdez to Fairbanks as early as 1899, which became successively
a stagecoach route in 1917, an asphalt-paved highway in 1957.
(Alaskan roads, constantly rebuilt, never attaining a final state,

are lined with bypassed segments of their past—every journey recapitulates the history of Alaskan transportation.) It was not till 1957 that it became possible, circuitously and by a bone-shaking, summer-only gravel road, to drive to the vicinity of Mount McKinley from Anchorage or Fairbanks. (On a good day you can *see* the mountain from Anchorage, and before 1957, from either end, there was the Alaska Railroad, the dedication of which was the other reason for Harding's unfortunate visit in 1923.) There was no direct road between the state's two chief cities until 1971, when the separated sections along the rail line were finally connected.

Even so, there are only two roads into or out of Anchorage, both linked to the mainland by causeways bordered by deep mud flats, and on weekends, going and coming, the traffic slows to the pace of the Lincoln Tunnel approaches in New York; Anchorage counts about as many vehicles as people—for a family, perhaps a light truck, an offroad hybrid of some kind, a snowmobile for winter, one or more dress-up cars for Sundays and other formal occasions—though it only seems that all of them manage to get on the two roads at once. The feeling, strongest to newcomers, fading with familiarity, is one of being arbitrarily shut in as in occupied Berlin or Vienna in the years after World War II —again, only two roads out, through one or two hundred miles of hostile territory where the penalties for small errors sometimes included machine-gunning by insecure Russian occupiers. Fairbanks, though smaller, has three ways out, one of which now stops dead at the Yukon River; the others lead to Anchorage and, 2,500 miles down the Alaska Highway, Seattle or the Montana border.

In both places, usable land is scarce and expensive: apart from statutory restrictions, it may be underlain by permafrost or other material no less chancy and, either way, likely to slither down the nearest gentle slope the first time the earth shakes, as it does; the scarcity of roads limits the development of land—and vice versa. Those who come from the comparative spaciousness that surrounds most American cities—the "urban sprawl" abhorrent to city planners but nevertheless a form of freedom—find themselves boxed into a narrow city lot or, of late, a plasterboard and cinder-block apartment. Oil workers, who a few years back moved their families north in hope of cheap and abundant land and an unfettered life, have since reset their sights, as the boom ended, for the backwoods openness of Arkansas or western Pennsylvania, where the deer and wild turkey, if you've a mind to hunt, wander past the back porch of your cabin and you have no need to fly hundreds of costly miles, with a chartered plane and a guide, in search of the elusive caribou.

I encountered one such family on a later trip north, packed up pell-mell for departure—house plants, suitcases, moose antlers, cardboard boxes, seven-year-old daughter and all—and booming down the Alaska Highway. For two days I'd battled a late-spring blizzard of pelting slush and gravel-filled ooze, mired trucks and broken-down campers, but as I got past that, my own car abruptly balked, slewed to a stop, and would not start, and the road, which had seemed a spring freshet of similarly belabored travelers, was suddenly a lonely and anxious place, nothing coming from either direction. So, after half an hour, when a dust cloud topped the next rise north and a car came out of it, I put on a smile and waved and invited myself aboard for the fifty-mile run to the nearest gas station. The man's wife roared across the gravel washboard at seventy, attentive and smiling steadily, and we talked a little, about the hazards of the road ahead, the distance to the next station after the one I was headed for, an oil company job left behind in Alaska, which has for its state motto "North to the Future." As I got out, waving my thanks, I noticed that in the russet mud caked on the back window of the station wagon one of them had reversed that motto: "South to Freedom." They had been there, tried it, and were going Out.

The problem is the paradox of starvation amid plenty, too many people crowded into too small space, surrounded by land that is abundant but out of reach—or public, with the effect that it cannot be possessed, a different kind of problem, to which we shall return. However viewed, the solution is flying. The first sign that a *cheechako* means to stay is flying lessons. (There are a dozen flight schools in Anchorage, and the local community college offers an assortment of practical courses.) The sign that he has succeeded is a down payment on a small plane. There are, it is said, proportionately more pilots and more private planes in Alaska than in any other state. In the Anchorage *Times*, the classified ads for secondhand vehicles are headed "airmotive," and used planes run about even with used cars. A quarter of the world's seaplanes are concentrated in Anchorage, and Lake Hood and connecting Lake Spenard, near the International Airport, are said to constitute the world's largest seaplane base; for the seaplanes and the conventional kind, the waiting list for tie-down spaces runs to several hundred, perhaps a couple of years, as in an old-fashioned and exclusive club. Yet these comparisons are relative. The actual numbers are less than they sound—perhaps 6,500 pilots, 2,500 private planes. For many Alaskans, merely getting from one place to another, which to Americans elsewhere is one of those rights inscribed in the Constitution somewhere, presents insuperable difficulties. It does not seem strange, therefore, to meet, as I did, a cabdriver in Anchorage, born in the oil town of Kenai a few miles down the Inlet, who had never been north of Palmer—a junction about as far off as Philadelphia from New York—and whose feeling for the rest of the Great Land was one of incurious passivity. If so circumscribed a view is not altogether common, it is because born Alaskans are not common either; more even than other Americans, most have come by stages from somewhere else and are still in the process, arriving, trying out, settling, moving on, departing, in a luxury of choice—that is the underlying sense of Alaska's approving self-description, "the last frontier." Perhaps, then, to be born to an apparently boundless land is to perceive the necessity of setting bounds and of living within them.

Yet, to qualify the Alaskan brag of bigness and puncture a few of the other clichés is to get no farther than Dr. Johnson's dislike for America and her tiresome revolution, brought up to date—that *it was not London.* When that has been done and the judgment made, what remains is nearly everything that matters, a whole reality still at large, to be captured and known. The outer skin of that reality is the discovery that Alaska is what America once was. Or if the state motto is true and Alaska is also the future, then it is a future made hopeful and anxious by the old, many-colored American dream of perfection, but capable of realization with its faults corrected; a dream, but still compelling.

With intelligence and determination and the luck of landing
where you get a view of the whole, you can see that reality, that
dream, in process of unfolding. So, a few days after that first arrival in Anchorage, I sought out a man so placed and talked
through an October afternoon in his office over a bookstore on a
side street near the town center, taking my first steps in learning
the Alaskan reality. At the end, it was the American past we
talked of, the nation-making, continent-spanning common impulse that repeopled the eastern seaboard with new immigrants
and carried earlier settlers across the Great Plains and the mountains to the Pacific.

"What we're seeing," he said, "is the whole westward movement compressed in my lifetime. From the East Coast to the
West Coast we saw people land who didn't understand the English language, so they went into the mines, the steel mills, they
did all the hard, dirty work, and they made all the sacrifices: for
one reason—their kid was going to be an attorney and a teacher,
become a good American. And the second part of that is that
there is no choice. If by the year two thousand we're going to hit
three hundred million people, that's ten cities of ten million each
and we're going to see it right here in Alaska, the spillover is
coming up here: *there's no turning back. . . .*"

The speaker, as it happened, was Emil Notti, first president of
Alaska Federation of Natives and as such a charter member of
that first generation of college-educated Native leaders, a man
still young who was living the process he described, whose children's destiny is to inhabit that new world cut off from the
Yukon village where he was born. So viewed, the destiny is one he
shares with other Alaskans, newcomers no less than sourdoughs;
their similarities on the whole are larger than their differences.
If, then, Alaska is a laboratory in which the whole national history is being reworked and remade, the first step in mastering
that reality is to think of it as a continent to itself, large and
apart, like the one to which it is attached. History is man in a
landscape. It begins in geography.

The surface area of Alaska figures to 586,412 square miles, one
of those numbers too big to take in. If you man-scaled it to the
distance you could reasonably cover in a day on foot, two
square miles to look at and explore, 1,200 acres, it would take

you nearly a thousand years. The area is about one fifth that of the continental United States, but the coastline, 6,640 miles measured on USGS charts, is greater by one third. East to west, four time zones divide the state, as many as in all the Lower 48, and the kink in the international dateline was put there by treaty so that all of Alaska would at least occur on the same day. The most easterly and most westerly point in the United States is found near the Rat Islands in the Aleutians, 180 degrees, halfway round the world, from Greenwich and 1,500 miles west of Honolulu. The huge Alaskan distances account for the awkwardness in referring to the rest of the United States, or anyplace else—Alaska is a world to itself. "The States" is still the easy and natural form, left over from the time when the nation and Territory were more tenuously connected than they now are. It is normal and polite to ask a visitor if he is new to "the country," if he is planning to leave "the country." Periodically, its citizens come forward to propose independence (among them, a few years back, Emil Notti, despairing of Native rights in the national context), and the movement does not seem Quixotic. Sovereign, the state arranges conferences and trade relations with the nations of the Arctic rim, neighbors as near in spirit and distance as the South 48. In the towns, the corresponding message on bumper stickers is "Leave Alaska for the Alaskans"; or ambiguously spray-painted on a sign where the Alaska Highway crosses the Canadian border, "Leave while you still can."

Geologically, two mountain systems cut Alaska in three: the northern extension of the Pacific Range, walling in the roadless Southeast and looping west down the Alaska Peninsula toward Asia in the thousand miles of sea-born peaks that form the Aleutians; and, in the north, the Rockies, bent westward by tectonic pressures and named the Brooks Range. Below the Alaska Range, and formed by the mountains, is a climate not very different from Seattle's, lightly refrigerated—mild, wet and foggy in summer, snow-deep in winter. North of the Brooks Range is Arctic desert, where snow and rain are measured in inches, not feet. Between is continental plain, a great almost-level bowl drained by the Yukon, as hot, sometimes, in summer as Kansas or the Dakotas, but colder in winter even than the Arctic plain. The geography is human and more complicated. In that scheme,

Southeast still stands apart, along with the Aleutians and the Arctic plain, homeland of the Inuit. The south coast, facing the Gulf of Alaska, broken by Cook Inlet and Prince William Sound and centered on Anchorage, is known as South Central. The west coast, fronting the Bering Sea, stands apart from the Yukon drainage and is subdivided by culture, the last region where many of the young of whatever branch of Eskimo still learn language and custom from their elders. Fairbanks, founded on a gold strike at the head of navigation of a Yukon tributary and kept going by Army and air bases and a railhead, centers Alaska in a loose grip; the upper Yukon has links—of language, tribe, history, transportation by water and trail and road—as close with Canada as with the rest of Alaska. Even in outline, the Alaskan continent is too big for an American state, too various to form a nation. But it is a country; the description fits.

The separateness of Alaska is a fact of geography reinforced by history. In an era in which American colonialism has become an axiom of academic logic, Alaska deserves study as a true instance: the only U.S. territory that was and in many ways still is textbook colonial. When, more recently, government by Congress has become a slogan, Alaska is an object lesson in effects and consequences.

Alaskan history began in Russia, in two epic centuries of expansion that explored, conquered, and possessed the immense Siberian mainland. The Russian movement east paralleled and coincided with the Western European drive across the Atlantic to seize and populate the North American continent—a parallel movement but in reverse, a mirror image, with differences in character and effects that went deeper than geography or national interest. For both, it was furs that furnished the immediate motive and assured the profit in acquiring those vast and thinly populated lands—a trade that in northern Europe supplied both luxury and practical necessity and in China provided the medium of exchange for the complex and profitable commerce in silks, spices, teas. European Russia had been the original source of furs in a trade that is probably prehistoric, but by the 16th century, as Europe prospered and its appetites grew, the overhunted Russian forests paid smaller dividends. Hence, by the time the

Europeans were finding new sources of their own across the Atlantic, the Russians had begun their push east, seeking virgin territory to supply the raw material of their trade.

In the vanguard of this eastward movement was a half-civilized south-Slavic tribe that through slow centuries of liberation had stood between Tartar tribute and dominion and the Russians: the Cossacks. Although the Cossacks, like the Tartars, are renowned as the great horsemen of the Russian steppes, they were also, in their Don homeland, the premier boatmen of an otherwise landbound nation. In trackless Siberia as in North America, a succession of great rivers cutting through taiga and tundra served as highways to the farthest reaches of the continent. Hence, in the Russian drive to the east, Cossack *promyshleniki*—outriders, advance guards, hunters, scouts—had exactly the role of the half-wild French-Indian *coureurs des bois* and *voyageurs* who led the way from the Atlantic to the continent's farthest north and west, though traveling the rivers in crude, heavy-built *bidar* rather than the sleek, bark-covered canoes the French had learned from the Indians. Hunting on their own, the Cossacks worked in small, light-traveling teams, *chunitzi*, bringing the catch in to a central *artel* commanded by a Russian *peredovchik* who was bound, in turn, to companies of fur merchants organized in far-off Moscow or Kiev. Where the Cossacks encountered more primitive tribes that could be subdued with the advantage of steel and gunpowder, they imposed the system the Russians had learned from the Tartars in the centuries of subjection: the defeated tribesmen hunted on the Cossacks' behalf and rendered periodic tribute in furs; as guarantees of payment and peaceable conduct, the hunters delivered their women and children to the artel as hostages, where other uses were found for them; slaves, their men serfs according to established Russian notions. The French and Scots in North America practiced trade rather than subjection and tribute, quite scrupulously, and were therefore dependent on a steady flow of European goods brought down long and difficult supply lines; violence in the American fur trade was more often that of rival traders quarreling among themselves than against the tribes. The Cossacks, in contrast, were virtually self-sufficient, had no general interest in the well-being of the Siberian tribes, and had the means of compelling them to produce their assigned quotas of skins.

By the turn of the 18th century, while the coureurs des bois were only beginning to penetrate the wild country west of the Great Lakes, the Russian system, fueled by brutal Cossack energy, had planted outposts across Siberia and reached the shore of the North Pacific. In the midst of wars to the west, north, and south, Russia's demonic giant of an emperor, Peter the Great, whose reign touched and reshaped every layer of the disorderly Russian reality, turned his attention to Far Eastern Siberia, and for similar purposes—to consolidate and expand what the promyshleniki had begun; and to assure control of Russia's land routes to China, the final source of the profit in furs. It was part of this purpose to discover what lay beyond the Siberian coast, still farther north and east—whether, for instance, the far northwest of the American continent, unknown, unexplored, was in fact connected to Siberia, as was thought possible. To this end, Peter set in motion an ambitious expedition commanded by the Danish sailor Vitus Bering, one of the many Europeans whose skills were recruited to fill the Russian void—of which, apart from the Cossacks in their little boats, seafaring was among the most obvious. Peter would be sixteen years dead before the effort bore fruit, but it was his outreaching will, finally, that discovered Alaska.

Taking advantage of the frozen earth, Bering in late winter of 1725 set out with a train of heavily laden sleighs from the swampy new capital of St. Petersburg. It took him two years to reach the forbidding shore of the Sea of Okhotsk, a long summer, with his Cossacks and Russians, to build the raw Siberian timber into a small ungainly ship, bravely named for the archangel Gabriel. In it, he crossed the Sea of Okhotsk and at last reached the actual Pacific, wintering on the Kamchatka Peninsula, a name still proverbial as a barren end of the world. After more than three years of preparation, when the ice cleared the next summer and the St. Gabriel at length sailed north, in mid-July 1728, the actual voyage was uneventful. After three weeks, Bering sighted a large uninhabited island which by custom he named for the nearest feast in the church calendar, St. Lawrence, actually about a hundred miles off the west Alaskan coast; thereafter nothing. After sailing cautiously as far north perhaps as the Arctic Circle and, through the shroud of summer cloud, seeing no further land in any direction, Bering concluded that his commission was fulfilled, that the two continents were not joined; and sailed back to Kamchatka. The return to St. Petersburg took him till March of 1730, five years after his setting forth, when he first heard of his master's death and the untidy succession that followed—Peter's wife, then a grandson, a niece, the beginning of thirty years of turmoil. In time Bering's name attached to the northern sea and the strait he had sailed through.

Nevertheless, within the decade the imported functionaries who kept things going in Peter's capital devised a new and more elaborate exploration, with scientific purposes. Hence by the spring of 1741 Bering had repeated his previous exertions: had recrossed Siberia, established a permanent base on Kamchatka, Petropavlovsk, built two ships with the same pair of names, *St. Peter* and *St. Paul;* and set forth to discover America. The expedition went badly. Misled by nonexistent islands shown on current European maps, Bering set a southerly course and missed the Aleutians altogether; in a storm, the two ships separated and never rejoined. Nevertheless, after six weeks of fruitless navigation, he caught a glimpse of the giant, snow-topped dead volcano which is still the air traveler's most likely first sight of Alaska, and gave it a name, St. Elias. (About the same time, Bering's second in the lost ship came within sight of the Alexander Archipelago four hundred miles to the southeast.) Sixty years old by now, worn, unwell, indecisive, fearful of weather and hostile natives, Bering was persuaded to put in at what was probably Kayak Island, east of Prince William Sound—but only for a few hours, to refill the ship's water casks. On board, as ship's doctor and the expedition's scientific ornament, was a remarkable and bad-tempered German naturalist, Georg Steller, who hectored a taciturn Cossack into rowing him ashore. While the crew collected water, Steller furiously gathered a boatload of specimens —plants, birds, animals, implements abandoned by frightened natives; wrote the only scientific description of a new and soon-extinct species of sea cow; recognized a local jay as kin to one he had seen years back in a book of American birds, imported to

St. Petersburg—and, with a leap of imagination, guessed that they had indeed found America and not an Asian promontory.

Bering, incurious, kept to his cabin; called the men back, and Steller with them, protesting the lost chance and making threats; lifted anchor and sailed away, west. The voyage back was worse than the setting out—autumn storms, men dying of scurvy—and ended in a wreck on an island off Kamchatka. In the wintering there, Bering's death was one among the many; in the spring, the survivors knocked together a boat from the ship's timbers and got back to Petropavlovsk. Nevertheless, in those months they had made the first step in the real but undocumented discovery of Alaska: the island, also named for Bering, swarmed with sea otter, an animal scarce on the Kamchatka coast, its fur more precious to the Chinese than jade. They would be back, hunting. The island and its furs, soon exhausted, led to the Aleutians; to Alaska.

From Kamchatka, Cossack chunitzi crossed these terrible seas in clumsy boats whose planks, with the scarcity of nails and any metal, were bound together with green-skin cords and, when the cold waves made up, often simply came apart. The chances of getting to Bering Island and back with a load of sea otter were not great but the reward was stronger—not simply the silver rubles the merchants in Petropavlovsk would give for the skins but a degree of freedom within that remote and desolate camp, perhaps even return to the Cossack homeland; and for men accustomed to death, that by thirty-degree salt water was not greatly feared.

By the 1750s, the swarms of otters on Bering Island had been exterminated, but the hunters, sailing the latitude three hundred miles east, as you can without compass, by sun and stars, had hit a new source, with seals too, in season—the tip of the Aleutian chain, which thereafter they had simply to follow inward, island to island, till they found the land whose name sounded like Alaesksu, Alyeska Alashka; the variants depending both on what Russians heard and what Aleuts said, the dialect differing from one island to the next. Merchants formed companies to regularize the commerce: bigger and solider boats, small ships; supplies packed overland from Moscow; fixed stipends for the hunters, a modest share in the profits for the peredovchiki, as in the North American trade. Where the Cossacks encountered groups of natives, they applied the Siberian system of subjection and tribute and set up camps that in places lasted years at a time, supplied by Aleut hunters, served by their women. A few bands resisted and with Aleut quickness and ingenuity fashioned armor of plates of hardened bone to ward off the musket balls, grasped at first sight the meaning of powder and guns, and seized both to turn on the invaders; but the Russians came back, burned or blew up the domed dugout Aleut houses, slaughtered the men; and the survivors served and interbred, adopted Russian names and Cossack ways. Reports of extreme cases sometimes filtered back to St. Petersburg—a shipload of Aleut girls, for instance, carried back to Kamchatka and, in the course of things, killed—and were answered with stern imperial decrees a year or two later; so one Petropavlovsk merchant whose promyshleniki were notorious for violence was tortured and exiled, his company dissolved.

As the Russians advanced, the fighting was more often between the servants of competing fur companies than against the now docile and serviceable Aleuts. One Kamchatka merchant, Gregory Shelikov, took a longer view of order and profit: a colony settled by Russian families, an addition to the imperial dominions, with schools, mission priests; and in 1783, on Kodiak Island south of the Alaska Peninsula, established the first permanent Russian post, which he named Three Saints for the ship that brought him there. Eight years later he sent out as manager Alexander Baranov (or Baranof), who would give the rest of his life to Alaska and who, alternating between drunken torpor, savage energy, and paternal benevolence, would effectively create Russian America and give it the essential form which it had to the end. In 1799 Shelikov's vision was rewarded: a twenty-year imperial charter transformed his partnership to the Russian American Company, in which the Emperor and his family took out stock; gave it a monopoly of the trade, ending the bloody rivalry among competing merchants; and conferred responsibilities for government and imperial law on the model of the Hudson's Bay Company in Canada, assisted by naval ships based at Petropavlovsk and detachments of Siberian troops. Baranov, now both company manager and governor of the colony, established posts along the Cook Inlet, a small shipyard farther east on the Gulf of Alaska, and built a palisaded headquarters at Sitka in the Alexander Archipelago. Destroyed by hostile Tlingits and rebuilt with the resounding name of Novoarkhangelsk—New Archangel, for Gabriel—it became a capital, though on a small scale: never more than a few hundred Russians and the half-breed civilized natives called Creoles, but with at various times schools, churches, a cathedral and resident bishop, a seminary, a lighthouse and an astronomical observatory, a hospital; and, at the top of a defensible rise, the log-built, barnlike seat of government known as Baranov's Castle.

The extent and degree of Russian control was debated at the time of the American annexation and has been, recurrently, since then. The Aleutians were secure, their original population thinned by disease and intermingling, the people transferred elsewhere as the Russians found uses for them; so also Kodiak Island, much of the Kenai Peninsula, mainland fringes around Cook Inlet, Prince William Sound, and the south coast generally on around into the Archipelago. Russians traded up the Copper River east of Prince William Sound but never reached its head-waters—it was among the important Indian trade routes, by which in time both Russian and British trade goods were carried over the mountains between Alaska and the Mackenzie valley. Russians (and, later, Sitka-educated Creoles) explored up the big rivers of southwest Alaska—Nushagak, Kuskokwim, Yukon, magical names—and set up a scattering of trading posts, as far inland on the Yukon as its junction with the Koyukuk; naval parties surveyed the huge bays and peninsulas of west-central Alaska and as far north as Point Barrow, about the limit, with the ice pack, for a summer's voyage under sail. On the other hand, although tribute was abolished and a schedule of wages decreed for native hunters and workers, even around New Archangel the Tlingits were restive, sometimes dangerous, to the end. (In 1852, they attacked and destroyed a fort a few miles southeast of the capital.) Among the Orthodox missionaries there was a sprinkling of martyrs and, by the same route, one acknowledged saint. As late as 1851, at Nulato, the most remote of the Yukon stations, the handful of Russians was unexpectedly eaten by neighboring tribesmen.

At the same time, Russian sovereignty was never seriously disputed. From Captain Cook onward, before the imperial charter conferred even a semblance of Russian government, British, Spanish, and French expeditions explored the Alaskan coast, scattering place names across the map, but scrupulously refrained from claiming territory. In 1824, with Mackenzie-based Hudson's Bay traders probing the upper Yukon and British and American settlers beginning to vie for the Pacific coastland that became British Columbia, a treaty with Great Britain settled the Alaskan boundaries about as they have remained; and they were confirmed the following year in a treaty with the United States.

Russian control was tenuous by reason of the impossible distances. Whether across Siberia to Kamchatka or by sea from the Baltic and around either of the great southern capes, communication with St. Petersburg was never greatly faster or more effective than in Bering's time. With supportive colonies in northern California and trade to China, governors after Baranov tried to solve Alaska's supply problems but never fully succeeded. For the difficulties and costs, Alaska in Russian hands remained a doubtful proposition. The company paid steady dividends (regularly about 13 per cent or more per year on invested capital) but ran up debts. In terms of income, the take in furs bordered on break-even and overall declined; periodically, by way of conservation, and to the dismay of natives relying on the trade, the Russians closed districts to hunting to allow time for the stocks to recover—a system called *zapusk*, holiday. Only seal furs from the Pribilof Islands—the great rookery of the whole North Pacific, to which the animals return each spring in multitudes, by thousand-mile migrations—provided regular and predictable income; the islands, uninhabited but known to Aleut tradition, had been discovered in 1786, colonized with Aleuts to harvest the skins, and it was there that the zapusk system was first practiced.

Increasingly, the Russians were in the awkward position of maintaining their fur supplies for the benefit of others. American companies based in Puget Sound disrupted the trade, and the natives learned to play the Russians against the interlopers for better prices. From about 1850 on, New England whalers invaded Alaskan waters in numbers, penetrating the Arctic and occasionally overwintering, until the profit in whaling was destroyed by reckless hunting, Confederate raiders during the Civil War, and easier sources of oil. (The era in large measure ruined the livelihood of the northern coastal Eskimos dependent on whaling; to this day in a few such villages a white American is simply "a Boston man.") Sea-borne sealers from Europe and the United States timed their voyages to cut off the animals in their migration to the Pribilofs: the seals were shot with rifles, the one in ten that did not sink recovered in boats; a method known as pelagic sealing that could destroy a whole generation of seals and grew more obdurate as the whales diminished. (The carcases, boiled down, were useful for oil as well as their skins.) In 1837, the Russians in desperation had rented the whole of the southeast archipelago to the Hudson's Bay Company on a long-term lease, for the payment of a few bales of skins a year. A few years later, the British from the Mackenzie, rediscovering old Indian routes through the mountains, established a post at Fort Yukon, a hundred miles inside the present Alaskan border. At the time of the Crimean War, the two companies agreed to treat each other's territories as neutral, an agreement approved by their governments, although a British squadron raided Petropavlovsk, which was administratively linked with Russian America.

In the aftermath of that war, imperial policy-makers reconsidered the prospects for their unprofitable American holdings—the colonies were not defensible, if they did not fall to the encroachments of British traders, they would be swarmed over by land-hungry Americans from Oregon Territory, as Texas, California, and the American Southwest had been a few years earlier—and concluded that the time had come to withdraw. Unlike the other nations that had shown interest in Alaska over the years, the United States had no rivalry with Russia for European or Asian power and was therefore the most suitable candidate.

As early as 1857, the possibility was delicately broached to President Buchanan by the Russian minister in Washington, Baron Édouard de Stoeckl, a Gallicized descendant of one of Peter the Great's German functionaries. Buchanan was not a man to take notice of delicate implications and in any case had more immediate concerns. When the Civil War intervened, however, and the British treated the Union with indifference or hostility, the Russians made friendly gestures, playing their own game, which included the harmless disposition of Alaska. With the war over and won, the baron found a sympathetic listener in William Henry Seward, Lincoln's Secretary of State, who stayed on under Andrew Johnson.

At this distance Seward's interest in Stoeckl's proposal looks obsessive, irrational—a recurring quality in the men whose minds have touched Alaska, as if something like destiny had seized and made use of them. An undersized New Yorker with a rooster's profile who entered politics as a certified Abolitionist Republican senator, Seward, as Secretary of State, nurtured large but contradictory ambitions. On the one hand, he hoped by acquiring strategic territory to provide his country with absolute defenses against European encroachment—in particular, so to encircle with the American flag the loosely held British provinces to the north that Canada would end in a natural union with the United States, an ambition as old in American policy as Benjamin Franklin. At the same time, however, in a nation sick of war and death, it was necessary to pursue these ends by pacific means. Thus, when Stoeckl came forward, Seward had already made offers for most of the island colonies in the Caribbean (Santo Domingo, Cuba, Puerto Rico, lesser places in French and Swedish hands) and had sought means of acquiring the Hawaiian Islands; Danish Greenland too, flanking the Canadian Arctic, had been on his mind. None of these projects had borne fruit. Stoeckl's proposal promised a success after many failures.

Other motives can be inferred. Congress was already on the course that would end in a trumped-up impeachment in 1868—a grand achievement might divert Congress, or at least the people, from more immediate problems. Traders based in the new state of California, with leverage in the national politics, had concluded that in American hands Alaska would be worth money, particularly the Pribilof seal rookeries. A telegraph company aimed at quick and profitable communication with Europe by way of Siberia and a cable across the short and shallow Bering Strait and in 1864 and 1865 had gone to the expense of sending surveyors to Alaska to work out a route.

Seward's interest sent the Russian minister back to St. Petersburg for instructions, where he was authorized to sell Alaska for
a minimum of $5 million (enough to cover the Russian American
Company's debts, mostly to the government), with an incentive
for bargaining in the form of a vague promise of a commission.
Back in Washington in February 1867, Stoeckl talked the price
up to $7 million, to be paid in gold, in London, within ten
months from the signing of a treaty. Seward threw in another
$200,000 for an assurance of clear title, but whether for the pos-
sessions of the Russian American Company or for the territory as
a whole was never put into words. On March 29, 1867, Stoeckl
stopped by Seward's home after dinner to say that the terms
were acceptable; the Secretary sent for clerks from the State Department to write up the treaty, and it was finished and signed at
4 A.M. the following morning.

There was at least one practical reason for such urgency: the
current session of Congress was about to end. Johnson annoyed
the members by proclaiming a special session to consider the
treaty of purchase, commencing April 1st; Seward turned to a
powerful ally, Charles Sumner, the ailing Massachusetts chairman of the Senate Foreign Relations Committee, who was persuaded that confirmation of the treaty was necessary to repay
Russian favor during the Civil War. Overnight, Sumner collected
and mastered all that Americans knew of Russian America and
delivered it in a three-hour oration on the floor of the Senate,
along with the suggestion for a new name—Alaska, which stuck.
(Others were proposed: Seward Land, Yukon, Sitka; a wit
offered Walrussia, for the comic animals Americans knew from
circuses and the vaudeville stage.) After desultory debate, the
treaty was approved in a closed session on April 9th with two dis-
senting votes; later, it was inferred that fur-trading money from
California had been more persuasive than the arguments of
statesmanship, but if so, the tracks were well covered.

At Seward's urging, the Smithsonian Institution dispatched a
revenue cutter to Alaska to find out something at firsthand about
the land the nation had acquired. A commissioner was appointed
to accept the transfer of sovereignty from the last Russian governor, Prince Dmitri Maksoutoff, and by way of garrison a detachment of troops sailed north from San Francisco under a stiff-

necked, disreputable Civil War veteran and Indian fighter, General Jefferson C. Davis, a name with only the middle initial for distinction from the Confederate leader. Boomers from California and Oregon arrived at Sitka, as Americans would call it, to open saloons, block out a town plan in salable city lots; Hayward Hutchinson, a Baltimore man partnered with a California senator and promoter, Cornelius Cole, turned up with cash to buy up the Russian American Company, which he did—stockades, warehouses, ships, flour barrels, sheepskin coats and all—for $65,000, quadrupling his investment by reselling the stock. On October 18th, Alaska Day, the Russian flag was hauled down in the yard in front of Baranov's Castle and the American run up.

While these distant ceremonies were transpiring, the Russian minister Stoeckl back in Washington began to feel a certain uneasiness; it appeared that the Americans were going to take the land but not pay for it. The difficulty was in the House—the same body of men that moved Mark Twain, as a young reporter, to incoherent outrage; of whom the young Henry Adams recorded the high-placed opinion that "You can't use tact with a congressman! A congressman is a hog! You must take a stick and beat him on the snout." These gentlemen at the time were engaged in providing the occasion for Johnson's impeachment and trial that came to a head the following winter. (When the President attempted to dismiss the irascible Secretary of War he inherited from Lincoln, Stanton with congressional encouragement locked himself in his office like a naughty child.) Hence, when Johnson conveyed the Alaskan treaty to Congress for the appropriation of the purchase price, the House waited a disdainful four months—and then in November voted a resolution against further money for Seward's territorial ambitions. The impeachment circus reached its three-ring climax in the Senate, and Johnson remained in office, by a single vote. The deadline for payment came and went—invalidating the treaty by its own language, though the symbolic American occupation of Alaska was already half a year old—but the appropriation bill languished in the House and the gold sat in the Treasury vaults. Derisive stories began to appear in newspapers—"Seward's Icebox," "Seward's Folly," and the other sallies of journalistic wit which have conditioned American attitudes toward Alaska ever since. Seward responded by inspiring favorable stories of his own. On the floor of the House, gentlemen on both sides debated, in a vacuum of general ignorance, the exact value of Alaska for homesteading and agriculture; someone suggested that if the national honor did indeed require payment, it would be best to *give* the Russians their millions and let them keep their worthless land; someone else wanted to deduct old Russian debts whose validity had long since been discredited in American courts. Seward and Stoeckl busied themselves in private talks with the opposition.

The obstruction is explained in part by congressional pique at having failed to send Johnson back to Tennessee, if not to jail; there was also (among men who had carried through a great war for, among other things, the liberation of black slaves) a decided resistance to one overgenerous clause of the treaty assuring the rights of citizenship to any civilized Alaskans who wanted it—a few hundred Russians and ten or eleven thousand dusky Aleuts, Creoles, and assorted Indians. (The rights of the perhaps thirty thousand Natives who did not live in or around the Russian settlements were left to be specified by Congress at some later time.) This attitude, which we would now call racism, would color congressional views of Alaska for another century.

There was a more compelling reason for the uproar over Alaska: in the political commerce of the time, it was a means of bargaining; resistance raised the value of the thing withheld, a congressman's vote. Hence, it was foregone that the appropriation would finally pass, as it did on July 14, 1868, the anniversary both of the destruction of the Bastille by the *sans-culottes* and of Bering's first faring forth in search of the intercontinental strait; the delay merely gave space for negotiation. "A congressional appropriation costs money," as Mark Twain observed at the time, quoting figures; favorable newspaper space could be had cheaply enough, but congressmen and senators were not inexpensive ("the high moral ones cost more," Twain drily noted), serviceable lobbyists came as high, and a charming and persuasive lady or two headed the list. All in all, only $7 million was actually delivered to the Russians and Stoeckl himself went home none the richer, letting out privately that Seward's odd $200,000 had melted away on congressional favors. (A committee, investigating, invited him to testify about the missing gold but he politely declined, pleading diplomatic custom.) Back in St. Petersburg, Stoeckl discovered that his crowning achievement was no more highly regarded there than in Washington. The government insulted him with a paltry commission of 25,000 rubles, stockholders raised an outcry, patriots were scandalized at the thought of surrendering even the most remote and barren portion of the Russian domain to foreigners for cash; and to this day the baron's name is generally omitted from Soviet histories and encyclopedias.

It was from these sour beginnings that Alaska's American history lurched forward toward the present.

The Americans who hopefully converged on Sitka in the fall of 1867 were soon disappointed. With a spontaneity as old in America as the Plymouth Plantation, they claimed land, began buying and selling, building, setting up businesses, established a town council and a school board, elected a mayor—and discovered that, legally, all of these activities were fictions. The ceding of Alaska had abolished such legal order as the Russian system conferred, but Congress provided nothing in its place. There was thus, for instance, no means of getting title to land, of forming a local government or collecting taxes to support it, of charging, trying, or punishing someone who had committed murder. Alaska had been attached to but not made part of the United States. It had become a kind of limbo, its status undefined, lawless in a quite literal sense. The Americans drifted elsewhere. The Russians, some by now Alaskan-born, having no personal links with the motherland, had been assured by the treaty of transportation home if they chose it and departed in shiploads. The Creoles and Natives, having no place else to go, remained; and General Davis's troops. If Alaska had a status, it was that of an occupied province.

Over the next seventeen years, apart from the original appropriation bill, Congress conferred two laws upon the province. One in 1869 extended U.S. customs regulations and provided an officer at Sitka to collect import duties, though the forgetful Congress often neglected to pay him. A second law, in 1870, confirmed the motive that had tipped the balance toward acquisition: it granted Hutchinson, now organized as the Alaska Commercial Company, a twenty-year monopoly on the seal trade in the Pribilof Islands, specifying an annual rent and royalties on furs and seal oil, with pious regulations on the Russian model for conserving the seals and providing for the welfare of the Russianized Aleut workers; but no means of enforcing these regulations. Besides the lucrative Pribilofs, Hutchinson's company gradually took over the old Russian trading posts, and through successors with altered names—the North American Commercial Company (1890) and finally the Northern Commercial Company, headquartered in Seattle—has remained an Alaskan presence to this day.

General Davis irritably bridled at efforts in Sitka to make him assume the law-giving and justicial functions of the departed Russian governor; his hands were full with more immediate responsibilities. The men pillaged the Russian cathedral and imposed their attentions on the women—Creoles, Indians, the remaining Russians—with a fine absence of discrimination; the officers supplemented their pay by smuggling in whiskey or, more profitably, molasses, the raw material for what the Tlingits called *hoochinoo*, whence a useful new word in English, *hooch*. It was therefore with some relief that, after ten years of this, the general and his detachment departed for the comparatively peaceful task of suppressing an uprising of the Nez Percé. The only official American presence remaining was then the impecunious customs officer, and when the Tlingits threatened a massacre, he appealed to the captain of a passing British warship for protection; ships of the U. S. Navy relieved the British, when in the area, and until 1884 their commanders constituted such government as there was.

In that year, responding to years of petitioning, Congress passed the Alaska Organic Act to establish a government of sorts: the country would not have the dignity or rights of a territory, but it would have an appointed governor, a judge, a body of laws. From the standpoint of those affected, it was an extraordinarily careless piece of legislation. The Organic Act, for example, applied the U.S. mining laws (miners could now stake claims, and many did, producing the famous series of gold strikes from Juneau to Nome and up and down the Yukon drainage) but not the general land laws (there was no way for anyone to get title to any land). The laws adopted were the statutes of Oregon—"so far as applicable," which was not very far. The gentlemen in Washington evidently supposed that since the two remote places were within a thousand miles of each other, they must have something in common, but Oregon law assumed a territory organized in counties and townships, inhabited by people who paid taxes; in Alaska, with no land laws, there could be no towns or counties and therefore no schools (Oregon required county superintendents) or any other form of local government; with no taxes and no taxpayers, there could be no juries (Oregon defined a juror as a taxpayer); and so on. The governor was obliged to inspect his domain, including the far-off Pribilofs, and report periodically to Congress, but Congress had not thought to provide any means of transportation. Such schooling as there was continued to be paid for largely by the Russian government, which felt some obligation toward its Orthodox churches and schools; or by occasional American missionaries. When Congress did bethink itself of schools for Alaska, it tended to forget about money for teachers' salaries or, if it remembered, voted the appropriation after the last ship of the season had sailed north from Seattle in August. Since trial by jury could be had only by taking ship for San Francisco, the Alaskan judges did not have much to do, and the office became a convenient dump for disreputable personages with political connections—such as the drunkard Ward McAllister, Jr., or E. J. Downe, a practicing thief who found new scope for his profession on the bench.

Congressional ineptitude—the inability to write laws that are consistent either with themselves or with the reality they are designed to regulate—is, of course, as familiar in the 1970s as in the 1880s, the everyday feedstock of newspapers, though today, like everything else, it costs more. In Alaska, however, it had a purposeful pattern. Periodically, in chorus with Oregon, Washington, and California, Alaskan voices swelled to demand notice in Congress and the various American rights: in the seventeen years that culminated in the first Organic Act; again in the period leading up to a second Organic Act in 1912, granting an elected territorial legislature, with limited powers and subject to congressional veto; and thence forward to statehood, forty-seven years later. And at each step other voices were to be heard, louder, reciting the same arguments that had obstructed the Russian payment—that Alaska was a barren, unpopulated, unproductive, frozen wasteland, unfit for white men, a bottomless drain on the national treasury. Such arguments, which served to justify delay in congressional action and drastic limitation when it came, were both circular and self-demonstrating: Alaska was inhospitable, and therefore not many settled, and therefore it was not necessary to grant them the forms of government, and therefore— It appears that the Alaskan population remained level at fifty or sixty thousand from the Russian period till the 1930s, though it is hard to be certain—a census was one of those amenities, along with maps, charts, lighthouses, schools, that Congress could not see much call for in Alaska. Even so, the population was greater than that of every other American state (except Oklahoma) at the time it was first organized as a territory. There was a difference, however. Whites in Alaska remained a distinct minority; most Alaskans were uncivilized natives, living the more or less traditional pattern of subsistence, and therefore members of that class of persons which in 1867 the United States had promised to define and regulate at some future date; a treaty provision which had vanished down the congressional memory hole.

Contemplating the comedy of Alaskan history, one naturally wonders, *cui bono?* The *bono* is not far to seek. The Alaska Commercial Company in the first twenty years of its monopoly regularly paid dividends of about $1 million per year (with a more modest return to the federal treasury), despite the ravages of foreign poachers and the tendency of the dutiful Aleuts to die off from undernourishment and disease. And so with the other industries that followed—salmon canning, large-scale mining, logging—on a still grander scale, amounting by 1920 to hundreds of millions of dollars. All these enterprises worked on the strict colonial principle of taking out resources and leaving behind as few pennies as possible lying on the ground. Where feasible—in the canneries, the Southeast forests—the workers were seasonal, arriving in spring, departing with freeze-up. The settling of Alaska was a matter about which company lobbyists in Washington and later in the territorial seat of Juneau grew passionate, and with them their legislative and journalistic allies. People would require government, their government would require taxes, a penny here, a penny there, perhaps even interest itself in laws to conserve and regulate resources, prevent the companies from taking out what was there for the taking. Interference!

So for decades the sealers, canners, miners, loggers expressed themselves, and in Congress they were heard. There were no other voices so audible or so assured. Their assurance rested on cash and was profitable.

Population was among the benefits promised by the Alaskan boosters of statehood; and it came. After the decades of stagnation, by 1960, the first national census following statehood, the count had passed 200,000, by 1970, 300,000; five years later, under the forced draft of oil, over 400,000. Anchorage, which like most Alaskan towns began as a mud-trampled tent camp (in its case, of workers recruited in 1915 to build the Alaska Railroad), has kept pace, with nearly half the total and has the air—some of the architecture, most of the urban problems—of a much bigger town, a metropolis. Yet all the numbers rank near the bottom on any American scale. Anchorage is not big by that standard, and although the map of Alaska is dotted with place names, most of them when you get there turn out to be a gas station-roadhouse-store with, perhaps, a cabin or two hidden among the spruce; or nothing at all, a vanished camp commemorated in a name. Alaska remains a country of few people on much land, and the reasons, the separateness, distinctness, have as much to do with history as with geography or climate.

This much is abstract. You learn the meaning by going there, looking, talking; by submitting to some of the risks of the Alaskan landscape.

If the fundamental decisions have always been made else-where—in St. Petersburg, Washington, San Francisco, Seattle, and now in Houston, Dallas, Los Angeles—the Alaskan character has been marked by a wary deference for outside authority that is founded on fact. When decisions must be made that affect Alaskan business (or the larger business of state government), it is natural to turn to Los Angeles or New York or Boston for guidance at the hands of high-priced teams of consultants; or to Congress. In the small population thinly spread, the pool of talent is not deep. A Fairbanks economist or an Anchorage engineer, recruited to Juneau, leaves a slot unfilled. A skill applied in one place means that it will not be available in another.

The obverse of the feelings that grow from this situation is a willful determination to show the world that "After a hundred years of malignant neglect, we can indeed deal with our own affairs, our own development, in a rational and reasoned manner"—the words of one high state official in Anchorage, Chuck Champion, but the sentiment of many. And since more than in any other American state the new population is not born but drawn ready-made from other places, it is also the determination of the immigrant, the colonizer, not to repeat errors that have been left behind: *California was once Alaska*—a formula often repeated where more likely than not the immigrating majority began as Californians, including many state officials and Jesse Carr, boss of the Teamsters and thereby, perhaps, the most powerful individual in the state; by which they mean the national byword of freeways, smog, traffic jams, and human indifference, yeast-grown cities with the cohesion of melted fudge and John Muir's glacier-carved valleys and alpine lakes turned into Disneylands or worse. Hence a new road in Anchorage or a winter ice fog in Fairbanks, compounded of too many cars operating in a saucer-shaped valley, touches the sensitive nerve of a rejected past; Alaska is a last chance. Yet it is a chance perceived through immigrants' memories of home. The Kenai Peninsula, with its pleasure boats and working fishermen, clamming and crabbing, weekend cabins and a climate moderated by deep-water currents, is described as Alaska's New England. Valdez, once the chief port for mainland Alaska and reawakened by the coming of oil, calls itself "the Switzerland of America," though it might be as sensibly said that Switzerland is the Colorado of Europe. It is not easy to see this country simply as it is, in itself.

"There are four or five major constitutional questions that are going to be resolved here," Al Mongin said, attempting to specify the roots of these Alaskan contradictions. The frame of reference came naturally from a lifetime of official historiography, in New York, Greenland, brought to Alaska to write a pipeline history that was canceled when the Department of

the Interior grew shy of scrutiny. I had sought him out in those
first fall days of arrival and now he talked through chapters in a
book that would not get written. We drank coffee in the living
room of his house, in a suburban part of Anchorage called Turnagain for an arm of the Inlet named by Cook; where ten years
earlier the Alaskan earth had dropped houses and cars and people into wounds slashed in the glacial muck and gravel and still,
rebuilt, when earth flexes the houses tremble.

"Congress," Al was saying, "has deemed that things in Alaska
are national, not state-wide. Now here's a concept: everything
that happens here is dealt with as if it were going through two
or more states. And it all stems from the condition under which
Alaska was purchased. The United States government didn't buy
Alaska, it bought the rights that Russia had. In the same way,
you'll find contracts that are written—concerning oil lands, for
instance—they only buy and sell the rights that we actually had,
without specifying, because they don't know yet what their
rights are. . . ." A quick mind, moving faster than words could
be listened to, heard, in a flux of articles to be written, books,
ideas, opinions: ". . . a unique entity . . . none of the other
states . . . and when the pressure gets too tough, they go back to
Congress, they say *change the law* and say what their rights are,
and then we can litigate within these narrow areas. Everything
that comes up here, more so in the last six years, we don't know
what the rights are, so we're going to have to deal with it more
as a matter of treaty than of the state constitution. . . ."

Ambiguities, historic contradictions; law has come but is not
yet firmly seated. Hence it is still possible as in past generations
for a determined man, among these few people beset by uncertainties, to seize the loose ends of the situation and make it serve
him: to land in Fairbanks or Anchorage with pennies in the
pocket and walk away a millionaire—though now the chance is
less likely for money than power, and power in the service of an
idea. Unlike the populous, settled societies that the other states
have become, Alaska does not impose a prolonged apprenticeship, and the gap between youth and power is not wide.
Thus a few years back Peter Scholes turned up in Anchorage, a
young and articulate man schooled in the Berkeley of the 60s,

grasped the situation; within days found a base in a tiny local organization, the Alaska Center for the Environment, and a friend in Chancy Croft, the presiding officer of the state Senate, an Anchorage lawyer who makes himself accessible; and was able to turn his hand to electing the next governor. Although the organization is small, it had leverage, which could be applied, and when a month of recounting determined a victory by 247 votes, it became one to which debts would be owed. The feeling was still exhilarating when I talked with him the following winter, in the antique, downtown clapboard cottage where the Center makes its headquarters:

"I've never known a place where you could become so involved in the daily decisions so rapidly. There's a lack of the kind of expertise that you find in California, but on the other hand, a much greater involvement. . . ." Yet he saw also, with the political instinct of his generation, that the things he had left California to fight for, the wild land and its preservation, were not and would not be controlled by Alaskans, but "The issue will be won in states like New York—urban conservationists wanting to save an image of a wild part of the United States." That sense of distance from the centers of power—and therefore of dependence on others' wills one can partly influence but not finally control—lies near the surface in a newcomer, but it is only a current form of the old predicament. And so—to try to live there and not simply go and come and look, as I had to do? "I sometimes get sort of scared," Peter concluded, "being in Alaska, I feel too dependent—on transportation, on ships transporting my food— and how really in terms of the real necessities of life, people in Alaska are parasitical . . . an increasingly dangerous position in the world that I see coming. If you see the whole trade structure breaking down, then you start to fear. . . ."

Those who consider themselves sourdoughs—even the ones working in tenth-floor Anchorage offices and dreaming of freeways to Nome—will snort at such uneasiness, but the image is real. To live there or anywhere in Alaska is to live as in a stockade, surrounded by half-conquered territory: one's relation to that land is grudging and provisional. Perhaps it is true that one is no less isolated and dependent everywhere, that this is the

American—the human—condition, but if so, the truth is muffled. Here, the land compels you to acknowledge it.

In the spring my youngest son flew up from school to join me. With most of my work out of the way, we would have time to explore the country together, hiking, climbing, canoeing. In the months alone I had been consciously collecting, saving, things and places that we could share.

The Kenai River was among these possibilities, cutting across the Moose Range about a third of the way down the peninsula south of Anchorage and accessible at several points from the Sterling Highway that leads to the fishing village of Homer, where, from the road, high up grassy ridges that look vertical, you see mountain goats and sheep ruminatively following their six-inch trails packed solid in the scree, browsing, and below in the narrow valleys cocoa-dark moose crop the meadows and willow scrub, placid as cattle. The river flows west from a narrow, elbow-shaped lake, also called Kenai (for Indians who once shared the peninsula with a coastal subgroup of Eskimos), fed by glaciers capping low mountains to the east. For four miles below the lake, the river is near the road, then in a sharp bend leaves it and enters a steep-walled canyon. Ten miles farther and the river deltas into another lake, leaves it, still dropping fast over rapids, and after ten more miles returns to the highway, whence it meanders thirty more miles through flat, alluvial moose land, to empty finally into the Cook Inlet at the oil town of Kenai. We thought of starting somewhere on Kenai Lake and paddling the forty miles to the last rapids and road connection— a day and a half, two days, without exertion—but I did not know much more of the river than I could learn from a map.

In mid-June, a few days after Matthew's arrival, we headed south from Anchorage to have a look, the canoe racked on top of the car, our gear sorted and repacked in the back. Below Kenai Lake, we pulled off often to study the river. From what you know elsewhere, you would expect the spring runoff by now to be on the ebb, the river dropping, but it looked brimful and still on the rise, crossed at each bend by heavy riffles that in places rose in good-sized standing waves; altogether, this section looked less mild than its reputation, but manageable enough—at each drop you could see a clear passage through, sometimes a choice. The banks opposite the road were low, muddy, and thickly wooded.

A half mile above Schooner Bend, the road crosses the river again and veers north, going its own way. We parked, studied the rapids at the bend through the field glasses: on the right, a narrow side channel formed by a gravel spit of an island, then on the left, the outside of the bend, a tumbling mass of foam-topped rolling waves sliding behind the island and going out of sight into the mouth of the canyon. Rough, what you could see of it, but not too long, and it looked as if, staying close to the island, you would be all right.

But the *canyon:* for the canoer it means steep banks and high, therefore caution if not fear, the end of choice—once in, you have no way out but to go on to the end. We climbed and walked along the rim to look. After a mile, the rock rose in a bare jutting knob overlooking a fairly straight half mile of river, and we settled ourselves to eat lunch, study the water a hundred feet below. The left bank, opposite, was lower, wooded and muddy. The river, constricted, poured between the banks in huge, spray-tossing sinuous combers, but at every drop we could see, there was a clear chute that looked smooth and deep, not much maneuvering, and generally left of the middle; and the left bank was not perfectly straight but knotted with little indents where you could see the current slow—eddies to get into, if you had to, to rest, where in a pinch you could get out. The roar of fast water on rock was muted by distance, the enclosing walls. Our outlook seemed to be about the midpoint, but it did not seem worth hiking to the end—probably, I thought, no worse farther than what we could see.

Then what about it? There were other streams nearby that I knew were easy; we did not have to attempt this one. Matthew agreed that we could probably do it; we had canoed rapids as heavy or worse before, though nothing as long. There were wild rivers farther north that I had hopes of flying to later in the summer, and since we had not worked together for more than a year, it seemed well to practice while we had the chance. So it was decided, though I kept some reservations: we would start well up the lake to get the rhythm; if we had trouble in the first big rapids, Schooner Bend, we could still pull out and stop before entering the canyon.

We camped, ate, slept. In the morning, we left the car at a campsite five miles up the lake, loaded the canoe and started out. We took it easy, paddling down the lake, practicing turns, trying to co-ordinate. Toward the end, you began to feel the tug of current—hold the paddles and it sucked the canoe along toward the lake mouth. Then the banks drew in and we had entered, moving faster, accelerating: the first broad riffle, I thought back to the course we had picked from the bank and we were through it, the bow riding up, beautiful and willed like the lift of a horse to a jump, flying; then another, others, revealed by each winding of the channel, the current getting stronger and running back in a heavy cross-chop, and the canoe, lightly loaded for our two days of travel, rose, settled, rolled in the cross waves, and we took some water. The morning was misty-white, spattering with rain, and seeing an opening in the trees on the right bank, I steered for it. We got out, tied up, dug the ponchos out of a pack, and agreed on an early lunch. We were getting near the first big one now. Before going on, we strapped the two big packs to the center thwart, stuffed ponchos between them—the rain was stopping again—and tied on life jackets. That left the tent and the small cooking pack loose in bow and stern, but there was nothing to tie them to.

We pushed on, dropping under the road bridge in a roar of turbulence around its concrete buttresses. As we came toward Schooner Bend, I doubted the course I had picked the day before, the waves looked fiercer down here at water level than through the field glasses. I shouted to Matthew and we dug in hard, trying to slip the canoe crosscurrent toward the inner channel, but the current had us now and would not let go. The canoe swung sideways, ground onto shallow cobbles above the island and stopped dead, lodged against a drifted log. We jumped out, lifted, cautiously floated the canoe back toward the outer channel, and were off, through, hugging the island through smooth, fast water, the waves of the rapids churning and tossing just to the left of sight. We had done it, all right, but not well, not the final test and checkpoint I'd intended, but we would not stop.

We had entered the canyon and the first rolling hillocks of

piled-up water, then a huge, low-lying squared-off boulder we had not seen from above, looming, coming toward us like a runaway truck and we swung to avoid it, slip past, running, no time now to think, look, calculate, hunt chutes and passages. More boulders, the waves continuous, building, coming in a dozen directions and clear across the channel, without a break: the canoe plowed into them, valiant, settled, half-filled; Matthew hunched, his paddle rigid, unresponding. Ahead to the left the current caromed against a high slab of rock, carrying us toward it, and I tried to steer—a little bend, eddying, beyond the rock, *get into it* —but the canoe, gunwale-heavy now with water, no longer answered, carried against the rock and held, precariously balanced against the current.

Don't be afraid! I shouted to my son. The habit of fatherhood, from infancy, meant to soothe, reassure, drive out fear because you will stand between him and the thing feared; meaning, now —put aside the fear that will damp the mind that will save you; meaning the small boatman's axiom that I had never had to test —*Stay with your boat, it will float you, save you, even full;* all which in that instant went unsaid.

Matthew, panicked, unhearing, stepped out and we were dumped, the canoe righted itself and we hung on, held for the moment with it against the rock, bow toward the bank and the little eddy I had tried for. Chest-deep, my feet touching cobbles on the channel floor, still gripping the paddle, and I tried to work the canoe forward—*get into it*—toward the eddy, safety. It was enough. The current lifted me, caught, carried me. I let go of the paddle and tried to swim.

There was another indent below the jutting slab of rock and I swam for it, struggling, gulping down water, choking, but could not get across the current and swept on past. I was tired, heavy in the water, and could not get out; it would not take long, the cold, though I did not feel it, would be enough. Once before I came close, badly scared, but was not dumped, but now my mind was calm, observing, ready, only curious as to what would come, and a question formed, a statement: *Then it is to be by water after all?* And, strangely, it was answered, with a loud, curt, instantaneous *No!*—the sort of answer you might give a troublesome child when your mind is on other things.

Then—and again it is strange, that a man, every man since the gift of consciousness, that this ragged, destructible skinful of protoplasm, sentience, and will should *in extremis* find words to bargain with the all-creating Father of the universe—I shouted: *Lord of heaven and earth, save my son for Thy Son's sake! Save my son!*

I began to remember in their ordered sequence the things you are supposed to do, which I had never had to practice. Relax, don't fight. Get on your back and float, and let the life jacket hold your head up to breathe. Point your feet downstream so you can see what's coming, watch for being run down by the canoe—it was behind me somewhere, I had not seen it pass. Give yourself to the current, let it take you.

A wave caught my hat and took it off, and it floated on ahead. Then another eddy and the current was taking me toward it, I dog-paddled a little, sculling, easing into it just ahead of another little point. I felt my feet touch, held on, waded, grabbed for a shrub leaning over the water. The paddle had followed me in, and I seized it, leaned, pushed myself up and out. I heard my son's voice roaring, a hundred yards back, two hundred, but could not hear the words. I tried to answer but could not shout, could not make the words form. I leaned on the paddle like a crutch and stumbled toward his voice.

I found him trying to get to me around the rounded hump of rock where we had lodged—he had not been carried off, had gotten out. The rock was worn, crumbling into the swirl of water ten feet below, and would not hold him; I hugged it, reached across, touched his hand, but could not get him across. He drew back from sight, climbed; I waited. Then he came to me and I hugged him, was hugged, as we had not since he was small: *My beloved son, you are saved.* My beloved son.

Matthew's paddle had gotten away from him, but he had caught the tent, floating loose, as it went past, and had it. That left the canoe, the packs if they stayed with it—the current would drop it somewhere in quieter water below the canyon—and we started downstream to look for it, supporting each other, grabbing at trees on the steep bank. In that instant, like angels dispatched on a mission, three men and a girl swept past in a big rubber raft, and, voice cracking, I shouted: *We've lost our canoe. Will you watch for it?*—and they agreed that they would. There was no time for other words, more thought.

A few minutes more and we came to them, they had given . thought, put in, produced a match-safe, and gotten a fire started. I was still not cold—the water, someone said, was about 35°—but Matthew's teeth were chattering. The girl found a heavy sweater for him, a change of pants, then a bottle of Jack Daniels, dry cigarettes for me. We passed the bottle, hugged the fire for warmth, talking now, voluble; a bearded leader who had rafted the river before, two young men and the girl from Anchorage.

The afternoon was getting late, cooling—my watch, I noticed, had stopped at three-fifteen; time to get on, the leader suggested, he could carry us, help us find the canoe, and we doused the fire and got in. Still with my paddle, I tried to help, but their game was the reverse of ours, to find and hit the biggest waves, and the clumsy raft rode them, safe as a baby's crib. It was, it seemed, the easy part of the canyon we had hiked and studied; below, where we swamped, the hard part began, with worse beyond.

The river slowed and went shallow, spreading into a lake, and the raft went aground. We climbed out, waded it off—and then a motorboat coming up from the lake with our canoe in tow by the painter, and we waved, shouted, claimed it; caught on a bar another mile down, half full of water but not much damaged, with the two packs still strapped to the thwart, the spare paddle tied in place.

The boat went off. The four rafters were headed for the edge of the delta, a trail they could follow back to the road. Matthew returned the clothes. We would be all right, I jauntily insisted, go along the lake a bit and look for a beach where we could camp. We had the packs, the tent. All right. We said thanks again, good-by, paddled off, and turned to wave.

In a mile we came to a high sandy beach backed by the marsh of the delta, piled with driftwood, and put in. It would do. The packs are meant to be waterproof (and are, in rain) but had broken open, the contents—sleeping bags, spare clothes, food, cam-

eras, film—damp or dripping. I rummaged for matches—without
the cooking pack there was no way of cooking but we could
make a fire, keep warm—but could not find them; they must
have been in the lost pack. There was still the butane lighter in
my shirt pocket with a soggy pack of cigarettes—a precaution I
have advised other canoeists but never put to the test. We
gathered driftwood, found a bundle of dry pine twigs for tinder,
knelt to try it; the lighter had dried, sparked, made a flame, and
the twigs lighted. Fire: we kept adding wood, building it up,
lengthening it, wrestling logs around it to spread our things on to
dry; I toasted cigarettes by the fire and managed to get one
lighted. For food we had a new box of granola that had stayed
dry, a bag of pemmican I had made in Pennsylvania and carried
for months, on principle, wet now, and with a taste of ferment,
but nourishing.

Toward midnight, as it must have been, sun low and turning
cold, it started to rain again, driving us into the tent. In the
morning we paddled another five miles down the lake to a camp
where we could reach the road and drove back to an Anchorage
hotel to regroup.

Washed and alive, warmed, fed, I pondered. God the all-
mighty, ultimate being—granted the little one knows, it seemed
inconceivable that He should take such pains, arrange so elaborate a demonstration of one's dependence and the limits of one's
ignorant freedom, one among the billions. Then if not oneself, for
one's son—sixteen, with the promise of dignity and strength
about him, perhaps even great things to be done, it would be ru-
inous, such a death, at this point in his life; so one had life for
his sake. Then another short step: if I lived for him, so also he
for me, neither could I, whatever worth, have lived past or forgiven such a loss; and though we did not talk of it much, we got
that far, admitted it, and his reasoning was as mine. But there
was another link in the chain, perhaps final: the pains taken, the
complex stratagem of salvation—that was fact; and the fact
suggested that He attached some value to the end, one was of
worth *because* He had cast a glance at one's foolish temerity and
the consequence. And it is best to accept that and leave it at that,
get on with the things that remain to be done, wherever they

come from, whatever their purpose. It is not worthy in the creature to bandy words with his Creator.

I also prayed: that is twice now, but may there be no third.

Like a pilot walking away from a crash and flying again lest he be unmanned by fear, my son and I replaced our lost gear and canoed elsewhere, farther north; dropped down the Gulkana in high water, where the rapids come minutes apart, shrouded in the glitter of the southerly sun, and boulders lining the narrow chutes grated the sides of the canoe and at one point grabbed the paddle out of my hands and passed us through, helpless, but did not spill us. It became possible for us to argue, shouting and vehement, yet remain what we were, father and son, grounded in love and the need of each for the other. From habit, whenever we came near a river, I studied it, canoed it in my mind—keep to the inside of the bend, then a hard turn, straighten out, point through the break in the ledge and let her sail, and eyes ahead for the next one coming: *don't be afraid, we can do it!* Matthew stood apart from these imaginings, cool and tolerant, his caution redoubled.

"These rivers!" Matthew said. "These rivers in Alaska—*they are all too fast.*"

Chapter 3

LIFTING THE CRUDE

> The essence of any project is to minimize the time between the moment you start committing capital and the moment you start receiving returns. Time is a very important base in that calculation.—W. J. Darch

Like the merchant in the parable who, learning of a pearl of great price, went and sold all that he had in order to acquire it, the Prudhoe Bay partners moved with the energy of obsession to possess their prize—to "lift the crude," as the operation is expressed, and deliver it where it could be turned into gold; into percentage points in the corporate balance sheets.

Although the tests from the first well showed an unprecedented volume of oil and earlier studies of the geology had suggested that if there were indeed oil in the formation the field would be a big one, it would take months more of drilling, analysis, and calculation to determine even roughly how much. The first step, therefore, was to consider methods of transporting crude oil from the Arctic coast to a place where it could be sold, estimating how much that would cost and whether the market thus reached could in fact absorb the supply; preliminaries to the fundamental question—*can we afford to produce it?* This was the assignment given Frank Therrell and other Humble Pipeline engineers in March 1968: an engineering-economic study. In Dallas, their counterparts at Arco were working on the same question, in parallel but independently.

The engineers were not entirely unprepared. As early as 1964,

perhaps in response to Richfield's offer of a split in its Arctic
Slope explorations in return for financing, Humble had done a
"real quick and dirty" study of a pipeline to the south Alaska
coast and apparently liked it well enough to put up money on
the prospect. In 1966, with the first dry hole still drilling sixty
miles south of Prudhoe Bay, the company dispatched Frank
Therrell to visit its Canadian subsidiary at Norman Wells, a
small field in patchy permafrost midway down the Mackenzie
River, producing since the early 20s, to inform himself about
Arctic pipelines. Halfway up the Great Bear River from the
Mackenzie, a small pipeline had been pumping barge-hauled oil
around the seven-mile Charles Rapids since the mid-30s. Across
from Norman Wells, spurred by the fear that Alaska's sea links
would be cut, the U. S. Army during World War II had laid
down 450 miles of road and ten-inch pipe to connect, at
Whitehorse in the Yukon, with the Alaska Highway and another
small pipeline run north from Southeast Alaska. The line from
Norman Wells, code-named CANOL, was never in fact used
(the war ended), but it answered a few of the engineer's questions; there were as yet no other sources to turn to.

From maps they could calculate gradients and river crossings,
both factors in selecting a pipeline route and estimating its cost;
permafrost, another factor, had been sketchily mapped by the
USGS in 1965. Despite a good deal of such general information,
the two teams in Houston and Dallas remained in the dark as to
fundamentals: the size of the Prudhoe Bay reserves and therefore the volume that could be produced and the size of the pipe
needed to carry it; and the actual character of the oil, which
might affect many aspects of design, starting with the chemistry
of the pipe steel itself, perhaps even where and how profitably
the oil could be sold. Nevertheless, the companies were urgent—
they wanted answers. Their engineers consulted, worked up
some hypothetical costs—from construction averages, from the
recent transalpine pipeline, the biggest yet with its 42-inch diameter, from another planned to cross the Andes—and harmonized
the two studies. In mid-April of 1968, barely a month after
Arco's fourth test, the executives gathered at Humble's headquarters in Houston, Frank Therrell displayed his map of North
America, and the engineers presented their findings.

Both teams had apparently given thought to several possible routes. Frank Therrell and his Humble colleagues, studying the maps, had, for instance, traced a pipeline that would run east from Prudhoe Bay and cross the wide Mackenzie delta, then move diagonally through Canada's rolling barren lands to the southern tip of Hudson Bay and on to the American East Coast. Arco showed continuing interest in a route that would follow the Mackenzie south and branch in southern Alberta to the Pacific Northwest and to some point in the Midwest, near either Minneapolis or Chicago, where the oil could be delivered into existing pipelines. Nevertheless, the route that received most attention was simply the shortest from Prudhoe Bay south, in as straight a line as the topography permitted, to the Gulf of Alaska—in the pipeline business, there are no economies of scale, and the costs of construction (which determine how much it will cost per barrel to pump oil through) are in direct proportion to the distance traversed. The principle is indeed so axiomatic within the oil industry that it has never been much discussed, although Charles Jones, who ran Richfield for thirty years and, more than any other one man, conceived the idea of scouting the Arctic Slope, hinted in his memoirs (*From the Rio Grande to the Arctic*) that a trans-Alaska route had been assumed from the beginning; and also that his company had done some early figuring of costs. From what they knew, the pipeline engineers believed most of the Alaskan pipeline could be buried as it is everywhere else but that on 5 or 10 per cent of the route it would have to be aboveground, where unstable permafrost posed risks of breakage and higher maintenance costs that would cancel the savings on buried construction—roughly the eighty miles from the Arctic coast to the foothills of the Brooks Range; since elevated construction, almost unheard-of up to this point, would cost at least double, there was a strong incentive to avoid it wherever possible.

On this basis, the engineers proposed a basic pipeline cost of $750,000 per mile (and $1.5 million for the aboveground portions); with allowances for all the things assumed elsewhere but nonexistent in Alaska—roads and airfields north of the Yukon, construction camps where there are no towns, power and telephone lines—these figures became the first formal estimate for the pipeline, $900 million, though it was ten months before it was published. Even at this earliest stage and before the extras that Alaska would necessitate had been added in, the cost was high. For reasons we shall examine at a later point, it would yield a pipeline tariff twenty or thirty cents more per barrel than the current U.S. average for similar distances. (There are more than 200,000 miles of American oil pipelines.) Even in the face of the current low market price for crude oil, however, the tariff was bearable, particularly over the twenty- or twenty-five-year life that the Prudhoe Bay field could be expected to have; as a matter of cost, and again for reasons we shall come to in their place, the tanker link, between the Gulf of Alaska and whatever market the oil was finally delivered to, would not have been a primary consideration.

And so in April 1968 the fundamental decision was made: the oil, however much it turned out to be, would be produced and would be moved south by pipeline and ship; and from that decision all else has followed. Among several possibilities, the actual south terminus of the pipeline was probably settled at the same time or soon after, although the companies did not appear to have made up their minds for several more months; but that decision too seemed self-evident, within the logic of the oil business and its costs.

While the pipeline engineers were doing the initial planning on which the two managements based their decision, others within the companies—chemical, petroleum, electrical engineers, physicists, mathematicians—were at work on the estimates that constituted the first steps in designing (as distinct from planning) the pipeline. Howard Koch at Arco, who would be responsible for the production layout of the Prudhoe Bay field, talked about how they proceeded—another stage in my education in oil:

"From the wells that you drill," Howard explained, "you draw maps of the thickness of this pay over the area, and you try to define what the thickness looks like on any given area, and then you integrate the total thicknesses over the area to give you the volume. Then you have to know some other things, like what fraction of the rock has void spaces in it—porosity. You get that through core data. So then you have the second part of the equa-

tion, which is how much void space you have down there. Then
the next thing, how much is oil and how much is water, and you
can get the water through measurement of the small samples. . . ."

From these calculations, they knew about how much oil was in
place. The final question, then, was how much could be extracted. And the answer: "The way we do that," Howard said, "is
to model the reservoir mathematically—in effect, *produce* the
reservoir and see what the ultimate recovery will be." The early
stages of this process, touched on in Chapter 1, in July produced
the electrifying announcement of five to ten billion barrels of
recoverable reserves; it was this computer exercise that refined
the figure to the 9.6 billion barrels that has been official ever
since. By the same process, since the computers were in fact living the field's life through to the end, in anticipation, the companies arrived at the eventual target for maximum production: 2
million barrels a day*—which at the time, be it remembered, was
about the deficit in U.S. consumption that had to be made up by
imports. The coincidence seems intended; although the oil companies were necessarily actuated by short-term and immediate
interests, the effect of that self-interest, at least for a time, would
be to bring back into balance the troubling national equation of
supply and demand.

By similar methods, the Humble team arrived at results like
those Howard Koch described, as did the Houston consultants,
DeGolyer and MacNaughton; and, later, British Petroleum and
the engineers of Alaska's Division of Oil and Gas.

Consequences began to flow from the basic decision. On July
29th, Humble transferred Frank Therrell and the rest of his pipeline group to Anchorage to begin the process known as alignment—exact surveying of the route, drilling of cores along the
way to determine soil characteristics, which they already knew
might be critical (as on the whole they are not, in more temperate lands). Anchorage, with a population of perhaps a hundred
thousand, was still rebuilding from the 1964 earthquake, and
office space was not abundant. The group set up shop in the

* Apparently an early estimate of maximum production and the basis for
the pipeline design; hedged since to from 1.5 to 1.7 million from the main
(Sadlerochit) formation, with suggestions that the difference might be made
up from other oil-bearing formations not yet confirmed as producible.

basement of an aging downtown hotel that had survived the cat-
aclysm. On the last day of that same month, Humble and Arco
jointly announced that they were appointing a consultant firm,
Pipeline Technologists, Inc., to study a route terminating at Valdez. The announcement underscored, for anyone caring to draw
conclusions, two characteristics of the oil business as it is now
conducted. For one, although consideration seems to have been
given earlier to the general configuration of an Alaskan pipeline,
the terminus (and hence, in some detail, the route by which the
pipe would get there) was chosen with little or no on-the-spot
information of the sort that company engineers could provide.
And, by entrusting to a fee-paid, free-lance consultant the task of
providing the analysis on which fundamentals of the project
would depend, the companies were taking a very narrow view of
cost, a myopia that would prevail to the end; they would act so
as to save dollars today, and *damn* the millions that the saving
might incur tomorrow.

At about the same time, the presidents of the three oil companies journeyed to Ottawa to discuss with Joe Greene, the
Minister of Energy, Mines, and Resources, the chances of running a piepline from Prudhoe Bay through Canada. (The third
company, BP, was dealt in by virtue of its landholdings, although it had yet to find anything but dry holes.) As the engineers had noticed when they looked at a globe, a Canadian route
was an obvious possibility; moreover, despite an off-on policy of
development, it seemed likely that the Canadians would eventually find oil in quantity in the Mackenzie delta, the Beaufort Sea,
and the Arctic islands still farther north—running the Canadian
oil in with that from Prudhoe Bay would spread the considerable
pipeline cost. Despite years of argumentation since then, only
one point of substance from these discussions has ever been
made public. Pierre Trudeau's government, lightheaded with its
recent election, intent on independence from the southern
colossus like a mosquito on the back of a steer, would insist on
Canadian "control" of any pipeline built through Canada. This
meant Canadian ownership and, in turn, more cash on the line
than the money markets in Toronto were accustomed to produce; the oil presidents felt a chill and resolved on what they
had already decided. Attention would continue to be paid to the

Canadian possibility, but it would not be a serious alternative to Alaska.

In August, BP and two companies with lesser landholdings shipped rigs to Prudhoe Bay for a winter of drilling to discover if they too had a share in the bonanza. From what was already known or could be guessed and the distribution of their leased tracts, the chances of course looked good, but in the oil business the only assurance that counts is the pay at the end of the drill bit.

At about the same time, the state government—with tentative patents on most of the oil land amounting to ownership—began to show a lively interest in Prudhoe Bay and to shower the oil companies with attentions which in the long run would prove to be of dubious benefit. The governor at the time was a forty-nine-year-old, recently elected Republican, Walter J. Hickel—Wally Hickel, in the Alaskan style—who in a brief span of years had fulfilled the Alaskan dream of turning the loose change in his pocket into millions, a lavish home in the unlucky Anchorage suburb of Turnagain, and, finally, a seat at the top of the heap in Juneau (the means in his case had been not gold or the newer oil but large-scale construction, including leading hotels in Anchorage and Fairbanks); a sequence that not uncommonly gives a man an exaggerated notion of his capacities and the feeling that because the gods of luck have once smiled they will do so forever. In any case, the governor caught the oil men's sense of urgency and resolved that the state should lend a hand.

The immediate problem in that late summer and fall of 1968 was transportation. In the absence of a road north of the Yukon, the companies had two unsatisfactory alternatives for getting their rigs to Prudhoe Bay—and the living quarters, trucks, and bulldozers, the steady flow of food, fuel, and other supplies: by barge either through Bering Strait and around Point Barrow or down the Mackenzie from Great Slave Lake, both fairly expensive at a hundred dollars or so per ton and requiring elaborate planning since it is only in August and not every year that the pack ice backs off far enough from the coast to allow barges to slip through and get out again before the ice closes in; or by rail to Fairbanks and thence north, at $270 per ton, in the work-horse Hercs. Urged by truckers demanding a share in the traffic and eager to speed the development of Prudhoe Bay oil—the sooner it began flowing, the sooner its royalties would replenish the state's generally depressed treasury—Mr. Hickel determined that the state should ease the transportation problem by cutting a winter trail across the federal lands that lay between the Yukon and the Arctic coast. In November the bulldozers tracked north from Fairbanks.

A winter trail was hardly a novelty in the Alaskan or Canadian Arctic—there had, in fact, been several along the general route to Prudhoe Bay, serving old mining camps. Where you can, you follow shallow streams that freeze solid; a big river like the Yukon is crossed with an ice bridge—packed brush and logs for reinforcement, snow-filled, made thick and solid with a spray of water pumped through the ice; overland in wooded country, you cut the stumps short and fill in with hard-packed snow. It makes a rough road, but negotiable for trucks in convoy, at low-gear speeds, or for tractors hauling flat-bed loads. Whether through haste, ignorance, or hubris, however, Mr. Hickel conducted his project differently.

Following what was still at this stage the expected pipeline route, marked by a line strung through the treetops from a light plane, the state bulldozers bladed off trees, stumps, and roots, stripping the surface heath and mosses to the bare earth, and by the following spring had cleared a road of sorts through to the coast. A convoy of 343 vehicles assembled in Odessa, Texas, made it through with many casualties to machines, at speeds, beyond the Yukon, little better than walking (that part of the run took two weeks). Since the freight cost turned out to be only $30 less per ton than flying, the road was never much used. It did, however, provide a public object lesson in how *not* to deal with permafrost: with the insulating cover of plant life stripped back, the underlying muck and gravel thawed, first the normal foot or two of every summer, then deeper and deeper, with nothing to keep off the melting sun, a process that may conceivably continue forever since the perennial vegetation has little chance of re-establishing itself (in the brief Arctic growing season there are no native annual grasses like those that patiently smooth earth's scars in more temperate regions). By summer, the trail for long stretches looked more like a canal than a road.

By then, however, Mr. Hickel had been summoned to Washington and the dizzy eminence of the Department of the Interior. His successor doggedly reopened the winter trail, using the same slash-and-cut methods as before to bypass sections rendered impassable by the natural succession of the climate, and in a pompous speech, full of references to "faith in the Great Land," he dedicated it to the former governor: the Walter J. Hickel Highway. There are not so many roads in Alaska that numbers are needed to keep track, and it is customary to give them names, though usually for some safely historical personage.

In late October, the three companies announced the formation of an organization to design and build the pipeline; the name they chose for it, the Trans-Alaska Pipeline System, yielded the unfortunate acronym by which it has generally been known ever since—TAPS. It was a name, not an entity. Where it could not make do with consultants, it would borrow people as needed from the sponsoring companies, which paid its bills, owned its assets, and, more important, exercised control through committees, meeting serially at the various headquarters or sometimes in Alaska, chosen to set policy and oversee operations in every phase from over-all management to publicity. Although TAPS would in time be elaborated in an organization having the appearance as well as the name of a conventional corporation, the initial pattern of power and command—enlarged to accommodate new members as fresh discoveries supplied the admission fee—would carry through without fundamental change. Moreover, since power as well as financial obligation would necessarily be in proportion to what each company had in the ground at Prudhoe Bay and no one of them possessed absolute dominance of the field, TAPS would be the focus of the shifting alliances and uneasy, self-interested co-operation of its partners. Law and the natural forces of the market push simultaneously in opposite directions, toward co-operation on the one hand but also toward the secretive silence and obstruction of hard competitiveness. Oil does not make for comfortable bedfellows, but TAPS was the bed in which the partners would have to lie.

Three weeks after announcing the formation of TAPS, on November 21st, Humble and Arco (but not BP) jointly applied for land at Valdez, on Prince William Sound, on which to build a pipeline terminal and tanker loading facility. Presumably by now, given the speed with which they had carried out the preliminary route planning, the companies had received at least some assurance from their engineering consultants and the Humble team in Anchorage that the location would be suitable or at any rate possible. The decision seems, however, to have been arrived at negatively, by eliminating other possibilities; the positive justifications that supported it were evidently argued from hasty and superficial study and included fundamental errors about the nature of Valdez geology, not corrected until years later, in the course of construction. The site chosen was on the south side of a narrow, sheltered fjord at the head of Prince William Sound, named Port Valdez by an 18th-century explorer to honor a Spanish dignitary; opposite was the moribund fishing village of Valdez, founded in 1898 as a port to service copper mining farther north and, later, the gold-strike town of Fairbanks. About two hundred acres of the thousand needed was part of the minute fraction of Alaska that had passed into private ownership and could therefore be bought. The rest was part of a national forest embracing the whole coast of Prince William Sound, but with the Forest Service policy of multiple use—recreation, hunting, logging, mining—getting permission to lease a few hundred acres did not seem to present any serious problem.

If the two companies had harbored any uncertainties as to how to get their oil out of Alaska or the basic pipeline route, they were now resolved. Against the years of passionate debate that followed, they would adhere undeviatingly, and with only minor and technical modifications generally originated by themselves, to the first fundamental decisions made. Such comparative inflexibility is curious but characteristic. To understand it and the habits of mind and action that lie behind—and, as I noted earlier, *we had better understand*—we need to step back for a moment and reflect on the history of oil, from which both derive. That history begins in western Pennsylvania, in the year 1859.

One cold, wet spring, hunting the meaning of oil in the bones of its past, I slipped off the Pennsylvania Turnpike and steered

crosscountry toward the northwest corner of the state, through
the outward curving arc of the Alleghenies. It is big country,
heaved up in massive, broad-backed ridges like a battlefield
where an army of titans has fallen, bearded in scrubby, new-
grown hardwood, deeply cross-hatched by knotted streams and
rivers draining south toward the Ohio; only the backward
reasonings of geology tell you what your eyes deny, that this too
was once a frontier between sea and land, as wave-smoothed
and level as the Texas or Alaska coast. The road crosses the
Allegheny River and assumes a name, the Colonel Edwin Drake
Highway, and points toward a place called Titusville. It was
here in a late summer on the eve of the Civil War, beside Oil
Creek in a narrow, steep-walled valley south of town, that the
man so honored made the world's first oil well. That first bears
qualification—oil had been found and produced, before and elsewhere; here it was deliberately sought and was drilled for, by
methods that have been used ever since, elaborated and refined
but not fundamentally changed. It was the beginning not merely
of the oil industry but of industry as we know it, cheaply and
abundantly powered, corporately structured under new laws
whose origins coincided with the origins of oil.

Oil had been known along the western face of the Alleghenies
—from western New York across Pennsylvania and into West
Virginia and Ohio—since the arrival of the first settlers around
the time of the Revolution, mostly as an inconvenient by-product
of digging wells to reach closed-in underground pools of brine
that could be boiled down to turn a profit as salt. The same necessity that drove New England whalers around Cape Horn, and
in the 1850s to the North Pacific and Alaska, found uses for the
dark liquid. Samuel Kier, a Pittsburgh salt merchant who had
tried bottling petroleum—rock oil, a literal translation of the
learned word—as a medicine that must be good for something,
found that when heated it could be separated into parts; among
them kerosene, a clear, sherry-yellow fuel that burned as clean in
lamps as costly whale oil and a good deal cheaper. Distillation,
the first step in refining; others in the area took up and improved
the process. It caught the interest of investors in New York and
Connecticut, who thought the oil might be worth producing for
itself and not simply as incidental to the salt trade. The country

northeast of Pittsburgh was an obvious prospect: good land in the narrow valleys but poor farming on the slopes, a zone of frontier left behind in the leap to rich homesteads farther west; before settlers came, the Indians had skimmed oil from pools along Oil Creek, collected it in shallow pits dug in its flood plain. Edwin Drake was hired to investigate, an out-of-work New Haven trainman (the rank accorded him in Titusville was fictitious, a title of respect). With the labor of a local blacksmith and the blacksmith's two sons, he tried digging for oil beside the creek. Finding nothing, he built a rig like those used earlier for brine, but steam engine-driven: a drill at the end of a cable, pounded up and down, the crushed earth and rock washed out with water. When the sides of the hole fell in, he drove lengths of cast-iron pipe down thirty-two feet to bedrock and went on drilling inside it—casing in, as oil men have since called it. Two weeks later, 69½ feet down, the bit hit oil, which bubbled up to within sight of the derrick floor. Drake attached a pump to get it out.

Not only the methods of drilling and extraction but the whole ethos of the oil trade—the raging hopes and reckless greed, the hatreds and the cold, cunning calculation—begins at Drake's well. He was not chary of explanation, and others in the area built rigs like his and had equal luck; or, since the Oil Creek reservoir was not deep, cheaply stomped their wells down with a man-powered spring-pole—a limber sapling angled over a forked post, from the end of which hung a rope with the drill bit, and rope slings by which two or three men could use their legs and human weight to drive the bit down through earth and rock. So while the war lasted, farmers probed their sloping quarter sections above the creek and did better by drilling than plowing; and when the war ended, demobilized troops occupied Pennsylvania's oil regions and in 1865, a few miles southeast of Titusville, discovered a new field. Up the hill from the find, in months, a city of fifteen thousand grew—mud streets, town lots, telegraph offices, boarding houses, banks, stores, hotels, brothels, churches, saloons, some of the buildings knocked together in five working days—and was called Pithole. Rigs sprouted across the fields and uncut woods like ferns in spring. The oil was barreled and hauled by wagon down the axle-deep roads to the nearest railhead and the Pittsburgh refineries, the wooden barrels standardized at the 42 gallons, weighing about three hundred pounds, which has been the American unit of oil production ever since. (Elsewhere and later, as in the Persian Gulf, oil was mostly transported by ship and its measure was tonnage.) A branch rail line followed, with big wooden casks on flatcars; then a six-mile, cast-iron pipe through which the oil flowed downhill, by gravity, at a dollar per barrel instead of the three dollars exacted by the teamsters.

Within months, the Pithole fever began to burn itself out. The most productive wells in the new field, Homestead and Frazier, ran low and had to be pumped. A succession of fires swept the rigs, caught the town itself, single houses, whole blocks; the first truth of oil is that it burns, and the vapors rising where it touches air will burn explosively. By 1870 the army of oil men, speculators, and hangers-on had moved elsewhere, though not far, and within a year or two the last buildings were gone, pulled down or burned. Today along Oil Creek and its tributaries, from Titusville to the Allegheny River, you can follow the melancholy early progress of oil in the place names left on the land: Titusville and Pithole, Petroleum Center, Plumer and Rouseville, Oleopolis and Oil City. There are small, early refineries, decrepit but functioning, at both ends of the route, Titusville and Oil City. Pithole is marked by grass-filled cellar pits, a dozen memorial street signs down the steep hillside, the work of a local newspaper publisher with a mind for history, and an embryonic museum. Of Oleopolis nothing remains but a name preserved in a few books and an approximate location on an oxbow bend of the Allegheny, near the point where Pithole Creek flows in.

If the short-lived cities of oil have come and gone, or have simply resumed their earlier pastoral somnolence, the symbols are everywhere along the streams and Drake's asphalt highway, up steep-climbing gravel roads and far back in the spring-budding woods, down single-track wagon or Jeep trails: pumps with their donkey-headed counterweights lifting and falling at the end of the walking beam in the slow rhythm of a heartbeat, sucking up the oil, some of them rigged in series by clanking lengths of steel rods to a single engine; cabin-size stock tanks to collect the oil for shipment, set down with deliberation like pieces on a checkerboard; shallow pits dug to take overflow or as a small producer's contrivance to collect a few barrels of oil a day, cheaper than steel tanks; grass-grown gathering lines threaded across fields and through woods, heaps of drill pipe. Some of the pumps are new or newly painted and bear famous oil names, such as Getty—serving wells that have been repeatedly abandoned, reworked, redrilled, renewed, since their first finding—but earth-brown rust is the predominant color; in the woods you chance on the remains of machines of still older model, left where they stood when the oil gave out and picked apart in the fifty or hundred years since, like insects in a child's inquisitive fingers, by trees that in that time have grown up from seedling to maturity. These remnants should be ugly and perhaps are to imagination—you remember the stripped hills and slashed fields, the hasty, pretentious little towns, the wagonloads of oil in barrels churning the earth to gumbo, strained at by heavy drays whipped to steaming exhaustion. What your eyes actually see, however, is less obtrusive: these signs of human energy and need have been absorbed into the landscape by the same natural forces that raise mountains and grind them flat again, fill and empty the salt seas. To speak of these things as scars on the earth is metaphor, exaggerated. Against the relentlessly shaping purpose of such forces, man's power is not so great.

What happened in Pennsylvania's oil regions was repeated in 1901 in the Spindletop field near Houston and the Texas Gulf Coast, again in the great East Texas field, near Dallas, in 1930; and in many lesser places in between. The first big strike was followed by a rush, and hundreds of new wells were drilled as near as possible to the discovery, with no regard for the geology of the oil-bearing formation underneath and limited only by the surface land rights, divided, as in Pennsylvania, among many small holders. Wild overproduction had two unfortunate results. First, more oil poured into the market in a short time than existing uses could absorb, and the price dropped below the point of profit—in Pennsylvania from $20 per barrel when Drake drilled his well to 10¢ three years later. The falling price, however, tended to stimulate still further the overproduction that caused it, and the whole field was soon played out, the great early wells dropping from uncontrollable thousands of barrels a day to nothing, or leaving oil in the ground that could only be gotten out by methods that were beyond the resources and the technical abilities of the owners. The two results were connected: there seemed to be no mid-point between high-priced scarcity and dirt-cheap abundance, with reckless waste; and at both ends of the cycle, individual consumers and the growing number of industries dependent on oil suffered, in the long run, no less than the producers.

As early as 1865, while the Pithole rocket flared across the sky in its flight to extinction, one man, at least, saw a solution to this dilemma: John D. Rockefeller, a young Cleveland clerk with a head for figures and a talent for keeping his own counsel. If you could dominate the transportation and refining of oil, he reasoned, you could assure yourself steady and predictable profits at the marketing end and incidentally control and conserve production, and it would not much matter who owned the actual wells. Rockefeller found partners to buy into a Cleveland refinery, acquired others with a secretiveness later oil men have inherited and found useful; secured favorable rates from railroads, such as the new Atlantic and Great Western, hauling oil from Pennsylvania, with penalties for rival refiners and unco-operative producers; and five years later formed his first stock corporation, which he called Standard Oil. Within twenty years he had built this beginning into a network of companies stretching from New York to California, many of them called Standard Oil, others, such as Atlantic Refining in Philadelphia (with roots in Pittsburgh and the oil regions, drawn east by the export trade), less obviously connected. Ostensibly the Rockefeller companies functioned independently in accordance with state corporation laws, but in fact control was in Rockefeller's hands, first through the Standard Oil Trust, then through exchange of stock with one of its components, Standard Oil of New Jersey. With this control went a virtual monopoly of the American oil business: 90 per cent of its refineries and pipelines; ownership of production was also substantial, but less important in the Rockefeller scheme of things.

Such dominance provoked resistance: state antitrust laws

aimed at giving substance to the regulatory intentions of state corporate charters; passionate denunciation from the standpoint of the small producers (Rockefeller's firm hand on production meant low prices for crude oil) or of government probity (Rockefeller contributions kept the legislatures docile); and, in 1890, a national Antitrust Act. It was sixteen years before a President, Theodore Roosevelt, dared aim this formidable weapon where it was intended, at Standard Oil; five years of lawsuits ended in the Supreme Court and the famous decision of 1911 dissolving the trust into its component parts. They include many of the names familiar today on the gasoline pumps: Socal (Standard Oil of California), American (Standard of Indiana), Sohio (Rockefeller's original Standard of Ohio), Mobil (descended from Socony, Standard of New York), Arco (Atlantic, married to Richfield); and Exxon (Standard of New Jersey, the Rockefeller keystone, left by the Court as a large company with limited production and thereby impelled into the international oil trade whence it emerged as the biggest of all international corporations, operating in the United States as Esso or Humble and since 1972 under its present name).

The alternative to the Rockefeller system was to control the oil at its source: limit and regulate production, sell the oil to the refiners and marketers at an agreed price that would ensure a return on investment. The Pennsylvania oil men tried more than once to accomplish this by forming producers' associations, but someone always broke ranks, secretly cutting the price in order to dispose of the crude that poured from the ground like an uncontrollable natural event, and the other small producers fell into line, earning Rockefeller's cool contempt for their lack of determination. If the many Pennsylvania producers could not hang together, then the solution was ownership of the oil fields by one or a few large companies; all it needed was money and luck, in quantity. The chance came with the discovery of the Spindletop field in 1901, the biggest yet. A Pennsylvania-Irish oil man, one Buckskin Joe Cullinan, picked up leases, and with outside money from New York laid the foundation of the Texas Oil Company, Texaco, which in time would rank second to Exxon in the hierarchy of American oil. What Cullinan did not get fell to a Pitts-

burgh banking family, the Mellons, with assets to rival Rockefeller's, and became the nucleus of another durable oil giant
which called itself the Gulf Oil Company since the obvious first
choice, Texas, was already taken.

Although it assured orderly development of the oil fields, a
steady supply, and a degree of conservation, such parceling up
of the resource raised the same objections as the Rockefeller system. It concentrated power in few hands, and the temptation to
use that power for narrow ends—to raise prices, exclude competition, rearrange governments for the benefit of oil—was not always resisted. A solution of sorts was to transfer part of this
power to a third party, representing government and the public.
The opportunity and indeed necessity came in the early 1930s in
Texas, which had aimed some of the earliest antitrust legislation
at Standard Oil and thereby soured Mr. Rockefeller on competing for Spindletop. The East Texas discovery in 1930, running at
a million barrels a day, produced a new oil glut in that first year
of the Great Depression and reduced the orderly system of production, marketing, and profit to a shambles. Crude prices tumbled as low as 20¢ a barrel, then 10¢, depending on the grade;
give-away schemes at the gasoline pumps would be followed by
bankruptcies, scarcity, and high prices. It was argued at the
time that this sequence of happenings was arranged by the oil
companies in order to overthrow existing conservation measures,
but the governor responded by calling out the National Guard to
shut down the wells; the legislature in 1932 empowered an existing regulatory body, the Texas Railroad Commission, to determine demand month by month and to limit the supply to what
was actually needed, prorationing production among the lease-
holders; the federal government encouraged the other oil-
producing states to control production by mutual agreement, and
in 1935 Congress made it illegal for producers to bypass such
regulation by smuggling oil into other states at cut prices.

One effect of these arrangements (and of newer recovery techniques made economic by secure prices) was to make the oil go
farther and longer. The East Texas reserves, for instance, which
provoked the state-administered production controls, were originally estimated at five billion barrels. In the thirty-seven years of

its life, the field had produced 4.4 billion barrels of crude, but at the end of 1977 the estimate of recoverable reserves still in the ground was 1.6 billion; which adds up to a billion-barrel bonus.

In time, the Texas principle, if not always the system, was adopted in most other producing states and to a degree at the national level; under the name of *maximum efficient recovery*, or MER, it became the means by which, as a requirement of state law, the companies could organize the production of shared oil fields—space the wells and set their production rates so as to extract the crude without excessive depletion, at the same time maintaining a steady level of prices—without violating the antitrust laws. The field, that is, would be developed as a *unit*—rather than scrambling after whatever could be pumped from their scattered tracts of land, its owners assigned responsibility to one of their number for rational and efficient operation, sharing costs and profits in proportion to their ownership. Although the mechanism derived from public authority, its practical effects were not very different from those of Mr. Rockefeller's trust—or of OPEC, the cartel of producing nations that came into being in 1960. The geography of oil changes but the dilemma remains: the functioning of industry—of society itself—depends both on the price of oil and its uninterrupted supply, and we have yet to find any means of assuring the one without interfering with the other.

The effects of this history, and of the differing styles of the men and companies that made it, persist in the oil industry today—though, given the characteristic secretiveness that is also part of that inheritance, it is not easy to say in what precise ways or to what degree. No one company, including Exxon, approaches the dominance of the industry once exercised by the Standard Oil Trust; whether Gulf, Texaco, and the progeny of Standard Oil can or do achieve the same ends by co-operation—and whether, if they did, that would be wrong in itself or harmful—is the theme of a long-running national debate. It is probably not a question that can be finally answered—the evidence is inconclusive, going both ways; and incomplete.

The reason for this uncertainty is the organization of the major companies. The companies are "integrated" in the sense that they are involved in all stages of the oil business from discovery and production to transportation to refining and marketing, but in most cases these stages of the total business are conducted by subsidiaries or affiliates—legally distinct corporate entities all of whose stock is owned by the parent corporation; the visible parent, then, constitutes a kind of federation of separate but related or interdependent businesses. There are functional arguments for this kind of organization and also, I think, distinct weaknesses, both matters that we shall return to. The result, however, is that it is virtually impossible, from the outside, to find out all the kinds of activity any one company is engaged in. Some of these subsidiary corporations, such as the pipeline companies, are easily identified, but others—which may include drilling, trucking, coal, uranium, and various lesser industries having no connection to energy or oil—are not at all obvious, and the annual reports of the parent companies are only very moderately enlightening about such matters; when you ask directly, as I have, for a comprehensive map of the corporate maze, you are met with vagueness, or a retreat behind the all-purpose oil-company barricade—that the information is "proprietary," a valuable trade secret like an industrial process.

What this means, apart from one's baffled curiosity, is that it is impossible to find out where, philosophically and practically, the power and the profit in oil actually reside: whether, like the original Rockefeller, the oil companies make their money by refining and selling cheaply bought crude; or whether the real source of profit is in the production itself. Within the United States, it *appears* that the major companies are production-oriented and look on every other aspect of the business as a complement to the necessity of disposing of crude, but not a primary base for profit. Except in the pipeline phase, where a profit is built into the rates by federal regulation, the balance sheets for shipping, refining, and marketing, if ever made public, would probably show results ranging from operational losses to break-even to very modest profits, the real money made on the crude itself. (Outside the United States, this situation appears to be what OPEC has dedicated itself to changing.) With few exceptions, ultimate control of policy within the companies is in the hands of men—chairmen, presidents, key executive vice-presidents—who rose from production, petroleum engineers who until quite recently were started out as drilling-rig roughnecks with their engineering degrees put away in a drawer somewhere for later use. Perhaps this balance is inevitable. There *are* no important technical secrets, no tricks of selling, to distinguish one company from another—none can be concealed, all are there for anyone to find; the distinction that counts, then, is the possession of crude and the ability to get it out of the ground.

If this much is approximately true, it explains what is otherwise strange about the oil business: the antic scramble after reservoir rights, in 1968 as in 1859, 1901, 1930, and all the other vintage years of oil; the mismatched partnerships of natural rivals for the sake of leases, exploration, production; the rise or fall of famous names of oil, their merger or disappearance, depending on the outcome of such transient alliances. Getting the point of view also casts light on another debate not yet resolved by academic economists: whether oil follows the pattern of other commodities, the supply rising or declining with the price, or is somehow an exception to the laws of supply and demand in a market economy. To common sense, it does not seem that the public—you and I, brother—has the option of doing without oil that it cannot afford, the textbook remedy in such an economy; we may have a choice between hamburger and steak, cornmeal and wheat flour, but not between empty stores and the oil that delivers food to their shelves and our tables—and history since World War II has only underscored this absence of alternatives, in the United States as in much of the world elsewhere. Less obviously but perhaps more seriously, it does not appear, on the other side, that the oil men have the corresponding option of refraining from production when the price is discouraging. Theory declares that such a course lies open, but each stage in the flow of oil—exploration, production, refining, distribution—moves at steady, continuous rates that can only be gradually altered over a period of years; shutting down a refinery, for instance, or any part of it, is as tedious a process as cooling the hearths in a steel mill—and explosively dangerous; and so in reverse, starting up, adding capacity. The oil business is compounded of human acts and therefore surely, one might suppose, subject to human reason: yet in fact and history all we know is that oil *must be found* and, once found, *must flow,* like a spring flood breaking the winter ice—and woe to him who would stand against that earth-opening force.

Corporate realignments followed the Prudhoe Bay discovery. On the last day of October 1968, three days after the official formation of TAPS, Arco announced a $1.6 billion agreement in principle to acquire Sinclair, whose faintheartedness had kept it from a share in the wealth; BP was to gain from Arco, for $300 million, the start of an American outlet by way of unwanted Sinclair gasoline stations and refineries in Pennsylvania and the Northeast. In early March of 1969, after scrutiny by the antitrust lawyers of the Department of Justice, BP got a few more stations for an additional $100 million. Just what Arco got for its exchange of stock is not clear, but one element was a Sinclair petrochemical subsidiary, filling a profitable slot in the integrated major oil company that Arco and its boss, Robert Anderson, aspired to become.

For BP, the pickings from the Sinclair corpse were only a step in the same direction. In March, with the Arco transaction completed and promising signs from its first successful Arctic Slope well, BP made an offer for Sohio. This original descendant of Rockefeller, still based in Cleveland, was a regional, Midwestern oil company, strong on refining and marketing, unlucky in its supply of crude. The merger supplemented BP's Sinclair acquisition in the Northeast; more important, it provided, ready-made, an American corporate structure equipped to borrow money to develop the BP holdings without exporting scarce sterling from Great Britain. It was a complicated arrangement. BP used its mostly unproven Prudhoe Bay leases to buy Sohio, on a sliding scale tied to the eventual rate of production; Sohio then used its theoretical interest in the Arctic Slope to acquire BP-U.S.A., its immediate owner; and contracted back to BP-Alaska the development and operation of its share in the Prudhoe Bay field. The Department of Justice took its time puzzling out this arrangement, even as you and I, but eventually gave its consent, and it was finally consummated on January 1, 1970.

While these corporate complexities crept toward conclusions and BP and then Socal and Mobil completed their first Prudhoe Bay wells, the state of Alaska made some news on its own account. On July 14, 1969, urged as desperately by those who already had land at Prudhoe Bay as by those who did not, the state announced that it was scheduling a new Arctic Slope lease auction, to be held in Anchorage in September. It would be the twenty-third sale since statehood, the fourth on the Arctic Slope: 179 tracts in the vicinity of Prudhoe Bay, 450,858 acres. It seemed unlikely that the land this time would go for the parsimonious $5 or $10 per acre that the previous sales had brought. In Juneau and Anchorage and Fairbanks, the talk was of $1 billion, a fine round sum that would surpass any previous federal offshore lease sale and come to six or seven times the current annual state budget. With such expectations, the state in August appointed the Bank of America to manage the take, turning, as usual, outside for expert assistance; then, thinking further, chartered a jet airliner (for $23,000) to fly the certified checks to San Francisco for deposit—since the interest on the 20 per cent down payment that had to accompany each bonus bid might amount to as much as $40,000 per day, it was a gesture worth making.

In early September the landmen, geologists, and accountants began filling up the Anchorage hotels. Some brought technicians to debug their rooms and made doubly sure by booking adjoining rooms on either side and above and below so that at least the central one in the block would remain secure. Just what value these precautions might have in practice is not clear: although most of the companies that wanted in on the play bid independently on at least some tracts, all placed most of their bets in partnership with their rivals, in dizzying combinations—Arco with Exxon, of course, but also with Home Oil and BP, BP with Gulf, Mobil with Phillips and/or Socal, and so on. Those that had already done seismic work or drilling on the Arctic Slope had a fairly clear idea of which tracts were likely to produce oil; the rest would have to guess from such public information as was available and whatever they could pick up through gossip or by other means. All had to gamble on how high to go to get promising land against the unknown bids of the competition—and on what, given the costs and uncertainties of drilling and the nascent pipeline plan, their bids would finally be worth if successful.

At 8 A.M. on September 10th, the interested parties gathered at

the Sidney Lawrence Auditorium for the ceremonial opening of
the sealed envelopes containing their bids and checks: a building
of creamy cinder block and corrugated steel in downtown Anchorage, resembling a high school auditorium without the
school, named for a local artist; whose 350 seats normally are occupied by audiences drawn by touring orchestras and an occasional theatrical or opera company. As the envelopes were
opened, the tracts were identified, the amounts read out and
posted on a long board set up across the stage. By five-thirty that
afternoon, when with a short break for lunch the work was done,
1,105 bids totaling more than $1.6 billion had been entered. The
state took a week to sort out the winners and winnow bids on a
few tracts it judged too low (nearly all had received offers, for
humor or luck, of a dollar or two per acre, in a few cases pennies). The final award was for 164 tracts, 412,453 acres, amounting to just over $900 million—an average of $2,182 per acre.
When the Prudhoe Bay oil formations were at length mapped—
the main Sadlerochit, the lesser Lisburne and Kuparuk—162
tracts leased in this and the three previous sales were shown to
lie over known oil. After considerable swapping, of which the
BP–Sohio deal was only the most visible, this land had shaken out
among sixteen companies. Arco and Exxon together wound up
with just under 42 per cent, most of it jointly held with each
other; BP–Sohio had 23 percent, all but half of one tract undivided. These combinations still left substantial acreage in the
hands of late-comers: among them, Amerada Hess with Getty
(archetypal Texans who had tossed the biggest block of cash
onto the auditorium stage), Amoco, Mobil, Phillips, Socal,
Union, and the comparatively small Hamilton Brothers company.

With the field carved up, the serious work of developing it—
and of finding means of turning it into income—could now
begin. Both goals presented difficulties, beginning with the size
of the field and the number of operators that had been drawn to
it.

In most parts of the world where oil is found, the means of
delivering it to market is dictated by geography. Oil from the
Middle East, North Africa, Nigeria, Indonesia, Venezuela all
gets where it's going by ship, sometimes starting through a pipe-

line link to a port. (The pipeline built from Saudi Arabia to the Mediterranean, once the pride of the former Arabian-American Oil Company—Aramco—was designed to shorten the distance to Europe in the era that preceded the first closing of the Suez Canal and the coming of the supertankers; since it passed through several Arab states, it provided their citizens with a convenient means of political expression—blowing it up.) Soviet Russia, with its huge new oil and gas fields scattered across the Siberian landmass, is just as necessarily pipeline country. As the Texas oil trade has shifted over the years from substantial exports to supplying the Midwest and Northeast and serving as a port of entry for imported oil, its natural mode has changed from tankers to pipelines, though a fair amount of coastal oil shipping still originates at its Gulf ports.

At Prudhoe Bay, on the other hand, if you study a globe as Frank Therrell and his colleagues did in the spring of 1968, you see as they did that there is perhaps, at least in theory, a choice: several possible land routes, depending on where it is most profitable or otherwise useful to pump the oil; but also, since Prudhoe Bay lies on an ocean, at least two potential sea routes—a potential limited, however, by the fact that both are blocked for most of the year by deep-frozen ice. If you then remember that the cost of transporting oil by tanker is a fraction of that by pipeline and that a first rule of the game is to minimize costs, you would have a powerful incentive for finding some way through the ice problem.

There were in fact alternatives to TAPS, actively studied, but in the years of debate, lawsuits, and accusation that followed, the two sides seem to have used the word at cross-purposes. For the opponents, the alternatives were *exclusive:* the choice of tankers or one of the several Canadian pipeline routes (there were finally hundreds if not thousands of variants) would achieve the primary objective of preventing the building of the Alaskan line. The oil companies, however, used the word *inclusively:* given the volume of oil and the likelihood of future discoveries, more than one route or mode of transportation would eventually be needed. Hence the prolonged look at possibilities —but the basic choice of the Alaskan pipeline and tanker system, made very early, seems never to have been questioned, and the alternatives, as they presented themselves, were understood and judged in relation to that.

The reasons for this choice—and for the kinds of alternatives put forward, the ways in which these were evaluated—are to be found in the deeply divergent interests of the oil companies brought into uneasy connection by the Prudhoe Bay discoveries. The companies are extremely reticent about allowing such conflicts to come before the public. With patience and attention to what *is* known, it is possible to reason out at least the main outlines.

For a start, there was the matter of unitizing the field, a by now accepted industry practice that was also required by state law: that is, each of the companies with land on the field would not develop its portion independently; rather, they would agree on one of their number to produce the field as a whole, or unit, for the greatest efficiency in investment and recovery, MER, sharing the costs and the profits in proportion to the oil (and gas) actually in place under the companies' land. Agreeing on an operator for the field—actually two, BP and Arco, since the size of the job exceeded the management capacities of a single company—was fairly easy. Deciding each company's share of the needed investment and eventual production was another matter. Although the data of engineering appear to be absolute, beyond argument, any two engineers can usually find different ways of interpreting them and arrive at quite different conclusions, and over the years every new well added something to knowledge of just what was in the ground, and where. Early on, it appeared that actual ownership of oil and gas would turn out rather different from the distribution of land: that to BP–Sohio, with a quarter of the tracts in the field, had fallen something over half the oil and just short of a third of the gas; while for Exxon and Arco together the situation was the reverse—about a third of the oil but substantially more than half the gas. The dominant trio (and the lesser owners) agreed to proceed on the basis of this tentative split, with provision for a final accounting once the percentages were settled. Actual negotiation of all the companies' shares and responsibilities took *nine years,* right down to the point at which the pipeline was scheduled to go into operation and the oil to flow; and produced two enormous contracts, the

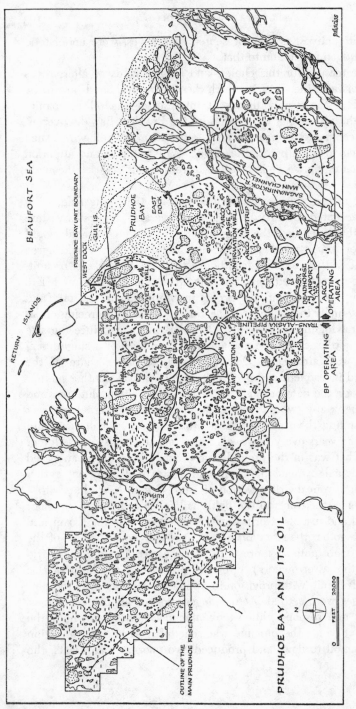

PRUDHOE BAY AND ITS OIL

N

0   FEET   20,000

palacios

It took eight years of drilling to define the main oil-bearing formation at Prudhoe Bay. This in turn was the basis for the Unit Agreement, not signed until April 1, 1977, in which the lease owners settled their shares in the wealth and expressed the strategy by which the field would be produced. Fewer than half the tracts auctioned at the $900 million 1969 lease sale finally paid off in oil and were included in the Prudhoe Bay Unit.

details of which the parties have refrained from making public.[*]
Along the way, Arco apparently tried to break the 1964 agreement, inherited from Richfield, by which Humble had put up
half the costs of the exploration that produced the original discovery; Humble responded with a suit to enforce the contract
and compel its partner to sign over half its interest. Humble
seems to have made its point, but the two companies avoided a
trial that would make their private arrangements public, and six
months later, in December 1969, they announced that they had
settled out of court with a "realignment" of the old agreement.

Among the three principals, there were more fundamental
differences, which could not be negotiated, though they might be
compromised, delicately balanced by force of capital and personality. The biggest of the three, Humble (or its parent, Standard
of New Jersey), had come out of the 1911 antitrust dissolution
with huge refining and marketing capacity but scant sources of
crude oil. It succeeded so well in making up that lack through
overseas exploration (in Venezuela, Saudi Arabia and, finally,
after breaking through the BP monopoly, in the Middle East
generally) that by 1970 it accounted for roughly 15 per cent of
the world's production and had, presumably, a corresponding
share in its proved reserves; sources of corporate income were
consistently about evenly divided between the United States and
the rest of the world. The company did not, therefore, feel quite
the same urgency as its partners about developing the Prudhoe
Bay find. It was prepared to take its time, look for ways of
delivering the oil elsewhere, as to its refining and chemical complex in northern New Jersey. Moreover, it had no keen interest in
aggrandizing potential rivals—Arco, an energetic comparative
upstart; BP, with reserves probably greater, world-wide, than
Exxon's, but so far, within the United States, of even less account
than Arco.

Arco's view of the situation was rather different from Humble's and the difference had less to do with its size than with the

---

[*] The percentages of participation in oil and gas production *were* published, however, in the spring of 1977, carried out to seven decimal places
(see Chapter 11). The boundaries of the Prudhoe Bay Unit and the
Sadlerochit reservoir, as finally agreed by the same process of negotiation,
are shown on the accompanying map, "Prudhoe Bay and Its Oil." Page 104.

personality of its recently elected chairman, Robert Anderson. Within the industry, Exxon has the reputation of getting first pick among the various competencies that oil requires. Within the company, however, these competent men seem to function with a decided leaning toward shared responsibility for important decisions, a system that is self-protective for the company as well as the individuals but also means that the perceptions brought to bear on a specific issue, the goals arrived at, are diffuse and quite generalized, the products of a committee or of a whole series of committees. Moreover, the continuity of men at the top, their views and programs, is not much greater than in the U.S. national government, with its programs tied to annual budgets and the regular upheaval of election and its corresponding inability to define and pursue long-term goals; at least since World War II, Exxon's leadership has tended to rotate at four- or five-year intervals, and the chairmanship of the parent company (and often the presidencies of the component subsidiaries) appears to be not a lever by which to apply the risks of new directions, fresh ideas, but a comfortable final platform from which to enter a deeply cushioned retirement. The Arco vice-presidents I have questioned on the subject claim to operate in the same kind of collective style as their competitors, and this may be true up to a point; but the company's fundamental policies since 1965 all bear the imprint of a single personality, that of Robert Anderson.

What Prudhoe Bay meant for Arco was growth, of course, but in particular directions. Arco's share of the oil—about 20 per cent, 400,000 barrels per day of the eventual maximum production of 2 million that was now projected—matched fairly well with its West Coast refining and marketing capacity. Along the East Coast, the company was supplied by production from Texas and the offshore Gulf of Mexico. Thus, by cutting back somewhat in the Midwest and expanding moderately on the West Coast, Arco would arrive at a uniquely secure and efficient position among big American oil companies: its supply of oil would balance its demand; moreover, this supply would come largely from domestic sources, independent of chancy imports; and, barring some new and unforeseeable coup of discovery, the company's over-all growth would not be in petroleum but in other

forms of energy and in minerals generally. Sooner than his competitors, the chairman seems to have seen and acted on the signs
that imported oil would become more expensive, less profitable,
scarcer, politically more dangerous, than it had been in the
heyday decades since World War II; and that the future of oil as
the primary source of U.S. energy was likely to be shorter than
its past. Or so I read the motives back from subsequent events.

BP's overriding reason for coming to Alaska in the first place
was to force its way into the American market, and that was
more important than just where the entrance came. On the credit
of its holdings at Prudhoe Bay, it was able to buy the skeleton of
a refining and marketing system on the East Coast (with some
cross-country links), then in the Midwest; it had at least a start
on the West Coast. Hence, BP's interests were compatible with
those of either of its two chief partners, and it might support one
or the other depending on the issue at hand. This balance was
reflected in a new division of responsibilities for TAPS reached
in September 1969, after BP had brought in its first wells and
had shown (with confirmation from DeGolyer and Mac-
Naughton) that it possessed at least half the Prudhoe Bay oil.
BP and Arco each agreed to own, and pay for, 37.5 per cent of
the pipeline; Humble was left with the remaining 25 per cent.

The furious urgency that infused the whole history of the
pipeline had a special meaning for BP. The sliding scale by
which it acquired Sohio was a protection for both parties against
an unlikely eventuality, as it seemed at the time—that Prudhoe
production might be indefinitely delayed or turn out a bust. BP
started with about 23 per cent of Sohio's stock—enough effectively to control corporate policy but far from outright ownership. For BP to gain absolute control in the form of half the
stock, Sohio's share of Prudhoe production from the leases it now
owned would have to reach a sustained rate of 450,000 b/d (barrels per day) by January 1, 1984. (This was net production, less
the state's 12.5 per cent royalty share, and since the Sohio share
was something over 50 per cent, it meant a total production of
close to 1 million barrels.) BP, as operator on Sohio's behalf, had
an incentive for achieving these rates with some dispatch: 75 per
cent of the profits on Sohio's share of the net sustained production beyond 600,000 b/d—provided that level was reached by

January 1, 1978. Thus, if BP failed the first deadline, it would lose profit, but if it missed the second it would have surrendered its Prudhoe leases without gaining the prize it sought, Sohio and a valuable chunk of the American trade. Both dates, of course, seemed comfortably remote at the outset, but became much less so as the project developed. The British company thus had a quite specific motive, over and beyond the general strategy of early return on investment, for getting the oil moving by the most expeditious route or method. Apart from the important matter of timing, however, among the three principals BP had no strong preference for any particular alternative to TAPS. Exxon had some interest in a different route, but it was not decisive, and perhaps could not be, in the character of the company. Only Arco would be strongly and essentially benefited by the route actually chosen. The three companies proceeded to act on the policies determined by these motives.

On October 28, 1968, the same day that the formation of TAPS became official, Arco announced the first big step in its own plan for the Prudhoe Bay oil: the construction of a new refinery at Cherry Point, on a peninsula a few miles west of Bellingham, Washington, and not far from the Canadian border, at the innermost meeting point of the straits of Juan de Fuca and Georgia which surround Vancouver Island. The new refinery would have a capacity of 100,000 b/d; with later expansion of its existing refinery at Los Angeles (185,000 b/d) the company would be able to absorb its probable share of the maximum production from Prudhoe Bay. To design the refinery, Arco chose a young engineer from Atlantic's Philadelphia headquarters refinery, Bob Steyer, and sent him west in early November. (Anderson by now had transferred the merged company's corporate center from Philadelphia to Los Angeles.) Bob was still there years later, as managing engineer, when I presented myself to spend a morning touring the facility and talking.

"We had as our target," Bob said, "a catch phrase, seven-one seven-one—meaning that we wanted to be in operation on July 1, 1971, which was to be the same date that the North Slope crude was to reach the West Coast. It was quite a rush for us to be ready when it got here. Obviously, we won. . . ."

The site is on a broad, deep-water seaway, part of the system focused on Seattle that constitutes perhaps the grandest port on earth, one of two natural ports in the United States capable of handling the current generation of deep-draft tankers (the other is Valdez); today also a place of great natural beauty, the farthest reach of American westward migration, where in the mild, damp climate, the sea grass grows thick, patched with scrub woodland, and local farmers eke a living from cutting hay, digging clams, fishing. From the start the refinery built here was intended as a showplace of environmental tact, fitted unobtrusively into the seascape, incorporating a number of novel techniques that later would be refined and expanded in the bigger oil terminal at Valdez (a plant, for instance, to process tanker ballast water, separating the residual oil so that it need not be discharged into the sea). The Cherry Point refinery was also designed specifically to handle Prudhoe Bay oil.

As we noted earlier, crude oil varies enormously in its characteristics—weight, the kinds of hydrocarbons it contains, the chemicals mixed with it. Prudhoe Bay crude is moderately heavy —it contains a fair proportion of the stuff from which asphalt or the heavy industrial fuel oils are made; to refine it profitably and efficiently, therefore, you need equipment that can process these "heavy ends," as they are called. It also contains a little less than 1 per cent of sulfur by weight, or about three pounds per barrel. Sulfur is among the mixed blessings of the oil business. In a pure form or converted to sulfuric acid, it is an indispensable raw material and tool in an infinite range of chemical and industrial processes. Mixed with oil, however, or transformed to sulfuric acid by heat, pressure, and contact with oxygen, it is bad stuff— destructive of machinery, environmentally offensive, foul-smelling, corrosive, unhealthy. Hence, if, like Bob Steyer, you are designing a refinery to accommodate crude you know has a significant sulfur content, you build in equipment to remove the sulfur; and you specify a chrome-alloy steel in the refining vessels that will resist corrosion. (The penalties for not doing so are unendurable—shutdowns, leaks, explosions.) That is what Arco did, and it is that, more than the environmental amenities, the pale, tasteful architecture, that makes the Cherry Point refinery special.

There are two other interesting things about this undertaking.

One is the timing—the rush toward production, the pipeline construction schedule already established, although it had not yet been made public and an immense number of essentials remained to be attended to, starting with obtaining permission for a route that would lie almost entirely on public land. Secondly, there were several refineries in the Puget Sound area that are designed for crude at that time still piped down from Alberta—generally lighter than the Prudhoe substance and lower in sulfur—and therefore unsuitable for Prudhoe Bay oil. Neither there nor anywhere else on the West Coast did any of the principals in the Alaska Slope discovery take steps either to expand their capacity or rebuild the equipment to process the expected deliveries; except Arco. It is one of the puzzles running through the whole project.

In December 1968, Humble came forward with the first clear signal as to *its* intentions for the Prudhoe Bay oil: a grand experiment in running a tanker through the narrow, ice-choked channels that lie between Canada's northern mainland and its Arctic islands—the long-sought Northwest Passage that had tantalized European explorers since the 16th century. The project was less Quixotic than it may at first sound. Archangel near the Arctic Circle has been an important Russian port for three hundred years, and though it freezes up for nearly six months of the year it is kept open by icebreakers; in the same way the northern Great Lakes with their oceanic storms are used through most winters by a huge traffic of tankers, barges, and ore carriers. What would be needed was a tanker built on the lines of an icebreaker: ice-cutting bow, heavy tempered-steel plating, powerful engines for ramming into ice—and for backing out again, as need be. Expensive but, given the inherent difference in transportation costs by tanker and pipeline, quite possibly feasible.

Moving fast, Humble found a vehicle for its experiment, a six-year-old American-built white elephant, the *Manhattan,* which could be redesigned for the job. Dating from the early days of the supertankers when naval architects were less confident of their specifications than they later became, it was overbuilt (heavier and stronger plates, more reinforcing, than were later found to be necessary) and overpowered (two steam-turbine engines of something over 21,000 hp each, designed for two smaller

ships that were canceled, driving separate shafts and propellers
—about four times the power, proportionately, used since in
bigger ships); fast and maneuverable but expensive to operate
and therefore chartered mostly for subsidized bulk shipments of
grain to India. Although far bigger ships are now quite usual in
the oil trade in most parts of the world, the *Manhattan* was nevertheless a giant when commissioned and by any previous standard: 940 feet long, with a draft of 50 feet and a carrying capacity
of about 115,000 dwt.* To fit the *Manhattan* for its task, a new,
icebreaking bow was added, rising from the keel at a shallow
30-degree angle, flaring outward at the sides to fend off ice; a
girdle of protective steel plates built out twelve feet all around
the original hull and extending seven feet above the water line,
fourteen feet below; laboratories, a helicopter landing pad, instruments to measure flexing of the hull, changes of displacement
as it plowed into ice. The work, urgent, pressed through the
spring and summer; by August 1969 the ship had been reas-
sembled at Chester, Pennsylvania, and on the 24th set sail, picking up along the way the Canadian and American icebreakers
that would shepherd her through. By mid-September the *Manhattan* had reached the vicinity of Prudhoe Bay and received a
symbolic barrel of oil, delivered by helicopter, for the return
voyage.

After three centuries of anticipation, the passage was uneventful but informative. Bram Mookhoek, later the chief of Exxon's
tanker operations—a tall, angular Dutchman bred beside the
North Sea, with the weathered features and steady eyes, the certainty of craft, of a master mariner—was aboard as designer and
gadfly. He worried that the ship was not having enough trouble
with ice to test its instruments:

"We passed an ice island, maybe three or four hundred thousand tons of ice, way up on the north part of Greenland, and we
were concerned that we didn't get any bending moment. I went
up to the bridge and told the captain, 'We want to ram that ice
island.' He says, 'You must be out of your mind. You mustn't

---

* dwt: dead-weight tons, the usual measure of size in tankers and other
cargo ships. It means the weight a ship can carry (in long tons—2,240
pounds), including about 4 per cent for its crew, fuel, and supplies. For the
unmodified *Manhattan* and Prudhoe Bay oil this would have been about
795,000 barrels.

know what the hell you're talking about.' I said, 'No, I think we know what we're talking about.' So we turned around, we found that ice island, we steamed around it. I said, 'Let's lay in at a right angle to that ice.' Five times at varying velocities we rammed it, and she climbed up, and of course we got some very nice readings, the bow went up, the stern went down—our instrumentation worked fine. Five times. Beautiful!"

From innumerable such experiments, Captain Mookhoek and his colleagues compiled the data from which to design a fleet of tankers to negotiate the Northwest Passage, first forty ships of 200,000 dwt, then fewer of 300,000 dwt, all immensely powerful, with as many as three plants of 65,000 horsepower each. The *Manhattan* came back for a second run in the spring of 1970 (and from a prong of ice that struck below the protective sheath of reinforcing steel picked up a sizable hole in a cargo tank). A conclusion was reached:

"I can only say in this case what the company has said," the captain reported, still disappointed. "That the pipeline had an economic edge." He remained convinced that the plan would work: *"In your lifetime you're going to see ships going through there."*

The *Manhattan* went back to hauling grain. The end, perhaps, was foreordained, despite the mariners' certainties.

From the early pipeline estimates, I have calculated that the transportation cost to Los Angeles would have come out to 52¢ a barrel, including the current U.S. tanker rates. There is some question as to whether those estimates were valid, but the oil-company management evidently relied on them and they were the basis for the decision Captain Mookhoek reported: the tariff in his ice breaking tankers to New York was predicted at 60¢ per barrel, then at $1 after the first voyage. The design work continued, with bigger tankers; under pressure from Arco, the emphasis shifted from the Northwest Passage to the Bering Strait and the West Coast. By either route, there was one great difficulty: the shallowness of the Beaufort Sea, which would require laying an undersea pipeline thirty or forty miles off Prudhoe Bay to an icebound loading dock in water deep enough for the big tankers to ride. Edward Teller of the Atomic Energy Commission proposed atom bombs to create a deep-water port at Prudhoe Bay, but the idea did not get very far.

Humble initially invested $40 million in its *Manhattan* project, another $10 million in the second voyage; these, by the way, were the published estimates, but gossip within the company guessed the real cost at $150 million, perhaps even double that. Whatever the true figure, the important question is why, if the experiment was economically doomed from the start, though technically feasible, it was undertaken in the first place. Despite two books published about it and many newspaper articles, it seems expensive publicity. It has been suggested that its purpose was tactical, to delay the fruition of the Arco and BP interests. Within the industry, it is thought to bear the imprint of the former governor of New York, Nelson Rockefeller, a man with a penchant for building monuments, and this is plausible. Mr. Rockefeller and his brother David, the chairman of the Chase Manhattan Bank (which has a special interest in oil), control the Rockefeller Brothers Fund, with offices in Rockefeller Center adjacent to Exxon's headquarters. The discreet lawyers who administer the Fund do not chat with inquiring authors, but it is probable that Exxon stock is the heart and soul of this enterprise; and that the stock is enough so that when an idea comes into a Rockefeller head, the Exxon management takes note and dutifully pursues.

Whatever the precise balance of interests among the three companies, Arco and BP contributed only $2 million apiece to the *Manhattan* experiment, and in early September, before the ship had completed its first voyage, Arco confirmed its commitment to West Coast delivery by ordering the tankers it would need to haul oil from Valdez. Both decisions seem to emphasize Arco's (and BP's) indifference to an all-tanker solution, but it seems equally possible that Humble had simply persuaded its junior partners (as it considered them) to accept a division of labor: Humble would pursue its tanker experiment while the others would undertake less costly investigations of alternatives to TAPS—land routes that would carry the oil somewhere other than the West Coast. These alternatives continued to be understood inclusively—as supplements to the intended Alaskan system, not replacements.

Arco took the lead in organizing preliminary engineering and economic studies of the kind that had preceded the TAPS decision, but all were commissioned to outside consultants rather than prepared by its own pipeline company or those of its colleagues. Despite the cool encouragement the three companies had received from the Canadian government, at least seven such studies were produced in 1968 and 1969. All were variants on a single idea: piping the oil east to the Mackenzie, then south to Edmonton whence it would continue east to a terminus at Minneapolis, Chicago, Pittsburgh, or perhaps even New York. One immediate question was getting to the Mackenzie—whether across the Arctic National Wildlife Range between Prudhoe Bay and the Canadian border (the act that created it made provision for a transportation corridor); or around the Wildlife Range, either offshore to the Mackenzie delta or south through the Brooks Range and then east. Beyond Edmonton there were existing pipelines pumping Alberta crude but without the capacity for the anticipated volume; effectively, one or more additional lines would have to be laid to take the oil wherever it was to go. In December 1968 or earlier (it was not announced for six months), Arco and five other companies formed a Canadian consortium on the lines of TAPS, which they named Mackenzie Valley Pipeline Research, Ltd. This body produced a series of studies of the possibilities and near Inuvik, the administrative center of the Canadian Arctic on the Mackenzie delta, built a test loop of pipe which functioned through two winters. No serious results came from all this activity, and again the conclusion seems foregone: pipeline construction costs and therefore the tariff on the oil carried are proportionate to the length of the line, and any Canadian route would produce a pipeline at least three times the length of TAPS—the transportation cost would be higher than the oil could sell for at the other end. The only thing that might make the route worth considering would be the possibility of spreading the costs over a substantial volume of oil from the Canadian Arctic; over the years, oil had been found in the Mackenzie delta, in the Beaufort Sea, and on the Arctic islands, but in the face of ambivalent Canadian policies toward Arctic exploration, it was not yet enough (and still is not).

Concurrently with these studies of alternative oil routes, the companies, more energetically, were considering what to do with the natural gas from Prudhoe Bay, and this was a rather different situation from the oil. Gas has always been somewhat incidental to the main business of oil, a by-product of exploration and production, in the past burned off as a dangerous nuisance where it could not be cheaply and readily sold; in addition, under U.S. policies which for years have fixed the price of gas at a fraction of the market price for equivalent quantities of coal or oil, it is a comparatively expensive and unprofitable nuisance. At Prudhoe Bay, however, the quantity of gas, 26 TCF (trillion cubic feet) was nearly as grand as the oil reserve; the oil besides contained substantial amounts of dissolved gas, about 300 cf per barrel, which would have to be disposed of somehow (Alaskan law prohibited flaring the gas and required that any not sold be pumped back into the ground for future use). Canadian Arctic gas discoveries had been considerably larger than the oil—enough to make a Canadian gas route feasible with the bonus of dual use to shave the tariff. Or so the companies evidently concluded, and that is the decision they have stubbornly maintained ever since. In June 1969, a project for a Mackenzie-Edmonton gas pipeline was formally announced, to run 2,080 miles at an estimated construction cost of $1.2 billion. For several reasons—inability to assemble the financial and construction resources for two such giant projects at the same time, perhaps a gamble that with delay the controlled U.S. price of gas would be allowed to rise to a more favorable level—the gas line would not start till the oil line was finished. In the meantime, as state law required, the companies would go on pumping their gas back into the ground. From an engineering standpoint, this can apparently continue indefinitely; in terms of cost, however, there are limits—the more gas produced with the oil, the more compressors, pumps, and injection wells that will be needed to put it back, and all of this machinery consumes, as fuel, a portion of the resource it is meant to conserve.

The $1.2 billion estimate for the gas line was apparently supposed to represent the complete cost as foreseen at the time, however shared among U.S. and Canadian companies. It is note-

worthy, therefore, that it amounts to a good deal less than half
the cost per mile then estimated for TAPS. A gas pipeline is indeed a different and simpler proposition than an oil line, at least
in the North—but not *that* different. Throughout, the companies
have tended to put their costs before the public in the most optimistic possible terms, at least until irrevocable decisions have
been made; then the realities begin to trickle out.

All through these early stages, the TAPS system remained so
attractive that the companies apparently did not give much attention to the fundamental question—whether the West Coast
could in fact absorb the quantities of oil projected for it. Or so
one concludes from events since 1968 and the impassioned denials by the various company presidents, repeated over the next
eight years.

Under a government system developed during the Korean war,
the U.S. oil industry is divided for administrative and statistical
purposes into five regions. In this system, the three Pacific Coast
states, with Arizona, Nevada, Alaska, and Hawaii, constitute a
district known as PAD V (an acronym meaning the fifth district
in the Petroleum Administration for Defense). The other four
districts—the rest of the country—are a distinct entity: broadly
speaking, oil produced in any of these states (or imported at any
of their ports) can be made available to any other state as need
requires. PAD V is different. For all practical purposes, there has
not been any means of getting oil produced in, say, Kansas,
Oklahoma, or Texas across the Rockies to California, PAD V's
principal consumer; or vice versa. For many years prior to 1968,
PAD V (and chiefly in California) had been producing about
half the oil it consumed, and production had tended to rise with
demand, the difference made up by imports from such places as
Indonesia: consumption of about 2 million b/d supplied fifty-
fifty from production and imports. In order to answer the fundamental question satisfactorily, therefore, it was necessary for the
companies to assume that PAD V's population and energy use
would rise so fast that the district could absorb the entire
Prudhoe Bay production when it began flowing at Valdez—and
dispense with imports into the bargain. The necessary difference

in the rate of increase was only 2 or 3 per cent per year, but that would have required *double* the average increase for recent years, both nationally and in PAD V. It was a large assumption.

I have been able to discover one sign that someone was thinking about the possibility of a West Coast surplus: a plan announced in January 1970 by a group of construction companies —British, German, Italian, none American—to build a 700,000 b/d pipeline across Panama. The Panama Canal locks will accommodate only quite small tankers, and with that limitation it is simply not possible to transport that daily volume of oil through the canal. Later, when TAPS officials were questioned as to what would become of a West Coast surplus if it in fact occurred, they talked vaguely of "Panama oil"—and still later gave not even that faint hint. Nothing so far has come of the Panama pipeline plan.

It is another piece of the puzzle which we shall have occasion to examine again.

Whatever unexpressed doubts may have lurked in anyone's mind, TAPS—that is, the committees of oil men of which it was composed—forged ahead as if guided by a divine revelation. In February 1969, Humble, Arco, and BP jointly announced their pipeline plan, to be completed by 1972 at a cost of $900 million. The estimate seems to be the preliminary one made a year earlier with the cost of the Valdez terminal added in, plus something for unpredictables; the completion date was no longer Bob Steyer's "seven-one seven-one"—the construction schedule had already slipped a year since the preceding fall.

A few weeks later, using mainly crews borrowed from Humble, TAPS began two programs essential to design of the pipeline: the drilling of core samples along the route to determine just where permafrost occurred, using small, portable drilling rigs imported from Texas and helicoptered into position or hauled behind tractors on sleds; and exact surveying of the proposed right of way, foot by foot and mile by mile, supervised from Anchorage by Frank Therrell. Although actual design was thus still in a very early stage, by April the oil men had matched the intended oil production to a pipe size big enough to carry it and felt confident enough to order the pipe. It was to be 48 inches in diameter, bigger than any yet used in a pipeline. Pipe in any size begins as a flat plate of steel which is then heated to a glowing temperature and coiled or, more commonly, rolled, into a tube of the required diameter, the seams closed by welding. As it happened, there was no plant within the backward and complacent American steel industry capable of producing 48-inch diameter pipe, although U. S. Steel offered to build one for the job —in a year or two. Three Japanese firms, however, came forward with an offer to fill the order at $50 less per ton than the current U.S. price, with deliveries to begin in the fall on specially constructed ocean barges. The oil men, naturally, accepted, and a contract was signed, though not announced till May 8, 1969: 500,000 tons of pipe in 30- and 40-foot lengths, at a total cost of $100 million; TAPS at the same time, for another $30 million, ordered the giant pumps for the first five of twelve planned pump stations that would be needed to move the oil through the pipe. If there had remained any faint possibility for fundamental change either in the pipeline route or in the means of transporting the oil, that possibility was now canceled. Moreover, since the pipe had been specified on the basis of quite sketchy knowledge of the conditions it would face along the route, the whole system would have to be designed to accommodate the pipe rather than the other way around. Finally, looking back from the present and pondering the very lively interest the Japanese government takes in the exports and imports on which the national survival depends, it does not seem likely that it offered such favorable terms purely from friendship for American oil companies; but just *what* the Japanese got in return is among those secrets of oil that have not yet come to light.

While the pipe contract was being drawn up, Arco, which at the start of the year began building a permanent Prudhoe Bay headquarters to replace the drilling camp beside its confirmation well, started work on a small refinery nearby, to turn the crude from Sag River State No. 1 into a usable fuel for trucks and drilling rigs. It is known as a topping plant: it separates the light components, or "tops," which with a minimum of further processing then constitute a low-octane fuel referred to locally as

Atlantic Supernox. (The heavy ends are pumped back into the ground.) By November the plant was functioning.

By April 1969, TAPS had assembled construction equipment worth, it was said, $50 million, perhaps pursuant to the original schedule aimed at starting work that spring and finishing in 1971. A number of more immediate jobs remained undone, among them the application for a right-of-way across the mostly federal land through which the pipeline would run (the right-of-way was wide enough so as not to depend on the exact survey still barely begun but could in a pinch be varied); this lengthy document, with descriptive maps, was delivered to the Anchorage office of the Department of the Interior on June 10th.

In July, without waiting for the completion of its long-pending land application, TAPS let the contract for the Valdez terminal to Fluor Ocean Surfaces, Inc. (The permit finally came through from the Forest Service on October 1st.) A month later the pipeline group did receive a permit to build a short section of gravel road from north of Fairbanks as far as the Yukon, and five days later, on August 20th, the construction contract was signed. The quick completion of this road made it possible to haul the facilities for seven construction camps as far as the river by freeze-up, then, with no public fanfare, tractor them north into position in the course of the winter so as to be ready for use in the spring. (Some components were built to fit the belly of a Herc and were flown into place.) The obliging Department of the Interior had agreed to rent the land for the construction camps for $200 per year. It did not seem to matter that the application for the pipeline right-of-way still hung fire or that TAPS had no formal authority to take its equipment overland to these rented campsites —from past experience the oil men could expect the Department of the Interior to come through, if slowly.

Delivery from Japan of the first 120,000 tons of pipe, begun in July, was completed soon after, in September, and the pipe was distributed to storage yards at Valdez, Fairbanks, Prudhoe Bay. Bidding was opened for construction of the pipeline and early in December Arco announced the contracts for the components of the two flow stations it would need in its half of the field—plants to gather the crude from its hundred or more wells operating at start-up, purify it, and start it on its way south through the pipeline. Despite a certain awkwardness in getting the various jobs in their proper and logical sequence, the project seemed to be going smoothly—no reason, one would have thought, not to hit the announced completion date in 1972.

Nevertheless, in the course of these months of preparation, there had been certain stirrings in Alaska that might have disquieted less confident men. Soon after the application for the terminal site was filed, Natives in the vicinity of Valdez protested to the Department of the Interior that the land was really theirs. That roadblock was eventually cleared, but as plans unfolded for the pipeline, similar mutterings began to be heard from other groups of Natives along the route. TAPS took the precaution of securing a waiver of rights from the recently organized Alaska Federation of Natives, of which Emil Notti was president. Five Native villages, led by the sixty-six residents of Stevens Village, on the Yukon a few miles northeast of the proposed pipeline crossing, disclaimed this waiver. In February, they filed suit to prevent the Department of the Interior from granting the right-of-way over nineteen miles of the 800-mile route; their argument was that since the land had never been sold or surrendered, it remained theirs by immemorial occupation and was not for the federal government to give away. The case moved rapidly through a federal district court in Washington, D.C. There on April 1, 1970, the same day the *Manhattan* headed out on its second voyage, Judge George Hart agreed that there was justice in the argument; and granted a temporary restraining order against the building of the pipeline and the year-round construction road north of the Yukon without which it could not be begun. The project's smooth forward progress came to an abrupt halt.

What the judge acknowledged in his brief order was a legal ambiguity left open in the original treaty with Russia in 1867, unresolved by Congress in the two Alaska organic acts and the statehood act, and in the few other legislative attentions the northern country had received in the course of a century. It was a tangle that would take four years to unravel, ending, where it began, in Congress. Considering how long it had been in the making, it is only surprising that it did not take another century.

Chapter 4

LAND, LAW, AND RIGHT

A decision had been made very early to build a Trans-Alaska
pipeline, for whatever short-term reasons. The years of litigation were simply devoted to rationalizing that decision.
                                        —Dennis M. Flannery

In March and April in Alaska, warming ocean currents coil
among the islands of the Southeast archipelago, changing the fog
and wet snow of Juneau's Pacific winter to the misty rain of
spring. Circling the coast of the Gulf of Alaska, the currents wash
the grimy snow from the streets and sidewalks of Anchorage, but
the nights still freeze, the surrounding mountains are layered in
snow, and down Cook Inlet rafts of winter ice run with the tides,
lifting on the gun-metal swells like the hulks of an abandoned
fleet; the deep thaw, renewing earth's life, engorging streams,
heaving pavements as if with an underground bombardment, is
still weeks off. Farther north, at Fairbanks and the Indian way
station of Nenana, bets are collecting in a pool on the breakup of
the Tanana River and a pole has been set up out on the four-
foot-thick mid-channel ice to mark the exact minute, which may
come in a couple of weeks now or not for a month or more,
depending on just how cold the winter went and the depth of insulating snow-cover. The ice lingers, but softening from the top,
beginning to crack, and the air—damp with melt and rotting
snow—smells of spring. Beyond the Yukon, the construction
camps still wait where they were hauled in and dumped the previous winter, but at Prudhoe Bay the drilling of development

wells continues methodically according to the agreed-on plan, paced to the pipeline schedule (Arco and BP each will have to complete sixty wells, averaging 10,000 b/d, by start-up, two hundred altogether to provide the production to fill the maximum capacity for which the pipe has been designed), though it no longer looks likely that the application will be granted and construction begin this year. The coast is still locked tight in sea ice, and the temperature, which in daylight reaches zero or better, drops low by night, but here too you feel the season changing. It is the light. Every day now the burnished coin of the sun lifts higher above the level white horizon of the Arctic plain, stays longer—the days by now are long at Fairbanks and Anchorage, thirteen or fourteen hours, but here the sun lasts two hours more, rising and setting in a swath of clear twilight. It is as if the pole is a fountain of light washing downward across the earth: before the end of April there will be no more true darkness, and by May the sun will have ceased to set. Light: all through May and June there will be more of it at Prudhoe Bay, in measured units of the sun's radiation, than at Fairbanks or Anchorage.

In Washington, decisions were being made that, so far as any human action can, would affect Alaska as profoundly as the recurring cycle of its weather, powered by the slow explosion of northern sunlight. In the Federal District Court, Judge George P. Hart, Jr., studied a map of Alaska and tried to relate it to the arguments before him: lawyers from the eminent firm of Arnold and Porter, speaking for the Koyukon Indians of Stevens Village, others from the Department of Justice, representing the Department of the Interior. The soft Potomac spring was already warmer than most Alaskan summer days. In Washington's squares, groups of girls from the offices occupied the grass, sunning their knees and eating sandwiches, while solitary young men sat upright and serious on the green-painted benches, reading newspapers; busloads of tourists circled Jefferson's domed memorial, viewing the first pink and white buds of the cherry trees. The judge rendered his decision: the injunction requested by the Natives—forbidding Interior to give permission for a pipeline on the federal land it administered in Alaska—was granted. April 1, 1970: the culmination of a century of uncertainty as to who owns the land of Alaska and in what sense, but also the start of a new process and therefore an intermediate step, a transition. With the confidence of a man secure in the court-room where his word *is law*, the judge also offered advice to the two groups of attorneys: Go away and talk among yourselves, agree, and come back in ten days' time with what you will accept; this is no matter for a court of law. A jurist's advice, weary of issues beyond reconciling, suits and their penalties, delays.

Circumstances did not permit the two parties to appear again in that relationship. On March 26th, when the pipeline permit seemed imminent, James Moorman, a young lawyer for the Wilderness Society, with two other environmental groups, sued the Secretary of the Interior with the same object but on different grounds—that granting the permit exceeded the Secretary's legal authority, that to do so violated the recently passed National Environmental Policy Act. The suit was denied, amended, united with the Natives' complaint, and on April 13th was argued before the same Judge Hart. Ten days later, on Shakespeare's birthday, he ruled for the environmentalists in a hand-corrected typewritten order that enjoined the Secretary from allowing the oil companies (which constituted TAPS) to build a construction road north of the Yukon, mine gravel for that purpose from public lands, or take steps toward constructing a pipeline anywhere along the route. The day before the judge's decision, crowds had gathered across the country to celebrate the first Earth Day and at such a gathering, at Fairbanks, the Alaskan Secretary of the Interior had assured the students and teachers of the University of Alaska that he would grant the necessary permit—but not without further and sufficient study.

The two court orders brought into brief conjunction interests that were at least divergent if not fundamentally opposed. The Natives' objective, for the moment given voice by the Stevens Villagers, was concrete, specific, and, if large, not unlimited: possession of the Alaskan land, in fact and in law—or as much of it as they could bargain from their fellow-Alaskans and the rest of the Americans, who throughout Alaska's history seem always to have looked on that land—whether with contempt, condescension, admiration—as a common possession, an attitude applied to no other part of the nation. Oil development and the pipeline (and for now the delay of both) were then simply a handy means to that end, but in the long run the Natives had no quarrel with either; on the contrary, if the land had been inseparable from the Native culture of the past, oil (and all the other forms of mineral wealth thought to lie below the surface of the land) was indispensable to the viable future that a few of the people's leaders had begun to envision. To that future also the land was, therefore, the necessary precondition.

The goals of the Wilderness Society and the other groups with which they made common cause were less simple—broader, more diffuse, less immediate, not, perhaps, subject to any natural limits. They came forward as the precursors of a new breed, beginning to name themselves—*environmentalists*, practicing *environmentalism, ecology;* words coined to express not so much a discipline having a specific content as an attitude. That attitude descends from ideas of conservation developed early in the century, perhaps ultimately from much older American virtues of thrifty self-sufficiency, moderation, balance, but now vastly enlarged. It has to do with man and nature, a word not much used or valued, but we have no other so inclusive for the thing meant: whether man and his works are part of nature or apart from it, intrusive. It is one of the fundamental questions, and fundamental consequences follow from one's answer. If in your view of the world man stands separated from nature by a great gulf which he cannot cross without doing harm, then the value you place on the merely human benefits from that commerce will not be high; and if you perceive that such harm follows necessarily because man's sense of his own good is scaled to his own narrow sense of time—today's sewage fouls tomorrow's streams, yesterday's cook fires have stripped today's land of its forests and bared the nurturing earth to the ravages of wind and runoff— then you will, if you can, compel the rhythms of human activity to match and be judged by the longer rhythms of nature, though the living earth is no more truly measured in human lifetimes, generations, than in years. It is a view that resembles that of the very earliest English colonists on the Atlantic seaboard, with their horror of the wilderness in which fate or luck or conviction had placed them, huddling for mutual comfort in temporary camps and makeshift villages, but leading to an opposite conclusion—not to tame or subdue the wild, remake the wilderness to the measure of its human use, but to leave it so far as might be untouched.

So far as Alaska and its pipeline are concerned, two possible courses of action follow from this frame of reference: either to prevent the pipeline absolutely; or to require that when finally built it be carried out in such a way as to guarantee no disturbance of the comparatively wild land through which it would pass. These are rather different goals, but they seem to have come together in those first court actions in the spring of 1970, and, in the subsequent years of litigation, their practical effects were much the same.

Today in Anchorage and Fairbanks there are two kinds of Natives, one visible, the other not. The visible kind is among the first differences a visitor notices, arriving from almost any other American city. Two blocks from the Anchorage City Hall, opposite the Holiday Inn, 4th Avenue is lined with the low, square-front buildings of an earlier Alaska, shored up from the 1964 earthquake: bars that open at eight in the morning and do not close again till 5 A.M. (the liquor stores shut two hours sooner), pawnshops sidelining in sexual artifacts, cheap up-a-flight hotels; a couple of blocks farther east and the storefronts give way to small cabins crowded onto subdivided lots with the look of a summer fishing camp on the bank of a remote river, built in the imposing shadow of the Alaska Native Hospital, which provides free care to Natives and a few others. Hence, if you stay at the Holiday Inn and your room happens to be on the 4th Avenue side, you become the audience of a continuously performing theater of the street. (The corresponding vantage point in Fairbanks is 2nd Avenue.) Dark young men with Mongol faces and lank hair, shiny-black, cluster outside the bars or crouch in groups along the curb, passing a bottle—villagers mostly, in town for a day or two and, knowing nowhere else to go, gathering in this reassurance of likeness and familiarity, bringing with them a little of that life of the village. Drunk—and that is the purpose, the drinks quick, one after the other till the state is reached—they sleep undisturbed against the storefronts or in good weather recline on the long open slope that descends on the other side of the hotel to the railroad and the port; or stretch out on the benches and the patches of grass in front of City Hall. Girls, undersized and probably very young, though there is no way of judging, cruise the sidewalks, indifferent, intent, and from time to time lead liquored-up white men to parked cars across the street or through empty lots to secluded rooms somewhere near. They do not eye the young men, who, drinking their money, are of no interest, but have the same sallow, impassive faces, the curtain of black hair streaming loose down the back, the bulky, shapeless down parkas, bright-colored and embroidered with geometric designs, over workaday blue jeans and boots. In the liquor stores as the night progresses the young men lean across the counters, drooling, spread a pocketful of dollars for the half-pint of whiskey or the bottle of sweet wine, the six-pack of beer, having trouble by now getting the words out. Women run the stores alone, unconcerned but fastidious—posters in the doors come and go, frequently torn down, warning: NO WHORES!

These are the visible Natives, though physically and apart from their demeanor, they might be the newer Japanese, heavyset Koreans. The rest in any Alaskan crowd are simply one more variant, perhaps with a Mediterranean cast, among the thousand American types, not distinguishable by looks, work, dress, name, though a Russian-sounding surname ending in -of probably means some fraction of Aleut; invisible, at any rate, to casual observation. The alliance of these two kinds of Natives—the enlarged meaning of the word—is among the consequences of the long struggle to possess the land, of which the April 1970 injunction was one incident along the way; but also a root cause of the eventual outcome.

Seward's hastily drawn treaty of cession left no doubt as to the sovereignty of Alaska—that, well established in international law, as we have seen, could be transferred without qualification. About most of the inhabitants—and therefore the actual title to the land they lived on, as distinct from sovereignty—the treaty was ambiguous. Those living in or near Sitka and the scattered settlements and trading posts along the south coast and up the Yukon were assured a choice of citizenship: American if they chose to remain, Russian, with passage home, if they did not; and were guaranteed "the free enjoyment of their liberty, property, and religion." From the last Russian census (1860) before the transfer, that provision probably applied to 800 actual Russians, 10,000 Russianized or interbred Aleuts, Creoles, and assorted Indians—the treaty did not attempt to distinguish; since the archives of Russian America have been closed to inquiring scholars since 1917, it has not been possible to determine exactly what areas of the country these people occupied or Russian government in some sense controlled. In any case the guarantee of citizenship did not mean much: it was, as we have seen, decades before Congress extended a semblance of government to Alaska, and decades more before it attempted to deal with such subsidiary questions as the application of laws and the ownership of land. In the meantime, the status of all the rest of the "inhabitants"—the forty or fifty thousand Indians and Eskimos beyond the ken of the Russians or known only from their occasional visits to outlying trading posts—remained unspecified: "the uncivilized tribes," as the treaty identified them, were to be "subject to such laws and regulations as the United States may, from time to time, adopt in regard to aboriginal tribes of that country" —meaning either that existing Indian laws would apply or, possibly, that the Alaskans would be dealt with as a special case. But whatever Seward may have had in mind, Congress did nothing.

The Organic Act of 1884, conferring a degree of government, made the ambiguity explicit, but also promised an eventual solution:

> The Indians or other persons . . . shall not be disturbed in the possession of any lands actually in their use or occupation or now claimed by them but the terms under which such persons may acquire title to such lands is [sic] reserved for future legislation by Congress.

No solution came. In time, the Alaskan Natives were granted citizenship (in 1924, along with other American Indians), but the more complicated question of their land remained and was among the stumbling-blocks to statehood: to make certain that the new state would not remain a charge on the nation (as most Americans since 1867, mistakenly, were persuaded it was), it was to be endowed with public land, eventually rounded out to 103 million acres; but since in the history of American law aboriginal land rights are subordinate only to the national sovereignty and superior to all other property rights unless set aside by treaty, purchase, or conquest, the lawyers in Congress could not see clearly how to guarantee the state title to its endowment. When finally passed in July 1958, the Statehood Act therefore adopted an ingenious legal fiction: it described itself as a "compact"—an agreement or treaty between two natural sovereignties —and paired the grant of land to the state with a self-contradictory assurance of Native rights. The "state and its people" (including, presumably, the Natives) were required to

> agree and declare that they forever disclaim all right and title . . . to any lands or other property (including fishing rights), the right or title to which may be held by any Indians, Eskimos, or Aleuts. . . ; that all such lands . . . shall be and remain under the absolute jurisdiction and control of the United States until disposed of under its authority, except to such extent as the Congress has prescribed or may hereafter prescribe. . . .*

Today, the leaders among Alaska's Natives can quote the words of these documents as readily as any schoolchild the Preamble to the Declaration of Independence, and all know at least their general sense. There were protests over land from the earliest years of the American occupation (from the obstreperous Tlingits in the Southeast), and they continued but went unheeded by Congress, the courts, the successive agencies charged with Native affairs. The people for decades felt no common interest. Their allegiances were local, to ancestral or tribal hunting and fishing territories which, as the basis for subsistence, required their own defensive hostility against neighboring bands

* This is the wording of the Statehood Act, Public Law 85-508, as it appears in the *United States Code*. To get it to this final state it was necessary for Congress, a year later, to produce a bundle of amendments more voluminous than the original Act, the Alaska Omnibus Act (Public Law 86-70, June 25, 1959), removing internal contradictions and reconciling conflicts with the considerable body of existing law that in one way or another had some bearing on Alaska; a characteristic congressional performance, it seems, whenever major legislation is at issue.

of differing language or culture, unless they came solely to trade. The cultures varied with the geography, from the permanent long-house villages of the seafaring Southeast groups—Tlingit, Haida, Tsimshian—to the nomadic Athabascans of central Alaska and the cosmopolitan Inuit in the north, with links of language, custom, trade across Arctic North America as far as Greenland. The dozen Athabascan languages of the interior are closely related but become increasingly unintelligible with distance from one territory to the next. The common language of the Eskimos evolved through five thousand years of separation to distinct languages native to the Arctic, southwest Alaska, and the coast around Prince William Sound—and Aleut, so different from the others that their common parentage is no more discernible than that of German and Hindi. And the three languages of the Southeast stand apart from the rest, with no family relationships, though there may once have been connections between Tlingit and Athabascan.

Divided by culture, language, and their own history, the Natives were subjected by the slow infusion of white settlement. The whites came at first as traders, prospectors, miners, as seasonal sealers, whalers, fishermen, canners, loggers—these transient uses of Alaskan lands and waters did not seem to conflict with the subsistence always in the past provided. But seals and whales grew scarce for the coastal people to live on, inland the traders' guns disrupted the once limitless supply of caribou. Starvation followed, recurred, and dependence on strange foods, brought in ships from the south. Those who lived acquired immunities to the accompanying diseases but not to tuberculosis (or influenza, penumonia)—there are those now, descended from parents who were among one or two survivors of a dozen children killed by that death, and it is still a commonplace of the distant villages. Notice of these consequences, from distant Washington and its agencies, came late and fumbling.

Grace Lincoln is a woman whose life was shaped by those impersonal attentions and by her own resistance to them. She is from Kotzebue, a place now with two thousand people, a city, at the end of a long spit of sand across the Bering Strait from Siberia and lashed since the last ice age by the Chukchi Sea;

whose layers of ancient habitation to archaeologists signify the beginnings of Eskimo culture and, more ancient by several ages of ice, the first entrance of man into the Western Hemisphere, a natural station in that passage. Five or six thousand years of communal life founded on whale and caribou in their seasons, supplemented by shore birds; when these failed and the people starved, her father had been among the first to build a herd of reindeer, Old World kin to the caribou but domesticable, an experiment early in the century by a missionary using nickels and dimes collected from American schoolchildren when Congress refused to fund it. Her face is formed in the broad planes of a textbook Eskimo, with the small nose and hooded eyes, the thick muscular body made for heavy work in hard weather. She became a public health nurse—knowing the closeness of family and village, the loneliness that would draw her back if she let it, chose the most distant place she knew for her training, Pennsylvania Hospital in Philadelphia; I met her in an upstairs office in Anchorage, in sight of the 4th Avenue bars, with questions about why the young men were down there in the street drinking their paydays, but it was her own life that we talked of, in Kotzebue.

"We had two different kinds of schools—this was in the Territorial days—one was the government school where the Native children went to school, the other was the Territorial school where the blessed ones went: the ones that had pale skins and blue eyes." The words as she says this have the lilting tune you hear everywhere in the North, of those not born to English, but although she understands her parents' Inupiaq, the language of the Inuit, it is hard, she had never learned to speak it: "We were forbidden to speak Eskimo—by the BIA. They thought that we would learn faster if we spoke nothing but English at home. They threatened our parents with jail if they spoke to us in our own tongue." There were common-sense reasons for the two school systems, one run from Washington by the Bureau of Indian Affairs, the agency of the Department of the Interior responsible for Native welfare, the other local to Alaska under the Territorial government (practically, till after World War II the BIA system excluded high school, though a few Native children were transported to Indian boarding schools in such exotic places as Oklahoma and Washington). English, at whatever cost, the teachers knew, was simply the necessary means to survival in the actual world.

In practice, the effect of these differences was contempt on the one side, shame on the other. Grace, with her head down, going to the inferior school as her parents directed, had the luck of meeting an anthropologist, one of several digging in the summers on Kotzebue's ancient coasts, guided by her father: "He put me through intensive kinds of things, just showing me the kind of skill that we had as a people and the fact that we survived in the harshest climate ever—I regained my pride just from his taking the time." It was the first step toward the scholarships and the decision, at the age of sixteen, to take the chance on distant Philadelphia: "The reason I went ahead and became an RN was just to prove that we could do anything we darn well pleased, that we weren't those dumb Eskimos and we weren't those dirty Eskimos who didn't know what to do or how to do it. My mother and father were *leaders* in their community. . . ."

For one who, from observation and study, knows Eskimos first of all for their quickness, adaptability, and dexterity, intuitive with machines, it is a useful reminder of a different attitude— and one not long past.

Grace Lincoln had lived her Kotzebue childhood at the end of a cycle of Native history. The old attitudes had begun to change, starting with the Natives' feelings about themselves. World War II brought the village people into contact with a different kind of white man, the soldiers, sailors, and pilots in the camps spread across Alaska for its defense against the Japanese or servicing the bases for the Lend-Lease bombers ferried to European Russia by the long but comparatively safe route across Siberia. Local Natives did civilian jobs around the camps. The war created a network of airstrips, the basis for a new and ready system of transportation; when an Alaskan airline extended regular service to Kotzebue, Grace Lincoln's father gave up his reindeer herd and went to work for it. Others, younger, joined up, acquired skills that later assured them responsible jobs within the white system and thereby leadership in the fight for land—such as Ray Christianson, a commercial pilot from the minute village of Eek at the mouth of the Kuskokwim River in southwest Alaska and a Yupik Eskimo, the most numerous of all the Native groups. After the war, there were opportunities for education on a scale never before attained: at the University of Alaska, last of the land-grant colleges or by reciprocity at the University of Washington; with statehood, at local high schools, at colleges linked to the university, at a new private university in Anchorage, Alaska Methodist. In the past, the occasional Native who managed to secure an education—such as the Tlingit lawyer William Paul, practicing in Seattle since the 1920s—had to go elsewhere to earn his living by it. Now for the first time significant numbers of Natives were formally educated, and statehood found uses for all there were— in local governments provided by the state constitution (confusingly called boroughs, with the size and general functions of counties elsewhere but the structure of municipalities), in the burgeoning state government or by a natural step in the state legislature; or in the new and aggressive organizations of Natives.

Statehood created an opportunity of another kind. As the state began picking the land Congress had promised, the curious contradiction written into the Statehood Act became an immediate practical problem—it was, all of it, land that one or another Native group supposed it already owned, from immemorial use, if not for the fishing mentioned by Congress then for hunting, trapping, berrying, and in many cases had already formally claimed. Emil Notti told me in outline what happened next, a story not made stale by probably frequent telling:

> The people started finding that the traditional hunting sites were being posted, you know, private property, no trespassing, and they got concerned. They were faced with a lot of budget problems, the state was, land was a commodity to meet the state budgets. And state government didn't give a damn about the people that were there, they were going to turn land into cash. The state started going in there and selecting lands, opening up lands for wilderness sites. Around Minto they started selling wilderness sites to sportsmen for cabin sites and hunting lodges. Then Dot Lake, they were selling wilderness sites at the New York World's Fair in 1964. They didn't give one damn, they went right out through Dot Lake and surveyed through a graveyard, no

consideration for the people that were there. And it was this
competition for land, between state government and the villages, that triggered the early meetings. . . .

The two early cases Emil Notti mentioned, both involving land
within reach of Fairbanks, were not the first, but they showed
what could be accomplished when a local chief organized his
people and teamed with a competent lawyer (at Minto, a young
white named Ted Stevens whose no-fee service carried him eventually to the Senate). What followed, finally, was a forceful
state-wide organization, the Alaska Federation of Natives, which
chose the thirty-year-old Notti as its first president, a young man
typical of the newest generation of Native leaders—educated,
with the name and looks and bearing of his Italian father who
had chosen the traditional life of hunting and fishing along with
his Indian wife; like Grace Lincoln, knowing no more of either
of his parents' languages than any second-generation immigrant.

The federation grew from a state-wide convention called in
the fall of 1966, formally proclaimed the following spring. The
federal emphasis was deliberate and political—a protection
against submergence of any of the by then numerous and mutually mistrusting groups, all drawn together by overriding interest
in their own land. There had been Native organizations in the
past—the Tanana Chiefs' Council in central Alaska, in Southeast
the Alaska Native Brotherhood dominated by the Tlingits—but
without much effect. Protests over land since the earliest years
had been successfully ignored, but now the state claims made action imperative and the Minto people had shown how it might
be carried out. Thus, each time the state submitted a land selection to the Department of the Interior's Bureau of Land Management for patenting, a local or regional Native group came forward with a counterclaim, often encouraged by another Interior
agency, the Bureau of Indian Affairs. By this process, the state
succeeded in getting tentative title to about a quarter of the
acreage promised by Congress, but the 42 Native claims added
up to 388,820,240 acres, nearly 24 million *more* than the state
contained (many of the claims overlapped and were therefore in
conflict with each other as well as with the state). In the fall of
1966, as the AFN was organizing itself, a bemused Secretary of

the Interior, Stewart Udall, ultimately responsible for making some sense of this tangle, ruled that no further state selections would be processed until the Native claims could be judged and either verified or refuted. Since that might take a while, white Alaskans groaned and made sour references to their unsatisfactory past relations with Washington. The new state of affairs was known as "the land freeze."*

Fortuitously, the AFN found a source of funds for its own campaign—and precedents on which to base it. Across Cook Inlet in the village of Tyonek lived the six hundred people of a Tanaina band called Moquawkie. When the oil companies in the 1960s discovered oil and gas offshore in the Inlet, the land around Tyonek became strategic. In Alaska it was peculiar: part of a 24,000-acre reservation set aside for the Moquawkies in 1915, swampy and therefore, in the eyes of the givers, worthless —but part of that tiny portion of Alaskan land to which anyone, even collectively, could show clear, private title. In the circumstances, that right was worth $14.7 million to the oil companies in lease payments and royalties, and the money in turn became the capital for a village corporation that took a substantial interest in Anchorage real estate—and could also find seed money for loans to promote land claims by other Natives.

About the same time in Southeast, a grievance of much longer standing came to a head. In 1902 the Department of Agriculture had set aside something over 16 million acres as Tongas National Forest, constituting nearly all of the archipelago—and the Tlingits once more duly protested that the land was theirs and not to be taken. (Another 2 million acres around Glacier Bay was taken later as a national monument.) In 1935, urged by William Paul, Congress gave the Indians permission to sue for compensation in the Court of Claims. Thirty-three more years passed and the

* In a message to Congress on March 6, 1968, President Johnson, Udall's boss, had dealt with the Native claims. The principles he set forth—that the villages should receive title to the land they occupied, with use of land needed for subsistence, and compensation for whatever rights they would relinquish—were those eventually enacted. The timing is interesting. The wonderful test reports from Prudhoe Bay were already in but as yet unpublicized, but in view of Mr. Johnson's aciduous concern for the oil interests in his home state of Texas, it seems highly possible that he knew by this time what was up.

government lawyers offered a derisory cash settlement of $7.5 million (it comes out to 40¢ per acre). Nevertheless, the Tlingits accepted. It was 1968 by now, the year of the Prudhoe Bay discovery—the money would have immediate uses; more important, a principle was established—Native land claims had been granted sufficient validity to form the basis for federal suits.

Native and environmental interests came together in the person of the Alaskan governor, Wally Hickel. The hotel builder had been zealous in securing state land, converting it to oil-lease cash, and assisting the development of production royalties by such ventures as the winter road to Prudhoe Bay; and correspondingly guarded in his relations with the Native voters—their claims, if they succeeded, might well be subtracted from the state endowment. (The state tried to thaw the freeze with a suit and in the federal court in Anchorage secured a favorable judgment, then a ruling of contempt of court against the Secretary of the Interior; but nothing came of it.) After his tour of Southeast, Brock Evans had returned to Anchorage to organize the first Alaskan chapter of the Sierra Club. Hence, when the newly elected President Nixon in December 1968 chose the governor as his Secretary of the Interior—and therefore the administrative authority for the pipeline question and the Native claims—the appointment provoked passionate protests from both parties.

Hickel's appearance before the Senate Interior Committee to defend his fitness for the office occasioned several neat moves on the political chessboard. Under grilling by the committee's chairman, Henry Jackson (with his traditional Washington state interest in things Alaskan), Mr. Hickel was obliged to promise, as a condition of senatorial consent, that he would not rescind the Udall land freeze without the committee's concurrence: he could not, that is, help the state to its land or the oil companies to their pipeline unless Senator Jackson approved. Later the corresponding House committee exacted a similar privilege. In effect, Congress (or at any rate the two chairmen) had taken a large step toward realizing the old congressional dream: to govern as well as legislate.

The day after the Secretary's appearance before Senator Jackson's committee, his predecessor Stewart Udall, as one of the retiring administration's last acts, announced a new and more formal version of the land freeze which Alaskans called "the deep freeze." Published in the *Federal Register* as Public Land Order 4582, it would have the force of law until the two chairmen thought otherwise: "All public lands in Alaska which are unreserved . . . are hereby withdrawn from all forms of appropriation and disposition under the public land laws. . . ." In a press release accompanying the text, Mr. Udall expressed the hope that "This action will give opportunity to Congress to consider how the legislative commitment that the natives shall not be disturbed in their traditional use and occupancy of the lands in Alaska should be implemented." Congress, however, was in no hurry to consider. The freeze looked to be a useful piece in the play, but it was not yet clear what it would be worth. The oil companies and their pipeline were a comparatively known quantity, but the Natives and their land were not and required assay; the state itself did not count for many votes.

Over the next three years, a succession of bills came forward in the two Houses. The earliest offered the Natives a flat $7 million in cash—congressional humor, suggested, presumably, by Seward's original offer. The AFN countered by securing a former Supreme Court justice as its lawyer, Arthur Goldberg (with a son, Robert, setting up practice in Anchorage), who proposed a settlement of 40 million acres and half a billion dollars. White Alaskans expressed outrage in their newspapers. The Stevens Village injunction suggested, however, that the proposal was serious; that without compromise the issue might take as long to settle in the courts as the Tlingit land claims.

Bills came and went, juggling money and land, governors and their staffs flew to Washington at public expense and departed, Emil Notti, Ray Christianson and other Native leaders, with borrowed money. We omit the comedy. Eventually the oil companies perceived and persuaded the state government that no settlement would be effective unless first of all acceptable to the Native leaders and the villagers who elected them and gave them power, and what the people wanted was title to their lands, not cash; and oil provided the lobbying skills to persuade the Administration and finally Congress itself.

In the fall of 1971 the Alaska Native Claims Settlement Act was passed by both Houses of Congress and signed on December 18th by a President glad for once of facing a problem less intractable than the Vietnam war.

The law thus enacted, Public Law 92-203, erected a smooth façade of compromise over the many conflicting solutions proposed by the two Interior committees, the Administration, the Alaska state government, the Natives themselves—with a gesture or two toward the environmentalists who had shown a lively interest in Alaskan land but very little in the Natives and their complicated claims. In essence, the bill's twenty-nine printed pages balanced two different kinds of concern: between money and land; and among the channels of power—regional, local, individual—that would flow from both. Collectively, the Natives would receive the 40 million acres they had demanded in the voice of Arthur Goldberg, and rather more cash than had seemed possible—altogether, $962.5 million, matching what the state had received from its Arctic Slope lease sales and the first published estimate of the cost of the pipeline. In exchange, aboriginal land rights (as distinct from the legal title to be conferred by the bill's various mechanisms) were "extinguished"; the state's earlier land selections were confirmed—and thereby the oil and other leases that depended on them; and Udall's deep freeze was at last revoked, enabling the state, the Natives, and anyone else so inclined to get on with cutting the Alaskan pie. As insurance against another century of lawsuits, Congress introduced the novel concept that the law's authority could not be challenged in the courts except by the state of Alaska and that within one year of its enactment. (This did not, however, prohibit suits over details of the Act, to determine exactly what it meant.) Cautiously, Congress declared that the Alaskan settlement was not to be interpreted as a precedent for Natives anywhere else in the United States—Manhattan, say, need not be handed back to any Indians who could be found to claim it. Once again, Alaska was exceptional.

The means by which all this was to be accomplished were ingenious. The largely unsurveyed map of Alaska was divided into squares, called townships, six miles on a side (36 square miles or 23,040 acres). The 220 Native villages with 25 or more inhabitants were allowed to select from three to seven of these townships, depending on population, so far as possible near their own locations. (National parks and of course any private land were out of bounds.) The villages would receive title to the surface rights in the land they chose, up to a total of 22 million acres, and from this, in turn, individual residents would get title to their homes or campsites; since the amount of land would not in most places be enough for subsistence—hunting, fishing, gathering—the villages were encouraged to form profit-making corporations that would find business uses for their land. Timber and mining rights (including oil) in the village lands were distributed among twelve regional corporations, along with surface and subsurface rights in another 16 million acres: that is, the state was divided into twelve regions corresponding to the Native organizations formed (or revived) in order to claim land; and a profit-making corporation was to be set up within each of these geographic units, receiving land from the total in proportion to its area. The regional corporations also got an additional 2 million acres in cemeteries and other historic sites, resolving another sensitive issue and making up the required total of 40 million.

For the purposes of the settlement, *Native* was loosely defined to match the Alaskan situation: anyone who could prove at least one Native grandparent (or who, lacking such proof, was so regarded) had the choice of enrolling in a village, a region, or both; and having done so received stock in the corporation, with voting rights and a stockholder's share in any profits. While participation in the village corporations was limited to residents, the regional corporations were open to anyone with at least an ancestral connection regardless of where he lived, and to allow for the several thousand Natives who had settled outside Alaska, provision was made for their forming a thirteenth, non-geographic corporation; in Southeast, the Tlingits and Haidas, having surrendered most of their land rights in the 1968 court settlement, were granted only a limited share in the land settlement but a full interest in the cash.

The money settlement was likewise distributed by a complicated series of formulas. The thirteen corporations were to receive $462.5 million according to their enrolled population, in installments spread over eleven years. The remaining $500 million would come from a 2 per cent royalty on future state mineral leases—bonuses, land rents, production royalties—until the total was paid off; past leases, as at Prudhoe Bay, would be exempt, along with offshore leases. The corporations, in turn, were required to share at least 10 per cent of this income with their shareholders and 45 to 50 per cent with their component village corporations (to get it, the villages had to produce an acceptable plan for using it); what remained was, in effect, capital for future ventures. To equalize the natural wealth, those corporations that made mineral leases on their own land were to distribute 70 per cent of the profits among the other corporations on the basis of enrollment. To answer attacks that had been made on the idea of the settlement, both individuals and the regional corporations were prohibited for twenty years from signing away their stock, income, or land; and the corporations were exempted from real estate taxes for the same period.

Besides the land from which the Natives would choose their 40 million acres, the Secretary of the Interior was required to set aside another 80 million as possible national parks, forests, wildlife refuges or wild and scenic rivers; and to submit plans for such uses for Congress to approve or reject within not more than seven years. In the meantime, the provision would give the environmentalists something to do, submitting counterproposals for what came to be called the "D-2 lands" (because specified in subsection (d) (2) of section 17 of the Act); but Congress would reserve the final say to itself. As a sop to the state, the Act created a ten-member Joint Federal Land Use Planning Commission, its membership equally divided between state and federal appointees, one of the former to be a certified Native.

At the time, the impression that filtered through the newspapers in the Lower 48 centered on the cash—that Congress was handing over a billion dollars, more or less, so the oil companies could get on with their pipeline; that the Natives had been bought off. The complex and abstract but fundamental questions of the land—everyone, it is thought, understands money—did not receive much notice.

Alaskans in general had a keener sense of what had happened —at a stroke, a large chunk of the country had passed into private ownership—and feared the obvious next step: that the same lands on which whites as much as Natives relied for meat and fish would be closed off by no-trespassing signs and fences and the dream of the Great Land's unhindered life, which most Alaskans nurture at least in imagination, would be at an end. Moreover, the regional corporations, once in full operation, would concentrate power on a scale beyond any yet reached by the rudimentary Alaskan economy. The 12 million acres of Doyon, Ltd., for instance, the mainly Athabascan body centered on the Yukon and headquartered in Fairbanks, made it one of the great private landowners of the world; the same land regarded as capital worth a cautious $1 or $2 billion, with the $120 million in cash it would receive in installments, made Doyon also a very substantial corporation by any standard.

The Anchorage *Times,* the state's leading paper, found it profitable to fan these anxieties. Combative Native remarks about the paramount rights of their private property were prominently quoted. Although some of the corporations were cautious about investment, others aggressively bought office buildings and basic Alaskan industries—construction, power, fish processing, transportation—and when they did rated headlines about "Natives." In fact, the corporate Native lands are so hedged by law and custom as to constitute a new kind of property, not strictly private or public; and whether the Native corporations, conforming to the demands of law and profitability, will behave differently from others, or be any more accessible to stockholders who happen to be Natives, is a question likely to take a generation or more to answer. The feeling conveyed by this journalistic shorthand was familiar to the Natives, new to the whites, bitter to both: exclusion. ("Neanderthals!" one commented. "They don't say Irishmen buy out—or Italians buy out: why *Natives?*") Yet if you value personal freedom within its necessary framework of general law, and if you are uneasy with the pervasive corporateness of American life—the necessity of belonging to *something* in mere self-defense—and if you have long memories for other corporate states: then I do not think you can regard the ingenious Alaskan arrangement altogether with rejoicing. It is at least a portent to be looked at steadily.

If the white Alaskans greeted the settlement anxiously, the Natives themselves were only moderately pleased. Sure, they had the 40 million acres, but it had *all* been theirs: looking at it from that end, they had handed over 335 million acres for about as

much as the state realized from the Prudhoe Bay leases; a poor
bargain. The Arctic Slope Native Association, representing the
Eskimos of the whole region beyond the Brooks Range, promptly
rejected the settlement and withdrew from the AFN; and presently sued the state and the oil companies for damages by reason
of trespass, in lieu of the oil land at Prudhoe Bay and in Pet 4,
from both of which it was excluded.

So whether you reject the bargain or determine to make what
you can of it, what remains is the inner sense of difference:

> The Indian, even if he has a good education and can make
> a good living, there's one difference: he's still brown and he
> stands off and he's not accepted in society, so he's not really
> part of the American scene, in a neighborhood. Is he still an
> Indian? But if they're brown or they're black, they're still a
> minority—so how does he fit in?

This is Emil Notti, speaking perhaps for himself but a to me
invisible Native, after the struggle in Washington pushed aside
from the AFN and given time to reflect as head of a research organization funded by the Ford Foundation, the Alaska Native
Foundation. He sat on the board of Doyon but was not hopeful
about what it could do to change the history of the future:

> All the enrollment does is give you a piece of paper that says
> you're a stockholder, no different than General Motors or
> RCA. I don't think the corporation's going to make one bit of
> difference in the outlook of the Native person. Look at a map
> of Alaska, pick any village or city, you'll find that it sits on
> the site of an original Indian village. Juneau, Sitka, Anchor-
> age, Kenai: as Alaska grows, I think we're just going to see
> history repeated over and over—mass movement of people
> going in, the city council is controlled then by non-Natives,
> the school board. So the Native's settin' there, all of a sudden
> —my God, what happened? All of his values about what's
> important—*we're going to put a minority committee to look
> after you.* I'm not very optimistic. . . .

When the oil companies did finally take a hand in the Native
claims controversy, their intervention was perhaps decisive, but

through most of the game they had stayed determinedly on the sidelines, regarding it as none of their affair. In the environmental debate that began as the Native question was inching toward its congressional conclusion and continued for two more years until Congress was again obliged to furnish a solution, the oil companies did not have that choice: from the start, they focused the disparate interests and anxieties of Congress, the Nixon administration, and numerous environmental organizations in temporary alliance—which sounds like a three-ring circus with the companies as performing bear; and so it was, at times.

Unlike a bridge, say, or a city office tower, a pipeline does not usually have to be designed in detail before it is begun. Rather, the engineers make a comparatively general plan based on a route sketched from topographic maps to avoid excessive gradients which would impose high pressures on the pipe itself and require extra pumps to move the oil up and over, expensive both to install and to operate; the volume of oil determines the pipe diameter and gauge, and if the oil is corrosive that difficulty is dealt with in the pipe steel specified. With such basic questions settled, local problems—wet, unstable soil, subsurface rock to blast through, river crossings—can generally be dealt with as they are uncovered in the course of construction, by on-the-spot engineering or minor realignment of the route; the range of possibilities is, after all, fairly narrow, and the pipeline builders have encountered them all repeatedly in the hundreds of thousands of miles of American pipe laid since the 1860s—not to mention the generally American-built lines in such places as the Alps, the Andes, the Arabian desert, the North Sea. Even with some discount for the hyperconfidence of the engineering mind and due allowance for the immense difference in scale, the process is not unlike the way you and I do our bits and pieces of household carpentry.

Alaska was rather different. Just how different—and whether the oil companies were sufficiently aware of the differences and willing or able to accommodate them in their pipeline—became the root question in the pipeline debate. To supply answers, they had to design the system in detail in advance and submit the design for comment, revision, and eventual approval, to the Department of the Interior and sundry other government agencies —and in effect to several congressional committees, the courts, and the general public (or at any rate that part of it represented by the environmental groups). Not the oil executives or their engineers but this balance of essentially political interests became the final arbiter of what would be done.

All this raises an interesting question: whether, if you were in the market for a pipeline, you would buy it from a pipeline builder or from a congressional committee, even one chaired by so august a personage as Henry M. Jackson or William Proxmire. The curious thing is that even now it is not absolutely clear that the common-sense answer is the right one—that the cobbler should stick to his last, the pipeline builder to pipelines, the legislator to legislating. As it turned out, however, after six years of preliminaries and all the impassioned debate, the pipeline design was in fact revised in detail almost every foot of the way as the backhoes and bulldozers disclosed new facts about the right-of-way that required engineering solutions. This raises a further question: whether—if engineering is determined by physical facts and is therefore not subject to debate once the facts are fully known—the pipeline would finally have been built any differently if the engineers had been left to their own devices; what, that is, we actually got in exchange for the six years of delay and the millions of dollars' worth of legal and engineering effort. Both sides in the debate are *certain* of the answer, but since it is the public, finally, that will bear every nickel of that cost, we had best consider the question for ourselves. *Our* answer is the one that counts.

It was (and is) a peculiarly American question. The first settlers got over their horror and homesickness and began to feel at home—and the change made independence inevitable. Two different feelings for the land succeeded and have run parallel through our history. There are those who with quickness see the land always as *something else*—the rusty, granular earth forged into iron, steel, the dark liquid fouling streams and salt deposits refined to power machines; means to personal wealth, freedom. Others of us look on the wild landscape and see not raw material but wonder, mystery, to be contemplated with reverence, in which we can perhaps modestly live, taking unobtrusive wood for fire and shelter, meat and roots and seeds for food, as they come. Knowing myself inescapably one of the contemplators—no landscape will ever turn into money at my command—I nevertheless believe the exploiters are also necessary to the human mix; that both have coexisted in varying balance throughout our history (nature in human as in geological affairs abhors dominance and seeks the level, the static); and that to suppress either, to deprive exploiter or contemplator of his necessary field, would impoverish the race of part of that pool of genes by which it adapts and survives. It is the *race* that is immortal.

In the 1960s the balance again shifted between these two sets of feelings about the land, these two sides of the American character, and those feelings having to do with conservation, protection, rooted in wonder, came together in an activist new ideology: environmentalism. The luck of the time and being in the right place provided voices to express it, issues to define it, and it became a force. Among those new voices, Brock Evans was representative. Bristly black hair and mustache, thick forearms of a serious climber: after college in the East, he got his first sight of the western mountains by way of a summer job at Glacier National Park in an interval of law school at Michigan and, with his law degree, returned, to Seattle, to take the bar exam and practice, for the sake of being near the mountains. The movement, as it was becoming, found its strength in the processes of law, and the young lawyer was soon giving more of his time to environmental causes—saving the mountains he climbed in east of Seattle from logging by converting national forest to national park—than to fee-paying clients: "I got very upset about the logging that I saw back there. I couldn't *believe* that anybody would be doing those things in such beautiful areas. . . ." What began as a young man's irritation at interfering powers—the Forest Service, the logging companies—and sense of desecration became a vocation when the aggressive Sierra Club chose him as its Northwest representative. This was the job that took him to Juneau in the summer of 1968, when the first Prudhoe Bay estimates were announced, and back in the fall and following spring to Anchorage and Fairbanks to organize the first tiny Alaskan chapter of the Sierra Club—two hundred members, divided, uncertain, formed for the purpose of resisting the oil companies' haste to begin their pipeline or at least ensure that a different point of view would be heard.

The wilderness lovers in the new chapter, being Alaskans, naturally included geologists and others with a practical interest in oil who took some persuading. Brock later recorded (in a staff memo preserved in the files of the Sierra Club's Seattle office) the arguments at the time and the strategy that grew from them: "I told them to ask themselves: do we really want this pipeline to come through? The answer they all gave was, of course, 'No.' Then if that is the case, I said . . . let us say we do not want it, let us say it is the wrong thing for Alaska—because that is, in fact, the truth." That was, however, too absolute a position to take in Alaska or anywhere else. What the Sierra Club professed, in common with the other environmental groups and consistently throughout the pipeline debate, was rather different and, on the face of it, moderate and rational. They were able to unite in "testimony to the effect that they were against the pipeline—now, because 'we need more study.'" It is an important nuance, and it explains, to a degree, the passions that inflamed the debate. The rational public arguments appeared to be the kind that could be debated, answered, compromised; but they were founded on incontrovertible conviction not openly or formally expressed and therefore beyond the reach of reasoned consideration. It was a serious contradiction: two different realities, or the same reality, differently perceived, and therefore no more reconcilable, finally, than any dispute over facts.

When the setting of the debate shifted to Congress, Brock Evans moved with it to become the Sierra Club lobbyist in Washington: fast talking in the style of the capital but still with the twang of the Northwest in his voice and sometimes eloquent; bringing the quickly reasoned advocacy of the courtroom to the cause and to the committees of Congress, where he would be attentively heard if not always heeded.

The newly elected Nixon administration had as keen a sense as Congress of the feeling that was abroad and adroitly turned it to advantage: by appointing as environmental advisor to the President a seasoned government lawyer, Russell Train, with a respectable interest in such matters; by following that expression of concern with the creation of a cabinet-level Committee on the

Environment, precursor to the agencies Congress would later authorize; and by reinforcing the lubberly Alaskan at the Department of the Interior with an interdepartmental task force charged with examining the proposed pipeline, with Russell Train as its chairman to ensure sensitivity to the environmental conscience. Gestures, of course, in a style that would become familiar: but Congress, too slow-moving to gain the initiative, could only respond with gestures of its own. Politically in 1969, the situation in which the pipeline would be debated was unprecedented: both a minority President (so far as the popular vote was concerned) and an opposition majority in both Houses of Congress; in the background, bloody mischief at the Chicago convention and a whiff of revolution that might yet turn out more real than romantic, 1860 if not 1917. It was a delicate situation in which to conduct the public business.

All considered, the pipeline plan progressed with surprising smoothness. The Nixon strategy was governed by the bargain made with Senator Jackson and his committee in order to secure approval of Wally Hickel's appointment: the immediate purpose in continuing the land freeze at the pleasure of the two Interior committees was to assure Congress its necessary leisure to cope with the Native claims; at the same time, this condition conferred on the committees as much control as they cared to exercise over any project on Alaskan public land requiring the permission which it was the responsibility of the Department of the Interior to administer—but only up to the point at which the senators and congressmen agreed to an exception to the freeze that would make the permit possible. To reach that point, it was necessary for the Administration to demonstrate to the interested public that it was at least as conscious as Congress of the environmental questions inherent in any large-scale development within the virtual wilderness of Arctic Alaska; and to proceed by deliberate stages small enough not to excite alarm and outcry. In the carrying out of this program, it appears that there was regular and detailed consultation between the Interior officials and the oil men sitting on the various TAPS committees; and between the oil companies' lobbyists and the congressional committees and their staffs. These exchanges do not, of course, form any part of the public record, so far as I know, and must be inferred but the style is so normal a part of the American way of government that it would be astonishing if the pipeline should have been an exception merely because it was big and controversial; on the contrary!

Although the oil companies announced their plan, budget, and schedule for the pipeline in February 1969, it was not till late that spring that they began the program of soil testing along the route that would be necessary to detailed design (see Chapter 3). Meanwhile, on May 1st, representatives of Arco, Humble, and BP appeared in Washington for an elaborate presentation of the pipeline plan before the congressional committees. A week later, as BP announced that it had brought in its own first Prudhoe Bay well, the three companies hired a Washington lawyer, Quinn O'Connell, who would represent TAPS to the end, and established a modest public relations office; the Administration matched these arrangements with the creation of its Interdepartmental Task Force headed by Russell Train, an idea "suggested" by President Nixon himself. And, strategically, the companies at the same time announced that they were buying the pipe in Japan (and had ordered the machinery for the first five pump stations). That agreement was by now probably a month old, but as things were progressing in Congress and the Administration, there seemed to be no reason for keeping it secret.

After these preparations, the actual application for the pipeline route was not completed until June 6th (the anniversary of D-Day, a recurring and perhaps deliberate parallel) and was delivered to the Anchorage office of the Bureau of Land Management on June 10th, with a request for a decision by July 1st. The application gave no information about the design of the pipeline system, perhaps because it had already been described in general terms in May and did not yet exist in any specific or technical form; but with characteristic secretiveness the oil companies were reluctant to divulge such information and, throughout, would do so only under compulsion, in laggardly fragments. The Task Force evidently knew what to expect and responded *the same day* with a letter from Russell Train raising 79 questions about the plan of the kind that were beginning to be on people's minds: prevention and clean-up of oil leaks, protection against earthquakes and thawing permafrost, reasons for choosing one or another terminal on the south Alaskan coast. (Curiously, Mr. Train professed official ignorance of the long-standing terminal choice, although the application for Valdez had been made by Arco and Humble in November 1968 and was in fact the trigger for Mr. Udall's deep-freeze order; perhaps so as to elicit now the objections to the other theoretically possible sites, such as Seward or Whittier on the Kenai Peninsula.)

The oil companies, in turn, were prepared, and on June 19th Mr. Train had his answers in twenty pages of single-spaced type-script from the current TAPS management committee chairman, R. G. Dulaney, an Arco pipeline executive and vice-president. These answers were more informative by implication than in substance. (The first recommendation for choosing Valdez as the terminal was that it was the shortest straight-line distance from Prudhoe Bay.) Like a good soldier, Mr. Dulaney did not presume to go beyond the letter of what was asked. To the touchy question of whether there were "any plans" for a pipeline "from the Prudhoe Bay area to Canada"—the delicate issue of alternatives—he could truthfully reply that the Trans-Alaska Pipeline System had no such plans, as of course it did not, by definition.

The TAPS application had candidly identified all the federal land it would need, including the 800-odd acres at Valdez administered by the Department of Agriculture as part of Chugach National Forest. Mr. Hickel prudently took one hurdle at a time. The first was the forty-mile segment connecting the state highway north of Fairbanks with the Yukon. The Secretary put the question to the senators at the end of July, Mr. Jackson responded favorably a week later, and within another week the road permit was granted and the construction contract let; no mention was made of the overland haulage *beyond* the Yukon, made possible by this link. With that camel swallowed, the gnat of Valdez presented no difficulty, and permission for the terminal site came through on October 1st.

Meanwhile, the Secretary was progressing toward granting the general right-of-way that would enable the pipeline to begin construction on schedule, in the spring of 1970. A formula was devised. The land freeze, which sooner or later would be superseded by Congress anyway, would not be revoked; Mr. Hickel merely proposed a "modification"—a provisional "transportation corridor" would be designated, a twelve-mile-wide slice of public land running from Prudhoe Bay south, within which the pipeline route would have ample room for maneuver. The permit for the right-of-way within this corridor would in turn be subject to contractual conditions, or "stipulations," assuring that the kinds of concerns expressed in Russell Train's 79 questions would in fact be dealt with. Under the stipulations, the design of the pipeline and its exact routing would be reviewed by the Secretary—actually, by an on-the-scene "authorized officer" acting for him—and approved or rejected, in detail; construction would be inspected as it proceeded and stopped if it did not conform to the approved plan or if new information required a change in design. From their initial draft, developed in early August, the stipulations were quite general in form, legal rather than technical, and were attacked on this account as weak and vague, though unless the Department of the Interior were to take over the corporate job of designing the pipeline it is hard to see how they could usefully have been made more specific. Actually, the intent was to give the Secretary and his staff of inspectors quite sweeping powers, comparatively unqualified by technical niceties. Despite several years of impassioned scrutiny by Congress and sundry environmentalists and a good deal of revision and reorganization, this principle prevailed, and no changes of substance were made. If the approach had a weakness, it was that, to be effective, it required an Authorized Officer of extraordinary technical competence and experience, possessing as well the force of character to assert his judgment against an array of oil company experts and executives committed to what was already turning into a multibillion-dollar project—a rare bird anywhere, particularly within the ranks of the Civil Service. But that was not a difficulty any of the debaters troubled themselves with.

At the end of August, Russell Train took the draft stipulations to Fairbanks for two days of public hearings (Brock Evans by then had his Sierra Club cadre organized to express themselves), and on September 30th Mr. Hickel transmitted a revision to the two congressional committees—38 pages of typescript assembled in a neat little booklet with the Department's symbol, a buffalo, on the cover. Intermittently till late December, the committees conducted hearings of their own, with Russell Train and R. G. Dulaney among the performers. The tone in the House committee was set by congressmen on the Henry Adams-Mark Twain model, senses dulled by advanced age and long continuance in office, who would have been quite comfortable in the Grant era: in particular the cantankerous chairman Wayne Aspinall, a Colorado Democrat, and the senior Republican, John Saylor, from Johnstown, Pennsylvania. Mr. Saylor was peeved by the announcement that the pipe would be bought in Japan and not from an American company, such as Bethlehem Steel—Bethlehem had a plant in Johnstown and was a prudent contributor. Mr. Dulaney explained that no American company, including Bethlehem, was capable of making pipe of the diameter required, but the argument did not penetrate: TAPS had made one big mistake and "You've just ruined yourselves as far as I'm concerned," John Saylor replied in homely accents. "I'm going to fight you every way I know how and the only way I know how is conservation." Or so the remark was remembered by one who heard it. Arco in September steered an order to Bethlehem Steel for tankers to haul its oil from Valdez—three 120,000 dwt supertankers that would about balance the cut-price Japanese pipe—but the congressman remained unpredictable.

The Senate proceedings were more to the point. On September 30th Senator Jackson, the chairman, sent the Secretary a series of questions in which he managed to touch two fundamentals: the width of the pipeline right-of-way; and the destination of the oil.

The Department's permit was governed by the Mineral Leasing Act of 1920, which included a puzzling provision—it limited the width of a pipeline right-of-way on public land to twenty-five feet on either side plus the width of the pipe. Just why Congress imposed such a limitation is not clear from the long-ago legislative history of the law, but apparently the congressmen who wrote the bill were irked at having been too generous with public land to the railroads and resolved to be the opposite with the oil companies. In any case, it had never been possible to build a pipeline within so narrow a strip of land, and though the permits may have been officially according to the law, the pipeline builders simply took what they needed with no questions asked, and when they were finished the pipe would, of course, fall within the legal fifty-foot limit. Even when completed, however, the TAPS proposal would take up a great deal more space: the 361-mile-construction road north of the Yukon, permanent access roads for maintenance all along the route, sites for airfields, camps, a microwave communications system, eventually a dozen pump stations each occupying the area of a small town, besides the pipeline itself; and nearly all to be built on public land. All this was specified in the original application. The planned "transportation corridor," besides acknowledging environmentalist demands for some over-all planning in the Alaskan wilderness, was evidently designed to accommodate the realities of the project and get around the arbitrary limitation of the statute. Mr. Hickel knew what the law said. Defensively, he insisted that the immediate issue was simply the "modification" of the land freeze by a new executive order creating the transportation corridor; the right-of-way permit would come later and separate, with due consultation.

Answering a very general question about the public benefits to justify the pipeline, the Secretary volunteered the oil company argument that would be standard throughout the years of debate: that West Coast oil consumption was growing so rapidly that it would absorb the entire Prudhoe Bay production by the time the pipeline could come on stream.

By mid-December, with Congress impatient for its holiday, the two committees had been sufficiently persuaded. TAPS submitted a revised application skirting the intractable question of the right-of-way width: a separate right-of-way for the pipeline itself; a series of applications for all the other land, to be granted by way of "special land use permits" (SLUPS). Following regular procedures, the head of the Anchorage BLM office in the January 1, 1970, *Federal Register* published the official notice that the transportation corridor was about to be established; and a week later the Secretary issued the requisite order, Public Land Order 4760. The permit for the all-weather haul road north of the Yukon would follow in order, replacing the unfortunate Hickel Highway, and then the pipeline permit itself—in time, presumably, for construction to begin on schedule in the spring.

Meanwhile, the competition between Congress and the Administration to demonstrate their respective sensitivities to environmental matters had borne curious fruit. Late in 1969, Congress passed the National Environmental Policy Act (NEPA, as it has since come to be known), requiring that every federal agency consider both the environmental effects of its actions and of alternatives to them, including that of taking no action at all; and document the results of these considerations in a form to elicit comment from other interested agencies and the public, all comment to be considered in turn and incorporated in a final version. Such a document is called an environmental impact statement—EIS in the jargon of environmentalism—and necessarily evolves through a series of stages, beginning with a DEIS (draft environmental impact statement) for the purpose of public comment and ending as a FEIS (final environmental impact statement). NEPA also provided for the creation of a Council on Environmental Quality (CEQ), replacing the Nixonian Cabinet Committee on the Environment, whose function would be to coordinate the governmental review of environmental impact statements. The President signed the bill on January 1, 1970, with a pious comment on its purposes, and in March, by executive order, created the required CEQ, with Russell Train in charge. In July, he went Congress one better by proclaiming a new Environmental Protection Agency (EPA) put together from existing agencies and units assembled from various departments; where the CEQ was essentially passive, a reviewing committee concerned with formulating policy, the EPA was a regulatory agency which would take an active and growing administrative role in environmental matters. A young government lawyer, William Ruckelshaus, was appointed its first administrator in November and, when the Watergate matter took him to the Department of Justice, was succeeded by Russell Train.

At the time, to the politicians in Congress and the White House, the simple and sweepingly stated purposes of NEPA and the agencies that followed from it seemed as unexceptionable as motherhood and the Fourth of July. There were consequences, however, which they did not foresee. Environmental impact: every human act occurs in, alters, and is governed by a complex, interconnected matrix of living and unliving matter whose ultimate inner processes touch every scientific discipline from molecular physics to mass psychology but are mostly unknown in any concrete and final sense—from the level of the single-celled organism upward, we do not really know what life is; and so on. Hence, to examine and describe in their totality the effects of even a quite specific and limited action brings into play the entire range of systematic human knowledge, and the result, even if it were the work of men of genius, cannot be more, in comparison with what remains unknown, than a crude and incomplete approximation of the reality. Moreover, the effects of any action are unknown in a different sense: they occur in the future, and if the future is one of the things that can truly be said to be indefinite, so too, at least philosophically, are the effects of a particular event. This is the realm of probability: beginning from knowledge of an event today and with many assumptions about the interrelationships of cause and effect, you can predict what will follow tomorrow with a degree of probability but no certainty; for the day after tomorrow, your prediction of effects introduces a second level of probability with geometrically diminishing certainty—and so *ad infinitum*. And finally, "alternatives": another infinite series, never complete, never final.

These are difficult philosophical issues that men have pondered for as long as we have knowledge of their thought. The effect of NEPA was to remove them from the abstract considerations of philosophy and deposit them in the courts and the committee rooms of Congress. Consider. Suppose you wish to travel from Philadelphia to New York. You present yourself at the train station to purchase a ticket, but the agent explains that the law requires certain formalities before he can oblige. What are the reasons for your journey? How will it affect the economy of Trenton, the shad in the Delaware where the railroad bridge crosses from Pennsylvania to New Jersey, the sea grasses in the north Jersey marshes? A committee of economists, biologists, and botanists provides answers of a sort; this takes a while. Now you can buy your ticket? Not just yet, if you don't mind. A bystander who has been observing these proceedings steps forward to accuse you and the ticket agent of conspiring to nullify the law. The committee's report is in the first place grossly inadequate and has not been publicly reviewed. Moreover, it has neglected to consider the alternatives: you could get to New York by car, on the Turnpike, or on foot; by ship down Delaware Bay and up the Jersey coast; by plane—and have you considered the interesting and historic route overland to Buffalo, thence by canal to Albany and down the Hudson to New York—? And that is only the beginning. Don't you *want* me to go to New York? you ask. That too is an alternative, the bystander replies, but his concern is disinterested, as a citizen he merely wants to see the law carried out, and by its nature it does not admit of qualification or exception. Ah! you think, if the law is the problem perhaps your congressman can do something about it, and you telephone. Why of course, he cheerfully tells you, you can expect some difficulties in putting a new concept into practice, but it's a *good law,* he voted for it. Nevertheless, he will see what he can do, there's an election coming as there always is. But it will take time—you know the complex procedures of Congress.

Fanciful, you say? Let us see.

When NEPA became law, the Department of the Interior was on the point of granting the TAPS permit for the necessary construction road north of the Yukon. Mr. Hickel paused to consider —any new law must be translated into regulations and procedures before it can be administered by the responsible departments or agencies; only particular cases, in turn, can make clear exactly what the regulations mean in practice. (As it turned out, it was *three and a half years* before the CEQ produced a set of procedures for developing environmental impact statements, and in the meantime officials like Wally Hickel were obliged to improvise—under the intent scrutiny of the federal courts.) On March 20, 1970, in an effort to keep the pipeline project progressing on its intended schedule, the Department of the Interior issued an eight-page EIS dealing with the road. Six days later, the three environmental groups brought their suit before a Washington district court and on April 23rd, as we have seen, Judge Hart granted an injunction against the issuance of the road permit.

The judge's ruling was based on two grounds. First, he agreed that the TAPS applications for rights-of-way—the statutory width of 54 feet for the pipeline itself plus 46 feet in special land-use permits and 200 feet for the haul road—comprised a single project which could not be so subdivided and were therefore

contrary to the Mineral Leasing Act. In addition, he found that
the hasty EIS put out in March did not comply with NEPA. And
finally, reasoning from the intent of NEPA, he greatly enlarged
the common-law concept of "standing"—that is, that the
members of the Wilderness Society, Friends of the Earth, and
the Environmental Defense Fund would be concretely affected
by the pipeline project and would in fact "suffer irreparable injury" if TAPS were built; and were thus qualified to take legal
action against it. The pipeline case was the first of significance to
be brought under the new law. It became the means, in effect,
for the courts, the lawmakers, and the public to find out just
what the law meant.

The state of Alaska responded to Judge Hart's ruling with an
ingenious solution of its own, separating the pipeline project into
strands slender enough to pass through the needle's eye of the
Mineral Leasing Act. The first step was a plan by which the state
would get permission for the haul road right-of-way, build the
road itself, and charge the cost to TAPS. The oil men considered
the idea and did not like it. The state tried a different tack: it
would still secure the permit, but it would then contract with
TAPS for the actual building of the road. This was apparently
acceptable; the *ad hoc* arrangements of TAPS were converted to
a corporation, the Alyeska Pipeline Service Company, with headquarters in Bellevue, Washington, across Lake Washington from
Seattle; and the oil companies, which still owned this new entity,
paid its bills, and determined its policies through the old committee system, made a start toward staffing it by appointing a president, E. L. Patton, a Humble executive with experience in building refineries (in northern California, Norway) in the face of
elaborate government regulation.

While the state pursued its will-o'-the-wisp, the Department of
the Interior began assembling an environmental impact statement that would deal with the entire pipeline project in a manner the courts might deem lawful. Again there seems to have
been consultation with the oil companies, and in December a
version of the EIS went to Alyeska for review and a significant
amount of editing—another reason, perhaps, why the new corporation had been formed. At this juncture, however, Mr. Hickel
fell out with the President—the war continued in Vietnam, the

promised peace did not come—and was cashiered and replaced by Rogers Morton, who proved a durable and more orderly bureaucrat. The land freeze was once more extended; the year turned to 1971 and the new Secretary, scheduling hearings on the edited DEIS, let it be known that he did not expect construction to start any time that year. The hearings lasted two days in Washington in mid-February, then shifted to Anchorage for a week at the end of the month. The Washington hearings alone generated 10,000 pages of testimony—most of it, mercifully, not spoken but submitted as formal written comment to be made part of the permanent record.

All through that summer of 1971 Secretary Morton (and eventually, it seems, the President) set targets for the completion of the impact statement, but they were not met. Complications for Alyeska in planning its pipeline were so also for the Interior committees attempting to estimate its environmental effects. In January 1970, the U. S. Geological Survey reported its doubts of the TAPS plan to bury the pipe for 95 per cent of the route, and journals such as *Audubon* were presently regaling their readers with horrific images of oil-filled pipe disappearing into thawed permafrost and fragile tundra—in the literature it is always "fragile"—awash in crude; the first awareness, probably, many Americans had of permafrost or the project involved with it. Alyeska, as it became, took a full year to replan the system so that only 52 per cent of the pipe would be buried—and in announcing that change also let out that the cost had risen from $900 million to a round $2 billion; changes that, coinciding with the completion of Mr. Morton's DEIS in January 1971, presumably made large chunks of it obsolete. In March 1971, the Corps of Engineers objected that the stipulations incorporated in the DEIS lacked technical detail of the kind it was competent to supply, and the report was promptly released by a Wisconsin congressman, Les Aspin, intent on replacing TAPS with a pipeline to the Midwest. Alyeska took out full-page newspaper ads arguing that Prudhoe Bay oil was needed immediately and that the alternatives now coming forward in increasing numbers could only delay it. On August 8, 1971, Alyeska at last provided the Department of the Interior with its Project Description—in three volumes accompanied by twenty-six volumes of appendixes. Possibly because it took time to read and further complicated the drafting of the EIS, Mr. Morton did not mention it publicly till early October.

Meanwhile, the adversaries in the lawsuit maneuvered toward trial. The Department of Justice, representing Secretary Morton, proposed shifting the case from Washington to Anchorage, a change of venue calculated to inhibit the plaintiffs' modest legal forces. Alyeska's oil-company owners petitioned to intervene on the side of the Department of the Interior. In November 1971, David Anderson, a Canadian MP, sued in Washington on the ground that the tanker leg of the TAPS system would endanger Canada's Pacific coast. All of these actions, with their accompanying briefs and arguments, took up time, but also made time for the matters of substance eventually to be tried.

The three environmental groups were represented by a recently formed nonprofit Washington law firm, the Center for Law and Social Policy, housed in an old-shoe-comfortable town house on N Street a few blocks from the White House; it bills its clients for expenses (Xeroxing, for instance, no small item) but not for services. Its running expenses are made up by foundation grants, mainly the Ford Foundation (which also put up money for the Alaska Federation of Natives in its pursuit of land); decisions on cases to be taken rest finally with a board of trustees that includes such distinguished advocates as Arthur Goldberg and Ramsey Clark. In the summer of 1971 James Moorman, who had secured the preliminary injunction, left for California to head the Sierra Club's legal defense fund. The Center persuaded Dennis Flannery to take on the case: first in his class at the University of Pennsylvania Law School, then clerk to the venturesome Chief Justice of the Supreme Court, Earl Warren, now three years into association with one of the great and remunerative Washington law firms, Wilmer, Cutler and Pickering. It was a risk to be looked at carefully—no chance of resuming that career if you failed—but also an opportunity:

> I approached the case with a feeling of trepidation—a David versus Goliath thing all the way through—David without even the sling-shot. You certainly have some thoughts about taking on a case of this visibility with the entire oil industry

grouped together and knowing that you'll be unable to match
them in the litigation with the resources or legal staffing. But
it was my view that the issue deserved to have a court decide
as a result of good advocacy on both sides. . . .

It was the lawyer's ultimate motive: the chance of making law
where it counts, in the courtroom.

The case evolved through hearings, motions, dissolutions and
reinstatements of injunctions, appeals—thousands of pages of
learned briefs and juridical opinions—and reached its terminus
in the Washington Court of Appeals in October 1972, which rendered its decision on February 9, 1973: in 70 double-column
pages of the *Federal Reporter*, with dissents, the judges
confirmed the prohibition against a right-of-way for the pipeline
but allowed the haul road, airstrips, construction camps, and
other uses of public land; and avoided the far more complicated
question of environmental effects and alternatives—and in more
than one place expressed the hope that Congress would take a
hand in resolving the issue. The oil companies carried the case to
the Supreme Court, which in April 1973 declined to review the
Appeals Court decision, on similar grounds—that the issue was a
congressional responsibility.

The question the courts finally declined to answer was one
Dennis Flannery and his colleagues at the Center had gone to
extraordinary efforts to make central to the case: whether the EIS
at length issued by the Department of the Interior was adequate
to its purpose, within the meaning of the law; whether, in the
same terms, it was final—for if it were judged not final but some
intermediate stage, then the whole process by which it had been
produced might be reopened, with new hearings, new drafts,
that could quite possibly continue indefinitely. The statement
was released on March 20, 1972: six fat volumes of duplicated
typescript with a three-volume supplement examining the "economic and security aspects" of the pipeline project; the Department noted with satisfaction that its preparation had consumed
175 man-years, at a cost of $11 million. Although Secretary Mor-
ton regarded his task as completed, the way at last cleared for
the pipeline permit, he cautiously waited six weeks to allow time
for comment. In that interval, the clerks at the Center managed

to cut the FEIS into manageable chunks, copy them, and distribute them around the country to a prepared team of waiting academics. From these experts by the beginning of May they had assembled four volumes of generally virulent comment, attacking the FEIS and the project itself root and branch. These further volumes, whose purpose was, of course, to support the Center's legal position and keep the case open, were transmitted to the Department of the Interior on May 4th, the day following Alyeska's delivery of its own less ambitious commentary.

I don't suppose anyone but the lawyers on both sides and a few harmlessly drudging authors have attempted to read these documents through. (Rogers Morton, for one, did not.) Measured against the lofty intentions of NEPA, they are about as satisfying as a sawdust hamburger—as, of course, one would expect any time a "task force" of bureaucrats is turned loose on the English language. There is, however, a more fundamental difficulty. Mr. Morton's FEIS does, I think, touch all the questions that human wit might raise about the pipeline, though in many cases—as the work of largely anonymous committees and subcommittees of government functionaries—not systematically or coherently. What is missing is the original research—*it does not exist*—without which most of these questions are not truly answerable; and any unifying intelligence. This huge document is essentially what the bureaucrats call a "literature search"—an assemblage and summary of other documents, published and unpublished, and in the nature of the case many of its sources were necessarily studies originated or inspired by Alyeska and its oil-company owners. It contains no direct or original research—all is at several removes from the reality examined. For a single example: will the tankers running in and out of Port Valdez harm the large and varied aquatic and shore life in that comparatively pristine environment? The answer depends on a large number of unknowns: whether the tankers will discharge oil in those waters and in what quantities, or will alter that environment in other ways; what that environment consists of, from the slime on the bottom of the fjord to the rock and glacial till along its shores; what forms of life exist there, in what numbers and in what relationship to one another; how they and their relationships are affected by oil; and so on. Probabilities: combining them, you arrive, if you are honest, at a kind of generalized possibility—"oily discharges from tankers, *if they occur, may* cause *some* deterioration of the marine environment"; an inconclusive all-purpose conclusion, numbingly repeated, with minor variations, in answer to most of the questions raised. It does not seem very good value for $11 million in public money, but—and we come to the real problem—it satisfies the law and is even *true;* it is only that, in the absence of real knowledge, it does not seem to have any practical utility.

Then was Mr. Morton's impact statement "adequate"? It does not appear that, by its nature, it could have been—and, therefore, that the judges rightly avoided judgment. But must we then blunder forward in ignorance, hopelessly? In building a pipeline as in every other human undertaking, we act from knowledge that is necessarily fragmentary, incomplete, contradictory, false in ways we may not be aware of. A healthy sense of the extent and nature of our ignorance induces caution as well as humility, both useful, but we must, finally, act. That is our human predicament, and it was perhaps in recognition of that truth that the judges refrained from expressing a conclusion about the adequacy of the FEIS.

The day the appeal from their decision was argued before the Supreme Court, March 9, 1973, Senator Jackson's Committee on Interior and Insular Affairs held its first hearing on legislative means of getting by the pipeline impasse. (A month later, the corresponding House committee got around to the same question.) Within a week, six bills had been introduced; another week and there were eight candidates before the House. All along, the environmentalists had proceeded on the assumption, often stated, that there was no limit to the consideration the nation could afford to give to the proper means of getting at its Alaskan oil—*if* it could be shown to be needed; that what was paramount was perfection of the means chosen so as to guarantee it would bring no harm to the Alaskan landscape. Oil company arguments to the contrary went unheeded or were dismissed as self-serving, as if the general good were not always and invariably a balancing of many self-interests; and Congress, generally, acquiesced.

In the years since the Prudhoe Bay discoveries, however, a

number of disquieting things had happened. By 1973 they had assumed so bold and threatening a pattern that even a congressman could not fail to notice and feel a certain torpid urgency. Between 1969 and 1973, U.S. demand for oil had risen from about 14 million b/d to 17 million, an increase of 21 per cent; at the same time, U.S. production, after creeping upward to the peak year of 1970 (11.3 million b/d), had begun the downward slide that has continued ever since. The proportion of U.S. demand supplied by imports rose accordingly, from 21.6 per cent in 1969 to 35.1 per cent in 1973.

During those same four years, signs came with increasing frequency, if anyone had cared to notice, that imported oil was not going to stay as dirt-cheap as it had been in the past, a threat against which American producers must somehow be protected; that a time might even come when it could not be bought at any price. In 1969, a revolutionary government in Bolivia, a minor producer, nationalized one small branch of an American company. Algeria followed suit in 1970, seizing six companies—a more serious portent. Libya, developed since World War II, claimed its BP unit in 1971 (and in the fall of 1973 all the others). In June 1972, Iraq, another revolutionary, abruptly took over the Iraq Petroleum Company, a consortium owned by BP, Exxon, and other major companies which since the mid-20s had been the prototype for similar ventures in other Middle East oil-producing countries; and in the fall, joined the four other Arab states around the Persian Gulf in an agreement by which they gained "shared ownership" in their production—a euphemistic step toward nationalization, which followed in due course. In 1971, Venezuela, the mainstay of U.S. imports, made a first parliamentary move toward nationalization; Iran in the spring of 1973 took to itself the rights and facilities of the group of companies that had come in after Mossadeq's abortive effort in 1951 (see Chapter 1)—BP, of course, but also Shell, Gulf, Exxon, Mobil, Texaco, Socal.

The immediate purpose of these novel arrangements was, of course, to secure to the countries a greater share of the profits in oil—by raising the price; by raising the royalties and taxes collected from those companies allowed to continue their operations in some form, whether as hirelings or in "partnership" with the respective governments; and in many cases by avoiding any compensation to the companies for their investments in the facilities seized. Price increases followed throughout the international oil trade, at accelerating rates and with increasing frequency. There were other motives. In Libya and Iraq, where the American and European managers and technicians were bodily removed, they were immediately replaced by Russians. (In these same years, the Soviet Union had made a succession of startling oil and gas discoveries in Siberia, each one of which dwarfed the finds at Prudhoe Bay, and had at the same time taken an aggressive interest in Middle East oil.) Meanwhile, the OAPEC nations* were feeling their way toward the possibilities of power founded on unified control of their substantial part of the world's oil production and reserves—a policy that would be put to its first ambiguous test in October 1973, in the new war against Israel.

One consequence of these events roused Congress. During the winter of 1972–73, for the first time ever, there were regional shortages of gas and fuel oil—not enough to go around. In the summer, shortages recurred, this time of gasoline and again spottily. Politicians in the regions affected looked for villains, suggested conspiracies, and, with no thought for the history from which these difficulties had grown, set about persuading their constituents that the shortages were "induced"—somehow arranged by the oil companies either to force up prices or to demonstrate once and for all the necessity of the Alaskan pipeline (and a dozen other less publicized oil schemes, similarly stalled). It is a tempting theory—from that standpoint the shortages were certainly convenient and in fact produced results—but it was supported by no evidence; and among the current generation of oil men, I for one have not found that chess-player's combination of craft, intelligence, and risky determination to conceive and carry out such a scheme. The Senate Judiciary Committee devoted most of the month of June to hearings on the subject but, predictably, accomplished nothing, though the investigation did provide a splendid forum in which to badger oil men.

* OAPEC: the Organization of Arab Petroleum Exporting Countries—the Arab members of OPEC, separately constituted.

The pipeline was at least an aspect of the situation specific and tractable enough for Congress to deal with, and it did, in hearings that continued through May and June of 1973. The cast was much the same as in the 1969 hearings and the testimony, depositions, and briefs of the long-running lawsuit that began in 1970. The issues they expressed had been refined in these years but not substantially changed: whether the Prudhoe Bay oil was in fact needed or could be indefinitely delayed; whether TAPS was a safe or economic method of transporting it; whether its underlying purpose was to deliver the oil not to the West Coast but to Japan; whether a pipeline through Canada (or some entirely different system) would carry the oil to more appropriate markets at less (or anyway defensible) cost.

The adversary system that prevailed in the hearings as in the courts—oil-company advocates on the one side, environmentalist detractors on the other, in shifting alliance with senators and congressmen—naturally produced extreme arguments. Thus, the oil companies offered projections of West Coast production and demand to show that all the Prudhoe oil would be absorbed there, though they mostly did not act on that conclusion; their antagonists manipulated similar projections to opposite results, arguing that most of the oil was destined for export. (Earlier, in July 1971, Alyeska's new president, Ed Patton, heedlessly told an interviewer that 500,000 b/d—a quarter of the peak Prudhoe production—would be surplus, and he continued to make vague references to "Panama oil," meaning that by some as yet undevised contrivance it would find its way to other American markets.) John Melcher, chairman of the House Subcommittee on Public Lands (a unit of the Interior committee), induced the oil companies to write letters affirming that they had "no plans" for or "no intention" of exporting any of the oil; these documents became part of the record and played a part in the pipeline bill that was eventually passed.

The Canadian alternative—TCP, the Trans-Canada Pipeline—was passionately argued but to no conclusion. The oil companies treated the idea as if they were being asked to build a pipeline through enemy territory in wartime; its proponents produced studies that drastically underestimated its costs and the time it would require and exaggerated its benefits. The generally ignored reality was that despite preliminary studies and a number of meetings with Canadian officials by oil men (and by Rogers Morton, among others in the Department of the Interior and the State Department), *no one* in Canada or the United States had actually proposed such a system; and Canadian receptiveness was not at all certain. The Canadian government was perhaps ready to consider a plan for transporting oil and gas from its Arctic fields at minimal cost to itself—but was at the beginning of a Native claims debate of its own that was destined to last a good deal longer than the one in Alaska; meanwhile, it would demonstrate independence from the "colossus to the south" by ending the export of Alberta oil and obstructing American imports where they passed through Canadian waters (as in some of the routes into Puget Sound or a new deep-water port planned for northern Maine). In attempting to stiffen these uncertainties with substance, the environmentalists insisted, as Brock Evans put it to the House Public Lands Subcommittee, that they had "never taken a stand in opposition to the idea that oil from Alaska's North Slope should some day be removed and used as a source of energy." It was only that there were "compelling arguments that it should remain in the ground"; and the "method proposed for transporting it . . . to the west coast of America will be . . . disastrous from an environmental standpoint." By rhetoric and the process of argumentation, the tissue of indeterminate probabilities had been transformed to certainty.

Other alternatives were advanced, with similarly righteous certitude, from various corners of the *academe:* a plan for refrigerating the oil and emulsifying it with salt water so that it could be piped below freezing, allaying the by now considerable anxiety as to what would happen if the oil's heat thawed the permafrost; a sort of aboveground, 800-mile tunnel, supported on pylons, through which the oil would be transported by fleets of tank trucks or some form of train, for the same purpose; a multitube pipe on the order of a giant Thermos flask; several more conventional railroad schemes, including one put forward by the premier of British Columbia and liked by a good many environmentalists, involving oversized tank cars and much automated loading, coupling, and switching, as in a modern subway; several variants on the submarine tanker idea proposed in December

1969 by General Dynamics—and ignoring, as it did, the impossibility of getting any large vessel within loading distance of the
Arctic coast; suggestions for tanker blimps, dirigibles, and airplanes, evidently made by people who had never experienced or
given much thought to Arctic weather. This list is hardly exhaustive, but it will give the reader a sense of what most of these alternatives had in common. In the first place, all depended on
technologies that did not yet exist to replace one that, even
granting all its possible fallibilities, was at least known from a
century of experience. Most suffered also in presentation from
bad arithmetic which converted misplaced decimal points and
too many zeros (a billion for a million, say) into forceful argument for economic feasibility. What counted finally was simply
that the oil companies were not prepared to invest money in any
of these possibilities, and it was, on the whole, *their* money that
was at issue. Congress was no less chary of the public purse. All
of these alternatives had, I suppose, at least some serious sup-
porters, but the conclusion seems inescapable that their function
in the debate was purely obstructive; and, therefore, their
strength was in numbers, not in any particular merit.

As it happened, it was a couple of years later, in the midst of a
mild patch in the Alaskan winter, before I began to get around
to most of these alternatives. In my room in Anchorage I sat up
reading an account of one of the railroad schemes in a 300-page
tract, *Cry Crisis!*, published, toward the end of the congressional
phase of the debate, by the Friends of the Earth—FOE, a Sierra
Club splinter group formed by David Brower when he felt the
Sierra Club, of which he had been executive director, was no
longer militant enough; and a partner, with the Wilderness Society, in the lawsuit. The book was an odd mix, written by Mr.
Brower and several associates: pretty pictures of summer Alaska
and bad logic; evidence combining the accurate with the fanciful
in a blend that would cost me weeks of labor to separate; strong
rhetoric about environmental values suffused with a sourness
about things human (caption for one of those aerial photos of
the Los Angeles city-mass intended as a prophecy for Alaska,
arousing repulsion: "We will have given in again to our old
habit—the conversion of the varied wealth of Earth into just
more human protoplasm"); and trouble with arithmetic. "Water

and air and land will be degraded steadily," the writers prophe-
sied, seeking cosmic images for their forebodings, "until it is too
late to do anything but swallow one last glass of diluted sewage,
inhale one last gasp of smog, and topple over into a strip mine or
reservoir." And therefore stop—cease and desist. The end was an
apocalypse, toward which the absolutist imagination tends:
earthquakes and tidal waves and million-barrel oil spills on the
Alaskan land and seas.

The account of the railroad system, however, seemed plausible, mostly quoted from other sources. I would have to learn
more about it—the mathematics of tank-car loadings, speeds,
gradients, scheduling, maintenance of diesel-electric engines,
rolling stock, roadbed and track; yet another mass of technology,
I thought a little wearily, that I would have to try to ingest. I
woke in the morning still thinking about it and over breakfast
read the day's headline story in the Anchorage *Daily News:* in
the night, a train of the Alaska Railroad had derailed, smashing
half a dozen tank cars and spilling a quantity of oil—unknown
but probably several thousands of barrels—over the snow somewhere between Anchorage and Mount McKinley. Cause not yet
accounted for: a frozen switch? sabotage? (the railroad adver-
tises regularly for information that will lead to arrests) or one of
the rambling and persistent moose? (The moose kill runs around
three hundred per year, almost one a day, and the railroad is
affectionately known as "the moose-gooser.") Perhaps, therefore,
the railroad alternative was one at least on which I would not
have to spend much more time.

The House ended its pipeline hearings on June 18, 1973. By
then, the Senate had ready, in broad outline, an act amending
the Mineral Leasing Act so as to authorize the Secretary of the
Interior to issue the necessary permits. (The corresponding
House bill took three more weeks to get out of committee.) Voting piece by piece on the Senate bill occupied a week in mid-
July, the final and crucial vote coming on July 17th on an
amendment introduced by Mike Gravel, the Democratic Senator
from Alaska and, like Mr. Hickel, a successful Anchorage
builder. During the hearings, both the oil companies and the environmentalists had promised to go on suing indefinitely until the

pipeline question was settled—unless Congress provided other-wise. Senator Gravel's amendment, therefore, proposed to make that impossible. It directed the Secretary to grant the permits; and it provided that

> The actions taken . . . shall be taken without further action under the National Environmental Policy Act of 1969; and . . . shall not be subject to judicial review under any law. . . .

This novel principle, refining a similar idea that came out of the Native claims settlement, was softened somewhat by a provision for a single suit, to be brought within sixty days, testing the law's constitutionality, but the vote still ended in a tie, 49–49. Spiro Agnew, not yet disgraced and still, as Vice-President, presiding in the Senate, cast the deciding vote and the bill was passed. The House bill went through on August 2 with less drama, the two versions were reconciled and repassed in November; and on November 16th the completed act was signed into law by the President as Public Law 93–153.

The environmental lobbyists had put their hopes in a different amendment brought in by Senator Walter Mondale—postponing the TAPS authorization until whenever the TCP alternative had been fully investigated—but it was defeated on the same day, 61–29. The final act nevertheless reflected many of the environmentalists' concerns. Besides amending the Mineral Leasing Act to allow the Secretary of the Interior to grant pretty much all the land that might be needed (for pipeline right-of-way, for pump stations, roads, construction camps, and so on), it affirmed the applicability of NEPA and the procedures that had grown around it and required technical and environmental studies in connection with any future right-of-way application (to be paid for by the applicants); public hearings and consultation with other government agencies and with the Senate and the House; and, responding to the oil companies' reluctance to make their plans known, submission of the detailed project descriptions without which this public analysis and review could not be meaningfully made. It also prohibited the export of oil carried in such a public right-of-way—unless the President could establish that export was "in the national interest" and was not overruled by Congress. And, finally, the act summarized the environmental stipulations developed for the Alaskan system and directed that similar provisions be included in any future pipeline right-of-way agreement.

All this was general law. The second half of the bill, including the Gravel amendment, dealt specifically with TAPS and is known as the Trans-Alaska Pipeline Authorization Act. In its preamble, under the heading "Congressional Findings," the Act laid down as national policy that

> The early development and delivery of oil and gas from Alaska's North Slope to domestic markets is in the national interest. . . .

and

> The earliest possible construction of a trans-Alaska oil pipeline from the North Slope of Alaska to Port Valdez . . . will best serve the national interest.

This was essentially what the oil men had been saying all along: that producing the Prudhoe field was urgent and that TAPS was the handiest means to that end. Their view was now law and, as expressed by Congress, would control all the actions that followed from the congressional decision.

At the same time, however, the environmentalist arguments—the dangers of oil spills and the uncertainty about the market to which TAPS would carry the oil—were duly acknowledged, though subordinate to this overriding urgency defined as national policy. Alyeska, as the pipeline operator, would be subject to stiff penalties for any damage done to the land by oil spills or otherwise—up to $50 million for any one incident. In addition, when the oil left the pipeline at Valdez, its owners and the tanker operators would be further liable for damages up to a total of $114 million: the first $14 million to be paid directly by the owners and operators; the balance accumulated from a fee of a nickel a barrel paid into a nonprofit corporation to be known as the Trans-Alaska Pipeline Liability Fund and administered by

Alyeska. (Once the pipeline was at full capacity, 2 million b/d, the limit of the fund would be reached in a little less than three years.) With an eye to the regional arguments that had surfaced in the course of the debate, the Authorization Act invited the President to negotiate with Canada for pipelines to carry oil and/or gas to the Midwest and East; and required him to "use any authority he may have to insure an equitable allocation . . . among all the regions and all the several states."

So the track was cleared at last, and every questioner had some kind of answer, though not necessarily the one he had sought.

"The oil pipeline is a battle that was fought and essentially lost." The voice is that of Jack Hession, who succeeded Brock Evans in the Sierra Club job in Alaska—a mature, youngish man, articulate and carefully reasonable; not, perhaps, the battler his predecessor might be, but not at all the squinty-eyed environmental radical, the Outsider, that is a stock figure in the Anchorage *Times*. (In that rhetoric, oil men from Texas or Oklahoma are never Outsiders.) I had wondered how he viewed the outcome, and his answer was the practical one for him at the time. The debate was over, the pipeline was being built, and there was not much he could do about it; his energies were absorbed in a dozen newer issues, less publicized but no less technical, among them the disposition of the public interest ("D-2") lands provided for in the Native Claims Settlement Act and the building of a line for Prudhoe Bay gas, which sooner or later, by some route and mode, would come as surely as the oil line, though the process of decision was hardly simplified by the Pipeline Act. Brock Evans, near the heart of the battle, had felt the same sense of defeat, but bitter and personal, and had gone back to the Northwest to hike, climb, and brood.

Both reactions are understandable but nonetheless surprising. The record of the courts and of Congress is one of many compromises on the pipeline and as many specific environmental goals achieved (at a cost we shall all eventually bear). It does not, at this distance, look like defeat—unless the only acceptable victory was the original goal of prohibiting the pipeline altogether and therefore the development of Arctic Alaska. That, reasonably, was in fact how it looked to Brock Evans after he returned to Washington from his weeks of solace in the Cascades:

> We were up against big labor, big business, at least half the media, all the power, all the money and all the resources. To get a forty-nine to forty-nine vote still, after all that, was hardly a defeat—it was a defeat with honor, anyhow. And we did get the tougher stipulations, we did get the tanker liability provision, there was some antitrust stuff in there—seeds of future actions. . . .

A battle, but perhaps not the war: *seeds of future actions.*

Dennis Flannery's view of his own part in the proceedings was unqualified: in the courts, as far as the case went, they had won against considerable odds, and when Congress took the issue beyond his reach he resumed his career at Wilmer, Cutler and Pickering and presently joined the long list of partners on the firm's letterhead. The congressional solution, however, new law for old, left him, as a lawyer, dissatisfied: the precise legal questions he had labored to formulate—the scope of NEPA and its practical meaning for the conduct of public affairs—were crudely answered but not resolved. As a lawyer also, he was conscious of irony in the way the oil companies had gotten, finally, what they sought:

> If, following the granting of the preliminary injunction in 1970, they had gone to Congress—as they should have—and requested an amendment to the Mineral Leasing Act, I think you might well have found at that point Congress taking some concern and interest in the whole project and perhaps assuring that there wouldn't have been the kind of slanted review process that went on. But they decided to take their chances in court. They lost in court. The Congress they then went to was a very different Congress with regard to environmental concerns. So the way things worked out . . . maybe their judgment, on purely pragmatic grounds, cannot be faulted. . . .

Ponder that outcome, brothers: those battles lost, those blunders toward victory. Almost, it is enough to persuade one of the subtle prescience and invincibility of oil.

Chapter 5

THE BUILD-UP

> It was mostly Exxon engineers who put this whole thing to-
> gether, out of Houston. They simply didn't know what in the
> hell they were doing, and once they had a thought, they
> thought they had all the answers.
>
> —Vic Fischer

"When the oil companies decide to organize a new company,
they take a guy out of their operations and they say to him,
'George, you're never going to make the top, boy, got to tell you
this. But you're a hell of a good man and you're too good to
waste in a definite slot here for the rest of your life. We're organizing a new company over here, gonna call it the Alyeska Pipeline: you go there and you'll get a hundred thousand a year and
we'll guarantee you still get your same stock options, get all your
other fringe benefits, and you're going to have a hell of a nice
deal, gonna be a *big man.*' And if George is halfway smart, he
cleans out his desk that afternoon."

The story is Bob Johnson's, one of the old-time Californians associated with another Californian in the management of the
Alaska Teamsters' local, Jesse Carr—who, under the modest title
of secretary-treasurer, is reputed the most powerful labor leader
in Alaska, maybe the most powerful individual of any kind; we
shall meet him again as we follow the history of the pipeline.
Bob himself as a boy turned from his well-off family to labor and
labor organizing and formed his ideas about both on picket lines
in the head-cracking 1930s; long hatreds, longer loyalties. He has

lived most of the Alaskan ups and downs, from comparative
affluence in trucking and construction to log-cabin poverty, surviving the cruel winters by manhandling iron on Dew Line
towers along the Arctic coast, where the cold blew down to
thirty below and filled your lungs with sharp pain when you
breathed in deep and the men worked twenty-minute shifts—
twenty minutes climbing the tower to fumble another bolt into
place, twenty minutes down to the safety of the warming hut:
talker, boozer, raconteur, ex-boozer, wit, one of several
Teamsters in various slots whose job is to turn Jesse Carr's muscular thought into persuasive testimony and public statement.
So: not a man whose views of oil are to be taken without the salt
of skepticism but with, nevertheless, an enemy's perceptiveness
that is worth attending to and pondering; the exact relationship
of the industry to the multiplicity of subsidiaries by which it gets
things done—where, in fact, the real power and authority reside
—is among the final questions raised by the pipeline and which
we had better try to answer.

Late in August 1970, the oil companies announced that the
TAPS partnership had been replaced by a Delaware corporation
to be known as the Alyeska Pipeline Service Company, the name
derived from the same Aleut word that Alaskans like to translate
as "the great land." The new arrangement had been evolving
through the cumbersome committee system for nearly a year, actively since January, when BP swapped its oil leases for eventual
control of Sohio. It took the partners till June 1970, six months,
to agree on a first step, the hiring of a president, E. L. Patton,
from the management of Standard Oil of New Jersey (i.e., Humble's parent, since renamed Exxon). The agreement by which
Alyeska was created, signed on August 27th, specified the oil
companies' percentages of ownership, which signified, in turn,
both what they would contribute to the cost of building the
pipeline and the proportion of its capacity they could claim once
the pipeline was in operation, either for their own production, or,
since the law would require the pipeline owners to function as
common carriers, that of any other Prudhoe producer tendering
oil for transportation to Valdez; ownership of the oil itself was a
quite separate matter not finally settled until shortly before the

pipeline was completed, in the general agreement for unitizing the field. So far as the pipeline was concerned, Arco and Sohio (the latter by virtue of its acquisition of BP-US—see Chapter 3) now had 28.08 per cent each, Humble 25.52 per cent; the balance, a little more than 18 per cent, was distributed among Mobile, Phillips, Union, and Amerada Hess. For their share of the oil production, the remaining dozen or so companies with some interest in the field would, in effect, be renting space in the pipeline from one or other of these owners. The existing committee system would continue, functioning like a board of directors in a conventional corporation but with the difference that there would be a separate management committee overseeing each aspect of Alyeska's work.

These decisions were made before the environmental suit obtained its first injunction, but the companies did not alter course, they remained unswervingly hopeful that the difficulty would be surmounted, the permit granted, and construction begin with the next season, the spring of 1971; or so they insisted to Ed Patton, perhaps to persuade him to take the job. In any case, whether construction was coming or only more lawsuits and hearings, the pipeline was becoming more than the various committee chairmen and their colleagues could see to in their spare time.

President and toward the end chairman of Alyeska, silver-haired and florid, almost handsome, with a strenuous, combative will at the center and a blunt dispassion that has offended plain Alaskans and others, Ed Patton was curious enough, when I asked him, to have wondered why he had been chosen after the months of searching. The oil men who had power in 1970 would be retired or dead by the time he finished the job; it was not a question that would ever be entirely answered. He had worked for Jersey Standard since 1938, barring five years in the Navy during World War II, from which analogies with the Normandy invasion informed his sense of the pipeline project. He got his first exposure to governmental and environmental attentions in 1959 when Esso sent him to Norway to manage the building of a refinery on the Oslofjord. Five years of that, a couple of years back in the New York corporate headquarters, and then the company sent him to San Francisco on a similar job: "It was my first encounter with the hysterical fringe—the days of rioting in

Berkeley and all that. But we did within five months get a permit to build the refinery. . . ." He remained detached: "I didn't build that refinery any more than I am personally building this pipeline: I built the management and the operating staff—trained them, mainly." At the beginning, the immediate job was to assemble the vice-presidents and department heads who, at a remove, would organize the work and select the contractors and subcontractors to carry it out.

At first, the seven owner companies intended to be represented in the Alyeska management in proportion to their ownership, but that turned out to be impracticable. The small owners would be overrepresented by a single executive; the big ones, due to the custom of subcontracting their construction, had few people with direct experience in project management, fewer they could spare, and none who had worked on anything so big—but it was becoming evident that in size, complexity, difficulty, and cost there were no precedents for the Alaska pipeline, not the Suez or Panama canals, not the transcontinental railroads, not even, perhaps, the pyramids of Egypt or the Great Wall of China. Nevertheless, at the start, the companies came through: "The first round of vice-presidents and division heads," Ed Patton explained when I interviewed him, "I personally recruited, I went around and talked to all the owners and asked them to submit nominations, and I interviewed these people and went over their backgrounds and their ambitions and made the decision and nobody challenged it. I put the slate up to the owners for election and they elected them." Humble and Arco were the main sources for this first level of management, with a sprinkling of men from BP and Sohio, but thereafter the companies were less generous, and in the final mix of executives only about one third were so provided, with specific loyalties and intentions of returning once the pipeline was built; the rest were "outsiders," generally construction specialists with not much to say about policy or over-all management, which the oil men controlled.

Despite the optimistic schedule, there was not much for these first recruits to manage. In August a first barge convoy of supplies and equipment was assembled and towed north from Seattle—the practical reason for placing the Alyeska headquarters at Bellevue, across Lake Washington from the port city, though the Alaskans were disappointed by the infrequent attentions that resulted. (About the same time, the pumps and compressors arrived at Prudhoe Bay for the flow stations needed in the Arco half of the field to assemble the crude from the many wells and deliver it to the pipeline; and were stored and eventually, as the design of the field developed, discarded and replaced.) In the fall, the second shipment of pipe was delivered and went into storage with the first lot in yards at Valdez, Fairbanks, and Prudhoe Bay. In December, to protect the pipe from rust and avoid the expense of wrapping it with corrosion-resistant tape before it was finally laid, Alyeska began the process of coating the seven hundred or so miles of pipe already on hand with a green epoxy that was sprayed on and heated to harden; an unsuccessful experiment—the coating stretched and cracked in the Alaskan weather, came loose and peeled like old house paint, and, underneath, the pipe went on rusting. (A stiff, papery chip of the stuff is on my desk now as I write, picked off a section of pipe at Valdez.) The material, Scotchkote 202, was made, as its name suggests, by the 3M Company (Minnesota Mining & Manufacturing), the same folks who bring you Scotch tape, but was soon abandoned. Alyeska, disappointed, brought suit to get its $24 million back and began hunting other ways of protecting the pipe against corrosion.

By the fall of 1970, besides most of the pipe and some of the construction equipment it would need, Alyeska had received the pumps and drivers for the five pump stations that would move the oil down the pipeline at the planned start-up rate of 600,000 b/d (eventually, at the peak rate of 2 million b/d, there would be twelve pump stations); at Prudhoe Bay, where Arco had run up its permanent base camp in six months and moved in during the summer, things had reached a similar state of readiness. Nevertheless, the oil companies' insistence that they could start building in the spring seems to have been bluff—or worse. Despite the quantities of hardware ordered and already delivered, the design of the pipeline system had not advanced much beyond the surveying of the route necessary to the right-of-way applications submitted in June 1969 and revised in December in an effort to get around the restriction of the Mineral Leasing Act. Moreover, when Ed Patton came on board in June 1970, the oil company engineers were engaged in a fundamental dispute with the Department of the Interior over their general plan for the pipeline. In particular, in a report that surfaced in January, the USGS challenged the engineers' assumption that 90 or 95 per cent of the pipe could be buried and proposed the reverse—that 90 per cent of the route should be aboveground (at twice the cost per mile) to avoid damage by and to the permafrost. This was a question that had to be settled, of course, in order for the Department to prepare its EIS and answer the courtroom arguments surrounding it, but the USGS did not have much more to go on than the oil companies' careful public announcements and discreet answers to specific inquiries, such as Russell Train's 79 questions; it was not till August 1971 that Alyeska provided a detailed description of the project, and with the supplements elicited by further questioning, it was December before it was complete.

The issue was more fundamental than a technical argument between two sets of engineers, even to such opposite conclusions. It touched, finally, the competence of the oil companies to do the work they had set themselves and their responsibility and authority for how their money was spent. "What we wanted," as Ed Patton put it, "was the right to protect our property as we saw it so you could justify that protection economically. But we never got that." Eventually, though, the technical disputes and the question of authority were compromised in the mechanism of surveillance confirmed and given legal standing by the Pipeline Authorization Act. In public, the oil men diplomatically adopted the line that the delay and the debate had been beneficial, that in the process the design had been "improved" (a view magnified in the environmental rhetoric into an admission that it would otherwise have been a "disaster"). In March 1973, for instance, Arco's president Thornton Bradshaw told a television audience: "Early in the game, environmentalists blocked us for very good reasons indeed. I think, early in the game, we did not know how to make an environmentally safe line." Privately, however, and particularly among the engineers who started early with TAPS and stayed to the end, they remained unpersuaded.

The engineering debate impinged on and paralleled the public arguments in the courts, Congress, and the newspapers, but it was mostly conducted in private, by men on the ground in Alaska, feeding information to drafting rooms in Houston and elsewhere, to laboratories in California and government offices in Washington.

The first step toward design had been to establish a feasible route for the pipeline from a study of maps, then refine that study with aerial photographs. The next step was the exact topographic plotting of the route on the ground: cross-section profiles within what would be the right-of-way, every physical feature of the terrain, measured up and down, fore and aft, laterally; stream and river approaches and their crossings; location of calculated bends in the pipe every few hundred feet to allow for expansion and contraction within the huge range of Alaskan temperatures, the 3,500 PIs of these angles (points of intersection) staked and marked with squares of yellow plastic for a second round of aerial photographs; and, in the course of this work, the center line of the pipe was staked out every 200 feet, much like the foundation of a new house, but with far greater exactitude. The data thus derived, to a required accuracy within one to five thousand (that is, an allowed margin of error of not more than 0.02 per cent), were written up, translated into hundreds of sectional maps, and, delivered to Houston, determined the design of the pipe. Moreover, the process was repeated many times: for the haul road north of the Yukon; for campsites, airstrips, access roads, gravel sites, pump stations, microwave towers; for every small variation in any of these determined by on-site engineering; and finally, as-built, for the benefit of the federal and state governments.

All this in the beginning was done by a few small surveying parties working under Frank Therrell—party chief and rodman, a couple of men with chainsaws, a pilot for the helicopter that carried them in and out. Two of them were Maurie Smith and Cliff Antonini, borrowed from BP and Humble, an unlikely team that stayed together and ended, when I met them, in charge of all Alyeska surveying, chafing at being confined to an office—men, as Maurie put it, with a liking for "desolate and wild and uninhabited country." Tall and sandy-haired, beginning to stoop, he had been a bomber and navigator during World War II, done government cartography afterward, then gone to work for BP, which sent him surveying to Tanganyika, Kenya, Libya, Yemen, the Persian Gulf, to New Guinea among tribes as yet unknown to anthropologists; still, after six years in Alaska, keeping his family in London and going home for long summer leaves, with an Englishman's inability to pronounce the Texas city and hero, Houston. Cliff, dark and compact, had done similar work for Humble in similar places and retained a Texas accent as decided as Maurie's British.

Together they had hiked the whole 800 miles of the route, several times over, probably, shouldering transit, tripod, and rod. Progress was measured in feet, 15 or 20,000 a day; in the endless days of the Arctic summer they worked twelve hours a day to make those distances and spent another four in the tents after dinner, writing up notes. In the wooded country from the Brooks Range south, the helicopters carrying them from the temporary camps to each day's starting point hunted a clearing to land in or, if there were none, let them down through the trees on ladders. The chainsaws cleared the survey route; one man died from a falling tree, several in helicopter crashes. Twice in the rush to finish, the company worked them into the winter, but they thought it a dead loss for what they accomplished—the light, brief and weak, the paralyzing cold. In Keystone Canyon north of Valdez, where the grade of the pipeline route exceeds 100 per cent, they were let down with ropes, like mountain climbers. Grizzlies, fearful for their cubs, chased them up trees, ripped the sides off helicopters, hunting food. A few were allowed .44 magnum revolvers ("like Hopalong Cassidy," Cliff, who'd carried one, said), shotguns with a ration of shells loaded with deer shot, but were under orders not to shoot. Foxes plagued them as they had every explorer since Bering. The fox was Maurie Smith's story. "A brown fox, very pretty little thing, just wandered around in circles, obviously crazy, but in those days we didn't know too much, we thought it was just a playful little fellow. Bit the party chief in the backside and they couldn't get it away, and the only fellow that had a sidearm was the helicopter pilot, who was away at the time. We'd been instructed, if we had to kill a fox, to cut its head off, box it up, bring it in and have it tested for rabies. They finally hit this fox on the head

with one of the survey rods, stunned him. The helicopter pilot
came back just then and he shot it, but unfortunately he shot the
damned thing through the head. . . ."

The line staked by Cliff, Maurie, and the other surveyors
forms a kind of laboratory specimen of Alaskan topography, as if
a surgeon's scalpel had extracted an 800-mile section of tissue for
examination under a microscope. For about a hundred miles
south from Prudhoe Bay, the line follows the multichanneled
braidings of the Sagavanirktok River, across the Arctic plain to
one of the river's sources, Galbraith Lake; by then, although the
river course is still essentially at sea level, the mountains have
closed in on either side in two ranges, the Philip Smith and En-
dicott, rising to 7,500 feet and forming part of the larger Brooks
system. Beyond the continental divide, the route joins another
river, the Dietrich, then a branch of the Koyukuk, a major tributary of the Yukon, and descends the southern highlands of the
Brooks Range to the Yukon, its approximate mid-point. This pattern is repeated in the southern half: rising uplands from the
Yukon as far as Fairbanks, then the flat flood plain of the Tanana
River into the heart of the Alaska Range and the descent to the
south, following the valley of the historic trade route provided
by the Copper River; and a final plunge through the coastal
Chugach Mountains to Valdez.

Geography defined the problems that the pipeline engineers
would have to solve. There were the three mountain ranges and
the corresponding passes through them: Atigun in the Brooks
Range, a steep, narrow canyon rising to a high point of 4,800
feet, so little known that it had never been named until the surveyors reached it; Isabel, 3,300 feet, in the Alaska Range,
Thompson, 2,700 feet, in the Chugach mountains. None is high
by Rocky Mountain standards, but in the climate the high-moun-
tain effects are intensified—furious winter storms, frosty nights
and occasional snow showers through the summer, treeless, the
bare rock and scree perpetually colored in the browns and grays
of autumn. There were the flood plains of five big rivers to follow, 800 river and stream crossings, from the giant Yukon to tiny,
intermittent trickles in the mountains and including, besides the
Yukon, a dozen the engineers counted as major. Every stream
provided specific variations on a basic set of problems: protect-

ing the pipe against the overwhelming force of bed-deep winter ice and spring-scouring flood; protecting the life of the stream itself, the salmon above all, from the channel-ripping erosion and siltation that, equally, make spawning impossible. All along the route, in the valleys and on the high slopes, there were other animals to consider, near the center of the interconnected system that is human and animal life in Alaska but about which not much that is basic was known—numbers, migrations, behavior, breeding, food sources, hence the effects of an infusion of pipeline builders and their machines: mountain sheep and goats, the moose nearly everywhere except the farthest north, several herds of caribou which the French-Canadians long ago, with wonder, called simply *la foule*, the mass, a flowing tide of life. There were other questions, possible hazards to be described, circumvented: glaciers around the Isabel and Thompson passes, within sight of the route, some of which within the last generation have unpredictably surged forward and retreated; four active seismic faults between these two passes, including the Denali, comparable to the San Andreas in California and running west to Mount McKinley, which in Tanana is also Denali (the word means simply "the big one"); and, on the coast, the grinding edge of two tectonic plates, the Pacific "rim of fire" which in 1964 produced the most violent earthquake yet recorded in North America.

These difficulties, while considerable, were comparatively specific and local. There was one overriding problem that would be encountered almost everywhere along the route, a unifying complication: permafrost. Even at Valdez, shielded by a coastal mountain range, caressed by warming ocean currents, it would be incautious, until the entire area has been drilled and tested, to assert that the ground is entirely free of this northern phenomenon. It was permafrost that became the focus of the engineers' debate and the public anxieties expressed by the environmentalists.

The word has made the substance sound more mysterious than it is in fact. By definition, permafrost is any ground—rock, gravel, sand, fine-ground silt—that has remained frozen for at least two years. On the Arctic Slope, where eons of cold have sunk the frost deeper every year than it can thaw in summer, the permafrost runs two thousand or more feet deep, unbroken by

rivers or ponds; south of Fairbanks it is increasingly thin and discontinuous, occurring in subsurface masses determined by local
climate and elevation. In summer, the sun penetrates a foot or
two to give life to an insulating cover of plants, the active layer;
below, the actual ground temperature remains constant in a
range from 14° F. in the north to the freezing line at the southern limit of permafrost.

This would be of no moment if oil did not everywhere come
out of the ground hot. At Prudhoe Bay it was the temperature of
a too hot bath, about 180° F. In pipes of the diameters previously laid, in temperate climates, the heat soon dissipates and
the oil drops to the temperature of the ground. What would happen to the oil in the 48-inch pipe chosen for the volume to be
pumped was unknown from experience. The engineers modeled
the effects mathematically by computer: the oil would enter the
line at something less than 180°, give off heat, gain some back
through friction and the pressure of the pumps, and arrive at
Valdez still hot, at about 160°. These predicted temperatures
were approximate and assumed that the pipeline would be
pumping at full capacity; in operation they would vary considerably with the volume of oil and the season, with important
effects on how the oil would behave (below 117°, for instance,
dissolved wax would begin to precipitate, clogging pumps,
building up inside the pipe like fat in aging arteries). The question before the designers at this stage, however, was more fundamental. Whatever its precise temperature, it was agreed that the
oil would enter the system hot and would remain so. What mattered, then, was the precise character of the ground in which the
pipe would be buried, and this at the start was known only in
general terms. You can make the experiment. Fill a bucket with
gravel and water, freeze it and it expands a little, shrinks again
when it thaws but, contained, it supports weight. The same
bucket filled with muck from a swamp or lake bottom or river
delta behaves quite differently: thawed, a weight dropped into it
sinks to the bottom or till it reaches equilibrium; the frozen,
thawing muck is, in the language of the soil engineers, thaw-un-
stable. This was the immediate problem faced by the pipeline
designers: they knew that most of the way was permafrost but
not of what kind.

To find out, TAPS in March 1969 drilled 300 holes along the route north of the Brooks Range, using five portable rigs towed on skids or truck beds behind bigwheeled D-6 Cats; in the six years that followed, using as many as 19 rigs, some of which had to be helicoptered into place, they drilled 4,000 more, south to Valdez, yielding 15,000 core samples shipped frozen to labs in Houston and San Francisco for analysis. From this sampling, solutions were calculated, argued. Where heat thawed unstable soil between sections of solid ground, the pipe would form a bridge, supported at either end. For the difference in cost between conventional burial and other solutions, the companies were prepared to take a chance—the net cost for maintenance and repair would be less; both could be cut by routing the line to avoid unstable permafrost. Hence, from the early drilling, the designers shifted the pipe east from Anaktuvuk Pass (the route of the Hickel Highway) to Atigun and thereby saved several miles of distance and corresponding millions in construction— this was the basis for the early estimate, for a line 789 miles long, 90 or 95 per cent buried. The USGS study in January 1970 disputed this reasoning; environmentalist papers published visions of sunken, broken pipe spewing hot crude across the tundra. The strength and flexibility of the pipe, about half an inch thick, became an issue: alloyed with vanadium for hardness' and elasticity, with niobium for resistance to corrosion. It had been specified, ordered, was being delivered, could not be changed. In July 1970, the month after Ed Patton was hired, pipe samples were shipped to a University of California lab in Berkeley to determine if they conformed to their specifications and the spot tests made at the Japanese mills.

Early, there had been another thought: that in the aboveground section of the pipe planned north of the Brooks Range, the oil would cool fast enough so that from there on the pipe could be safely buried. Prudhoe crude has a fairly low pour point, about 15° F., below which it gels—turns too thick to pump; the trick, then, would be to move it in the narrow interval between pour point and freezing, which, if it could be managed, would cause no disturbance to the permafrost. To find out how the pipe itself would stand the cold, a thousand-foot loop was set up at Barrow in the winter of 1968, with the University of Alaska contracted to monitor the test. The test succeeded—the pipe was unaffected—but the plan failed: the computer simulation mentioned earlier, carried through at the same time, showed that the volume of oil in the 48-inch pipe would retain most of its heat and come out at Valdez not much cooler than it began; cooling it before it entered the system and holding it within the range of temperature at which it could still be pumped (for instance, by storing it in tanks at Prudhoe Bay) was ruled out by cost.

In January 1971, with the Berkeley tests completed, Alyeska changed its tune: a little more than half the line would be buried, and the cost of the project had more than doubled, to $2 billion. Aboveground pipe had been and remained a solution to avoid: cost, always cost. It had to be supported by uprights drilled into the permafrost, collectively (under the new plan), more feet down than all the wells needed to produce the Prudhoe field; and raising further questions about permafrost—whether, for instance, the pressure of freeze and thaw would push the supports out of the ground. In addition, the aboveground pipe would be vulnerable to winter shutdowns, for maintenance or accidents, when the oil could solidify and be difficult to start again. To allow for that possibility, the pipe would have to be insulated, with four inches of plastic foam batting stuck to folded sheets of galvanized steel, a contrivance that would allow two or three weeks of midwinter shutdown before the oil became unpumpable. But this was only part of the objection.

In the course of the debates, Ed Patton and others made sweeping defenses of the quality of the pipe—in particular, that it was a special alloy that would remain strong, flexible, and weldable in Alaska's extreme temperatures, which range from .a winter record of —80° F. recorded at the Prospect Creek pipeline camp in 1971 to as much as +100° F. at Fairbanks—but were chary of supporting details. Any steel will behave strangely in extreme cold unless heavily alloyed with comparatively rare and expensive metals such as vanadium: drop it on the frozen ground and it shatters like glass. In time, it came out that the Japanese alloys had been specified for operating temperatures only down to —20° F., and this was a further argument against putting the pipe aboveground: even sheathed in insulation it would turn brittle in the course of a prolonged winter shutdown. The drastic change of plan in January 1971 was then not merely a yielding to mother-hen caution in the Department of the Interior. It confirmed that the pipe was not so *very* special after all: along much of the route, from what was so far known, the risk of burying the pipe outweighed the risk (and cost) of elevating it on aboveground supports. Frank Therrell and the other engineers became acutely sensitive to these opposite constraints on their design, which were imposed by the pipe steel:

> It began to be delivered in August of sixty-nine, so from that point the design and the operating conditions had to be predicated on what pipe you had. So when we got more soils information plus some design criteria that we had to meet, then the pipe would only do so much. So that restrained us to the point where, in some of these areas where normally you'd have pipe that would take more flexing with the stress and all that it would be under, you could have buried some of the pipe, but with us locked in on this particular pipe, that forced us aboveground.

Even so, remembering, he disputed design decisions that had been imposed and, more than that, the dilution of the engineers' authority in the process of decision:

> The people we are dealing with in the government agencies and all are not that familiar with the pipeline, and we just could not gain any credit for these types of things that have been done in the pipeline for years. So consequently we ended up with a superconservative design. Some of the people—"We're going to build this pipeline, but we want you to guarantee us that it's not going to fail." Fail-safe: so you have no alternative but to design it supersafe, essentially ignoring what could be taken care of from the maintenance standpoint of minor settlements here and there. . . .

But the pipeline's integrity, as it came to be called, *could not* be considered in isolation: it touched matters beyond the concern and competence of engineers. Among these were the caribou.

Not much, as I said, was known about caribou; or, more accurately, the kind of intimate knowledge by which, for millennia, Eskimos and northern Indians have survived through skill in hunting them was not accessible to engineers, biologists, and environmentalists, lacking that necessity. Now, from studies commissioned by TAPS and the oil companies or occasioned by their pipeline, we know a little more, but the knowledge is still imperfect and uncertain.

The caribou is the American cousin of the reindeer herded in northern Scandinavia by the Lapps: a big, hump-shouldered, milk chocolate colored deer with complex, many-pointed, backward-curving antlers grown by both bucks and does. In Alaska there are a dozen native herds that have been identified and named. The biggest is the Arctic, ranging from Prudhoe Bay west to the shores of the Chukchi Sea, south into the Brooks Range, numbering in good years several hundred thousand—the numbers fluctuate, the estimates vary. East of Prudhoe Bay and across into the Yukon Territory as far as the Mackenzie delta is the Porcupine Herd, named for a Yukon tributary and averaging, perhaps, a hundred thousand animals. Below the Yukon two smaller herds in narrower territories, the Delta and Nelchina, impinge on the pipeline, from a few thousand in size to tens of thousands.

The herds are migrant, in spring and summer seeking open ground for browse and calving, returning to sparsely wooded uplands in winter. In obedience to this pattern, the Arctic and Porcupine herds follow circular migration routes centered west and east of Prudhoe Bay and the pipeline, which seems to lie on the border between their two territories; in summer there is some mingling, interchange of fragments of the two herds, depending, perhaps, on weather and food, the mosses and lichens of the Arctic plains, but no large passages across the pipeline right-of-way. South of the Yukon, however, the pipeline would cut through the range of the two lesser herds.

The pipeline posed two different threats to these herds. Buried in unstable permafrost, its heat would create a zone of thaw that would eventually be fifty feet wide, a belt of quagmire, depending on the nature of the soil, in which the pipe itself might be safe, as the engineers calculated, but which the animals presumably could not get across. On the other hand, if the pipe were laid aboveground on supports to answer that problem, it was uncertain whether the caribou would cross it. To find out, biologists conducted a number of experiments: sections of elevated pipe raised high enough off the ground so that in theory the animals could pass under it; simulated pipelines on which passage over was provided by steep ramps built up of gravel. These studies, supported by observations of caribou and half-wild imported reindeer around working pipelines in western Alaska, confirmed during actual construction of TAPS, revealed that hatred of enclosures is among the deepest of caribou instincts: with rare exceptions—solitary males that in midsummer might get under the elevated pipe where shade and breeze would provide some relief from the torment of mosquitoes and biting flies—the caribou would not cross under or over the pipeline; newborn calves, separated from their mothers by the pipeline, would trot beside it, voicing and answering their distinctive, identifying cries, but would not cross.

To the Eskimos and a few white hunters, it was not much of a revelation. Scattered across the North wherever the herds migrate there are pairs of fences made of brush or stone, sometimes many miles long, between which the animals advanced in a narrowing funnel, like cattle down a slaughter chute, to the final opening where a few hunters could pick them off with leisured certainty. (The big-horn Dall sheep, in their steep mountain pastures where man is their only predator, behave in a similar way: with nerve and the sure feet to follow their almost invisible trails across the plunging scree, you can approach, keeping down, to within a few yards; but outlined against a skyline ridge, your human form stirs fear and they scatter uneasily and move off.) This was in any case a serious discovery. If the herds could not reach their immemorial calving grounds, where food, weather, and a seasonal scarcity of predators combined to assure the new generation's survival, the herds would thin, die out. This in turn was a matter not merely of one species, even in numbers so great as to constitute one of earth's natural wonders, or of the few men who still depend on it for sustenance: remove one link in the chain of life, whose foraging and dung controls plant growth, whose flesh feeds several predators besides man, and the whole chain alters, in unpredictable and perhaps disastrous ways. It was not a tolerable risk.

Concurrently with their caribou studies, the biologists learned things too about the moose, the other large and much-hunted Alaskan meat animal, migratory also by recurring routes though not in the same swarming life-tide. Moose are forest creatures, habituated to ducking under branches, crashing, long-legged and high-rumped, through heavy brush, indifferent to barriers; around Anchorage and other Alaskan towns, in season, they follow their accustomed trails with stupid persistence, crossing in places four-lane roads because that is the way to the willow shrubs and waterside meadow grasses they feed on. Their behavior in relation to the pipeline was therefore the opposite of the caribou. They had no inhibitions about stooping to get under elevated pipe, at times, to judge from the tracks, getting down on their knees to scramble beneath pipe that had been set too low. This difference was a source of confusion to oil men unable to distinguish caribou tracks from moose; several jauntily assured me that caribou and the pipeline were no problem—the animals simply went under. As with most other realities connected with the pipeline, this one cost me a good deal of skeptical determination to untangle.

It took the biologists several years, patiently tracking and counting the herds from light planes and helicopters, then on the ground, to find out something about the behavior of moose and caribou and their migrations, which is why not more is known; but in this case there was time—and money. What they learned led to opposite solutions. Where unstable permafrost required aboveground pipe, they settled on a minimum clearance of eight feet to accommodate the moose, allowing a couple of feet for winter snow-cover. For the caribou in similar areas, the elevated pipe was reversed and run underground, still insulated, so as to leave a minimum opening of sixty feet, the frozen muck removed in and around the trench and replaced with gravel, the ditch itself surrounded with sheets of Styrofoam. There are twenty-four such crossings along the route, which the designers call "sag bends." Where the nature of the surrounding soil made this solution impractical, it was necessary to bury the pipe and refrigerate the ground around it to keep it from thawing, using brine chilled in cooling plants to one side of the line and pumped through underground pipes—a costly arrangement applied altogether on about three and a half miles of the route. At Prudhoe Bay itself, where the oil is gathered from the wells through a maze of smaller-bore pipes, all laid aboveground, a different design was adopted: gravel ramps, twenty altogether, with snow fences to guide the animals to and over them. Whether any of these solutions will in fact work is matter for further years of careful observation. The Prudhoe Bay design was tested early and disproved, and the fringe groups of caribou seen there seem already to be clearing out, though whether the ramps or the general bustle of human activity is the reason—there may be several natural explanations—is not known; since neither of the two northern herds apparently has to cross the area to survive, the loss is probably not a vital one.

It will be noticed that gravel figured largely in these designs, as an insulator of permafrost: the freeze would build up through it (a process known as aggradation, the opposite of degradation, the thawing of permafrost), and the gravel would then become a new active layer, but stable; or so the theory. For this purpose (and in general, to prevent erosion), gravel was needed around the pipe, wherever it was buried, but also as a base for the thirty-nine construction camps* that were eventually built and, north of the Yukon, eight airfields; and, of course, as the foundation and surface for the 361-mile haul road from the Yukon to Prudhoe Bay and for the work pad—in effect, a second road— that had to be spread along the entire route of the pipeline, from which all the various machines would operate through the successive stages of pipe laying. Depending on the soil beneath and the weight of buildings and equipment to be carried on top, the gravel was laid down three to five feet thick. In the course of construction the builders were able to limit the gravel layer by starting with insulating pads of Styrofoam, but the total quantity required by the project was still enormous: altogether, more than 65 million cubic yards. That is one of those numbers too big for

---

* Twenty camps (shown on the endpaper map) served the pipeline itself through most of the construction period, six others more briefly; plus one camp for each of the twelve pump stations and the big camp at the Valdez terminal.

the human mind to make sense of. A householder knows gravel a yard or two at a time, making a driveway or patio, running about a ton and a half to the yard, many wheelbarrow loads; four or five yards fills a heavily built medium-size dump truck. Put the whole 65 million yards required by the pipeline project in a single heap and it would form a cone a mile across its base, 250 feet high—not an Everest or a Mount McKinley but a lot of gravel nonetheless.

Part of the early planning, therefore, was finding places near the pipeline route where gravel could be mined in quantity. It is one of the resources Alaska has in abundance. Eventually it was supplied from 470 locations—"material sites," in the jargon of the pipeline; where none were found near or big enough, it was necessary to crush and screen rock to the required, uniform sizes for its various uses. Most of these sites were on federal land, and the Bureau of Land Management improved its budget by selling what was needed at a contract price that came out to about 18¢ per yard. The haul road and three of the airfields, accounting for a third of the total, were exceptions: Alyeska would build them, use them for the duration of construction, and then turn them over to the state; and therefore, as ostensibly state projects, the gravel of which they were built was, under an old law, free.

In terms of engineering and design—what kinds of stress steel pipe will stand before it breaks—permafrost and extremes of temperature are related problems, comparatively limited and solvable. Along much of the pipeline route and particularly from the Alaska Range south to Valdez, these problems were enormously complicated by another factor: the earthquake zone. Earthquakes are among those aspects of earth's life of which man has much experience but little understanding. We know a good deal about their effects but are not much beyond guessing at their causes. Their occurrence remains essentially unpredictable.

The theory of earthquakes is that they result from pressures generated by the heat and flow of probably molten rock deep within the earth. Where that pressure finds a weak point in the earth's crust, perhaps at the intersection of two tectonic plates, it forces the crust apart—a fault—or in places spews molten rock through it—a volcano; around the Pacific rim, the two go together, geological fault and flowing lava. The fault is not only a break but, if it is active, a line of continuing tension, like a tightly wound spring: as the inner pressures build, the earth moves on either side of the fault, up or down, sideways, the tension is released with explosive force, and the surface of the earth dances in rolling, rebounding waves, like a whip that has been snapped. These are earthquakes.

There are several systems for estimating and comparing the force of earthquakes, all approximate. The commonest is known as the Richter Scale, a sequence of numbers from 1 to 10—each higher number represents a tenfold greater release of energy (some seismologists say thirty times). A quake rated below 3 on the Richter Scale may be felt but will probably not cause damage. A value of 3.5 means slight damage—cracks in plaster walls, cups rattled on a shelf, knocked off; at 6, whole buildings may collapse. The earthquake that destroyed Guatemala City in February 1976 was rated 7.5 on the Richter Scale. The Alaskan quake of 1964 was 8.5.

Everyone who lived through it has stories to tell. It happened about five-thirty in the afternoon of Good Friday, March 27, 1964—Easter was early that year, in most places in South-Central Alaska there was still a thin cover of snow on the ground, on streets and sidewalks. The shaking, wave motions you could see (or so they were remembered), knocked men down and they struggled on hands and knees toward doorways, the safety of open space outdoors; cars on the roads slithered sideways into ditches. In Anchorage, heedlessly built on the gravelly outflow from several epochs of vanished glaciers, the streets heaved, opened, sank, and along 4th Avenue, the main street, the shabby one and two-story bars, stores, and pawnshops, wood-frame and cinder-block, dropped into basements, roofs level with what had been the street; 3rd Avenue, a block nearer the Knik Arm of Cook Inlet, slid down the hillside toward the railroad line and the water; the suburban area called Turnagain trembled like pudding and its expensive new houses collapsed, broken-backed, and were swallowed up. The tremors and aftershocks lasted minutes but seemed hours. Those who survived and stayed (there were 120 deaths) are sensitized to those vibrations and when they feel them are afraid; the earth still moves, often, though not yet again with such violence.

The destruction on land was only part of the effect, perhaps not the greatest. The center of the quake was somewhere in the northern Gulf of Alaska, near the island-shielded entrance to Prince William Sound. From there a whiplash wave spread across the Pacific, 70 feet high, moving at 50 miles per hour: not a tidal wave but a *tsunami*, to borrow the Japanese word since we have none better, scientifically a seismic sea wave; lesser waves followed. The wave traveled quickly up the Sound, funneled through the narrow mouth of the fjord of Port Valdez, and caught twenty-six men and women on the town dock where a freighter was unloading; the town itself, set low at the head of the fjord between two glacial streams, was wiped clean. Seward, the Kenai Peninsula railhead just west of Prince William Sound, was likewise swept away. Farther south, the fishing villages and towns on the east side of Kodiak Island were destroyed, the boats carried far inland, but with few deaths—those who used their radios had five hours' warning that the wave was coming; one man in early evening noticed his cattle nervously making for high ground as with a common impulse, having felt or heard something beyond human sense, and followed them and was saved. The next day, 1,200 miles to the southeast, a family camped for the weekend on a Washington beach was caught by the same wave, their bodies never found.

Part of the planning of the pipeline, then, was to locate the epicenters of past earthquakes, record their intensities, and plot both in relation to the intended route. The maps that resulted are almost clear north of the Yukon, increasingly spotted with seismic events from Fairbanks south, black with dangerous possibilities in the coastal arc from Valdez to the Alaska Peninsula. The maximum intensities correspond: from generally 4 or 5 on the Richter Scale around Fairbanks, disturbing but probably not greatly hazardous, to 8.5 at Valdez.

The Stipulations that evolved through the pipeline debate and the drafting of the Pipeline Act required that the pipeline itself and its dozen pump stations, the terminal facilities at Valdez, be built to withstand the maximum known earthquake in each area, stringently specified but with an out—"where technically feasible." For the low and light buildings that would be needed, this was a comparatively easy though not inexpensive requirement—

the techniques had been developing for fifty years, since Frank Lloyd Wright designed the first earthquake-proof hotel in the aftermath of the 1923 Tokyo earthquake; from calculation and experience—pipelines in California that had come intact through quite severe earthquakes—it appeared that the buried sections of pipe would probably be safe also. The aboveground pipe was another matter.

From an early stage in which the pipe would be held in a semicircular truss mounted on a steel post set in the permafrost, the aboveground design developed to allow free movement from side to side, the distances figured to accommodate the earthquake forces specified in the Stipulations (and incidentally the contractions and expansions of extreme cold and heat). The pipe would be carried by a steel-beam framework resting in turn on a Teflon-coated crossbeam supported between pairs of 18-inch uprights drilled forty or fifty feet into the permafrost. The pipe would thus be free to slide from three or four feet either side to as much as ten feet, the length of the cross beams and the positioning of the vertical supports specified to match the earthquake standard*; the pipe itself would be protected by rubber bumpers welded to the vertical supports and by crushable aluminum honeycomb enclosed in a Styrofoam-insulated fiberglass jacket. Every few hundred feet, the pipe was to be "anchored"—held rigidly in place by an arrangement of beams, steel rods, and turnbuckles bolted between four uprights: between these anchors, the pipe could, when the earth shook, snap like a whip, sideways, up and down, without breaking; or so it was calculated. The vertical support members (VSMs) were to be set in 24-inch holes and held by a slurry of sand and water that would pour and, when it froze in the permafrost, set like cement. In most places, to prevent thawing, the VSMs were fitted with sealed ammonia-filled tubes extending to the bottom of the hole; if the permafrost warmed, the ammonia would absorb the heat and carry it to the top where it would dissipate to the air

* This variation was due to engineering exactitude but also to a decision to save a little money by making the crossbeams as short as possible— the whole rig was said to be worth the price of a Cadillac, say $13,000 apiece. It was a costly and characteristic saving, making it necessary to fit each set of supports exactly in the course of construction and sort the several sizes of crossbeams to their required locations.

through radiator flanges. A similar heat-transfer arrangement was planned for the sag-bend caribou crossings south of the Yukon, where the temperature of the permafrost is barely below freezing.

This elaborate sequence of designs, which I have only sketched, took time. The finished engineering drawings, which are on my desk now as I write, are dated May 1975, *after* pipeline construction had actually begun. The date is perhaps evidence for the improvements made possible by the delay.

The hundreds of streams and rivers the pipeline would encounter presented two different and perhaps opposite problems: protecting the pipe from the stream; and protecting the stream from the pipe—and the effects of construction. Any change in a river course is likely to affect the life it sustains in ways not predictable and, once brought to pass, not easily reversed. Straighten the channel and you speed the flow, making it impossible for fish to reach their spawning grounds, with multiple and complex effects on the life that depends, in turn, on theirs; conversely, a bend or obstruction introduced into the channel may slow the current so that runoff silt is no longer carried on down but dropped—with similar results for the life of the stream. At the same time, the simple force of flowing water, above all in the spate of spring flood, is immense, almost imponderable, year by year breaking up bedrock, capriciously lifting and carrying off or burying snagged trees, boulders. Add to these hydraulic forces the fact that nearly all the Alaskan streams are underlain by permafrost, whether thick and continuous or in isolated and unpredictable masses, and the design for crossing them with a four-foot pipe laid so as to resist destruction becomes a task of considerable delicacy.

The task was further complicated by the fact that oil weighs less than water: as we know, it floats. In the case of Prudhoe Bay oil, a cubic foot weighs about 55.4 pounds, against 62.4 pounds for a cubic foot of water. Hence, a running foot of oil-filled pipe weighs 696.5 pounds, about 90 pounds less than with the same amount of water. The pipe itself is surprisingly light—about 250–310 pounds per foot, depending on the gauge of the steel: pipe and oil together would lift from the bed of a strong stream like a water-logged boat. There are two solutions to this difficulty: excavate the stream bottom in order to bury the pipe deep enough so as not to be affected by the current; or carry the pipe above on a bridge. Since bridges cost money, TAPS initially intended all its stream crossings to be buried, including the Yukon, and produced core samples to justify the intention. The state of Alaska, however, wanted a bridge over the Yukon and by doing the work itself—the bridge would support the pipe but also in theory be a link in the state highway system—could secure most of the cost from federal funds. Further study of the stream bottoms, continuing right through pipeline construction, induced further caution. Eventually, it was decided that twelve more rivers, all wide and powerful, would have to be bridged, and these structures, like the pipe, were ordered from Japan— prefabricated and preassembled, then dismantled and barged across the Pacific in sections.

At the hundreds of lesser crossings, the pipe would indeed be buried—and weighted with a concrete jacket to keep it from floating; or, where the force of the current required it, held down by sectional precast concrete saddles, each about six feet wide and high, eight feet long, and weighing nine tons. The same arrangement would be needed in the long sections of the line— north of the Brooks Range, on either side of the Alaska Range— where it paralleled the flood plains of rivers that might violently alter course from year to year, each, in effect, a prolonged stream crossing. Both the bridge foundations and the buried, weighted stream crossings were to be further protected by dikes to control stream flow and erosion.

Taken together, these arrangements have a makeshift quality: in the conditions faced, they are near the limits of technology, and whether they would in fact do what they were meant to do would not be known except by trying them, perhaps through several years of pipeline operation. That was not, however, among the questions that received attention in the course of the pipeline debate.

Once in the ground, the pipe would be subject to a less dramatic but more pervasive threat than earthquake, permafrost, or spring flood: corrosion. In a representative year in the United States, there are perhaps 500 pipeline breaks in which 50 or more barrels of oil are spilled. That adds up to a minute percentage of the hundreds of millions of barrels actually moved by pipeline, but it is still a good deal of oil—enough, say, to heat an average house through a thousand winters. Most of these breaks are probably not due to any large accidents, human or otherwise, but to the natural wearing away of pipe steel. Particularly in wet ground (but there is always some moisture), there are weak electrical currents by which the metal from the pipe is transferred to whatever minerals are present, the same electrolytic process used in chroming the bright work on a car. The steel becomes pitted, the pressure of the oil inside breaks through, and the oil spurts. With the volume of oil to be carried in the Alaska pipeline, even a comparatively small spill would necessarily be a large event—like everything else connected with the project— and in that environment perhaps disastrous. To complicate things further, Alaska seems to be an extreme case of this universal phenomenon: it appears that the aurora borealis, which makes the winter nights glorious with its sheets and coils of many-colored light, induces powerful ground currents— measured along the pipeline route at anywhere from 30 to 200 amperes, surging at times to 1,000 (1 ampere, the unit for the rate of flow of electricity, is enough to light a hundred-watt bulb); if, as is thought, the intensity of the northern lights is connected with the solar storms called sunspots and these are in a rising cycle, then protecting the pipe from electrical corrosion will be one of those problems likely to get worse before it gets better. To provide against possible disaster, the designers planned what was called cathodic protection: half-inch zinc rods laid on either side of the buried pipe which would then function as anodes in the natural electrolysis—the zinc would transfer to the steel pipe rather than the other way around; at intervals, test leads connect from the zinc rod to the surface so that it can be periodically checked—if the zinc wears away so that it breaks and no longer carries a current, it can be dug up and replaced.

Nevertheless, given the common experience elsewhere, it was necessary to assume that in its expected twenty- or twenty-five-year service life the pipe *would break*. The real design effort therefore had to be toward devising means of detecting and limiting the breaks when they occurred and cleaning up the consequences.

The system devised was a mixture of hypersophisticated and simple. Essentially, it depended on continuous measurement of the volume and pressure of oil in the pipe, from Prudhoe Bay through the pump stations to the storage tanks at Valdez. All of these measurements would be signaled to a control center at Valdez, where they would be analyzed by a pair of computers. If the computers detected even a quite small drop in pressure (15 psi—pounds per square inch—between 2 and 5 per cent of the operating pressure, which varies with the gradient), it would alert the operator in charge, who could then close electrically powered valves at the pump stations or others spaced every ten or fifteen miles along the line—or all of them at once—in a matter of three or four minutes. Since the system, to work, depended on communications in both directions throughout the line, and therefore on electricity, several alternatives were provided: microwave (built and owned by RCA-Alaska) supplemented by satellite-relayed VHF radio; propane convertors to supply power to operate the valves and radio receivers, backed up by storage batteries (with heating for winter) and, in the southern half, by electricity from public power lines; in a pinch, the valves can be closed by hand. In addition, on uphill grades and again every ten or fifteen miles, one-way check valves were planned, so that the oil could flow through them, south toward Valdez, but not run back; if the pipe broke at the bottom of a slope, the oil ahead would stop moving but only that behind the check valve would run backward and out the break.

In the course of the debate, the system developed and the valves multiplied, with grumbling from the engineers about overdesign (and from the oil men about cost). Although ingenious and sensitive, there were limits to what the system could do. The pipe when full will hold about 9 million barrels of oil, more than 11,000 barrels per mile moving at between seven and eight miles per hour. From the placement of the valves, the sensitivity of the instrumentation, the time it would take to respond to danger signals, it was estimated that in a major break—the pipe torn in two by an earthquake or explosion, say—the oil spill that resulted might run from 15,000 barrels to in places as much as 50,000; small percentages, in fact, but large volumes. Hence, the design included plans for dealing with big spills: teams of men and equipment based at the pump stations to bulldoze dikes around the oil (and permanent dikes around the fault zones), machines to suck the oil up or gather oil-soaked gravel, earth, or snow for burning, which is all, so far, we know about disposing it.* Actually, such large and probably unlikely spills are comparatively easy to detect and master. Leaks too small to alarm the controls at Valdez, particularly from below-ground, winter-covered pipe, might go undetected for months until cumulatively they were big; as was demonstrated repeatedly once construction began. By way of defense the designers planned periodic inspections of the pipe by "pigs"—cylindrical containers of instruments moved through the pipe with the oil between pump stations—to record irregularities of shape or sound (the hiss of a leak). Planes would fly overhead on a weekly schedule, unrolling infrared film that would photograph the heat of leaking oil. And, finally, men would drive or walk the line, looking for the dark stain of oil on the ground. Men, bored perhaps, isolated, resisting routine and control, with every degree of moral attention and dutifulness: that is what the engineering abstractions and invariables come down to.

Designing and building a pipeline, even with all the qualifications imposed by Alaska, is a commonplace, but a pipeline terminal and oil port is a rare undertaking; most have been fitted to landscapes as old as man's knowledge of the sea, adapted from those ancient uses gradually, in the course of a century's history of oil. The difference between the two is like the difference between working in wood and working in steel. A carpenter works in feet and inches and his mistakes disappear behind decorative moldings and are not functional. A machinist is restricted to tolerances of a hundredth or thousandth of an inch and if he gauges wrongly, the thing he makes will not work and is rejected. Planning the pipeline terminal at Valdez was inherently more

---

* Several species of bacteria consume oil, and their action is assumed in ballast-water treatment plants and in biological sewage treatment generally. (All sewage contains oil.) They are comparatively slow-acting, however, especially so in a northern climate where all life processes are inhibited; attempts have been made to develop oil-consuming strains that thrive at low temperatures, but so far without notable success.

difficult than designing the pipeline it served, and the actual site furnished complications at least as great as those along the pipeline route.

The choice of Valdez was arbitrary, except for its effect on cost—it made the shortest route from Prudhoe Bay. After the fact, as we have seen (see Chapter 1), local considerations argued against other possible terminals. By comparison, Valdez offers a number of advantages in the handling of big ships. The fjord known as Port Valdez runs east and west, three and a half miles across, a little less than fourteen miles long. It is circled by a flat shelf a mile or so wide, forty or fifty feet above the water, with the look of an ancient shoreline. Behind it in all directions dark mountains, lightly wooded, rise sheer to three or four thousand feet, slashed by steep streams and veiled in waterfalls, hiding, from the sea level of the Port, endless ranges of still higher mountains, all connected to the coastal Chugach system. Wet winds, warmed by the North Pacific, come over the mountains to the south, swirl within the confines of the Port, rise along the northern slopes but do not cross, and deposit their moisture. The climate at Valdez itself is mild, even—pacific: hovering around freezing in winter, rising to the fifties or sixties in summer. It is also wet: sixty or a hundred inches of rain in any year, two or three hundred inches of heavy, sticky, clogging snow. From July on into fall and winter, the clear days can be counted on the fingers of one hand, and even when it is not raining or snowing the gray-white cloud mass hangs low on the mountain shoulders and you see the peaks rise up through them; in a half-dozen visits at every season, I have myself hit only part of one day that could be called sunny.

The mountaintops themselves are heavy with permanent ice and snow. In the immediate vicinity there are twenty glaciers large and distinct enough to be named, dozens of glacial fragments of the ice mass that in the last ice age covered the entire area thousands of feet deep. Since the turn of the century, all but one have been in general retreat; only the Columbia, the twentieth and biggest, is still advancing, calving great bergs into the open waters of Valdez Arm and Prince William Sound, beyond the mouth of Port Valdez. Twenty-five miles northeast of Valdez, where road and pipeline cut through the mountains, you can sample the mountaintop climate—officially an average hundred inches of rain, five or six hundred of snow, though the accumulation is almost too great to measure and those who still live at Valdez talk of a record fall in the early 50s that in ten days went to sixty feet. Thompson Pass is a climax. Beyond, the weather changes abruptly to dry cold and the comparatively hot summers of interior Alaska.

The beauty of the place was not, of course, the reason it was chosen. The Port itself is ice-free—the winters are never cold enough to freeze it over. It is big enough to afford ample maneuvering room for even the biggest tankers. It is also deep—the surrounding rim of mountains plunges to depths of 450 feet, in places as much as 700. For two miles at its entrance, it shrinks to a width of two thousand yards or less (but the Narrows is still shorter, wider—and deeper—than the approaches to most of the world's great ports), then opens out again to the sheltered, open basin of Prince William Sound and a long, straight run between big islands to the Pacific. Within the Port, a pilot has almost no current to contend with, and strong winds are rare, though sometimes in winter a gale blows down the fjord from the mountains so hard that to cross one of the mud-and-gravel streets of new Valdez, safely rebuilt since the earthquake, you must go on hands and knees.

How the tankers would manage, however, only indirectly concerned Alyeska's engineers. Their responsibility ended with the terminal itself and the means of loading the ships. From that standpoint also, Valdez looked good. The wide shelf on the south side of the Port, opposite the town of Valdez, provided space for the oil tanks, power plant, and other facilities the port would require—and it was well above the height of the tsunami that destroyed the old town, the only one known to have occurred. Early geophysical studies showed that the tanks could be built on earthquake-proof bedrock only a few feet below the surface and were much quoted during the pipeline debates.

As it turned out, these studies were wishful—or worse—but by the time that came out the basic decisions had been made and written into law, and construction was well along. The problem deserves a close look.

In 1975, when the bulldozers began folding back the skin of

earth and glacial débris overlying the terminal site, they did not hit bedrock six feet down, as the studies predicted, or ten feet or twenty feet. Eventually the machines had to scrape away up to *sixty feet* of overburden over an area of nearly 150 acres—about 15 million cubic yards to be picked up, moved elsewhere, and somehow disposed of. It was a remarkable error. *How could it have happened?* A few months later, I sat in a snack bar at the Valdez construction camp, talking over that question with George James, who was looking out for the state's interest at the terminal—the state had been allowed to claim the land that was being used. The answer, George thought, was pretty simple:

> They had some surveys here at the terminal that were done to aerial photos and they went up and they dug a few little pits, and they said—O.K., bedrock is at six feet down. It was another facility designed by somebody that was Outside that hadn't really recognized what the problems were. They do not accept—it's *very difficult* for someone to envision a thirty-foot snowfall. It's even more difficult for them to accept a hundred and twenty feet, which is Thompson Pass. And what is very interesting is for them to have taken surveys through some of these areas with snow on the ground. *How much snow?* And the guy says, "Well, there's never been more than six feet in the Sierra Nevadas,* so that's probably what it is." *And that six feet contributed to some hundred-foot errors!*

Even if George's explanation is more symbolic than literal, I think it is true: a marine architect as well as an engineer, he had designed and built tanker terminals and pipelines on Cook Inlet before taking on the state assignment, and that experience in fact contributed a number of refinements to the work at Valdez. And unlike everyone connected with Alyeska, he had no motive for concealing Alyeska errors.

There were other difficulties at Valdez revealed in the course of construction, but this one, if large, is representative. Among those the designers knew about beforehand were some connected to the site's advantages. The Port itself, for example:

---

* Fluor, which did the design and construction work at Valdez and some of the preliminary investigations, is a Los Angeles firm.

deep enough to accommodate the biggest tanker yet built or planned—and too deep, in fact, for any ship to anchor safely anywhere inside it; too deep also, alongside the terminal, for conventional docks set on piles driven into the sea floor. The docks needed for loading oil are light, however—they have not much to carry but pipes, loading arms, pumps, their own weight —and three of those planned to handle the traffic at start-up could be anchored laterally into onshore rock or the steeply shelving bottom near the shore; a fourth berth would be similarly anchored but floating. (Space was left for a fifth berth to be built once Prudhoe Bay reached full production.)

The pristine character of Valdez and the waters to which it gives access meant another kind of complication. Since the decline of the inland copper mining that brought the town into being, there had not been much traffic except small commercial fishing boats and an occasional freighter or oil barge with supplies for the town's thousand or so permanent residents. The local fishery is rich and connects with the Gulf of Alaska, the source of much of Alaska's wealth since the Russian days. Just how the multitudinous life forms of these waters relate to and depend on each other, how each or all might be affected by changes in their environment—even quite small oil or other discharges, slight rises in water temperature that might result—are among those things about which little is truly known and were therefore the subject of much strong rhetoric in the lawsuits and the congressional hearings and debates. A tanker is a one-way ship: it carries oil from source to destination and travels back empty. To keep from capsizing it must therefore be ballasted on the return voyage—sea water is pumped into its oil tanks, discharged when it is safely near its mooring. Some of the oil, however, clings to the sides of the tanks when they are unloaded, mixes with the water ballast, and goes overboard when the ballast is lightened for reloading—estimates of how much vary from 0.5 to 1.5 per cent. In a moderate-size supertanker of 165,000 dwt, with a capacity of over a million barrels of oil, this film might in theory amount to as much as 17,000 barrels. (In practice, no tanker would probably ever be ballasted to capacity.) With the world-wide growth in the oil trade and the tanker traffic that carries it, this routine discharge—apart from accidents and the thousand varieties of human error—adds up to huge quantities of oil every year: not, perhaps, the biggest part of the total but, being generally near shore, a very visible part; and though the exact effects of oil along a continental shelf, the nursery and source of most commercial fishing, may be argued, it seems only common sense that they are harmful and perhaps greatly so. And hence the rhetoric.

The answer was serious enough to be written into the Right-of-Way Agreement signed by Alyeska's oil-company owners: a prohibition against any discharge of ballast by tankers serving the terminal; and a requirement for Alyeska to receive and clean all ballast water from the inbound ships. What followed was a system like that designed for Arco's small Cherry Point refinery completed in 1971 (see pages 108-9) but much enlarged. At start-up, there would be three big tanks (two more later) to receive the ballast pumped from the tanker berths—250 feet across, 48 feet high, with a capacity of 430,000 barrels each. There the ballast would stand for several hours while the oil floated to the top, to be skimmed off and mixed with the supply coming down the pipe from Prudhoe Bay; the remaining water would then move to what was effectively a sewage plant for chemical treatment and further skimming until it met the contractual standard (ten parts of oil per million of water, and there were complaints it was too much); whence it would move by stages into Port Valdez. The final test of purity before the ballast left the plant would be a fish tank stocked with local species: if the fish died from a sample or the water failed any of the earlier tests, it would be pumped back to the holding tanks and the cycle would begin again.

Storage tanks for the oil are the essential facility of an oil terminal: enough capacity to absorb the ebb and flow of oil shipped through the pipe, to supply a mixed fleet of tankers controlled by weather and contracts. The tanks planned for Valdez would hold about eight days' production from Prudhoe Bay—18 tanks at the start, rising to 32 at full capacity. Each would be the same diameter as the ballast-water tanks, 250 feet, but taller, 62 feet—510,000 barrels each. They were made of 8-$\times$-30-foot sheets of steel, 1¼ inches thick at the bottom, tapering toward the top as the pressure to be contained diminished—nine or ten tons each; the steel plates were bent to shape at the mills in Japan (but a Chicago firm got the contract for welding them into place at Valdez).

Again and more seriously, spilled oil was a concern. The tanks were grouped in pairs surrounded by a reinforced-concrete wall high and strong enough to contain somewhat more than their entire capacity. Snow was a problem—the weight. The roof of each tank, more than an acre, would have to support ten or eleven feet of snow at any one time, until the oil's heat melted it. To facilitate runoff, a roof design in the form of a shallow cone was used, adapted from one George James had tried on Cook Inlet, supported by interior pillars; the steel-plate walls rested on rings of concrete poured on the leveled foundation rock.

And a final necessity. Inside the tanks, unless filled to the top, light vapors rise from the crude and, mixed with oxygen, are explosive. The usual precaution is to float the tank walls so they rise and fall with the oil contained and leave no space inside for air, but that solution was precluded by the earthquake threat. Or, where natural gas is produced as a by-product of oil, it can be pumped into the void at the top of the tank and, if sealed from air, will not ignite; but there would be no gas—the oil men were postponing the production of their gas and going to expense to extract it from the oil, return it to the ground (and had apparently decided at the beginning that the gas, when produced, would be piped through Canada, not Alaska). A system was devised to exclude dangerous oxygen: burned-out gases from the stacks of the power plant, needed to run the pumps, the control center with its elaborate communications system, would be retained, cleaned, compressed, and pumped into the tank tops; and the tanks and their charge would be safe from hazard—or, anyway, perhaps, from the casual smoker up on the roof in winter, clearing the snow.

Since early in 1969, the oil companies had been insisting that, if they got the right-of-way permit, they were ready to start building the pipeline in the spring; then next year's spring, the next. Construction seasons came and went. The question turned to Congress and the Pipeline Authorization Act passed its decisive vote in July 1973 but produced no visible reaction from

Alyeska—no schedules announced, contracts let, men hired to do the work. The design of the pipeline and terminal, in the hands of consultants relying on sketchy information, was, as we have seen, still far from complete. Admittedly, after four years of unintended delay, caution might have been in order: despite the slender deciding vote in July, there were months of congressional procedures to be gotten through before the entire and final act could be passed and, in November, signed by the President; there was also a tangle of court action to unwind before the Department of the Interior could issue its authorization—it took until January 23, 1974, to draft the agreement and get it signed. And in pipeline work, no design can be final until the entire route has been dug and earth's last surprise revealed. Nevertheless, with these allowances made, when the red light finally turned green Alyeska was unready in more fundamental ways.

Vic Fischer is an economist who after government work in Washington turned to teaching at the University of Alaska and for several years has headed the Institute of Social, Economic and Government Research within the university: a compact, bespectacled chain-smoking man, thinning dark hair, muscular forearms coming out of rolled-up shirt sleeves—an Alaskan intellectual, content in the state's frontier simplicities, connected to the world Outside through the fluent catch and parry of inquiry and discourse. Professionally and as an interested Alaskan, he had watched the pipeline's progress from the beginning. In 1969, at Fairbanks, on the eve of the state's big Prudhoe lease sale, he organized a conference on the future of Alaska—scientists, politicians, oil men, conservationists—that raised a number of the awkward and fundamental questions TAPS had shown no interest in considering and which would form the substance of the pipeline debate. In 1971, for the Secretary of the Interior, he and two colleagues prepared an independent assessment of the economic issues surrounding the pipeline, published as *Alaska Pipeline Report*. About that time, work had taken him to Bellevue for a visit to Alyeska's offices, and—

I was just *flabbergasted* by the extent to which Alyeska— Patton and his vice-presidents—had no concept of how they would build the pipeline—the programming of construction, getting the job done. By then the pipe had been ordered, camps had been established at various places, and we saw, when the permit was finally given—Alyeska *still* wasn't ready!

The programming of construction: there is a logic to building a pipeline as to any construction project, a necessary sequence of steps each of which must be gotten through before the next can be taken. So: survey and stake the route, assemble the men and equipment and pipe, open the ditch, weld the pipe joints together in sections, lower the pipe in, cover it; and so on. Each step involves specialized men and machines, distinct crews more or less self-contained. Since money and time, men and their equipment are not unlimited, it is necessary to divide the job into units, move the crews through them in their proper sequence; and since time is part of the equation and the job spreads physically along an 800-mile route, you further divide it into sections so that at any one moment the logical sequence of steps is occurring simultaneously at many points. The principles are simple enough; getting the whole multiplicity of steps in the right order—programming them—is a matter of considerable complexity, the more so as the project increases in size and cost.

Ed Patton and his men were generalists, managers: their experience of the oil business, like his, had included among other things overseeing a variety of construction projects but no intimacy, apparently, with the building of a pipeline. Watched from above by the oil companies' management committees, they would in turn organize construction through a series of specialized consultants, contractors, subcontractors, all in turn with their own dependencies and suppliers—eventually hundreds, perhaps thousands, of such relationships; although more complicated, this was the same kind of arrangement the oil companies had adopted in designing the pipeline (as distinct from the early feasibility studies and general planning). In both cases, the relation of the Alyeska managers to the actualities of construction, the physical happenings along the line, was and remained strangely remote, indirect, through multiple layers of contractors and subcontractors, each watched from above, watching in turn and in touch with, managing, the next level in the sequence. Those within Alyeska whose specialized competence connected directly with the work—and gave them, therefore, immediate and final responsibility for what happened—were few.

One thing apparently on the Alyeska managers' minds in that summer and fall of 1973 was cost. In October they were able to announce a new estimate—the pipe, they said, would now cost between $3.1 and $3.5 billion—or perhaps as much as $1 billion more; disquieting uncertainties. When Alyeska applied to its owners for $28 million with which to take the first steps toward starting construction in the spring, the oil companies were close-fisted: Arco and Sohio between them contributed $5 million; the other major interest, Exxon (as it was now named), declined to pay just then, and the four companies with minor holdings followed suit. This late reluctance is puzzling, though it may have been a sign of private negotiations aimed at reconciling a number of conflicting interests. BP and Sohio were concerned that their merger agreement would be enfeebled if the pipeline failed to reach capacity by the specified date; there seems also to have been some anxiety about how each of the participants would raise its share of the cost (and hence about the reliability of Alyeska's steadily rising estimates). What came out of this bargaining was a new agreement among the owner companies, the main terms of which came out piecemeal and months later, in June and July 1974. Originally, the pipeline had been scheduled to start at a capacity of 600,000 b/d, rise in about two years to 1.2 million b/d, and reach its 2 million maximum at an unstated later time. Now those first two phases were to be combined, so that the initial capacity would be 1.2 million b/d, a change effected by hastening the construction of pump stations—eight instead of five would be ready at start-up. Ownership was redistributed at the same time. Arco's and Exxon's shares were reduced (to 21 to 20 per cent respectively) and those of the four minor participants in proportion. Sohio's share, on the other hand, was raised, and part of it was then sold to BP, so that Sohio wound up, on the books, with exactly 33.34 per cent—and BP with 15.84 per cent; this exactitude is all the more curious when we recall that Sohio already owned, on paper, the company to which it transferred its share in the pipeline and was owned in turn (see Chapter 3). The practical effect of this shuffle was that Sohio, with just short of half the pipeline but financially the weakest of the partners, would be the major borrower of money to finance it.

After grappling with its cost estimates, Alyeska took a first step toward construction by appointing the Bechtel Corporation as its general contractor for the pipeline. "They called in Bechtel and they didn't even know what to ask Bechtel to do," Vic Fischer caustically observed. "Alyeska itself wasn't doing anything—except some macrodecisions that didn't have any reality." (Actually, it was help of the most obvious kind which the Alyeska managers, conscious, perhaps, of their limitations, had been demanding for years, but the oil-company owners had refused to authorize it until the Pipeline Act was final beyond recall.) Bechtel, based in San Francisco, is a big, privately held, international construction firm with a lofty regard for itself. (It hires its executives from among former cabinet members.) Bechtel would not, however, build anything. Its role would be that of CMC—construction management contractor, a supervisor of all the other contractors, responsible to Alyeska. Other contracts followed, in an accelerating tempo.

As the project evolved, the first step was the building of the long-delayed haul road beyond the Yukon. In the winter, another ice bridge was laid across the river and 33 tons of supplies were tractored north to the seven construction camps that had been in cold storage since 1970. In February, a Pennsylvania firm, Michael Baker, Jr., got the contract for the engineering of the road (from an office in Fairbanks it had been doing odd jobs of mapping and surveying for three or four years), and a few days later Bechtel was authorized to begin the "mobilization" of the work force—although the road-building contracts were not let till April 5th, for $185 million. Bidding on the pipeline was opened at the end of March, and contracts were announced on June 12th. By now the route had been divided into five sections ranging in length from 127 miles to 225 (the longest from the continental divide in the Brooks Range north to Prudhoe Bay): the whole line was too much for any one company, and each of the five contractors was in turn an association of two or more firms formed for the purpose.

Since the fall of 1973, Alyeska had been assembling men with a sense of the specifics of construction rather than the generalities. Among them was Frank Moolin, a round-faced, muscular engineer who had managed the building of a big refinery in Singapore, done structural engineering for BART, the unlucky San Francisco subway system, and, when that was finished, was hired by Arco and assigned to Alyeska. (He would shortly be promoted to senior project manager—Alyeska's finger on the pulse, distrusting computers, driving up and down the work pad keeping track of progress in a pocket notebook.) What Frank Moolin found when he joined Alyeska in September 1973 was much the same situation that had astonished Vic Fischer two years earlier—the absence of most of the necessities for any large construction project:

> We started to build the job before we had everything we needed to build the job—the beds and the governmental relations, equipment, spare parts, all those sorts of things that are normally available when you start any major undertaking. . . . You need communications, you need warehouses, you need camps, you need beds. We didn't have the construction equipment—we taxed the ability of the industry in the Lower Forty-eight to supply it. . . .

At first, good-naturedly, Alyeska tried to buy its supplies locally, with the result that stocks of things like batteries, pillowcases, four-wheel-drive light trucks were cleaned out in Fairbanks and Anchorage and took months to replace. None of the major contractors had enough equipment of their own and were therefore competing among themselves for bulldozers, earth movers, graders throughout the United States. Alyeska took the first of a series of decisions that the situation seemed to require, assuming its contractors' responsibilities, with a considerable effect on its costs: it would supply all the equipment on the job, stockpile the spare parts, provide maintenance, making no disinction between what it owned and what the contractors brought with them. This amounted to about 15,000 pieces of equipment (eventually 18,000) worth $400 million, of which about one third belonged to the contractors, while Alyeska had bought another third and leased the rest; and correspondingly huge inventories of everything from nuts and bolts to truck tires and complete engines in dozens of sizes. The mechanism for this large undertaking was, naturally, a subsidiary, the Alyeska Equipment Company, a separate consortium organized by Alyeska's owners.

As the contracts and purchase orders poured forth, Alyeska needed a base for the day-to-day management of the project and found it in some disused barracks and acreage at Fort Wainwright, adjoining Fairbanks, which in April 1974 it rented from the Army and began refurbishing. (Ed Patton and his executives by now had transferred from Bellevue to Anchorage but would go no closer.) On all sides—among contractors, suppliers, the ordinary people of Alaska looking for jobs, new business, within Alyeska—there was a kind of wartime exhilaration, a sense of commitment to action that had at last silenced, if it did not answer, all the indeterminable questions of motive and probability. From his new offices in Anchorage, Ed Patton observed what was happening through memories of his five years in the Navy in World War II:

> This thing has to be viewed almost like the Normandy invasion. There was a whole lot more material than they needed, you knew you had to get those guys there and you knew that once you'd committed yourself to that invasion, you had to go whole hog—to save as many lives as you could. You had to get your foothold and you had to expand this foothold, and then you had to go into a mature operation of winding up the war. Now our foothold was obtained in 1974, when we got the road built. . . .

Work officially began on April 29, 1974, four weeks after the contracts for the haul road had been signed, when the first Cat bladed the first scoop of half-thawed earth. Clearing, grading, gravel mining and hauling in millions of cubic yards, compacting, bridge building, and culvert setting: by the end of September all the gravel was down, and by mid-November the job was finished, another Yukon ice bridge had been built, and a small party of Alyeska officials had made a ceremonial run to Prudhoe Bay in a four-wheel-drive station wagon.

In October, Alyeska issued its most precise and detailed cost estimate so far. The price of the pipe, it said, now added up to $5.982 billion.

Chapter 6

ON THE LINE

> It's a production line in reverse: instead of the product moving by you, you move by the product, and it requires all these things to drop into place. If one thing is missing, that whole crew sits on its ass, and you go out there and you see hundreds of guys sprawled around.—Frank Moolin

It is about five in the morning and we are near the Yukon, a hundred and some miles north of Fairbanks but still another long day and morning from Prudhoe Bay. In the night the sun in a haze of golden twilight slipped to the hilltops that gird the south bank of the river but did not set and now it is on the rise again, a polished disc of copper, bright and warm, the sky light blue and clear. It is the bad time of a night without sleep: if sleep takes you it will take you now, but I resist. Except along the highest ridges this rolling country is still wooded—tawny aspens in clumps, straight thin pines hugging their drooping branches, widely and regularly spaced like pegs stuck in a board; among them and along the edge of the deep ditch beside the road the earth is dotted with the flowers of the Alaskan spring, blue of cornflower and red of poppy, translucent pink wild onion, rare puffs of Alaska cotton: in places the raw ditch itself is sprayed with the bright green of new grass, unnatural among the dusky colors of the subarctic and already washing out—the beginning of the restoration and revegetation program imposed by the Stipulations under which Alyeska is building its pipeline. As momentary sleep half-seels my eyes, I see visions: red spheres glowing back among the trees, the shapes of low buildings that would mean a work camp. I force my eyes open, the shapes dissolve, and there are only trees. I try to light a cigarette but my hand shakes with the vibration of the truck, a ten-wheel Kenworth with a sleeping compartment behind, tail-heavy with the oil tank it is towing, the front wheels riding up and chattering over the washboard ruts of the worn gravel road—I cannot make the flame connect with the end of the cigarette. It is not easy to talk, not just the roar of the diesel under the square blue hood in front; your voice shakes.

Beside me in the cab Bill Granger works the truck down through its dozen gears and stops and you feel the thrust of the oil in the tank, sloshing forward and rebounding, an awkward five- or six-thousand-gallon partial load of low-grade oil, running north to lay the dust clouds on the summer road; that will be his job for the next month or so, hauling oil between the northern camps. The Kenworth is only a few months old and looks new, but the clutch is shot and Bill has been learning to shift without it, doing better now, with less grinding of gears. He tests the brakes with a pneumatic hiss: we are at the top of another long grade, 10 or 12 per cent, this one with sharp curves, doubling back, running down toward the Yukon. We are past the Steese and Elliott highways out of Fairbanks, some of that a mining-camp trail laid down in the 1920s, which the state does not exert itself to keep up—rough and worn, barely wide enough for two of these rigs to pass, and in the winter, with the snow graded into the ditches it is easy to go off; on the steepest grades Cats were stationed halfway up to catch the trucks as they spun out and tow-chain them to the top. Bill remembers and as we pass points out the slopes where tractor trailers and an occasional car or supervisor's light truck have gone over, rolled, burned.

Eighty-five miles north of Fairbanks the road improves—a little wider, better maintained: the Yukon Highway as it is now named, built in the fall of 1969 anticipating the pipeline and an Alyeska responsibility, but technically open to the public; a sign where it begins warns those who insist on seeing the river of heavy truck traffic, wide loads, dust clouds, and archly concludes, "Drive *Carefully* and Have a *Pleasant* Trip!" In the night a woman's voice has been among the voices coming over Bill's

CB radio and we finally catch her at Livengood, a gold camp evolved into a base for a state road crew, where the Elliott Highway swings west to Eureka and Manly Hot Springs on the Tanana, seventy-five miles off: a man and woman in a Toyota station wagon going for a visit with relatives (this road is not among those kept open in winter). We pull over to drink a thermos cup of coffee, talk. There have been no other little cars, and Bill is grateful: one more anxiety added to the others.

Bill tries his brakes, turns to me: "If they lose pressure, a line breaks or something—if they don't hold—I'll tell you to jump, O.K.? And you get out that door fast—you only got a second or two before this rig's running too fast to do it." He is quiet, matter-of-fact, merely conscientious—I am a passenger, a guest, one more anxiety. I try the door handle; the passenger seat is a five-foot climb over the right-side fuel tank—dive for the ditch, tuck your head, roll; a rough landing if you have to do it. He eases the brakes off and the truck moves forward and down the hill in its lowest gear.

Transportation is the first necessity of any big construction project, and never more so than in Alaska where, except for the mountains of gravel and crushed stone, all the raw materials, every piece of equipment and the fuel to run it, every plate of fabricated metal, all the food for the workers, had to be produced elsewhere and brought here. The word for this process is the properly military term *logistics:* an adapted French word that means first of all the lodging of troops and therefore the gathering and distribution of the food and supplies they need to fight, but also with an implication of doing these things in a logical sequence—delivering the necessities to the right places at the right time and in the right order; management. In both size and complexity, the logistics of the Trans-Alaska Pipeline was the root difficulty from which all others derived. My passenger's run from Fairbanks to Prudhoe Bay with Bill Granger represented one tiny link in this immense chain.

Altogether, the project required about 3.5 million tons of supplies and equipment, a lading that, if moved by truck, would amount to 180,000 twenty-ton truckloads and form a convoy stretching bumper-to-bumper from Chicago to New York: the total transportation bill came to around $1.5 billion, perhaps one fifth of the total cash cost of the pipeline, depending on just how that is calculated. As a matter of cost, at one fourth the rate for over-the-road haulage when the choice existed, sea-going barges, not trucks, were used whenever possible, starting with the pipe itself—120 bargeloads delivered from Japan on specially built barges from the fall of 1969 on and distributed to storage yards at Valdez, Fairbanks, and Prudhoe Bay. (The pipe destined for Fairbanks was unloaded at Seward or Whittier on the Kenai Peninsula and hauled north by rail; for the purpose, the Alaska Railroad rounded up 130 flatcars in the Lower 48 and barged them up from Seattle.) To get around the scarce and expensive Alaskan labor and the short season for outdoor construction, Alyeska, when it could, assembled the components of pump stations and the Valdez terminal in the south—at Seattle by preference but also at more distant sites ranging from Chattanooga to Houston to Los Angeles and San Francisco—and brought them north by barge; for the larger production installations at Prudhoe Bay the oil companies adopted a similar strategy and were therefore at the mercy of the two weeks toward the end of August when, in most years but not all, the pack ice moves far enough off Point Barrow to let tugs and barges through and back again.

Barge shipment, though cheap, is slow on route, slower to organize and carry through. In emergencies and whenever time was a vital factor, delivery by air was the solution—the workhorse Hercs, at ten times the barge rate per ton from Seattle. In the first two construction years, 1974 and 1975, air cargo ran at better than 80,000 tons per year.

Within Alaska, the main transportation job, getting the right materials to the right places at the right times, was necessarily done by truck. At the peak, the job included nearly 300,000 gallons of oil a day to heat and power the camps from which the pipeline, the pump stations, and the terminal were built; correspondingly huge quantities of fuel and lubricants to run the trucks and construction equipment (the standard for Arctic gasoline is —60° F.—a little of this special blend was produced at Arco's Prudhoe Bay topping plant but most had to be brought in from Outside); most of the hundred or so tons of food the men consumed every day; and, of course, all the pipe. To save time and money in moving the pipe, Alyeska prewelded about three quarters of what was actually used: in buildings erected in the pipe yards at Valdez and Fairbanks, forty-foot lengths were welded into eighty-foot "double-joints" by automatic welding machines, immune to the Alaskan weather and the unpredictable temperament of pipeline welders. The resulting eighty-foot lengths were then hauled—three at a time, thirty-six tons, fourteen thousand such loads altogether—on a rig adapted from the logging trade in the Pacific Northwest: at the front, the joints of pipe were supported on a cradle mounted behind the truck cab, at the rear by a ten-wheel dolly connected by wires to the tractor so that, in theory, the rear set of wheels would track around curves as the driver steered—an arrangement that in the killing winter cold proved as shaky as it sounds. In the earliest stages, haulage beyond the Yukon (see Chapter 3) was limited to the four or five months of a winter ice bridge. By 1974, when construction officially began, the summer months were filled out by shallow-draft river barges and tugs at the crossing, with gaps for break-up in late spring and freeze-up in the fall. The following year, the season was extended by bigger barges winched across on cables, riding the surface of the river on cushions of air, out of reach of drifting ice and snags; by the fall of 1975 the river was spanned by a 2,300-foot bridge which eventually would carry the pipeline itself as well as trucks, a two-year joint project of Alyeska and the state but actually built by the state highway department so as to recover the $27-million cost from federal funds.

It was the importance of trucking in the building of the pipeline that made Jesse Carr and his Teamsters, as it is said, the most powerful force in Alaska, and not only because Teamsters drive trucks and can, if they choose, refrain, but because the union's members—in full they are the International Brotherhood of Teamsters, Chauffeurs, Warehousemen and Helpers—do most of the other unskilled and semiskilled jobs that concern logistics. It was for the same reason that, watching the pipeline's progress, I wanted to have a look for myself at the reality of Alaskan trucking. By the time I reached that point, in early June of 1976, the intention turned out to be difficult to fulfill.

Alyeska was becoming restive under the scrutiny of newsmen,

photographers, and congressional investigators (see Chapter
9)—not to mention writers. The camps and the road beyond the
Yukon, though built on public land, were private property in the
company's view, and it was not happy with people going there
for no other purpose than looking. After months of persuasion I
was offered a kind of Catch-22: I would be given a pass for the
haul road if a truck line would agree to let me ride one of its
trucks; the trucker, in turn, was amenable if I could provide
some assurance that Alyeska approved. I went out to the Sourdough Freight yard to introduce myself to the manager (one of
two lines still under contract to the pipeline company—there had
been twenty); back into Fairbanks to the Alyeska office to secure
the necessary piece of paper. This was in the form of a military-
style travel order but carried fewer privileges: it would get me
past checkpoints on the haul road; it forbade me to enter the
camps or eat their grub, on pain of detention and removal by
the security guards. There would be a truck heading north at
seven the next morning and, nervous of missing it, I did not
sleep much and was up early to pack up my camp and fit some
necessaries into a pack—sleeping bag and a small tent, a bag of
pemmican and some sandwiches, apples, water, camera, film,
and maps; even in summer in Alaska it is well to allow for the
possibility of being stuck somewhere, and in this case I had been
instructed to look out for myself.

I got to the yard within a minute or two of the time, but the
expected order had not come through from Alyeska, nor had a
Sourdough driver bid the job. It was late afternoon before the
two came together and the dispatcher introduced me to the man
I would be riding with: Bill Granger, after a year of driving fifth
in seniority among eighty others and therefore with all the hours
he wanted, eighty or more a week, up to 5,000 a year, which, at
the current $12.48 an hour, with several rules for time and a half,
is maybe $70,000; and, at thirty, lacking wife and kids to give
form to ambition, no goals more specific than a once-in-a-while
Hawaii vacation, some Fairbanks land for investment, and an
early retirement under a Teamsters' plan based on hours per
year. No man, if he can, works except of necessity. Before coming to Alaska he had driven lumber trucks in Washington; before

that, with his father, fished for salmon till the catch declined below a cash living.

We went out into the yard to inspect the unfamiliar Kenworth. The tanker was already hitched, run in winter with four inches of foam insulation, now starting to peel like a molting skin; Bill knocked the tires with a mallet, squatted to look and, finding one that was delaminating, backed the rig into the garage for a replacement. He loaded a suitcase into the sleeper behind the cab, I hefted my pack, and he plugged in his radio—Sourdough supplied what the men thought an inferior make, and they provided their own. Under his Alaskan flannel he wore a T-shirt someone had given him with a motto silk-screened across the chest: CB'ER'S DO IT ON THE AIR. We were ready to start.

At Fox, a few miles north of Fairbanks, we stopped for a check of Bill's manifest: a motherly Alyeska woman in round glasses who threw up her hands at my unfamiliar pass and waved us on. At the Hillside Cafe, new-built by a German proprietor, he pulled into a raw clearing among half a dozen Sourdough trucks and we ate at a truckers' table, five dollars' worth of hamburger, sauce, french fries, and coffee, and, for me, a bottle of beer. Before we left Bill bought six quarts of Seagram's from a pocket roll of twenties. There were rules against liquor in the camps, and it would be a while before he was back this way.

Beyond the Yukon bridge, at 6 A.M., we turn off for a security check, and the Nana sergeant, with dark glasses, badge, and powder-blue uniform, has never seen a paper like mine. (Nana is an acronym for Northwest Alaska Native Association, which under minority preference obtained the security contract for the northern half of the pipeline; its employees, more or less Eskimo, are reputed simple but honest.) The sergeant worries: he produces a week-old duplicated order against intruders, signed by Ed Patton; maybe he will have to send me back to Fairbanks. I suggest that he get Ed Patton on his radio—I offer to talk to him; I mention oil men in Anchorage, Houston, Dallas, Los Angeles, New York. The sergeant thinks it over, returns my paper, and waves me back to the truck. I am waking up.

At Old Man Camp, six miles below the Arctic Circle, Bill pulls off again to refuel and eat breakfast. Playing by Alyeska's rules, I sit in the cab and eat an apple, wander through the yard looking at stacks of truck tires, whole engines on skids; Bill's white-haired father is a mechanic, working in the camp, and Bill greets him profanely, like a trucker, though with me he is carefully clean-mouthed. A bearded college-boy Bechtel supervisor finds us and gives Bill a pep talk—for God, for country, and for Yale—about meeting his schedule, and Bill coolly reminds him of the contractual rest period, not yet taken. Thirty miles farther, at the top of another long grade, we pull into a truck park which on a public road would be called a scenic overlook: it faces a west-ward valleylike fold of hills laced with streams and tiny ponds a thousand feet down, spread out like a topographic map. Tundra now, underlain by permafrost, the thin active layer carpeted with low shrubs that from this height and distance look like velvety lawn—beautiful walking because from the hilltops you can see your route for miles in all directions, though the chest-high brush is more strenuous than it looks; we are past the treeline, not yet in sight of the mountains. Bill crawls into the compartment behind the cab to sleep. With my pack I hike down the slope, find a level bench, and set up the tent—the heat is bringing out mosquitoes, in the night it was cool enough to use the Arctic heater in the cab but since breakfast we have been running the air conditioner. Now, glowing under the high sunlight, the tent, even open at both ends, is a sweat bath, and I strip. As we pulled off, a pair of red foxes watched, vanished over the rim of the slope, and I mean not to sleep, only rest for a little and then go and look for them: intensely curious animals and, except for being often rabid, harmless; here they are drawn by truckers' leavings and, despite penalties that include loss of job, are probably fed. Sweating, skin cooled by a faint stir of breeze, I close my eyes against the glow of the tent roof. . . . It is four or five hours before I wake to Bill's voice, calling: Time to go on. The foxes have not come back.

As we follow the road north, I find that the country is too big and too new for my eyes, I see it at large and in general. Bill's sight from the months of making this run week by week has adjusted to the scale and he sees first and points out details as we pass: a bear shape outlined on a distant ridge, a place where an undersized culvert has collapsed and the road has washed out in the spring runoff, another where a long string of pipe was marked for rewelding, then buried before any repairs could be made—the pipeline builders have no secrets from the truckers. A grizzly ambles beside the road where the ground has been staked and cleared for the gravel work pad from which the pipe will be laid: a tawny-golden hump-shouldered two-year-old, five hundred pounds of adolescent playfulness and unpredictable ferocity. The bear moves off, trotting, looking back nervously over his shoulder: this thing, which he has perhaps never seen before, is too big to challenge. Later there are three of them, honey-colored and huge, digging for scraps in the smoking layers of a camp garbage dump. They are playing, they stand up, come together as in a dance, shove each other over backward, bounce as if on a trampoline, roll. There are dozens of bear stories: bears that break into camp kitchens and demolish a tableful of fresh-baked pies, get into an office and have to be fire-hosed out. Those that persist or turn savage are shot with tranquilizers, trussed, and helicoptered elsewhere but come back—their range is hundreds of square miles. For emergencies the camp managers keep a gun or two, locked up, with shells loaded with deer shot, but they are not used; private guns are among those things forbidden by the camp rules.

We have entered the mountains. Actually, we have been among them for hours now but from the south by this route the approach is gradual, you see them at a distance, ahead and on either side, as a darker, snow-washed upward continuation of the foothills. Then as road and pipeline climb the narrowing valleys, Koyukuk and Dietrich, the mountains close in, and suddenly, as it seems, at Dietrich Camp a hundred miles north of the Arctic Circle they are all around and near enough to hike to. Bill parks and goes into the camp office to recheck his routing and I wander around the yard, matching what I can see with the names on my map: Sukakpak and Wiehl and Snowden, to the north the Endecotts, a whole range, rising like a wall. Sheltered in the stream beds there have still been pines and along the road an occasional wide-eaved Alaskan cabin, roof weighted with live sod, nestled among them, signifying mining, trapping, to which their owners will perhaps return when the pipeline is finished; the rivers have been brimful with spring melt roaring over boulders and in my mind I have been canoeing them, hazardously and with frequent spills, but climbing toward their source along the continental divide we are beyond that now, the streams have spread and shallowed through many channels and are choked with spongy ice. *Rugged* is the adjective invariably applied to the Brooks Range but that is a journalist's word, seeing these mountain folds from a jet plane twenty thousand feet up. Down here the look is different: austere, unyielding perhaps, not threatening. The near peaks, four or five thousand feet, are carved and sheered in steeply tilted planes and knife-edge ridges, shaded in a narrow range of color from cast-iron gray to deep brown, creased with lingering snow. Below, their shoulders are rounded and softened by loose scree in the same somber colors, thinly covered by lichens and grasses in autumnal gray-greens and browns. The steep-walled narrow valleys between remind me of New Hampshire, but in skeletal outline, as it might have been ten or twenty thousand years ago when the ice sheet withdrew, before the trees came back. These mountains are still young, in process of formation.

Bill comes back from his errand and we return to the road, still climbing toward Atigun Pass and the continental divide, the high point on the pipeline route, 4,800 feet. The road winds steeply now, clinging to the mountainsides, carried on a corniche of the same dark rubble of which the mountains are made, and in places the grade reaches 22 per cent. There have been wrecks along here and a dozen deaths, not trucks but light planes and helicopters undone by winter gales and white-out and the long climb through thinning air; Bill points out the places where they lay through the winter and have only this spring been removed. Over the top and gearing down the widening valley of the Sagavanirktok River that we will follow to Prudhoe Bay, Bill slows the truck and stops, points: a solitary gray wolf, looking scrawny in his half-shed winter coat, heavily muscled in his hindquarters, broad-faced and narrow-jawed, not doglike. He trots beside us when Bill eases off the brake, stops when we stop, and the intelligent eyes appraise us. He is a familiar at this point, hoping to be fed. We drive on.

By midnight we are out onto the coastal plain, patched with snow and shallow ponds, and the mountains are an abrupt, steep-rising, snow-topped wall behind us, but we are still eighty miles short of Prudhoe Bay. The next camp and its airstrip are visible miles off, like a mirage in the desert, and we steer for it— Happy Valley its unlikely name, which the men transform to Giggly Gulch. At the office, Bill shows his manifest, is assigned a room, and hands me the key—he is more comfortable, he says, in the sleeper—and I accept. He will see me at six for breakfast; in the chow line here, badge checks are made only at dinnertime and I will get by.

From the outside the camp looks like a trailer park set down in the middle of a gravel-covered shopping-mall parking lot: dozens of trailer-sized modules laid out in the geometry of dominoes on a board, linked by enclosed wooden walkways, raised off the frozen ground on pilings; this was one of the seven original camps built in 1970, with low ceilings in the modules, scaled to fit the bellies of the Hercs. My room is the standard eight by fifteen feet, single bed at either end, a table and chair between under a window that looks out on the soft late-afternoon light of the Arctic summer night. I treat myself to a shower but am too keyed up to sleep. I find my way to the dining hall where coffee and soft drinks, sweet rolls and doughnuts are set out on a long table and help myself. The feeling is of a raw and utilitarian college campus where all the courses are practical, the men and a few women mostly young and hairy, on their own, with defined and circumscribed purposes. On a corridor wall a verbose handbill signed by the Alaskan Committee for Democratic Use of Resources denounces a forthcoming sale of offshore oil leases. Most of the pipeliners work a single ten- or twelve-hour shift, but a few of the bedroom doors are marked for day sleepers, mechanics, oilers, and camp maintenance people who work through the night.

In the morning at breakfast my plate is heaped with bacon, scrambled eggs, sausage, fresh toast, one installment in the four or five thousand calories a day that is normal in the camps. Serving the food—and washing dishes, making beds, sweeping—is the job of a *bull cook*, an Alaskan term for a lowly job that comes down from the mining-camp days: capable of cooking a bull but nothing else. Alyeska public relations lore abounds in chefs with *cordon bleu* diplomas from London and Paris, though

I have eaten one way or other in most of the camps and have not met them; but the food is there, in plenty and variety. The cost for food and housing has been quoted to me at anywhere from $25 per day to nearly $200; like other Alyeska costs it is elastic, varying with the speaker and his purpose.

The final leg to Prudhoe Bay is a straight and easy two and a half hours. Dust clouds swirl at intervals for miles ahead, enveloping other trucks, road crews; beside the road, completed pipe marches on its stiltlike supports, like a causeway crossing a frozen seascape. As we leave the camp, a ptarmigan flies up clumsily from the tundra beside the road, pure white still in its winter dress. Later, in a level, ice-lined bend of the Sag, there are a pair of caribou, grazing, heads down and white-rumped, indifferent to the machines rumbling along the road above them.

The outer limit of the Prudhoe Bay field is marked by a lift gate across the road and a guardhouse, and we stop. The flat horizon is rimmed with faint signs of the work of development: drilling rigs at wide intervals, a squat clump of buildings that may be the permanent Arco base camp, a little nearer, to the southwest, the control tower of the state airport, which is called Deadhorse. This time it is Bill's papers that puzzle the Nana security man and he takes the time for several radio calls to verify routing and destination. While we wait, I jauntily offer to deliver the oil and pick Bill up on the way back, and they laugh; I have learned by now enough respect for the skill to know that I could as well fly the rig to Prudhoe Bay as drive it.

The Sourdough manager in Fairbanks had suggested that I would prefer to fly back—others who had gone as passengers mostly did not want to make the return trip by truck—but I insisted I would make the run both ways and he did not argue. Now that is looking a little complicated. Bill will not be starting back immediately and then will only go part way. Connecting with another truck bound for Fairbanks may be difficult; and I am getting short of time—my son is due in Anchorage in two days and I must be there to meet him. It is thirty-nine hours since we pulled out from Fairbanks and, allowing for stops to eat and sleep, we have covered the 502 miles at a speed of about 20 miles per hour. I will have to fly back.

Bill drops me off at Deadhorse. We say good-by, shake hands, and I take a last self-conscious picture of him, standing beside the truck. The midmorning jet is leaving in less than an hour and I check my pack, sit down to wait, drink coffee, and page through a two-day-old *News-Miner*. In another hour and a half I am on the ground in Fairbanks and telephoning home to wish my son a good flight north.

On November 18, 1974, as a new ice bridge was being laid across the Yukon, the haul road was officially complete: 35 million cubic yards of gravel, twenty bridges, in 154 days; 361 miles at a cost of something over $500,000 per mile. Henceforth, as in construction projects in most other parts of the world, supplies and equipment could be trucked where needed, subject only, until the bridge could be finished, to the awkward barge or ice-bridge crossing of the Yukon. The road had another meaning as well: for the men who would work there, the Arctic world beyond the river had become in one jump an appreciably less hazardous place to be than in the early days of route planning and surveying. Those working on the pipeline would still be indoctrinated at the start in Arctic survival and outfitted with $500 worth of suitable clothing and equipment bought from Alyeska (and Alyeska's own people would be put through a more elaborate winter survival program at a mountainous ski resort a few miles south of Anchorage—Frank Therrell, as an old hand on the project, was among the instructors), but they would have little call for such skills. It was a fact about which those on both sides of the pipeline debate could agree, though drawing opposite conclusions. For Robert Larson who, as Alyeska's safety manager, was responsible for the lives of all connected with the project, the change was simply one that made his job a little easier:

We have a rather well-defined construction zone, a road north where there wasn't one before, traffic going by with regularity. It is still possible for people to get stranded, because of white-out conditions, for instance, but the period that you've got to tough it out is now hours rather than days or weeks. In essence, the wilderness is gone; we're working in a crowd out there.

Although indispensable, the haul road was no more than a preliminary first step toward "getting the job to the point," as Frank Moolin put it, "where it was run like a normal job." As such, it was excluded from the Right-of-Way Agreement and Stipulations that governed the pipeline itself—in effect, Congress and the Department of the Interior had rejected the argument on which the original injunction had been based, that road and pipeline were one and inseparable: the Right-of-Way Agreement for the pipeline was signed on January 23, 1974, by the Secretary of the Interior and the seven oil-company owners; the road, which in theory would eventually be designated a state highway, was covered by a separate and simpler contract between the Department and the state.

Although the pipeline agreement runs to 30 closely printed pages and another 54 pages of exhibits, including the Stipulations, its essence is a single brief clause:

> Permittees shall submit construction (including design) plans, a quality assurance program, and other related documents as deemed necessary . . . for review and approval. . . .

The agent in charge of this process of review—and for seeing that the plans were carried out as approved—was to be an official known as the Authorized Officer, acting on behalf of the Secretary of the Interior. His powers to control construction would be sweeping and absolute:

> The Authorized Officer may at any time order the temporary suspension of any or all construction, operation, maintenance or termination activities of Permittees. . . .

In January, Andrew P. Rollings, Jr., a retired brigadier general in the U. S. Army Corps of Engineers, was appointed to the job and set about establishing an office in Anchorage. A few days later the Alaskan governor created a parallel position, the State Pipeline Coordinator, and moved a youthful engineer from Juneau to fill it: Chuck Champion, responsible for what happened on the state-owned land south of Fairbanks and north of the

Brooks Range, and for the state's rather different interests and objectives in the federal right-of-way elsewhere in Alaska. Under the agreement, the costs of approval and surveillance by the two pipeline offices would be borne by Alyeska's owners, along with the expenses the Department of the Interior reckoned it had already been put to since the pipeline proposal first came up, exactly $12,253,730.

The real start of pipeline construction was thus a flood of paper—plans, engineering drawings, written specifications by the hundreds—that in late May 1974 began flowing to the two pipeline offices from the Alyeska building on South Bragaw Street in suburban northeast Anchorage. Reversing this flow would take time—each step in construction would be authorized or modified by a Notice to Proceed (NTP in the evolving jargon), followed by a work order directed to one or more of Alyeska's contractors—but pipelaying was not scheduled to start until the next construction season, the spring of 1975. Meanwhile, a number of more general steps had to be taken, in anticipation. Double-jointing started at Valdez and Fairbanks within days of the first applications, and by January the welded pipe was moving out; so did work on construction camps for the southern half of the route. In June Alyeska let the contracts for drilling the holes in which to set the vertical supports that would carry the aboveground pipe.

All this sounds complicated, but Frank Moolin, when I suggested as much in his plywood-paneled office in a former barracks at Fort Wainwright, demurred. You do not in fact have to be a graduate engineer to figure out the common-sense sequence of pipeline building: deliver the pipe to the work site, open the ditch, weld the pipe joints, wrap the pipe with corrosion-preventing tape, lower it in, cover the ditch with gravel and earth. The complicating factor in this simple procedure was Alaska itself, and not only in the matter of logistics: it was necessary for the engineers, pipeliners, and construction men to abandon their accustomed ways and learn to do things differently. For a start, the pipe would actually be laid from a gravel work pad running almost the entire length of the pipeline route, in effect a second, parallel road twice as wide (54 feet) as the haul road and connected to it and to the public highways in the south by a further 120 miles of access roads—more millions of yards of gravel to be dug, transported, spread, compacted. This was done to protect the permafrost, "the environment," of course; but for the more immediate reason that in most places in summer the heavy construction equipment simply could not move without a four- or five-foot bed of gravel underneath. In the north, to minimize the quantities, the gravel went down on insulating sheets of polystyrene foam, the same stuff a backpacker carries in lieu of a mattress. In a few places, where the underlying bog would not support equipment even when topped with layers of gravel, the pipe was laid in winter from a pad of compacted snow—and when, one winter, the snowfall was light, it was necessary to provide artificial snow, using snow-making machines borrowed from ski slopes in the Lower 48.

The aboveground pipe was a further complication. Holes for the vertical supports had to be drilled in pairs, laterally spaced according to the designers' specifications and at intervals of about sixty feet. The VSMs, 18-inch-diameter steel pipes, were then lowered into the holes, and since the exact soil conditions could not be known until the holes had been drilled, it was frequently necessary to weld on additional pipe to make them long enough (and sometimes to cut them short). To hold the VSMs in place, the 24-inch holes were filled with a sand-and-water slurry poured like premixed concrete, which in time would freeze solid in the permafrost. Horizontal support beams could then be welded between the VSMs, the pipe, welded in long strings, was lifted high and lowered into place, and the sheets of metal-covered insulation could be wrapped around it and riveted.

In either construction mode, aboveground or buried, each of these steps was subject to multiple inspections: by Alyeska's contractors, checked in turn by Alyeska's own inspectors (this was the meaning of the quality assurance program—QA—required in the pipeline agreement); and by field representatives of the federal and state pipeline offices (see Chapter 8). In addition to the visual inspection and other physical checks, the Stipulations specified that all pipeline welds be X-rayed. Elsewhere, the U.S. pipeline code, which has the force of law, requires a spot check of welds by X ray, about 10 per cent. Here, the standard was 100 per cent, and it proved an almost insurmountable complication (see Chapter 10).

The sheer size of the project multiplied these difficulties. As we noted earlier, there were simply not enough construction equipment and spare parts to go around, a problem Alyeska solved by the costly expedient of taking the buying, leasing, and maintenance out of its contractors' hands. Although the Japanese firms were able to supply virtually all the structural steel pretty much as needed, the precision machinery at the heart of the system was mostly manufactured in the United States: 151 valves of various types inside the pipe itself, dozens of pumps and generators for the pump stations (and for the flow stations at Prudhoe Bay); and so on. Delivery dates receded once the contracts were signed: "Suppliers were promising us the moon," Frank Moolin sourly remembered. "Well, that's not the way it happened." So also with more usual equipment, such as the trailerlike modules from which the construction camps were assembled. Some efforts were made to insulate them against the Alaskan winter and finish them with fire-resistant materials, but the real problem was quantity. Altogether, Alyeska had to provide housing and food services—and lighting and heating, power, water, sewage treatment, communications—for about 25,000 workers spread among as many as 39 camps ranging in size from 300-man Atigun to the Isabel Pass camp in the Alaska Range, with about 1,600 beds; the dozen camps at the pump stations, each with a capacity of about 250 each; and the giant camp at the Valdez terminal, built up slowly over a period of 22 months, to a final size in early 1976 of about 3,500.

As it happened, the shortages and prolonged deliveries Alyeska encountered in 1974 as it attempted to set to work were vexed by a booming inflation set off, as it seems, by OPEC's experiments with the price of oil. Steel, for instance, went up by about 50 per cent in a period of eight months; the cost of energy, we are learning, is basic to most other costs. In September, the Federal Energy Administration offered some help by ruling that Alyeska's orders for equipment and supplies were subject to the Defense Priorities Act, which in 1950, at the outbreak of the Korean war, established a means of allocating manufacturing capacity but had not been invoked since then. In effect the ruling once more confirmed what Congress had declared in the Pipeline Authorization Act, that Alaskan oil was vital to the national interest. But although delivery schedules improved, the prices did not.

These preliminaries were sufficiently advanced so that in late March of 1975 actual pipeline work could begin, weeks ahead of the Alaskan season. The place chosen was the Tonsina River, about 70 miles north of Valdez. By April 19th the first 1,400-foot string of pipe had been welded together, lowered into its ditch angled below the stream bed, and covered over; and presently floated and had to be reset. It was a preview of most of the trials that were to come. The project by this time had passed through a series of what Frank Moolin called "constraints"—establishing construction camps (or anyway the rudiments), securing NTPs, gathering equipment and parts. What remained was "negative geotechnical surprises"—finding, when you opened the ground, that the soil was not what you expected or had predicted; another engineering euphemism. The practical effect of such a surprise was, repeatedly, the wrong equipment in the wrong place at the wrong time, crews of men (at a base rate of about $10 each per hour) drawing time while engineers and surveillance officers argued out a change in design.

When the "surprise" was unstable permafrost, as it usually was, there were two alternatives. The simplest change was known as realignment—moving the pipe route far enough east or west so as to avoid the patch of unsuitable soil. This was done so often in the course of construction that alignment within 200 feet on either side of the surveyed center line was approved on the spot by the state and federal surveillance officers; a greater change required new plans and, often, negotiations between Alyeska and the supervisory and technical staffs of the two pipeline offices in Anchorage. Although the right-of-way officially occupied by the pipeline was small—still the 54-foot band specified in the old Mineral Leasing Act—the actual acreage was thus considerably greater: a construction zone at least 400 feet wide and stretching from one end of the pipeline to the other. South of the point at which the first pipe had been laid under the Tonsina, the route was shifted a mile to the west of the original plan

for a distance of about five miles—not in that case to avoid unstable soil but to reduce the number of stream crossings.

The other solution was to change the "construction mode"—
that is, move the pipe from below ground to above, a change invariably resisted by Alyeska's managers on the grounds of cost.
It was disruptive as well as expensive—coming as it did unpredictably in the course of construction, it halted work until additional VSMs, support beams, and insulation could be delivered
to the site or, in some cases, ordered from the suppliers. Simply
drilling the holes for the VSMs became a major undertaking. Initially, when the designers were thinking in terms of no more
than 80 miles of aboveground pipe, Alyeska had contracted with
a single drilling company to do the work. As the length increased, Exxon's drilling department stepped in and supervised
the design and testing of a variety of more or less experimental
systems proposed by a large number of Texas drilling contractors. Although drilling is perhaps the most routine activity there
is in the oil business, the VSM holes went beyond the industry's
experience in several ways: they would be large in diameter, 24
inches, requiring a correspondingly powerful rig (at Prudhoe
Bay, where the wells are unusually big, the oil comes out
through 13⅜-inch pipe, supported near the surface inside
20-inch conductor casing); the rig, however, had to be
sufficiently portable to be moved around on the back of a truck;
the soils in which the holes would be drilled ranged from frozen
muck to solid rock; and the sheer amount of drilling to be done
was huge. By 1974, as construction was beginning, six contractors had been selected that could meet these requirements, using
three basic types of rigs: a large-diameter auger, a kind of per-
cussion pile driver, and the rotary drill-pipe-and-bit system normal in oil-well drilling.

At that point, the plan still called for 381 miles of the route, a
little less than half, to be aboveground. In the course of construction nearly fifty miles was added, bringing the total to about
425 miles. That translates to 76,000 VSMs and, at an average
depth of 50 feet, almost 4 million feet of hole. That, in turn, is
the equivalent of 400 oil wells of the 10,000-foot depth drilled at
Prudhoe Bay, where Arco and BP between them took eight years
to complete the hundred-odd wells that would be needed when

production began. By contrast, Alyeska would have just two years in which to accomplish four times that amount of drilling. It was a big order.

Early pipeline work, possibly even the first 1,400 feet at the Tonsina, raised disquieting questions about two fundamental aspects of the project: welding and the quality-control program. As early as March 1975, the State Pipeline Coordinator, Chuck Champion, was informing Alyeska of his anxieties about both matters, and his criticisms were echoed in the course of the summer by his federal counterpart, Andrew Rollins.

Quality control over most phases of pipeline construction is a quite straightforward matter: if the pipe ditch, say, is dug deeper or shallower than the design calls for, you can measure the error exactly and correct it without much difficulty. Welding is different: an art, not an exact science, and whether a particular weld is any good or not is a matter of judgment about which two experts may reach opposite conclusions; the only way of determining who is right is to pull it apart and see which breaks first, the weld or the original metal, and under what force. The 100 per cent X-ray standard imposed by the Right-of-Way Agreement redoubled this difficulty. To X-ray a pipe weld, you wrap it with a twelve-and-a-half-foot strip of photographic film, sealed off from light, expose the film to a source of radiation inside the pipe and develop it like any other film. If these things are done correctly, you then have a negative on which the weld shows as a pale, textured band against the uniform dark gray of the pipe steel; an irregularity, such as a void left by a gas bubble trapped inside the weld metal, should show as a dark spot against the lighter tone of the weld. Interpreting the film—determining that such dark spots represent defects in the weld rather than in the film or its processing and that, if so, they have practical significance for its strength and durability—is as problematical as making a visual judgment on the quality of the weld itself; an anxious art. The X-ray work, which like everything else was to be watched over by quality control and quality assurance, was divided between two companies: Ketchbaw Industries south of the Yukon, Exam to the north. Their contracts were announced on March 26th—one day before the work began beside the Tonsina.

On the Alyeska organization chart, the quality-control staff was part of the production department charged with all the steps involved in actually assembling and laying the pipe. Moreover, the authority of the individual QC within that department was quite limited. As originally conceived, it seems to have been thought that he would fulfill his function by being present during critical operations such as welding and ticking off an elaborate checklist of steps designed to minimize personal judgment—clean and bevel the ends of the two joints to be welded, bring them together and clamp them in exact alignment, heat them to the prescribed temperature, and so on. If in the course of this he spotted what he judged an improperly executed weld—or, as happened, a whole series of such errors continued day after day—there was nothing directly he could do about it. Instead, he could make a report to the manager on the work site or on up through the chain to the project manager—to a superior, at whatever level, who was in fact administratively responsible for what the QC thought was wrong and who, in deciding what, if anything, to do about it, would have to weigh such factors as the cost, the production schedule, and his own position as manager. It was this anomalous arrangement, apparently, that the two surveillance groups questioned. Abruptly at the end of May, Alyeska relieved Bechtel of contract management and placed all of those responsibilities in the hands of its own expanding staff—except for the supervision of quality control, which Bechtel would continue to manage; there were references to economy and efficiency, but in fact no concrete reasons were given for this drastic change, either publicly or to inquisitive writers. In mid-July, Alyeska granted the quality-control people authority to order its contractors to redo work that did not conform to specifications or, if that failed, to stop the work altogether; but the QCs remained within the production department. Five weeks later, concluding the procedures outlined in the Right-of-Way Agreement, Mr. Rollins formally approved Alyeska's quality-control program. By then close to 250 miles of pipe had been completed.

In early October, by grace of Alyeska, I flew north from Fairbanks in a leased yellow Apache, four passenger seats on either

side, with a random group who, as I did, all had some interest in
looking at the pipeline work: an elderly editor from a pipeline
trade journal; a bearded ex-pipe hauler who by driving the
Prudhoe run through the winter had put together the stake to
outfit himself with cameras and was now free-lancing; a slender
young man from New York with ambitions of making a movie,
who in the meantime was hacking in Fairbanks; at the front, the
pilot and copilot; and Beverly Wilson, the Alyeska woman in
charge of such outings, who till February had taught English
and journalism in the high school at Ketchikan, in Southeast (in
her travels up and down the line she encountered former students to whom she was still, respectfully, Miss Wilson), and who
since then had bought two houses in Anchorage, into one of
which she had moved over the weekend. We were outfitted, according to the company rules, with a duffel bag stuffed with
sleeping bags and white plastic hard hats, and I had, besides, a
small pack with my own notions of necessaries for an emergency,
but the weather was still mild.

The Yukon came in sight at a point where the channel cuts a
great loop between steep hills, slows and spreads across a broad
mud flat named Sightas Island on the maps. The pipe crossing
was to be immediately upriver from this bend, and as the plane
slipped toward its landing, the last preformed concrete section of
the state bridge was being lowered into place by a crane
mounted on a barge tethered in mid-river; a few more days now
and the temporary plank roadway would be complete, the
bridge would be carrying its first trucks, and the pair of captive
hovercraft that had ferried the traffic through the summer would
be pulled up on the bank and stored. The plane circled once and
dropped to the landing strip at Five Mile Camp, no more than a
widening in the adjacent haul road and separated from the
trucks by gates at either end, let down as the plane was due.

In a supervisor's yellow wagon from the camp pool we drove
north, pulling over to the work pad at intervals to look at the
successive stages of the job: aboveground here, the VSMs mostly
in place, some with their heat-conductor pipes installed; beside
them, the pipe joints laid out like a giant string of sausage, bent
to their predetermined curves and waiting for the side-boom
cranes that would lift them into place for welding. When we

reached a work site and parked, Beverly shepherded us along the pad, sticking close to make sure no one got in the way of a machine or hindered the work by talking to the men. A joint was being heated from the inside by a corona of gas jets, a spider torch, and the other crews stood around waiting their turn: shaggy young men like college boys, several with cameras around their necks who came over to admire ours, wonder if one of us was famous. Getting toward lunchtime: a welder's helper had rigged a gas jet to warm a foil-wrapped slab of beef saved from dinner—the cold sack lunches furnished on the job had been a matter of recent dispute.

In the night a first fine sifting of snow had fallen and stuck, the day was sunny and warm, around freezing, and in front of the camp office and along the work pad there were fresh tracks of bears, though only the cautious blacks, not grizzlies; one we paused to look at beside the road, driving north, a big fellow, staring back, curious and a little menacing, and we did not get out or linger. At the work site, where the access road ended in a steep drop over the edge of the gravel pad, the men had boyishly built a snowman in the shape of a crossing guard, stick arms outstretched, cigarette butt between his lips, and on his head the dark-green regulation hard hat of the execution contractor for this section of the line, Associated-Green.

First snow: the season for most work was about over. In another month or so the temperatures would be hitting minus 35 from here on north, the cut-off point for outside work, when the metal of the machines turns too brittle to be trusted, though in places Alyeska would risk pushing late into the cold to make up time. By now the work force was down to about 20,000 from its summer peak of 23,000; by Christmas it would dwindle to 6 or 7,000, leaving only those doing inside jobs at Valdez and the pump stations or readying the machines for the spring.

By the end of October Alyeska reckoned that the pipeline itself was 50 per cent complete and that after falling weeks behind early in the summer it was essentially on schedule and could therefore be completed, as planned, in another year. This calculation was made by subdividing the work into measurable tasks, then totting up the percentages of completion for each—the chief use the company found for its computers ("a giant adding machine," Frank Moolin disparagingly considered it). Thus: 330 miles of pipe finished, 227 miles of ditch excavated, 390 miles of pipe welded and ready for the spring, 42,755 VSMs set; the work pad was 90 per cent complete—21.5 million yards of gravel had been mined and spread. By the same process, the project as a whole was figured to be only 35 per cent complete: averaged with the pipeline, the work at the pump stations and, particularly, at the Valdez terminal was thus seriously behind. Nevertheless, a rapid-fire series of Alyeska press releases expressed a modest optimism about the chances of bringing the job up to its over-all goal within a year. At the end of September, one more piece in the jigsaw dropped into place. Since late July, 47 barges and their tugs, the entire summer convoy destined for Prudhoe Bay, had been anchored west of Point Barrow, waiting for the pack ice to move far enough out so that they could get through, but the ice that year was late; the biggest peacetime private sealift in the history of the world, it was said, another of the project's superlatives and, with its qualifiers, probably true. Then, briefly, the wind changed, blew from the south, and 25 of the barges, loaded with essentials for the field, made a dash for it; the rest were towed back through the Bering Strait to ports on the south coast—800 miles from their destination, but at least in Alaska.

As it happened, I flew up to Prudhoe Bay on October 2nd as the last of the barges straggled in, and an Arco PR man, Mike Webb, intent on pictures, invited me out by helicopter for a look. The Arctic coastal fall had settled in: heavy mist, thin snow on the ground, turning cold. At Deadhorse, the Anchorage jet came in fast, the runway lights and control tower hidden in ice fog till the last moment, missed the landing, and lifted north in a long circuit, out of sight of land, to try again; I crossed fingers against an unlucky third attempt, but this time we made it down. At the airstrip across from the Arco base camp we ducked under the rotors and climbed in. As I strapped myself into the seat next to the pilot I had just time to consider that I am of an age, a generation, never to have liked airplanes much and this would be my first venture into a helicopter. Inside the Plexiglas canopy, it was like being in an open Jeep on a mountaintop: the feeling of spaciousness, of being already suspended in space and the limit-

less firmament pressing in; underneath, the machine sat on
baggy, half-inflated pontoons that supposedly would support it if
it had to be set down on water. Then we were lifting and the
sensation was of floating balance, buoyant equilibrium, and the
helicopter hung on its blades like a ball on a string, slipping sideways and fast on slanting strata of the gray Arctic air that its
movement made visible, birdlike, with none of the labored heavi-
ness of a plane, and I stopped thinking about it.

We were out over the bay. Mike behind me ran through his
rolls of film, leaning at the open door. To the north the wintering
light shaded into darkness. Below and east, the 400-foot, seago-
ing barges were spaced at random, like pieces on a chessboard
played by incomprehensible rules, each carrying a multistory
steel module, representing a part of a flow station or power
plant, resting on metal skids welded to the deck. The helicopter
dipped and descended for some closer shots. The sea was gray-
green, marbled white with broken ice kept open in a narrow
channel leading into the new dock built for the sealift in the
summer, on the west side of the bay, where now banks of
floodlights had been turned on, their firefly glow swallowed in
the immensity of the Arctic twilight. Below us, a hundred feet
down, a pair of stubby tugs shoved an elephantine barge toward
the dock, propellers churning the heavy water to froth. As the
barges moved east from Point Barrow, the weather had turned
cold enough to freeze the water four inches thick with new ice—
Coast Guard icebreakers led the way, opening a path; by now,
and here, the ice was nearly a foot. Most of the small cargo had
already been lightered off and landed. For the modules, the
barges would be sunk solidly on the shallow sea floor so their
loads could be transferred to smaller barges; for some, the solution would be a mile-long causeway built out to reach them in
the course of the winter. Either way, these barges would stay,
frozen in, till spring.

More than any part of the pipeline, the modules were essential
to the schedule. Delays there might be made up. Here, these
components had been put together in Seattle so that they could
be moved into place and finished inside during the winter. If
they had not gotten through, no expenditure of labor or money
could have replaced them for another year, and producing the

field—and bringing the pipeline on stream—would have been delayed accordingly. But they arrived, and the work could proceed.

Despite the flowers of heroic achievement plucked from among the nettles of failure, as the project neared the end of its second full year fundamental difficulties were coming into view about which neither the oil companies nor the hopeful Alyeska press releases had much to say. During the summer, the pipeline work force had risen to 23,000. In contrast, while TAPS was being argued, the oil companies had insisted the job could be done with perhaps 5,000 men, and more or less neutral economists like Vic Fischer concurred; even allowing for self-serving hyperbole, this amounts to an error of 350 per cent. And, since wages were the biggest single item in the cost of building the pipeline, Alyeska's estimates were still on the rise—to $6.375 billion in the course of the summer, and this seemingly precise figure made no allowance for the interest paid by the oil-company owners, also part of the cost, on the money they borrowed to pay Alyeska's bills. Chuck Champion was still complaining about the quality of the work this sum, however computed, was buying: in particular, about the number of welds rejected, welds that would have to be expensively redone. From a comparison of Alyeska's own end-of-the-year figures for the number of joints welded and the number of miles of pipe completed, it appears that rewelding must have accounted for anywhere from 10 to 25 per cent of the total number of welds, and in fact, as became apparent later (see Chapter 10), this mathematical estimate is overly cautious. By contrast, in Texas, where conditions are not so very different from Alaska in summer, pipeline contractors figure on having to redo no more than 2 per cent of their welds, and a 5 per cent reject rate is considered the maximum tolerable.

The anxieties were not confined to the welding of the pipe or to the quality-control and X-ray programs that backed it up. No one could visit a work site without being struck by the gangs of men standing idle, crews waiting for one job to finish so they could begin; the forty-six tasks into which the work divided, each performed by a distinct crew, about 400 crews altogether, were in theory linked in tight succession but in practice often out of step. It was expensive time. In the peak summer months, the crews were averaging better than eighty hours per week—ten to twelve hours a day, seven days a week, week after week without a break, paid at the union scale of time and a half after eight hours in a day or forty in a week. Hence, in the fall and winter as the Alyeska executives (and the oil men watching over their shoulders) planned for the next year of construction and pondered their chancy schedule and swelling costs, the difficulties centered on "productivity" and "cost-effectiveness": they did not seem to be getting from the men the quantities of work they had budgeted and paid for. To improve this equation, Frank Moolin developed a new plan under which the crews would be limited to a normal work week, their performance measured against hourly goals set by management. And then—

> Only after these crews get their productivity up to the point where it approaches or exceeds the target will we give them more hours. Then we say, if we give you more hours and we see your productivity drop, we're going to take them away. Now last year it was impossible to do that because in many cases they didn't have the equipment they needed to do the job. . . .

Carrot and stick: hours worked and therefore pay. It would make a lively summer, I thought, if the plan was actually carried out—20,000 men sitting around the camps at the end of their eight-hour day, on their Saturdays and Sundays off; when Moolin's plan was announced, the pipeliners who heard of it let it be known that they simply would not go out of the hiring hall for as little as forty hours of work—hours could be bargained in both directions. This was late January 1976. I promised Frank to stop by in the summer and see how the plan was going. "Well, if it doesn't work," he jauntily assured me, "you'll find some other guy sitting in this office!"

The new productivity plan, of course, represented a manager's view of what was wrong with the project—or might, anyway, be improved. It does not appear that any one of them ever turned the same kind of intent scrutiny on the management of Alyeska or the contractual system under which it labored. To raise the question is to answer it: managers are no more prone to critical self-scrutiny than most other humans.

In mid-February, a couple of weeks after I talked with him, Frank Moolin's professional peers expressed their approval of the way the project was going by honoring him: he was named Construction's Man of the Year by a McGraw-Hill trade journal, *Engineering News-Record*. Besides honor, the award meant a couple of days' respite from the ten- and twelve-hour days, six and seven days a week, that he, like everyone else, had been putting in. He flew to New York to attend the dinner at which the award was presented and deliver a speech of acceptance.

Chapter 7

SIGNING ON

> This contract is a bucket of shit.—Jesse L. Carr

In January 1974, about the time the oil companies were putting their names to the Right-of-Way Agreement that would permit the long-delayed Trans-Alaska pipeline to start building, negotiations got under way in Washington on a parallel and no less necessary contract: the Project Agreement governing the relations between Alyeska and its contractors on the one side and, on the other, the men and women who would supply the actual labor, spoken for by their unions. The Alyeska negotiating team was headed by a lawyer, William J. Curtin, a senior partner in the Washington firm of Morgan, Lewis & Bockius. The workers were represented by a committee drawn from fourteen international unions and chaired by Robert A. Georgine, an aggressive young executive (later president) of the Building and Construction Trades Department of the AFL-CIO. In the weeks of negotiation that followed, Alyeska had a single paramount objective—to make sure the project would be carried to its conclusion without the delay of a strike—for which it would give ground on many lesser issues. There were national as well as corporate motives for this urgency. Although the effects of the OAPEC oil embargo were muted in the United States, in Europe they were acute and in England, what neither wars nor successive Labour Governments had accomplished, had cut the work week to three days and turned off the light and heat in the little shops; unpleasant portents. Echoing Congress a few months earlier, the negotiators in their preamble agreed "that the national energy crisis adds another dimension to the urgency and importance of this project."

In March, with a sense of accomplishment on behalf of the nation and their members, the international representatives journeyed to Alaska to present a draft of the Project Agreement to the local unions who would have to live with and enforce it. After suffering their approbation—"what a wonderful bunch of people these oil companies were, their generosity, their benevolence," as one man remembered it—Jesse Carr, the secretary-treasurer of Local 959 of the International Brotherhood of Teamsters, Chauffeurs, Warehousemen and Helpers of America, made a speech. "This is a *bucket of shit,*" he said. "I don't know what in the hell you're all standing here saying this stuff for, because it's lousy." And: "Well, I'll tell you something, I'm *not signing it.* And nobody in the International's going to sign it for me!"

In the early 1950s, when Jesse Carr, ex-Marine, ex-trucker, left California, settled in Alaska, and went to work for Local 959, the union had about five hundred names on its rolls, few assets, and a lot of debts. By 1974, as pipeline construction was about to begin, the membership had grown to eight or ten thousand—and in the next three years would more than double. Its jurisdiction had become state-wide, and its membership and affiliates included not only the obvious drivers and helpers but trades and professions ranging from longshoremen and oil-field roughnecks to meatcutters and grocery clerks, telephone operators, firemen, policemen, and school principals—five or six hundred such groups (the union is vague about numbers), some, like the police, associates for the sake of sharing the benefits, with many more clamoring to be represented. Among the benefits are high wages and a complex of employer-funded trusts that provide generous pensions, comprehensive health care and legal services and give the union a remarkable and rapidly growing concentration of economic power within the state. These benefits are the main reason also why everything in Alaska from steak to housing to pipelines costs more than anywhere else in the United States. The two, union and boss, have grown up together by tough bargaining. Indeed, those who have faced him across the table do not consider that Jesse Carr negotiates—he makes offers which the employer accepts. Although he has sometimes given his word, take it or leave it, not to call a strike, he had not in his more than twenty years signed a contract containing a no-strike provision.

The Anchorage meeting ended, the negotiators went back to Washington and started over, this time with Jesse Carr at the table and his own team from Local 959. What came out—the Project Agreement dated April 29, 1974, elaborated in a series of clarifying letters from Mr. Curtin and from Alyeska's president E. L. Patton—was a good deal more like the standard Alaska Teamsters construction contract than it began. It was actually signed on May 7, about a week after the Cats rumbled north to start clearing the right-of-way for the haul road beyond the Yukon.

The no-strike provision was still present, as Article VII of the finished Project Agreement and several times repeated elsewhere in the document. An employee who for whatever reason refused to work or engaged in a slowdown could be fired; a local's obligation was fulfilled by telling him not to strike, and so too the international's responsibility for the local. Disputes at any of these levels were to be settled by a third-party arbitrator empowered to make an award if he found that a strike or slowdown had occurred, though it would take court action to enforce it. This moderately clear-cut arrangement was qualified by Article X, which specified that no one should be required to operate equipment or work under conditions that he deemed unsafe. The proposition is less self-evident than it sounds: anyone with a sharp eye and sufficient motive can find something wrong with a truck or forklift and "red tag" it—declare it out of service until repaired—and the work stops; as Jesse Carr once observed in the course of negotiations with a local contractor, "I don't *need* to strike you. I saw five safety violations on my way in to your office!" Hence, while acknowledging the idea, Alyeska stuck to the principle that an unsafe machine could not be used as a pretext for a strike, though a four-stage grievance procedure was provided for settling disputes (with the time limits set for each step, it might take a couple of months to carry out). Several clarifications were added, including one taken from the standard

Teamsters contract that permitted union officials to visit the work camps and job sites at will and call their members together for safety meetings on company time; but Alyeska's judgment on the safety of its equipment remained final—so far as the contract was concerned, a man could not refuse an order, say, to drive a truck with brakes he thought were no good. In the urgency of the work and the difficulty of keeping the machines running through the hard winters, such orders were given, and in at least one case a man died carrying one out, lacking the capacity to refuse.

Another issue hotly argued by the Teamsters boss was Alyeska's work schedule. Although it was implied but not specified, Alyeska, for the same reason it insisted on the no-strike clause, wanted a seven-day week of ten- or twelve-hour days to make the most of the long Arctic summer daylight. The Teamsters demanded a six-day week, even with longer days to make the same total, with one week off in seven. (The worker would collect forty hours' pay—but only after he returned to the job.) The issue was compromised: a man could, if he wanted, take a week's leave without pay after nine or thirteen weeks and would be transported free to Anchorage, Fairbanks, or Valdez, with a job kept open for him to come back to—"R and R" in the military expression, rest and recreation; after twenty-six weeks, the limit of the Arctic construction season for many outdoor jobs, he was required to take a break, at least in theory.

What the unions got in return for these concessions was of course money. Once hired and transported to the camps, the men were guaranteed forty hours' pay per week even if, as might happen, the weather made work impossible; the intended long days and weeks would earn one and a half times the basic scale —time and a half for hours beyond eight a day or forty a week. The pay scales would be those in the existing Alaskan contracts of the various locals, which were incorporated in the Project Agreement (along with their other provisions so far as they did not conflict). For the Teamsters, for instance, these were the hourly rates in a new contract made with the employers' association, the Alaska chapter of the Associated General Contractors of America, that took effect on July 1, 1974: an average of about $10.50 for the 67 job classifications covered, rising in specified increments every six months—about 15 per cent in January 1975, then $1 per hour, then 50¢; this contract was to remain in force till June 30, 1977, when, if it kept on schedule, the pipeline work would be finished.

It was time and a half that produced the huge pay checks that excited newspaper readers. At the rate in effect in the summer of 1975, a bus driver—hauling men from the camp to the work site in the morning, waiting in the bus all day for them, driving them back at night—was getting $11.98 an hour for his first forty hours, $479, and $17.97 for the rest, another $790, or over $1,200 a week, if the crew he was busing worked a 12-hour shift 7 days a week ("seven twelves," about the top); at these rates, Teamsters accountants figured, a single man with no dependents could not afford to work more than 39 weeks in a year—after that, his taxes, federal and state, would consume his additional earnings. With deductions for taxes, social security, and 40¢ an hour for union dues, he would actually see about $600 a week—and a tax rebate of $6,000 or $7,000 at the end, if he remembered to quit in time. (A few when they filled out their employment forms had the wit to put fanciful numbers in the space where they were asked for dependents—three 9s, or 1, followed by a row of zeros —and in the Alyeska computers were therefore immune to withholding taxes; the IRS noticed eventually and brought charges that were written up fully in the Anchorage *Times*.) On top of actual wages, Alyeska for 84 hours would be paying something over $250 a week into six Teamsters trust funds—$2.10 an hour for two different pension plans, lesser amounts for medical care, industrial training and advancement, prepaid legal services. Alaska, as we said, is an expensive place.

Bus driving was, of course, softer than most pipeline work— and therefore much sought by those who could stand extremes of boredom or had a lot of reading to catch up on; in the final year of construction, Alyeska attempted to find something more for its drivers to do during the day than sit, by making them part of the crews they drove—they would in theory have to get out and stir themselves. The quantity of money if not the quantity of work was, however, typical of the project. A bull cook in a camp, making beds and sweeping out rooms and subject to the Hotel and Restaurant Employees Union, would work fewer hours at a

lower rate and come out appreciably behind. At the other end of
the scale would be a journeyman welder, with an hourly rate that
rose from about $15 in 1975 to $16.34 one year later, say anywhere from $35,000 to $60,000 for a season's work, depending on
how much welding time the weather allowed him; a member of
a select craft, with problems of his own besides taxes—the one
indispensable skill on a job where most others could be (and
were often) self-taught in short order and were therefore replace-
able.

At whatever rate, these wages were, in the old term, *found:*
Alyeska would furnish food, beds, transportation to the work
camp, Arctic clothing (at cost, according to a clarification). There
was room for dickering about all these matters: number of
square feet per man (60), whether members of the same craft
would be housed together (only by accident), how often linen
would be changed (once a week), whether hot lunches would
be provided (if possible but not necessarily). Within the camps,
Alyeska intended Boy Scout rules against drinking, gambling,
fighting, stealing, enforced by searches of rooms and of inbound
and outgoing baggage, but prudently did not put them into the
Project Agreement (a clarification made the rules, when promulgated, subject to the grievance procedure). Jesse Carr addressed
the notion that everyone hired should live in a camp—and stay
there when not actually working: "That provision that the men
have to stay in camp. Tell me, I know you're going to have
guards, you say that—what about Doberman pinscher dogs,
too?" ("Well, we didn't necessarily plan on using Dobermans,"
someone answered.) A clarification allowed anyone living in Valdez or Fairbanks or one of the villages in between to go home at
night, with transportation provided; whence, among other consequences, a thriving nighttime trade in both towns and at several
roadhouses along the Richardson Highway.

Besides the fourteen unions and their members, who under the
prevailing closed-shop Alaskan construction contracts would
have first crack at all pipeline work, several groups followed the
negotiations with interest. The Natives, for instance. An unkept
promise of jobs had been the trigger that set off the lawsuit and
injunction led by the people of Stevens Village, the immediate
occasion for the Native Claims Settlement and the years of envi-

ronmental debate that followed (or so the villagers thought—
that there had been a promise and that the oil men were evading
it—and that was what counted; see Chapter 4). The Pipeline
Authorization Act had therefore required the Secretary of the Interior to take "affirmative action" to ensure that Natives (and
other minorities) benefited from the pipeline. His action took the
form of a four-paragraph section in the Right-of-Way Agreement
which obliged Alyeska to establish a training program for Natives and "do everything practicable to secure" their employment. This condition was in turn acknowledged in the Project
Agreement: the contractual norm for Alaskan construction, hiring through the union hall, would be modified accordingly.

Alyeska dutifully sent a form letter to the two hundred or so
Native villages, inviting applications from any who wanted to
get in on the money. It also contracted with the Alaska Federation of Natives to establish, at Alyeska's expense, a manpower
department for the same purpose. The AFN compiled lists of
interested Natives and referred them to the state labor offices,
but nothing much happened—the AFN did not know much about
Alaska construction or its unions; as a matter of contract, a
builder may not fill his crew from outside the union hall unless
the hall is empty—which is to say that he *never* has recourse to
the state labor exchange. The AFN began feeding its lists
through existing union training programs—typically, a couple of
weeks in Anchorage or Fairbanks, with room and board and $60
a week thrown in—and a narrow stream of Natives emerged at
the other end, moved through the hiring halls and out to the
work camps. Just how many actually went through the program
is problematical. The AFN counted a thousand the first year,
four thousand or more in the next two, but since one man might
be hired, quit, and be rehired by a different contractor, at a
different camp, more than once in the course of a summer—when
he had enough money for a fling in Fairbanks, when he tired of
the work or grew homesick for his village—it is not clear how
many individuals these numbers meant. Alyeska guessed 10 per
cent of the work force at any one time, perhaps two thousand in
the peak months. It did not *look* that many in the camps and on
the job sites, but since a Native officially may be one with no
more than a single Native grandparent, it is not a question that

can be answered by eyes alone. From time to time, AFN officials went out to look for themselves and came back to Anchorage to make denunciatory speeches, as unpersuaded by what they had seen as I was.

Under a recently enacted state law, Alaskans in general constituted another distinctive group that, like the Natives, was to receive preference in oil industry and pipeline hiring. The statute was meant to change a condition as old in Alaska as Baranov and the promyshleniki: that much of the cash work in the country is done by outsiders who come for a season, for a year or two, and depart, taking their earnings with them. Moreover, Alaska is a place of perennially high unemployment, generally about double the national average, according to official figures, and far higher in the villages; in good years, when the work force expands, the percentage of those unemployed stays about the same, 10 or 12 per cent—the difference, presumably, made up by people from Outside who come and go. At the same time, it is not clear how far Alaskan statistics are comparable with those compiled elsewhere: whether a man who works a road crew or crabbing boat in summer and in the fall retires to a cabin to get in the meat for a winter of trapping is unemployed in the same sense as a laid-off assembly-line worker in Detroit or Los Angeles. Much of the work to be had is inherently seasonal, and while the salmon run and the caribou migrate, the land does still provide a living for those who know how to take it; but whatever the precise reality, unemployment is among the unresolved constants of Alaskan politics.

In any case, the principle, known as Alaska Hire, was the kind that seems self-evident to legislators, it was written into the Project Agreement, and the unions signed: that, other things being equal, someone who resided in Alaska should be preferred for a pipeline job over someone who did not. It was easier to express than to carry out. Residence, in the first place—the state said a year, supported by an Alaska driver's license, and later was flexible about the time if it recognized an "intention" to stay, of which the state Department of Labor would issue a certificate. It was like the old philosophical problem of proving one's existence, impossible (*I reside, therefore I am*). Alyeska was diligent about applying Alaska Hire to its own employees, or at least the

visible ones—all but one of its public relations people, for instance. Most pipeline jobs, however, were filled by contractors
supplied from the union hiring halls, and the unions raised practical, not philosophical, objections: that, however defined, they
could not tell a resident from a nonresident; that, if they could,
to do so could subject them to lawsuits and other unpleasantness
on charges of discrimination. The principle is doubtful on several
constitutional grounds, but the obvious objection is common
sense: Americans are accustomed to moving freely from one
state to another, and that freedom is meaningless if, when they
do, they are prohibited from working, no matter how humane
the intent.

Alaska Hire gave occasion for rhetorical exchanges among
state, union, and Alyeska officials but had few concrete effects.
In the course of construction, Alyeska defensively estimated that
60 per cent of the pipeline workers were in some sense Alaskan.
When the state's leading bank, the Alaska National, studied what
happened to pipeline pay checks drawn on itself, it found the
opposite: two thirds of the checks were deposited Outside,
where the pipeliners still had their homes. Having gained a foothold in this transient state—a job, a business, a home—not many
certified Alaskans were prepared to give it up for the sake of two
or three years on the pipeline, even at $12 an hour and time and
a half.

The pipeline was to be built under one other well-meaning restriction: women, according to the Right-of-Way and Project
agreements, were among the groups to be protected against discrimination. To prove that it was fulfilling this imperative,
Alyeska maintained that women, like Natives, constituted 10 per
cent of the work force. There were press releases and inspired
newspaper and magazine articles about women fork-lift operators, women radio operators, women welders (the welder was always the same woman, an Athabascan bull cook who struck up
an acquaintance with a welder, was taught welding after hours
and qualified as a helper; two birds with one stone); officials
such as Gerald Ford and Henry Kissinger who toured the work
sites under Alyeska supervision came away impressed by the
number of women they saw doing construction work. There
were in fact enough women to make difficulties about plumbing

in the early camps, and at Chandalar, just south of the continental divide, I noticed a cartoon stuck up in a hallway showing
a young man standing at a urinal in the accustomed posture and
a woman beside him, raising her skirt—caption, "You really *have*
come a long way, baby." In practice, the sexes sorted out as one
would expect: the men were out on the job, running the machines; the few women were in the camps, clerking room assignments, sweeping up.

Given the interlocking obligations under which it labored—
closed shop, Natives, Alaskans, women, not to mention race,
religion, national origin, and mere skill—Alyeska produced and
distributed a leaflet called *The Truth about Pipeline Jobs in
Alaska.* The gist was repeated by labor exchanges as far off as
Massachusetts: that jobs were scarce and if you got one the pay,
though high, would be sponged up in Alaskan rent and food.
Some evidently took these warnings at face value, but many did
not. Alaska allures, as in the gold-rush days.

As I sit at the splintery camp table in front of my tent finishing
an after-dinner bottle of beer, a red squirrel scrabbles in the
sand and duff at my feet, hunting pine nuts and berries, scraps of
human food, small and neat as a chipmunk, bold almost to
tameness. White belly under the dusky red, white-rimmed eye-
lids: he dodges playfully behind the trunk of a jack pine, peers
around at me, runs down again, and resumes his hunt. In my
wool shirt I feel warm—the temperature is about 60. The light
now at ten o'clock on this early June evening is like midday and
looks as if it will never end, though in fact the sun sets at this
season around eleven-thirty and there will be almost three hours
of darkness before full light returns.

The camp is fitted into a wooded bend of the slow-moving
Chena River at the west end of Fairbanks, bounded on one side
by Alaska Route 3, heavy with traffic moving toward the suburb
of College, where the University of Alaska is, and Anchorage 350
miles to the south. Straight west is the airport, north across the
river the final miles of the Alaska Railroad. The camp and its
sixty campsites are screened by pines and birches. Except for the
planes skimming the trees every few minutes, coming in or taking off, you hear but do not see these things: steady whish of

cars on the road, hiss and rumble of the jets, whistles of shunting freight trains. At the site across the dusty camp road from mine, two small children querulously refuse to go into their tent and sleep, are spanked, wail for a minute or two, and subside; farther off you hear country rock from a radio tuned to the local station and turned low—despite the prolonged sunlight this is the end of a working day and another is coming in the morning, early. For those with jobs or looking for jobs, it is therefore a lucky camp, the only one anywhere near Fairbanks; the alternative is a $40- or $50-a-day room at one of the motels—or a sleeping bag under a Chena bridge, downtown. (At Anchorage the only public camp is twenty miles out on the road to Fairbanks.) According to a notice at the camp entrance, the fee is $2 a day for a place to park a camper or put up a tent, with a two-week limit, but for now the camp is not yet officially open and no one is collecting; there is no water either—the river is not to be trusted —and the only place to get it, a jugful at a time, is a tap mercifully opened at the edge of the campus, a mile down the road. (There is water also at the gas stations but not drinkable, they tell you—something wrong with the city water system over the winter and not yet fixed.)

The camp is full—the site I took near the entrance seems to be the last one open: campers and vans and battered pickups with cracked and broken windows and rags stuffed in place of gas caps, a few trailers, one surplus postal truck still painted red, white, and blue, but all except the trailers have set up tents. They come from few places, California, Arizona, surprisingly Minnesota, with Alaska predominant; but nearly all are new or at least potential Alaskans—as new anyway as their new license plates on the old cars and trucks—and like those who preceded them, or indeed most Americans since Jamestown and Plymouth, they start in a temporary camp. A few I have met even wintered here till they could find their way through the hiring halls to the pipeline, and with wood to be had for fires and an Arctic sleeping bag, you can do it. The pipeline is the lure, the chance of getting out of the camp and settling—or going home again for a new start—and the chance outweighs the discouraging Alyeska propaganda, if any of it reached them. And if pipeline jobs turn out scarce, there is work enough to be had for anyone young and ready—$5 an hour pumping gas and greasing cars, $15 or $20 doing rough carpentry for nonunion local builders, though without the bonus of long hours and the huge drafts of overtime that the pipeline pays; by now the pipeline has sopped up most of the casual labor that formerly did these jobs.

So they come, many with their wives, a few, like my neighbor across the way, with a child or two. Naturally, since jobs and money are the incentives, they come the cheap way, down the Alaska Highway, that narrow, winding, washboard-rutted axle breaker which some consider an adventure, still after thirty years left mostly unpaved by the penny-wise Canadians; a national insult or a national disgrace, depending on one's viewpoint. In 1975, the Immigration Service counted 141,069 people arriving in Alaska by this route; by now, in June 1976, they are entering at a rate of perhaps 500 a day, in 200 or so cars and trucks. Some of these are long-haul truckers coming and going on their more or less scheduled runs; as many as a third may be tourists, recognizable as elderly couples and retired—they have the time for a trip that will include two week-long chunks of wearisome driving from the nearest U.S. border crossing—towing their slow-moving, expensive new trailers. The rest are pipeline immigrants, or hoping, at least, to give it a try. Since the project started, more than a hundred thousand of them have stayed long enough to figure in census estimates. That means a one-third increase in the state population within three or four years, but nearly all of it is concentrated in two small areas: Anchorage, which has swelled to about 180,000, and Fairbanks, which has doubled to over 40,000. And to anyone finding his way around either town, these numbers seem cautious. It is the thing—the thronging traffic, the strangers' faces, the crowding and the prices—that older Alaskans feared at the outset and now complain of: *Los Angeles!*

From inside a Sears, Roebuck tent in the Chena River camp—or a YMCA hostel, a log-walled, $10-a-night flophouse, or any of the other halfway stations—it looks different, of course: the world is before you and its chances; the trick is to find your way in. The first step is a union hiring hall. The Hotel and Restaurant Employees' is easiest—dull work and steady, inside the camps, not well paid. The Laborers' pay scales are better. The Teamsters are the biggest and control the most jobs but have a

lot of members lined up ahead of newcomers. You go to the hall
and get your name on a list, in order. At the Teamsters, you get
preference if you have worked 400 hours on a union job in
Alaska any time in the last two years—you go on the A list.
There is a B list for those who worked between 80 and 399
hours, a C list for those with fewer hours who qualify as Alaskan
residents. (They take your word for it.) All three lists by definition and contract are for members of Local 959, if they have kept
up their dues. If you are young and on the loose and have no
connection with this local or any other, you land on the D list—
everyone else—and pay a nominal $10 a month, called dobbie
(pronounced *dough*-be) dues, for being registered: and then
you wait, sit on a bench and drink coffee, read magazines and
wait for your name to be called.

The contractors put in cryptic lists of jobs to be filled—semis
and 20-yard dumps, low beds and bull lifts, each at a different
hourly scale listed in your hip-pocket copy of the current contract—and the dispatcher reads them out, the jobs, the hours,
and the camps where they are based. If you are present and your
name is at the head of the right list—A list first, then B, and so
on—you bid the job: persuade the dispatcher you are qualified,
and after a day of Alyeska orientation you are on your way by
chartered plane to Five Mile or Dietrich or Happy Valley. From
the D list it may take many weeks to reach this point. Maybe it
is winter by now, fewer jobs, down to seven or eight thousand
and most of them tough, in the cold mountains beyond the
Yukon, but there are fewer to compete for them, most have made
their hours and gone home for Christmas, gone to Las Vegas or
Honolulu for R and R, and you get your chance. It does not
much matter if you know the job or have ever done it, you pick
up the words in the weeks of waiting around the hall, and the
contractor supervisor who hires you is no better judge than the
union dispatcher; you will learn when you get there and, if not,
you will at least pick up enough hours to rise to a higher list the
next time around.

One among the thousands of young men who went through
this process was a son of old friends in Pennsylvania, and when
he was home during a second midwinter break I spent an evening with him talking about it. Dick had gone to Alaska one

summer with his family and then—finished high school, little thought yet for college—liked it well enough to buy a second-hand camper and go back on his own, to Fairbanks. He parked in the Chena River camp, found odds and ends of carpentering, played soccer for a local semipro team. In May 1974, in Fairbanks and Anchorage, Alyeska had started its own hiring and by the end of July had moved into its construction headquarters at Fort Wainwright, but until fall, when the weather turned chilly and most of his friends had vanished to the work camps, Dick had not considered working on the pipeline. Now he did and, still green about jobs and unions and pipeline work, presented himself at Fort Wainwright to apply: "Everywhere else, if you want a job you tell the guy what a great worker you are or whatever, you talk to him and ask him for a job. So I went out and talked to Mister Alyeska. I didn't know anything." A secretary gave him a form to fill out. While he did this, a tall Texan appeared, looked him over—bushy hair and young beard, work shirt and stub-toed boots—and took the form away: the jobs were clerical, typing and filing, that was all Alyeska had to offer. (To an Alaskan, even a new one, all Outsiders connected with oil—broad-brimmed hats, pointy-toed boots, and double-knit suits—are either Texans or Okies; the happiest sight you can see, it is said, is a Texan leaving, with an Okie under each arm.) A Native stood by, laughing.

"Yeah, I'm the token Eskimo."

"What do you do?"

"Nothing."

"Well, that sounds like a pretty good job."

Dick found his way to the Laborers' hall, got on the D list, and waited. A month or so later when a requisition came through for tapers he had risen high enough on the list to bid. You ever done this kind of work, kid? Yeah, sure, I taped pipe two summers running, in Chester County, where I come from, I'm a good taper. Sure, kid, sure. *Scamming*, they called it: the art of persuasion. Dick was sent to Old Man, an Associated-Green camp in Section 4, about ten miles below the Arctic Circle. The first day, he looked around for someone in the crew who knew the job and met an old black man from the Gulf coast, worked pipelines for years and still had all his fingers; and watched him and began to learn.

The work is easy enough but heavy, for buried pipe the last step before lowering-in and backfilling. The pipe joints, welded in strings, are set on skids of imported oak 4×4s, lifted in slings carried by side-boom Cats, wire-brushed and heated to the temperature specified for the tape, primed with a dark oily liquid. The tape itself is self-adhesive, like household cellophane tape, but 9 or 12 inches wide, in rolls that weigh 75 or 120 pounds. The rolls are lifted onto the spindle of the tape machine and unwound, round and round the pipe. The final step is inspection. If the tape breaks, does not stick, forms a bubble, the defect—a "holiday"—is cut out with a pocketknife and a new piece stuck down; tape-patching, another subtrade. The inspectors at this stage were still the Bechtel college boys, making about $700 a week, which sounds great for a young engineer working his first construction job but was $300 less than the Laborers' whose quality they were supposed to control; and the tapers therefore looked down on the QCs and judged that they knew even less in a practical way about doing the job than they did themselves.

Early in the project, as part of the process of scheduling time and estimating cost, Alyeska—or Bechtel, still the contract management contractor, on Alyeska's behalf—analyzed the work into a series of discrete activities, each measurable and simplified to anonymity. Programming (see Chapter 5): breaking the big job into a sequence of small ones, as on an assembly line, would make it possible to limit and control the levels of skill and judgment needed for each, set precise goals, measure the results, and each of these purposes, in turn, could be expressed in a computer program and therefore predicted and monitored. This plan was complicated by the restrictions imposed by the Right-of-Way Agreement and the Alaska Hire law: the work force would be an unpredictable mix of more or less experienced pipeliners drawn from the Lower 48—the same core of men who follow the trade from Saudi Arabia to the Andes—and of Alaskans and Natives. Early on, therefore, Alyeska made some assumptions about productivity: the standard would be a Gulf workday, 100 per cent—the quantity of work to be expected from a Texas or Louisiana pipeliner; an Alaskan day—white Alaskan—was 80 per cent; a Native day was 50 per cent. (These percentages reached the councils of the local unions and occasioned some sourness—they suggested, despite Alyeska's protestations about affirmative action and adherence to Alaska Hire, a strong motive for filling the jobs from Outside.) The theory seems to have worked well enough in the comparatively simple phase of building the haul road but broke down once actual pipeline work started in the spring of 1975: the schedules slipped behind, the work force ballooned—and with it, cost; and disturbing questions arose concerning the quality of the work as well as the quantity (see Chapter 10). At the end of May 1975, Bechtel and its computers were accordingly eased out of most of their responsibilities.

In taking over the management of the project, Alyeska seems not to have questioned its basic assumptions. The computer printouts continued to yield their precise percentages of work completed, of goals attained or missed, and these were duly reported in weekly press releases. The job descriptions delivered to the hiring halls became ever more minute and specific ("lowboy driver to load operating engineer equipment aboard trailer, qualified on 594 boom Cat . . ."). Filling such jobs was intended to be an impersonal task of matching experience and qualifications to a surpervisor's requirements, computer work, the men themselves interchangeable, like spare parts for the machines they were called to operate—necessarily, perhaps, since in most of the trades the periodic leaves and the chance of rising through the hiring lists to a better job or more agreeable camp encouraged a high turnover which, even if as low as 1 per cent a day, would replace or redistribute the entire work force in the course of a summer construction season. (Since it was estimated that altogether about 70,000 individuals were employed, it looks as if, numerically at least, the work force did in fact change completely every year of the three years of construction.) When Alyeska labor specialists visited the union offices to complain about the match of workers to job specifications, they did so in an abstract jargon that further soured the business managers and dispatchers: they talked about "the products" the unions were supplying, by which they meant what was coming out of the hiring halls—the men.

The assumption behind these attitudes seems to have been that if the job could be programmed in small enough steps then the effects of differences in individual competence would be nullified, and the quality of the work would take care of itself. The quality-control program required in the Right-of-Way Agreement, which remained a Bechtel responsibility until early in the second year of pipeline work, nearly a year after the company had been removed from general management of the project, was on the same principle reduced to a series of checklists. This extension of programming still took no account of individual judgment—whether the QC completing the checklist knew enough of taping or ditching or welding to tell good work from bad. And similarly the work itself: except for the welders, who were exceptional in most other ways as well, the system could identify which crews were productive and might know something of how well they were doing as a group, but within a crew of ten or twenty men it had no way of distinguishing individual competence or productivity; nor, given the constantly changing roster, was such information thought worth collecting.

This situation produced a curious set of attitudes toward the work at hand. Among the Alyeska managers, planted in their offices in Anchorage and Fairbanks ten or eleven hours a day, six and seven days a week, there was a kind of dogged determination to get the job over with, get back to the oil companies they were borrowed from and resume their careers; seen from that vantage point, TAPS was a position of unwonted public exposure—a man might ruin himself in the course of the assignment but not in the long run do himself much good. For the men in the camps and on the line, the motives were personal and more immediate: educate the kids, retire young to the Alaskan dream of land somewhere, a cabin and a boat, or back home to Washington, Arkansas, or upstate New York, maybe if you were young and had never considered the prospect, college, a law degree and an upward start in life—all the modest forms of independence that a rapid accumulation of cash puts in a man's mind. What I did not find at any level was any strong commitment to the work itself, any sense of its importance or of the value of one's own part in it. It was too fragmented for that, the parts too hedged around and controlled from outside: like being

in a war, but with no justifying feeling of being caught up in a
large and worthy common purpose.

On the jukebox in the bar at the North Star Motel the same
record plays over and over, loud, filling the square, dark, windowless dining room next door: Helen Reddy singing, *That's no
way to treat a lady, No way to treat your baby* in her country-
woman's accents, the tune and the words going round in circles.
At the bar a couple of middle-aged welders are getting loud, but
the matter is nothing more serious than who buys the next round
of $2 drinks and whether to have them here or somewhere else.
Two nights ago a young fellow, crazy-drunk, ran out onto 15th
Avenue and shot a man with a pistol, apparently at random, and
since then I have been having dinner here, hoping, short of
gunshot, to see something of the high pipeline times that have
been in the newspapers Outside: the big-time hookers that have
added Anchorage to the Las Vegas-Seattle-Honolulu-Los Angeles circuit, the hundred-dollar-a-card poker; and so on. But
now in October 1975, with the pipeline work slowing, it is a
tame crowd, welders on their week of R and R with wives flown
in from Seattle, younger men with their girls; the place is homey,
the rooms $10 or $15 a day less than downtown Anchorage—
even a welder's money is not unlimited. I finish the $7.50 prime
rib, as it is called, drink my coffee, and head for my room two
blocks west on 15th.

The reverse of the newspaper image is conveyed in a weekly
tabloid, *The Camp Follower,* put out by the Alyeska public relations department and distributed free in the camps: after-hours
guitar groups and intercamp volleyball games, an incamp wed-
ding (a Teamster and a Laborer), the Athabaskan bull cook
who became a welder's helper. The reality is somewhere between the two images: periods of frantic teamwork surrounded
by stretches of inactivity, as in the Army. "I know a hell of a lot
of people working on the pipeline," remarked Vic Fischer, the
University of Alaska economist, "who have quit because they just
can't stand not doing anything for as many hours as they end up
not doing anything." If you are an Alaskan, you have done it
yourself or have friends or kindred who have: "My son, a
member of the Operating Engineers, works on the pipeline as a

service oiler, out there for a year now in many different camps, and there have been literally periods of a week at a time when he did exactly nothing because there was no rig for him to work on." Some of the early camps in the Brooks Range were small, a few hundred beds, and remained so, held together perhaps by a common talent for extreme weather; the pump-station camps were also small and coherent, much of the work done at close quarters, indoors. The general experience, however, is of camps running to thousands, too big to cohere, their inhabitants too transient, too divided by small loyalties to form naturally into groups that reach across the barriers of work crew, craft, union—jurisdictional rivalries have been suppressed by the Project Agreement and the simple abundance of jobs, but the instincts remain—or feel the large commitments that are born of social groups; and the ten- or twelve-hour workday, the two hours more for washing up and eating, does not leave much time for reflection or for social relations. The reality is boredom and isolation, as in a prisoner-of-war camp: the bosses are the enemy, represented by the camp managers, the security guards (though they too are unionized from sergeants down, Laborers), the suit-and-tie visitors who come through periodically on inspections, from Fairbanks, Anchorage, and beyond; resistance—the one unifying emotion that most seem to share.

Perhaps anticipating this state of mind, Alyeska drew up a definite and quite detailed set of rules for camp behavior, and it was apparently some intimation of their content that produced one of Jesse Carr's bellows of protest in the course of the Project Agreement negotiations. What finally came out was in fact normal enough in northern construction or mining and common sense for any group of men working in the isolation of remote and more or less wild country; a few of the rules were imposed by the right-of-way Stipulations. The difference was the size of the project and the camps supporting it, the constant turnover that made it unlikely for the men to stay long enough in place ever to know each other or their supervisors well. The camp rules had therefore to be formal and impersonal; they became, in effect, part of the labor contract, subject, as we noted, to the grievance procedure that applied to every other clause in the actual contract.

Along with some tips on Arctic survival and a once-over-lightly physical (a number of things, such as hereditary illnesses, that would make a difference in long-term employment were of no interest to Alyeska's doctors), a new hire got his first exposure to official camp policy in the course of his day's orientation at Fort Wainwright, before shipping out: a morning of lectures from which, if he were like most, he came away with the impression that "*everything* is subject to termination." Well, not quite everything, but: alcohol, marijuana and other drugs, guns; gambling, drunkenness, horseplay, and fighting; loafing or sleeping on the job; insubordination or refusing to accept a work assignment; smoking in bed; feeding or molesting the wild life, hunting, fishing or trapping on the right-of-way or around the camps (in deference to Natives, conservationists, and the Stipulations); private cars or trucks (at those camps that could be reached from public roads). These and similar prohibitions were all listed in a twelve-page, breast-pocket-size booklet the new man was handed at the start of orientation. The arguments had to do not with one rule or another—though negative in form, they were reasonable, unobjectionable—but with how the whole package was to be enforced.

Alyeska's first thought about enforcement was a multiple right of search—of inbound and outbound baggage, camp bedrooms, truck cabs and sleepers at the Yukon crossing point. That intention was accommodated to the unions' sense of their members' privacy and personal dignity. Before a man gets on the plane for camp, his suitcase or duffel bag will be X-rayed for guns, the same system used at most American airports for the same purpose, and a trained dog may sniff it for marijuana; the guards may ask to look inside and, if he refuses, the bags will simply be left behind. As a defense against thievery, all outgoing baggage is to be searched—but the owner may first go into a room apart and unload anything not his, no questions asked. (The sleeping rooms and lockers were equipped with locks and keys, and no guard entered unless summoned or without first knocking and asking permission.) It was in this modified form that the rules were printed in the little booklets. In practice, they were further compromised. The working standard was outward conformity and a decent discretion. A man might carry in a case of beer or a half dozen bottles of whiskey buried in his duffel or wrapped in a blanket and go unchallenged, but if he stupidly displayed it he would ride the next plane back to Fairbanks. In consequence there was a good deal of gossip, some repeated in the newspapers, about dormitory hallways reeking with marijuana smoke —and after-hours boozing, big-money gambling, girl bull cooks who banked more money from the infirmary beds than from their union-scale jobs—but in the camps I slept in my middle-aged nose detected no pot and, like most other inmates, I was too tired and occupied for games and girls and drink. There were, however, at least two camp deaths, of men in their late twenties, from overdosed drugs, and one of an older man frozen in a drunken stupor, a common enough way, in Alaska, for a drinker to go.

Even in their relaxed and compromised form, the rules presupposed an elaborate security force. In 1974, in the easy time of building the haul road, it began with three men hired away from the Alaska state police, a force with the same function in the Native villages—law, protection, a paternal justice—as the Mounties in the Canadian Northwest, but without the royal tradition and the glamorous uniforms for state occasions. There was not much at first the three could do or needed to do but travel from camp to camp and show their uniforms, signs of Alyeska's determination to have order. With the start of pipeline work in 1975, the force was expanded by contracts with the two security organizations: Nana north of the Yukon, a new venture of one of the Native corporations (see page 217), Wackenhut to the south, a long-established detective agency that, responding to the times, had branched into guard service and industrial security. Each had its own chief and hierarchy of captains, lieutenants, and sergeants. By the summer of 1976, the entire force—Nana, Wackenhut, Alyeska's own security department, clerks, and secretaries—had grown to nearly two hundred, about the same size as the thinly spread complement of state police.

The twenty-two Alyeska officers were put through a week's training organized by the state police academy and came out as deputies, legally empowered to make arrests; most were ex-troopers—their number in 1976 matched the vacancies in the roster of the state police, whose pay, while very good, did not

quite compete with Alyeska's—or experienced policemen from
Outside. At their head was the former chief of the Fairbanks
police. (The Anchorage chief had also worked on the pipeline,
though not in security, assembling the cash to buy a retirement
home Outside.) The rest of the force received a simpler training
provided by the two security contractors—safety, lifesaving,
crowd control—along with the standard Alyeska orientation. At
five of the northern camps—Deadhorse, Galbraith Lake, Prospect and, on either side of the Yukon, Five Mile and Livengood
—they were reinforced by state troopers rotated every thirty
days, fed and housed by Alyeska but still drawing their salaries
from the state; elsewhere, a hard-pressed security man could,
like any other citizen, call in the troopers from the local barracks.
Since there were no lockups in the camps, there was nothing a
trooper or commissioned deputy could do with an arrested man
but sit down with him somewhere and lead him off to jail in
Fairbanks on the first plane out.

Against the tedium of stop-and-go work and the long hours,
the controlled blandness of camp life, you find ways of asserting
your identity—"just so you don't feel like you're a grub the
whole time," as my young taper friend remembered it. For the
weeks in camp perhaps you go to the trouble of fitting an elaborate stereo system into the hundred-pound baggage limit, to set
up in your bedroom. Or for the hour of shuffling through the
evening chow line and another hour of after-dinner beer with
friends, you dress painstakingly, according to your private notions of flash, which may be a red satin suit, flower-patterned
shirt, and hundred-dollar kangaroo-hide shoes. The food is at
least plentiful, with a choice of several soups, hot meats, vegeta-
bles, and at some camps good, interspersed with steak, lobster, in
summer outdoor barbecues, though as the project progressed
and the work force grew, these costly treats became less frequent. After dinner, besides talk and beer, there is the nightly
movie in the recreation hall organized by the project recreation
director or his assistant, closed-circuit television spliced with 60-
and 90-second spots on safety and the proper use of Arctic clothing instead of commercials (they are produced for Alyeska by
the Oregon Museum of Science and Industry); there is also in

most camps a roomful of exercise equipment, but time and fatigue keep it from getting much use.

If you find nothing else to do before turning into bed at night, perhaps you get up on a chair and hold a cigarette to the smoke detector in the ceiling of your room, to see if it reacts; supposedly it will not. In the isolation of the camps and above all in winter, fire is a camp manager's great anxiety, but Alyeska's approach to fire protection is curiously cost conscious and pragmatic. Early on, it was calculated that in the general permafrost and the deep cold of winter a system of reservoirs, water mains, and hydrants would be unreliable—and unjustifiably expensive for temporary camps. Hence, although smoke detectors are in all the rooms and automatic chemical extinguishers in the kitchens, there are no sprinklers. Three lines of defense meet fire if it comes: clear the men; attack with hand extinguishers; if that fails, bring in bulldozers and remove and destroy the buildings where it has spread—that was one of the reasons the camps were put together in modules. The men charged with these responsibilities in the camps are volunteers, as in a village fire department, supervised by a small staff of Alyeska professionals in Fairbanks and Anchorage; in June 1975, when Alyeska assumed the project management, it changed its contracts to require that the general contractors provide fire chiefs—one for each of the five sections into which the job was divided, covering several camps each.

The system was a recurring irritation to the Teamsters, which includes Alaskan firemen among its members; at one point a business agent presented applications and résumés from the entire Anchorage force, demanding that they be hired and a full-time team set up in each camp, but Alyeska remained adamant. In the northern camps in 1974 there were in fact several fires, one at Dietrich in the Brooks Range in mid-December that destroyed the kitchen, dining room, and recreation hall before it was controlled, but surprisingly the volunteer system worked as intended—the fires were stopped, no lives were lost. These were thought to have started in camp heating plants run on high-paraffin fuel from the Arco topping plant at Prudhoe Bay. The furnaces were redesigned. When another serious fire came a year later, this time at Prospect, south of the mountains, there was talk of arson, but an investigation produced no evidence—or none, at any rate, that the security men would admit.

There were other forms of excitement, of self-assertion, besides playing games with the fire-control system or joining the brigade. These centered often enough on food so that Alyeska's attention to diet, chefs, and catering contracts does not seem misplaced. Maybe dinner is Italian again—Spiro-Getty you call it, making a joke about the disgraced vice-president and the oil man—and you empty your tray on the floor. Maybe the guy next to you does the same and adds the sauces and condiments arrayed at the center of the long camp table and in a moment it is not funny: several hundred men dumping their food, banging their trays, and the jars and bottles starting to fly. "We found just having tablecloths on the tables makes all the difference," an Alyeska manager observed. "They tend not to throw things on the floor. At home you have a tablecloth, you don't throw things on the floor."

In the fall of 1975 there was restiveness among the welders in two sections of the line between Fairbanks and Isabel, a big camp in the broad pass through the Alaska Range. It came to a head in late October: gangs of angry men hemming in the few camp guards, fistfights, six actual arrests and more attempted; at Fort Wainwright, doubling as a work camp and the construction headquarters for the whole project, a troop of MPs appeared with truncheons, masks, and canisters of tear gas but did not use them—no one afterward would admit calling them in. The word spread quickly among the nearly 1,800 welders and helpers on the job at the time, and many of them stopped work. This was not a strike in the meaning of the Project Agreement: officially, the welders' union was urging its members to stay on the job, but at the same time it produced a list of grievances ranging from hazardous working conditions to inadequate clothes washers in the camps and too few privies at the work sites. Alyeska countered by insisting that the men go back to work as a condition for arbitration, and most went, the next day. (In the north, where work by now was slowing anyway with the season, fifty men demanded and got immediate R and R.) What it came down to was food. Alyeska had agreed at the outset to provide hot lunches where possible—in practice, those working

in or near the camps got them and everyone else carried sandwiches to the work sites. After mutterings about discrimination,
the hot lunches were abandoned, but the welders started helping themselves to steaks from the camp freezers, which they
could then grill on acetylene torches during the lunch break.
When someone noticed and tried to stop them, they reacted with
outrage.

Through most of that summer and fall, Tonsina, ninety miles
north of Valdez, was among the unhappy camps. A mile or two
north in a clearing beside the Richardson Highway stood a
roadhouse dating from the stagecoach days, a big barn of a place
with, at the time, new owners and large ambitions—a busy bar
that opened in late afternoon when the workday ended, girls and
all-night games in the rooms upstairs. In the camp, animosities
grew, welders on one side, Teamsters on the other; in July, a
couple of welders, brothers, took aside a bus driver who happened to be a Teamster shop steward and gave him a beating,
and men from both groups went on a smashing rampage through
the camp. Alyeska consulted with the unions, removed all three
men, and let it be known that they would not be allowed back;
the steward brought a million dollars' worth of damage suits—
against Alyeska, the Wackenhut security guards, the Section 1
contractor, Morrison-Knudsen. Later there was another outburst
—a camp helicopter demolished by a truck, another truck driven
off a bluff, window smashing in the dorms, $200,000 in damage,
it was said—and the troopers were called back and made a couple of further arrests. Alyeska began compiling a computer list of
troublemakers it did not want on the job, "people that would
cost us money in just flat transportation from one camp to another while they're being fired and rehired, fired and rehired," as
Bob Koslick, the company's second-in-command for security, put
it. By the summer of 1976, the list had accumulated 2,000 names,
but except for a very few about whom Alyeska had taken the
trouble to present evidence in court and secure injunctions, it
was not effective. The unions on principle did not recognize a
black list; anyone on it could appeal and most often win—or, if
terminated on one job, simply go back to the hiring hall and ship
out to another camp, with a different contractor or subcontractor; fired and rehired.

Incidents like those at Isabel and Tonsina were, for all one could see, random and personal, unpredictable kinks in the smooth flow of work that Alyeska had sought to assure by its contracts, rules, and private police; inevitable too, perhaps, given a work force of 23,000 strangers assembled for as many private reasons—and if the Jack London and Robert Service Alaska ever had any reality, it is even surprising that the troubles were not greater. Other kinds of incidents, concurrent with these, were purposeful. The purposes were often obscure or were deliberately hidden, but the common element is clear enough: power—the three-cornered maneuverings by which Alyeska, the state, and the unions each sought advantage over the others.

Late in the winter of 1974, as the trucks hauled north from the pipe yards at Valdez and Fairbanks, carrying the double joints, VSMs, insulation, and all the other materials that had to be distributed along the line for pipe work to start in the spring, there was a series of ugly accidents. Several of these were on the Elliot Highway, the old, narrow, gravel road connecting Fairbanks with the newer haul road built between Livengood and the Yukon and kept up by Alyeska. Jesse Carr counted six of his truckers dead in the course of that winter* and attacked Alyeska for setting schedules faster than the road would bear. (The speed limit now is 35 mph, lower on the downgrades, but when I drove it in Bill Granger's oil truck in summer, we averaged about 20.) Alyeska countered that the Teamsters were sending out youngsters inexperienced on Alaska winter roads, with no training on the heavy rigs they were driving.

While the Teamsters were arguing schedules with Alyeska, they were remonstrating with Walt Parker, the new state commissioner of highways, about the condition of the road. In November, Jay Hammond, a bearded bush pilot and professional hunter, had been certified governor after multiple recounts (see Chapter 2). Mr. Hammond was nominally a Republican, however, so Jesse Carr had opposed him, and Teamster complaints were coolly received in Juneau—and the state had, perhaps, its

* A colleague of Jesse's told me there were eight deaths. Neither total tallies with circumstantial lists of project fatalities I collected from three different sources—unless one assumes that the Teamster figure include incidents unconnected with the pipeline; or are mere rhetoric. See Chapter 9.

own reasons for letting the roads slide (see Chapter 9). In February, Jesse asserted himself—called safety meetings for every trucker in the state, including those driving for Alyeska, its contractors, and the oil companies, and for four days nothing carried by truck, which was everything, moved; it was like a general strike, but legitimate under all the contracts. "I've got the hammer to shut it all down," Jesse said. "*Safety*," an Alaskan Teamster official explained in another context, "*is the hammer*." And then: "Well, oddly enough," Bob Johnson recalled with satisfaction, later that year, "we got Mr. Parker's attention then, and they begun flaggin' the edge of the road, and they begun running the flag trucks on ahead. But we still had our problems."

Among the problems were further accidents and some deaths, attributed to faulty upkeep of equipment, and Teamsters agents visited the camps to red-tag trucks and call safety meetings, though most now lasted a day or less. Alyeska reflexively took the view that none of these actions was justified, that all had some other purpose and were unsupported by agreed procedures, which included written reports by the drivers on the condition of their equipment; and was seconded by Mr. Hammond's commissioner of labor, to whom the complaints also went. (Among the Teamsters' employees at the time were two former commissioners of labor, four deputy commissioners, two commissioners of commerce—most of them Democrats.) All through and perhaps among the actual issues, the Alaska Hire law focused contention. The state brought injunctions to compel observance; Alyeska's contractors got around it by requisitioning men by name, as they could by law and contract, and when pressed the Teamsters simply closed the hiring halls, maintaining that Alaska Hire conflicted with the Constitution and with national law—and in fact charges of discrimination were brought against the Teamsters before the National Labor Relations Board. In December, as the 1975 season ended, Jesse Carr and his aides under threat of subpoena flew to Washington to tell the House Interior committee that the local had not obstructed the project, and the pipeline, at least, was back on schedule; and were heard with sympathy.

The welders noticed the effectiveness of safety. They did not hold safety meetings, but there were "wobbles"—one day the three hundred welders in a section of the line completed 68 joints and the next day it dropped to eight and stayed at that level till a complaint was resolved. It was Frank Moolin's job to listen, answer, and keep count in his pocket notebook: "We know what it was, a petty dispute, it manifested itself by the welders' saying that the ditch was not a safe ditch and they would not work on the pipe because it appeared as if the ditch might slough. We may lose two or three days' worth of production, and then, what happened, the ditch conditions didn't change at all—but we reached an understanding. . . ."

Late in 1976, as Alyeska neared the end of pipe work and began shutting down camps and selling off surplus equipment, there was a series of large-scale work stoppages. These were simply wildcat strikes and were so called, no longer veiled in safety disputes, but as such they left the unions clear of penalties under the Project Agreement—officially, at least, they were not encouraging the men to walk out. Again, the real issues are unclear, but in each case the occasion was a question of the jurisdiction of the three most numerous locals, the Operating Engineers, the Laborers—and the Teamsters: which would control the chief share in the dwindling work. There was also a more general objective: the current Alaskan labor contract, which had set wage rates since the start of the project (see page 242), would expire in the summer of 1977 and was now by its own terms subject to renegotiation; the strikes that came may therefore be understood as the opening salvos of labor rhetoric, anticipating the contract talks that were to come.

This contest naturally centered at the northern and southern ends of the line where for various untoward reasons (see Chapter 10) the work had fallen seriously behind and now was driven in a desperate race against the winter shutdown—there were accordingly plenty of jobs still. Thus, in Section 5 at the end of August a thousand Operating Engineers went back to their camp and stayed there for three days, until Alyeska secured a court order requiring them to resume work; the ostensible question was whether they or the Teamsters were to service their equipment. A few weeks later, as the cold closed in and the work still was not finished, the Teamsters were embroiled again, this time with the Laborers over which would have charge of removing equipment from the work sites. The Teamsters made their point by the simple expedient of refusing to drive the buses that carried Laborers from the camps to work, and for nearly a week, from Fairbanks north to Prudhoe Bay, while Alyeska went through the legal process of proving that a strike was occurring as defined in the Project Agreement and securing the necessary order to stop it, no work was done at the job sites. The issue simmered till mid-December when at Valdez, where work was still possible, 250 Teamsters reported sick, all at once. This time Alyeska did not trouble to dicker with the arbitrator and the courts, but on the second day of the sick-out brought in a fleet of buses and, under the nervous escort of a squad of state troopers, loaded the men aboard and hauled them back to Anchorage. The job was about over, and men, even Teamsters in numbers, were becoming dispensable.

It is mid-October 1975 and somewhere on the line between Five Mile camp and Old Man, the first two beyond the Yukon, I find myself standing beside a big well-stuffed man, moon-faced and ruddy, who like me is looking on, arms folded on his chest. Conversationally I ask something about the work—a gang of helpers is heating the end of a length of pipe with a spider torch, preparatory to fitting the internal clamp that will hold the two joints in position for welding. The half-dozen words have tagged me, like radioactive isotopes, as something he does not have much use for, another of these goddamn pointy-headed commie-pinko Eastern Establishment writer fellas, and instead of answering, with a grin he tells me a story in down-home accents: the one about the sheriff who, finding a young black man at the bottom of a pond weighted with chains and dead, arrests him. *Dat nigguh so dumb,* the sheriff says, *he steal mo' chains den he cn swim wif!* It is a nasty enough story in itself but if you have a memory for recent history it is like a bare-knuckle fist in the mouth. It originated in 1964 with the lynching of three civil rights workers in a Mississippi village called Philadelphia. The two white boys, so far as appeared, were merely killed, but the young black, James Cheney, died with an ingenuity worthy of Vikings or Arabs: castrated and otherwise cut with knives, beaten with chains, burned, drowned, and if not then certainly dead, buried alive. *Yes, sir!* I think. *And you, sir, are a credit to your race!* But I move off, keeping the thought to myself.

The storyteller was a welder, and the welders as a group, if not this particular specimen, deserve closer acquaintance, for a number of reasons. With the Teamsters, they provided much of the melodrama that enlivened the building of the pipeline and for that time imposed their special qualities on the work and on the life of the camps and of Alaska at large; when an Alaskan made his defensive jokes about generic "Texans," it was generally, whether he knew it or not, the welders he was talking about. Among the sixteen unions and twenty-seven locals that supplied the labor by which the pipeline was built, the Teamsters and the Welders constituted opposite polarities, mutually repelling: the Teamsters, more so in Alaska than in any other state, with their large membership and carefully cultivated political connections, their huge investment trusts (more than $100 million in the pension fund alone), their hands on every wheel in the economy, collectively powerful though individually helpless; the Welders few in numbers, a conscious and jealously guarded, close-knit elite among all the crafts, each member as nearly irreplaceable as any one worker can be, with a worldwide choice of jobs and few local loyalties except to each other. The two positions have their obvious anxieties. It is not surprising, therefore, that a good many of the incidental labor squabbles, of which the riots at Tonsina were merely large instances, concerned Teamsters and Welders.

Welding is a simple enough process in principle: heat two pieces of metal to a high enough temperature and, if they are sufficiently like, are properly cleaned beforehand of all foreign matter and machined to an exact fit, they fuse—their atoms mingle and the two pieces become one, a single, continuous unit. In one form or other, this skill figures in several metalworking trades. What is different about pipeline welding is the degree of risk that may follow from inferior work, and therefore the standard to which the pipeline welder must qualify and against which his work is judged. That standard is a narrow one, with little room for variations in quality: if a single weld fails out of the tens of thousands that hold the pipe together, the pipe breaks at that point and the entire system fails, at immense cost in lost rev-

enue, in repair and cleanup, with other losses that may be incalculable. On the Trans-Alaska Pipeline, the welders were the only
ones whose work was signed (stamped with an indentifying
seal) and therefore could be and was exactly monitored, any
flaws traced to their makers. Among the construction trades they
are like artists, with that intense personal and physical connection with their work, and their union is like a guild.

The guild to which most pipeline welders belong is the ver-
bosely named United Association of Journeymen and Appren-
tices of the Plumbing and Pipe Fitting Industry of the United
States, Local 798—its members are 798ers for short. Its headquarters is in Tulsa; if its members are called to work in a remote
part of the world, it may set up a local office, as it did in Fairbanks for the TAPS project.

If this union takes you in, you begin an apprenticeship that
may last years. As a welder's helper, you sandblast the pipe end
to clean it, run the spider torch, position the pipe and set the
clamps, stand by to hand up tools and, as each layer of metal is
laid down in the joint (six or seven such layers, or beads, in the
.462- and .562-inch pipe used by TAPS), wire-brush it and, if
need be, grind away surface irregularities with a rotary wheel
and polish the finished bead with a buffer. And you watch and
learn and if you are lucky in the welders you work with, they
teach you. It is an intimate relationship, each dependent on the
other's skill; welder and helper may stick together, travel from
job to job like knight and squire, a team.

To rise from apprentice-helper to journeyman welder, you pass
a destructive test, and a similar test is repeated every time you
apply for a job. John Ratterman, who wound up in charge of
Alyeska public relations, had been a helper as a kid in Oklahoma:

> Everyone who applied for a job took the test. They gave him
> two pieces of pipe of the kind we were working on and he
> welded them together and then they cut out of that four
> things that they called tickets, that go across the weld.
> They'd take each ticket in a little portable machine that has
> claws on it, grips the metal on either side of the weld, and
> then with a crank and a geared mechanical thing they pull it

apart. And if the weld breaks on any of those four tickets, the guy fails his qualification test and does not get the job: it has to break in the metal before it breaks in the weld. All welders who had welding cards—and five percent or less actually passed . . .

That such a test must be repeated for every job is due to several variables—in the welding technique specified, the character of the pipe, the filler rod melted into the joint (in electric arc welding, the commonest type), the changing code of welding. The welder himself varies. The work requires both a steady hand and sustained intense concentration and therefore pulses with the emotions of the man who does it—if he thinks his wife is neglecting him or he has a running quarrel with a Teamster bus driver or is keyed up and anxious about a test on unfamiliar equipment or has simply been out late drinking the night before, the quality of the work suffers. Like a gift for music or painting, a welder's talent can be recognized, cultivated, trained, but not compelled; some have what others for all their will and effort never acquire. Like an art in another way, a welder's skill is not a fixed and final state but a process, continually evolving as it is practiced, and every weld, like a writer's books, will be examined, tested, and perhaps condemned—with the difference that the standard is no critic's idiosyncratic judgment but absolute; whether, finally, the weld breaks or holds.

These are the common conditions of the welder's trade everywhere which make the work both proud and nervous. On the TAPS job, they were, like everything else, complicated and intensified. To certify welder qualifications, Alyeska contracted for test facilities at Tulsa and Fairbanks that would simulate the quality-control procedures it had established. Each welder who applied was required to execute a complex series of passes to join two pieces of the 48-inch pipe, and after each pass the bead laid down was inspected visually. The completed joint was again inspected and, if accepted, was X-rayed and the film examined. Finally, fourteen samples were cut from the joint—not the four "tickets" John Ratterman remembered from twenty years back— and subjected to eight different destructive tests (some were done twice): the basic one of being pulled apart, but also bending and twisting in all dimensions, cracking, nicking, hammerlike impacting. Only if he succeeded at every stage of this procedure would the hopeful welder be certified for the job, and, conversely, if he failed at any point, the test was over, no appeal or excuses, and he had only to collect his day or two of testing pay and catch the next flight out.

Something less than 800 welders got through the two test centers, of whom about half, along with 1,200 or more helpers, would be actually in the field, welding, at any one time during the two years of pipeline work. Among them they were responsible for the roughly 60,000 joints it took to put the pipe together (in addition to the 42,000 automatic welds performed in the double-jointing operation—see page 215); each joint was made up of six, or seven, circumferential beads, each of which, in turn, was divided among two or more welders. Every step in this procedure was subject to controls similar to those applied in the original qualification test: inspection of each bead; X-ray examination of the completed joint; and, periodically, the cutting out of a joint which was then shipped to a laboratory independent of the two testing centers for the same series of destructive tests. When we consider that the purpose of all this scrutiny was, among other things, to weed out those welders who did not consistently perform up to standard under the pressures of schedule and the exasperations of Alaska and its weather (with some allowance for an occasional off day), what is surprising is not that the Alyeska welders formed a temperamental bunch within a trade notorious for temperament—welders against the world!— but that to the end they remained, most of them, more or less sane.

October again and I am back in Anchorage, talking with Jake Lestenkof about the pipeline training program he runs for the Alaska Federation of Natives. He is an Aleut, started life on the Pribilof Islands clubbing seal pups for the present fur concessionaire, as his ancestors did for its American predecessors and the Russian American Company before them. He is a slender, thin-faced, tawny-skinned man with a narrow mustache, some variety of Russian you might say, or anyway Czech; and intelligent, later the executive vice-president of the AFN—Aleut history seems to be the one subject on which he cannot speak coherently. We have touched on most of the unions he works with, but when I mention the welders he turns thoughtful, cautious. 798, he explains, is a local out of Oklahoma, very close to the oil industry, partners with the oil men for years and years all over the world. "I think," he ventures, "they may still think that they're in a foreign country, up here in Alaska." He says it carefully, not liking to think about it, puzzled. But are there no Native welders? I ingenuously ask. "There has been welding training going on in Fairbanks all summer, we have a number of Native helpers, in the union. The bad part about it, they don't seem to stay too long, once they're on the job site." He ponders further, and finally: "It seems to be a very difficult bunch to work with."

Amen, brother. Amen.

I spent the summer of 1976 roaming the pipeline with my son, dutifully hunting evidence of the new productivity plan Frank Moolin had promulgated in January—more work, shorter work days, fewer workers—and thought I saw signs. In the late afternoons before dinner there were young men running in shorts and sweat shirts along the access roads to the camps, with time and energy to jog off the heavy diet or renew appetite. Volley-ball nets had been set up in corners of the gravel pads on which the camps were built and there were games in progress. The atmosphere was that of an overlarge summer camp or a utilitarian college campus in the spring. In the late evenings when the movies finished, the men gathered in the corridors outside the dining halls where coffee and cold drinks, pastries, had been set out on long tables, ice cream in freezer chests, relaxed and quiet, chatting.

On the work sites, it seemed, the crews had become teams, each member knowing his job and responding to the rest. The insulators on the aboveground pipe, for instance: the suction cups of the unfolding machine grip the galvanized-shiny 23-foot panel of insulation and open it, the boom lifts it from the flatbed and gently lowers it around the pipe, and the men skip along the platform beside it, doping the weep holes and inserting the pop rivets that will hold it in place, with the quickness and precision of dancers, a pleasure to the eye, and move on to the next panel. By four or five in the afternoon, although even in the south the full day of summer would last another eight or ten hours, the work site was still and the men had collected in one place, tossing a ball and talking, waiting for the bus back to camp. There were, I noticed, a few girls among them now. High up on the seat of a Cat a young man lolls bare-chested in the sun and a girl gambols toward him across the churned earth, a sandwich in a plastic bag in one hand, a paper cup of water in the other. She arches upward on one toe, like a ballerina in profile, reaching the sandwich, he leans and takes it, and she sprays the water at him, dances off, turning to blow kisses, and all of this charmingly, with the calculated grace men have always worshiped shining through the jeans and clumsy boots, the open-necked blouse: she is a nymph sprung from the sparse Alaskan forest, returning to it, beckoning; there is a bow of red ribbon tied to the peak of her hard hat. Yes, an old construction man observed, looking on, there *are* a few women on the job this summer, kind of mascots, you might say, morale-builders; but he seems as pleased and tolerant as I am.

So in July in Fairbanks when I went out to Fort Wainwright for a visit with Frank Moolin, I greeted him cheerily—"You're still here, I see!"—and with a smile he agreed that he was. The productivity plan, however, had not gone quite as intended. He had indeed started the crews out in February with shorter hours —not, to be sure, the contractual 40 hours' minimum mentioned in the written directive, but 60 or so, which was something. By now, though, the work force had once more swelled to last year's size, about 23,000, and they were getting paid for 78 hours a week, an hour or two less than the 1975 average. It was a very modest improvement, and it was qualified: last year's figure did not count travel time from the camps to the work sites and back, that had been paid separately at a flat rate; now the 78 hours included actual time on the buses, which is to say it was worth time and a half—and that, it occurred to me, was why I had noticed the crews at day's end taking time to clean up and wait. They knew how to work the new rule, all right.

The numbers in the work force and the hours they were working were abstractions, of course. What they meant was the hard, irreducible reality of money. The announcement of the productivity plan in January had coincided with a new estimate of the cost of the pipeline system—it had taken another billion-dollar leap, and the Alyeska accountants and managers now figured it would take $7 billion to finish the job; one of the objects of the productivity plan was to make that total stick. As the summer advanced and it became apparent that the plan was not having much effect, the managers tossed another 10 per cent onto the table: on July 1st, a week before I stopped by Frank Moolin's office for another chat, Alyeska let it be known that the estimate had been revised upward again, this time to $7.7 billion.

The reason for the soaring job time and its attendant cost, Frank figured, and from his standpoint it was a safe enough bet, was the troubles they had been having with the welding. These troubles are complicated and interesting enough so that we shall consider them separately, along with some other reasons, at a later point (see Chapter 10), but the gist was that Alyeska was having difficulty matching up X rays with welds: within the terms of the Stipulations, it was unable to prove that all of the 60,000 field welds were acceptable. From a secret audit of the X rays, completed in the spring (but it did not stay secret for very long, another interesting matter we shall come to), it appeared that about 4,000 of the weld X rays were what Alyeska delicately termed "questionable": that meant there were no X rays, or if there were they showed defects in the welds that seemed not to have been repaired. What this meant in turn was a lot of unanticipated work, more men, more hours: not just X rays; many of the "questionable" welds were buried, some under stream beds; so far as they were proved in fact defective they would have to be cut out and replaced. Under the welding code, the rule is that the piece of pipe welded in to correct a defect must have a length one and a half times the diameter of the pipe, or in this case at least six feet; the replacement pipe is called a "pup." If it turned out, as it might, that it was necessary to insert a pup in place of every one of the 4,000 unverified welds, that would add up to something over four miles of additional pipe, which as it happened was more than Alyeska had on hand in the heavy-grade, .562-inch thick, that was required. A rush order for that amount of pipe, about 3,600 tons, had accordingly been sent off and the Japanese were exerting themselves, but the barges were not expected before the end of the month. Meanwhile, Alyeska was digging bell holes at the points where the doubtful welds were supposed to be—in places they came so frequently that from the air the line looked like a battlefield after an artillery bombardment—and with crossed fingers was taking new X rays.

About half the questionable welds, nearly 2,000 of them, were concentrated in the northern section of the line, which is to say that perhaps 25 per cent of the field welds completed there were affected. The Alyeska project manager for that section had therefore been fired, that very day as it happened, and when the additional pipe came in Frank would be sent up to take charge of the repairs, though he did not mention that possibility at the time and probably did not know of it himself; his superiors in Anchorage were beginning to look closely over the shoulder of Engineering's Man of the Year. At the same time, a new job, chairman, had been created at the top of Alyeska, though just what it meant was not yet clear, since there was no board to be chairman of, and Ed Patton had been moved into it. In his place as president, BP's W. J. Darch had been inserted, hastily summoned from large projects in Libya and Abu Dhabi, the change so abrupt that no one within Alyeska seemed to know anything about him except that he was English (actually, Welsh, a distinction lost on most Americans). While their oil-company overlords insisted that the change of management was a coincidence, had in fact been long planned in anticipation of the completion of construction and the beginning of the operational phase of the pipeline, everyone else was naturally skeptical, inferring large and perhaps sinister meanings.

Confronted with this sea of troubles, Frank Moolin remained genially confident that the line would be completed on schedule or at any rate in time for start-up in the summer of 1977. The probability for completion in May, he said, was 50 per cent, 75 per cent in June, and a round 100 per cent certainty in July—the oil would begin its slow progress south from Prudhoe Bay and, with time for final cleaning and testing and the filling of the Valdez tanks, would be in the tankers and on its way wherever it was going sometime in the fall. I don't suppose these cheerful percentages actually came from the computers; it was simply the way he had grown used to looking at things, habitual.

In the face of such smiling confidence, there was not much more to talk about. We were both curious as to what Mr. Darch might have in mind, but since neither of us knew, or knew much about him, there was not much to say about that either. At the time, defensively, everyone seemed to be looking for funny things to say about him.

"It's sort of ironic," Frank considered, offering one of his own. "Here it is, the Bicentennial, and we're working for the British right now."

"Undoing the Revolution!" I suggested.

"They won after all!" he countered.

"Come home, America, all is forgiven!"

Chapter 8

THE WATCH ON THE LINE

> Our point has always been that we've done our job one hundred per cent correctly if you've never heard of us, see?
> —Andrew P. Rollins, Jr.

The Right-of-Way Agreement under which the pipeline was built is prefaced by a careful statement of intent. Among other interlocking purposes to which the signers bound themselves,

> The parties shall balance environmental amenities and values with economic practicalities and technical capabilities, so as to be consistent with applicable national policies.

The sentence, which sounds like a bland, unarguable generality, became the standard by which the Agreement would be interpreted and, as such, it determined the courses of action by which all concerned would attempt to do their various jobs—builders to build, surveillance officers to monitor construction, oil men to profit; and not only for the three years it would take to build the pipeline but for the life of the system, as long as it lasts. It deserves, therefore, serious attention.

The "parties" to this purpose, as to the entire Agreement, were, on the one hand, the seven oil companies whose control of Prudhoe oil made it necessary for them to create a system to bring it to market (later augmented by BP to make an eighth pipeline owner—see page 206); and, on the other, no less than "the United States of America"—all of us, as we are named in our government's contracts—represented by the Secretary of the Interior. A contract is a weapon sharpened at both ends: its obligations are mutual. The debate had cast the public in the role of onlooker, anxiously protecting itself from the exercise of pipeline technology for economic ends. The preamble evenly distributed these contradictory realities. The Secretary and his delegates could not confine themselves to "environmental amenities," they were obliged to give equal weight to "economic practicalities"— the cost, which is time and schedule as well as money—and "technical capabilities," what men actually know how to do as distinct from what they may sometime know or think they have some chance of knowing. For the men who would have to translate purpose into the reality of construction conforming to "national policies"—producing the oil with the expedition demanded by Congress, but harmlessly (see page 168)—it was a hard standard, a knife edge, sharp. We may quarrel with "environmental amenities"; is the air we breathe, the water we drink, an "amenity," a frosting on the cake of life? But the phrase, like the whole contract, is meant to be sweeping, and there is perhaps no other so full and so brief.

Together with the Stipulations (see Chapter 4) which legally formed part of the contract, the Agreement outlined the procedures by which these purposes were to be accomplished. In essence, Alyeska was obliged to submit its plans for review and approval by the Authorized Officer acting on behalf of the Secretary of the Interior. Written approval of the plans became in effect an order to Alyeska specifying how the work was to be performed. The Stipulations refined this procedure. Alyeska was to submit first a preliminary design for each "construction segment"—any part of the job that could be described and carried out as a unit, from a single pipeline stream crossing to a complete pump station to the aboveground and below-ground pipe for each of the six linear sections* into which the project was

* Six linear sections. Elsewhere, I have referred to *five* sections. This was so for construction purposes, but the pipeline was planned in six sections, and Alyeska apparently submitted six general applications to the Authorized Officer; when the contracts were let, however, a single combine, Arctic Constructors, undertook the two northernmost sections, covering about 225 miles (75 to 100 miles more than the other four contractors), and thereafter within Alyeska they were referred to as "Section 5/6."

divided. Further, in reviewing the plans, the Authorized Officer might request additional information or suggest changes, either as a precondition for approval or later, in the course of construction. The Agreement provided a language in which to discuss these matters: the Authorized Officer's approval was called a Notice to Proceed, an NTP; the plan on which it was based was an NTPA—a Notice to Proceed Application.

The Agreement allowed no deviation from the NTPs in the course of construction except with the written approval of the Authorized Officer. For assurance that this would be so, his representatives were allowed access to the work sites at all times, with room, board, and office space to be furnished by Alyeska in the camps. To ensure that the project was carried out in the approved form, they had the power to stop work that did not conform or that appeared to endanger health or the environment, with provision for Alyeska to appeal to the Secretary from orders it considered unreasonable; but, given the mutual commitment to completing the job with economy and speed, it was not a power that would be capriciously exercised. Besides stopping the work, the Authorized Officer might also undertake at the owners' expense any requirements of the Agreement and Stipulations not performed by Alyeska, other than the actual building of the pipeline system. All costs of reviewing plans and monitoring construction would be borne by Alyeska, with bills submitted quarterly.

These provisions were all in the nature of general principles. Technical specifics were referred to existing federal regulations for pipelines. Where there was no such standard, the Agreement and Stipulations created it: the X-raying of all pipe welds and the treatment of tanker ballast water before discharge at Valdez, mentioned earlier; prohibition of interference with big-game and fish migrations, the calving of caribou and moose, the spawning of fish; control of the tankers so as to penalize the dumping of ballast water en route; reporting of oil spills, with plans and equipment for cleaning them up. Whether damaged by spilled oil or other cause in the course of construction and operation of the pipeline, Alyeska would be obliged to "rehabilitate" any fish or wildlife habitat or other "natural resource"—another vague but all-embracing phrase.

All this occupied 44 double-column pages in the printed version of the Agreement and Stipulations that served as a manual of conduct for all concerned in the pipeline, but the ideas themselves are straightforward enough. There was, however, an immediate complication: not all of the right-of-way was on federal land, and the state of Alaska was intent on asserting authority over its own. Its authority was in turn ambiguous; in the process of settling the Native claims, the state had been able to patent only a small part of the acreage granted at statehood, and the rest represented tentative intentions, with a good deal of sorting out still to be done among state claims and those of the Native villages and regional corporations. Hence, in January 1974, before the federal Agreement could become effective, it was necessary for the state and the Department of the Interior to agree on how much of the right-of-way would be subject to each; in May the oil companies signed a separate lease with the state, complete with stipulations paralleling the federal ones, covering about 240 miles of the route (including 39 miles on private land, most of it belonging to Native corporations). Under the three complementary agreements, all plans would be submitted both to the Authorized Officer and to his state counterpart, known as the Pipeline Coordinator; the federal official might give his authorization for work anywhere on the line, but that of the Pipeline Coordinator was also required on state land. Since reviewing the plans was meaningless without control over their execution, both offices would theoretically monitor construction for the whole length of the route, but in practice, the state concentrated its efforts at Valdez and on the southern third of the line; to avoid duplication of effort, the Authorized Officer provided most of the detailed engineering work from his own staff and consultants.

There are two things to be noticed about the regulations under which Alyeska and its overseers were to operate. In the first place, there is the innocent belief that, to shape private motives to a predetermined standard of public good, we have only to declare the framework of law in which they are to operate, with suitable penalties for failure to comply—that we can, if you like, ordain and control future events. Secondly, whether the regulations are regarded as loose and vague or broad and uncompromising—both views are possible and perhaps even true, depending on how you look at them—they would make extraordinary demands on the two men and their handful of deputies finally responsible for enforcing them: the technical knowledge to make accurate distinctions, day by day, among a multitude of possibilities; the force of character to impose one's judgment on an army of competing motives, intentions, interests. These are large assumptions.

Andrew Rollins, retired from the Corps of Engineers and newly designated as Authorized Officer, was the man on whom these assumptions fell. In January, with the Right-of-Way Agreement safely signed, he put aside his general's rank and arrived in Anchorage to begin assembling a staff and, like everyone else in town at the time, look for office space. By the end of the month he had signed a contract with an organization called Mechanics Research, Inc., to serve as consultants on engineering and environmental matters (and it, in turn, subcontracted these responsibilities to two other firms, Gulf Interstate Engineering, and Ecology and Environment, Inc.).

The staff of Rollins's Alaska Pipeline Office grew to about a hundred people, including the consulting engineers and biologists supplied by the three contractors. Of these, about two thirds stayed in Anchorage, processing NTPAs; the rest were assigned to the field. Two monitors—AOFRs, Authorized Officer's Field Representatives—were assigned to each of the six sections of the pipeline, spelling each other ten days on and four days off, the schedules arranged so that someone would be present at all times; there was also an AOFR in residence at Valdez, and another who traveled the twelve pump stations; and regular assistance from the contracted technicians. In this considerable group, only Mr. Rollins and his secretary were classed as permanent employees as the Civil Service defines permanence; the rest were temporary, either as consultants or else borrowed from various federal agencies, to which, if all went well, they would return when the job was over.

Most organizations, regardless of size, take their tone from whoever is at the top, whoever has power: their style—the way things get done, their feelings about the job. The style is the man. Accordingly, when the opportunity offered, I went around to present myself to Andrew Rollins in order to learn what I could about his people and the curious business of pipeline surveillance. The quarters he had found were at 8th Avenue and F Street a little off-center from downtown Anchorage, a two-story structure of bronze framing, glass, and gray-painted cement block —any commercial office building of the 1960s, cut down to suburban quarter scale and translated to Alaska, new-built but haggard; inside, warrened into cubicles for the staff, according to their degree.

I arrived in the midst of an unusual occurrence. Behind the conference desk in the big second-floor office at one end of the building, the AO was going over the words and timing of a press release with Bob Gastrock, the APO's young public information officer who had taken a flier from a government job in Washington but had not yet had many opportunities for writing. "Our point," Mr. Rollins explained with a glance over his half glasses, "has always been that we've done our job one hundred per cent correctly if you've never heard of us, see?" A lesson in style: the release was a departure and therefore uncomfortable. Its occasion was a pipeline bridging of the Gulkana River in place of an intended understream crossing, a change Alyeska had proposed for engineering and environmental reasons in which the APO concurred and which it was now obliged to defend—"Our blessing," Mr. Rollins observed, "is that I'm five thousand miles from Washington." The river is fast, shallow, broad, and winding, with many rock ledges, good for three or four days of lively canoeing if you don't mind getting wet; clear and clean because its source is a lake, not a glacier, where in June going downstream you share the currents with thick-bodied red salmon coming up and with your paddle try to emulate their grace and pertinacity; and because for most of its length the Gulkana is near the Richardson Highway and the pipeline, a symbol accessible to Alaskans no less than to environmentalists Outside and in Congress. And therefore the press release and the unaccustomed visibility. He finished reading and Bob Gastrock departed with his approved text. I slid a chair up to the desk to talk.

The military rank, even in retirement, suggested rigidity and an iron-fisted imagination, and that thought had been embroidered by magazine writers. The man I sat opposite still signed himself formally, A. P. Rollins, but had become Andy to Alaskans and friends and, in the present context, impatient with the old title. Engineer by inclination and training, manager of undertakings on the large scale favored by the Corps, canny enough as a bureaucrat to have lasted a career: the mix left no room for soldierly stiffness. In his Alaskan flannel shirt sleeves and quick, emphatic speech, he was a conscious civilian, responding to the job he had taken on, letting his style be seen: gray-haired and getting lean, the thrusting jaw unwrinkled, sixtyish at a guess but not old—the Army, the Corps, turns a man loose before all the life is sucked dry; a fighter when fight is needed, or that is the effect he wants, but by nerves, not temperament, and too well schooled in the ways of government for frontal assault, he will proceed by indirection, unobtrusively. If there is rigidity behind the smooth, gray face, it is not military but managerial, in an older bureaucratic style, a determination to do no more and no less than the law, the Right-of-Way Agreement and its Stipulations, require, and keep his feelings to himself; and that impassive correctness he will also demand of those who work for him and will closely watch, like the commas and periods in Gastrock's press release.

We talked through the work remaining on the pipeline system, defining the work of the APO. By now—this was the second winter of construction, one full season still to go—there were only four or five big plans still to be worked out, submitted, approved: the oil-spill liability fund, collected, a nickel a barrel, from whoever bought the oil (see page 168); the tanker traffic control system in and out of Valdez, but that was a job for the Coast Guard, he had only to mesh their certification with his own approval for the start of operations.

"We got forty-five days," he said, "before we really get cranked up for this summer's work, because once we get in the summer it's going to be the day-by-day field problems—we had 'em last summer and they called in every day and they've encountered something, *well*, they're going to have to make a field change. And then be ready, and people in the field, to answer the problems when we come up with a remoding situation: we get in there and we dig out and we find that we've got frozen

material that we didn't anticipate, so we'll have to go high instead of low—or they encounter a condition that wasn't intended, so they've got to make a field change. . . ." A lifetime of
work in places as remote as Alaska had not worn the Southern
idiom and softness from his speech, Virginia or southeast Ken-
tucky to my ear. The voice slowed, concluding: "No, I don't see
Alyeska with a real tough problem. I think Alyeska's present
schedule is completely attainable."

Frank Moolin had been saying the pipe would be complete
by the end of the season, November at the latest; the terminal
and the pump stations were another matter, still running behind,
but not at least his problem. I wondered how that sounded to
the AO. November?

"Well, *you* say that, *I* didn't say November, I said their *present
schedule*. Moolin better *shut up* because his boss said not to talk
to people about that, the owner companies are goosey about
that, they don't want the word to get out about that. Goddamn
it, Alyeska's official position on that is still *next July!*" As he
talked, he thumped the polished surface of the desk, rhythmi-
cally underscoring the words, and the insistence was puzzling:
Alyeska was indeed turning circumspect in what it would or
would not say, but the schedule looked to be the least of their
worries. He and Frank Moolin, it occurred to me, would have a
lot to do together, but the law that bound them, like an old married couple, did not oblige them to like each other. Talking one
time with Moolin about a new scheme for verifying welds, I
suggested the AO would have strong views about it, and "He has
opinions about everything else, why should that be any
different?" he shot back, in his own half-Irish style of gab.
"Whether they're called for or not, he has opinions." Now Rollins
was giving him an answer of sorts. Moolin might be in charge of
building the pipe but he did so under several layers of management. The Authorized Officer was at the top of his particular
pyramid, it was Moolin's bosses he talked to, and, so long as he
walked the narrow path of the Stipulations and kept his head
down, was answerable only to Washington; and that, as he
remarked at the outset, was a long way off.

We worked on through the rest of the construction schedule.
The end was what the Stipulations called rehabilitation: recon-

touring the gravel pits and equipment dumps, the stream cross-ings and edges of the work pad, seeding the bare earth with grass, in places setting sprouted stubs of willow. I wondered. None of this was yet visible, but the grasses, where they had been planted experimentally, had worked in reverse: voles and other rodents found them, flourished, and stripped them bare be-fore the native plants could establish themselves in accordance with the plan. There was nothing to be done about the work pad, an 800-mile gravel road paralleling the line, which would stay, for maintenance and repair; nor about the crushed-stone fill covering the buried pipe. And aboveground—

Mr. Rollins caught at the idea, warming: "You are never going to be able to fly over Alaska and not see that there's a pipeline. *It will always be visible*—and the biggest visibility is going to be —and the people begin to notice what they're doing, they're going to be *absolutely shocked* at what an eyesore this God-damned elevated pipeline is! It's a *monster!* It's the Goddamned-est-looking thing I've ever seen in my life, mile after mile of this thing snaking over—" He broke off, quieted, slowed, and concluded: "The fact of that thing being there is a monstrous im-pact on the terrain."

I demurred: I was not so sure of the unloveliness of the pipe, of its being, in a strictly aesthetic sense, beyond appreciations; and—telephone poles, railroad tracks, bridges, high-tension power lines, why not pipelines?—the earth envelops them and makes them part of herself, the eye adjusts and they blend with the natural scene. For myself, I had flown the line, looked at pieces of it, but not yet walked it, which is the only way to see; and was therefore still thinking. But again the vehemence sur-prised me and I wondered, if it was real, what it meant. I was, of course, a writer, and writers are simple fellows who may know what things look like but not how they work; environmentalists by nature, not technicians. I was also one of that large public that had made the fuss about the pipeline and created the pres-ent situation in which the safe and legal answer to every doubt-ful stretch of frozen subsoil was to elevate another mile of pipe on stilts, "go high instead of low," and the AO, whatever his pri-vate views, was here to see that, by God, I got what I thought I wanted—and would pay for it too, as the oil men were beginning to say.

It was the environmentalist that he seemed to be addressing. Several groups had sent representatives to observe—"watching us to see if we're doing a good job"—and Mr. Rollins had invited them in, stood at a blackboard, chalk in hand, giving an old-style briefing on the work. They had annoyed him by discovering ironies and a general's posture beneath the open-neck civilian shirt:

"They've been up here twice on trips and I've gotten mad at people on 'em, people that always fuss at things and never found time to be at the exit interview—come up here on a trip and cain't sit in on the exit interview and report to us on the things but get home and write their reports and put it in the paper. Course," he concluded, with irony of his own, "I was raised *a little different way.* . . ."

Although the state lease was not signed till three months later, C. A. Champion, who is known as Chuck in the newspapers and everywhere else in Alaska, went to work as State Pipeline Coordinator at the end of January, about a week after his opposite number at the Alaska Pipeline Office. He was around forty at the time, a civil engineer working in Juneau in the Soils Engineering Division of the state's Department of Environmental Conservation, before that in Anchorage, doing similar work for an Alaska-based oil company—one of the small group of Alaskans, therefore, whose professional experience was directly applicable to the pipeline; a bureaucrat too but politic enough to survive the change of governors that came with the election that fall, set apart from Andrew Rollins by a short generation and its several wars. There was no place on Chuck's organizational chart for a public relations officer. He had no inhibitions about talking to the press (or inquisitive authors) in his own words and voice.

The SPCO—State Pipeline Coordinator's Office—shared a second floor with an insurance company on 7th Avenue a couple of blocks from the federal establishment. The office seemed spacious: fewer people, about thirty, half of them, counting Chuck and his deputy, Cy Price, in Anchorage, the rest on the line as field surveillance officers and their alternatives or deputies

(FSOs—the state has its own language, distinct from the federal). In the field the state month divided differently from the federal. An FSO worked fifteen days at a stretch, then took a six-day break while a deputy filled in; which sounds equivalent to the federal schedule, but the difference in style is worth noting—there was only one man in charge in each section of the line, his responsibility undiluted, and early on Chuck let it be known that it was to be exercised.

I first met Chuck Champion on a working Saturday morning complicated by his having charge of several half-grown children and their friends—they were at a gym class, were due somewhere else, and he would be coming and going, giving his wife a break. Broad enough to seem less tall than he is, which looks to be around six feet, he sprang from the desk chair and athletically prowled the room while we talked—a halfback, you would guess, back in California where he grew up, with basketball and track in season, though the games have left his face smooth, open, and unmarked. His speech is sprinkled with aggressive images he liked seeing quoted in the papers—punching someone out, cutting someone up, to impel agreement; the habit of a man who in youth would always have been big enough, imposing enough, for the threat alone to suffice, rarely tested. Now, you would say, he has reached the age of ambition, training and experience behind him, large possibilities ahead—a very visible man in the small world of Alaska, a term or two hence he could be governor if he wants, if the job goes well, if he remains visible, but for now the ambition did not seem overt, and it was the job alone that occupied him.

"One of the things that we face on this pipeline," Chuck Champion was saying by way of introduction, "is that we have to draw the line between the environmental amenities and the constructibility of the damned thing. You can imagine how long I would last if I held Alyeska to zero environmental impact—they'd just boot me out the door!"

It is the same pairing, environment and feasibility, set forth in the Right-of-Way Agreement, but the state cannot afford the judicious passivity of the federal officials, its interest in the pipeline—and therefore Champion's—is dynamic. Since the discovery of oil at Prudhoe Bay in 1968, successive state govern-

ments had made shift to balance deficit budgets against the expectation of prosperity paid for by oil royalties and taxes, but
these hopes could not be realized, these problems solved, till the
oil began pumping. To this extent, then, the Alaskan interest was
the same as the oil men's: completing the pipe, lifting the crude,
was the first concern; Alaskans would still have to live with their
environment, whatever happened to it in the process, but it is an
"amenity," and concern for it necessarily stopped short of serious
interference with the project and its schedule. Moreover, where
it counted, the Alaskan interest in the pipeline was as practical
as the oil companies'—if it failed through bad design or careless
construction, it would not merely foul the land and streams with
spilled oil, it would cost the state money in cleanup and lost production. Hence by the time pipeline construction began, the
FSOs entrusted with surveillance were not environmentalists but
engineers and builders experienced in the lesser oil fields and
pipelines on the Kenai Peninsula and around the Cook Inlet,
men who, like their boss, would take a positive and assertive role
in getting the pipe laid, on time and to specifications.

It did not seem, however, that the two aims need conflict. "I
feel," Chuck was saying, "that in order to have environmental
protection—we must have development, I concur with that, and
we must have environmental protection—we are going to have
to pay more for development. I don't want to stop the world and
I don't want to rape it, and I'm willing to pay whatever extra it
costs for gasoline or for my house or for oil, in order to protect
the environment." This was early in 1976, at the start of the final
year of construction. As the project moved toward conclusion
and the costs continued to mount, the state had second and third
thoughts about how much it was prepared to pay: the higher the
cost of construction, the less it would realize in revenues (see
Chapter 11). Hence, while the federal monitors had only an incidental interest in cost so long as the general will of Congress was
fulfilled, Mr. Champion and his FSOs kept track with growing
anxiety.

There was a more general motive for this activist approach,
shaped by Alaskan history. "The facts are," Chuck insisted, "that
this oil belongs to the people of the state of Alaska: it is state
lands, it is state's oil, and the development of that oil," he con-

cluded, slipping back into the jargon, "should be used to maxi-
mize the socioeconomic benefits to the people of the state of
Alaska." This was the Alaska politician and booster talking, but
what he was saying was true enough of the title to the land
under which the Prudhoe field was found, conferred by statehood, though what the land would be worth without the oil is
another question altogether; so far as the oil itself is concerned,
however, what the state actually owned was its 12½ per cent
royalty on production—which it could, of course, take out in
kind if it could find anything to do with it, any way of transporting it to market and disposing of it once there.* Even if the
state's proprietorship had been indisputable fact, however, its
rights had been much qualified by the intervention of Congress
and the possessive attentions of Americans in the other states;
one more incident, as it looked in Alaska, in the long history of
tutelage from Outside. Things in Alaska are national, not state-
wide, as the historian Al Mongin remarked when I began my
investigations.

Hence, overseeing the pipeline (and the oil companies behind
it) had from the start an importance for Alaska beyond immediate self-interest: however divided by federal authority in the person of the Department of the Interior, it was a chance for the
state to assert once and for all its command of its own destiny.
Alaskans of whatever stripe shared and approved this motive,
and—by temperament, office, ambition—the State Pipeline Coordinator more than most:

"We've somehow gathered the reputation," Chuck Champion
explained, "of not being very capable of handling our own
affairs. This is one opportunity for the state to show people that
we can indeed deal with our own environmental affairs as well
as our own development, in a rational and reasoned manner.
That we no longer need to have Big Brother—that is, the Department of the Interior—making our decisions for us. We've
had a hundred years and more of malignant neglect. . . ."

Since the state's most pressing interest was to fulfill these am-

* The Pipeline Act had taken account of this possibility in requiring that
each of the TAPS owners operate its share of the line as a common carrier:
accept oil from whatever source and transport it at the same tariff for all
producers; the same privilege applied, of course, to small Prudhoe lease-
holders that did not go into the pipeline project, as well as to the state.

bitions without actually delaying the pipeline any more than it already had been, the function of the SPCO was to simplify the process of approval and surveillance, so far as possible:

"Because of the unique nature of this project, because of the limited Arctic knowledge—and there was a lot of supersitition on the part of a lot of people about what the Arctic is all about, all the scare stories—because so much of the land is owned either by the state or the federal government, because there were so many government agencies that would have been involved, and the controversial nature, nation-wide, of the possible environmental impacts: it would have been impossible to deal with the issue effectively. So Alyeska and both the state and the federal government agreed that we would furnish Alyeska one-stop shopping for governmental services—one agency for the state, one agency for the feds, through which they would channel all governmental relations, and this would preclude everybody acting individually."

"One-stop shopping" was not quite what Alyeska got, however. From Alyeska's standpoint, of course, even two agencies to satisfy was a complication, but at least the SPCO would in large measure depend on its federal counterpart for technical advice and be to that extent subordinate; in any case, that was how the hand had been dealt, and the arrangement looked workable enough on the face of it. There was a catch. The broad authority conferred on the APO by the Pipeline Act and the Right-of-Way Agreement was subject to serious qualification: in going about the business of oversight, the Authorized Officer had no power to override any existing state or national law or public policy. Given the physical extent of the project, the many kinds of technology on which it drew, it touched a complex body of law and was subject, in turn, to a correspondingly large number of government agencies, each with its own responsibilities for enforcement, its own self-sustaining ambitions.

Toward the end of construction, Alyeska's headman, Ed Patton, complained in a speech that building the pipeline had required authorizations from 59 different government agencies. The figure is impressive, though, like most of Mr. Patton's statements of fact, it was supported by no evidence. Nevertheless, the number of agencies with a legal interest in the work and some voice in how it was done was considerable; enough so that if you were trying to build a shed in your back yard and found yourself subject to as many competing and sometimes contradictory bureaucracies, you probably would not bother. Besides the APO and SPCO, at least a dozen government bodies had serious and sometimes crucial authority over the pipeline. Some of these were concerned with limited aspects of the project more or less continuously, from beginning to end; others came into play only at certain stages of the job.

For a start, protection of wildlife along the line, of which much was made in the Stipulations, fell largely to a new body created for the purpose in which the state and federal interests were about equally represented, the Joint State/Federal Fish and Wildlife Advisory Team (usually given as JFWAT, pronounced "jeff-wot"). Its staff was made up basically of biologists, but in many cases hyphenated with various branches of engineering and with an eye toward recreation—hunting, fishing, boating; about thirty altogether. It was their job in the first place to identify the migration routes of caribou and moose along the pipeline, the streams where fish live and spawn, and to approve the designs of Alyeska's engineers by which these might be preserved. It was a tedious undertaking on a sketchy base of knowledge, honorably performed—tracking and counting the herds through the hard winters and insect-infested summers, all the work of hunting without the rewards. *Advisory* was, however, the key term: JFWAT could gather its data (at Alyeska's expense), endeavor to persuade the pipeline builders and the two pipeline offices of the conclusions to be drawn, but had no power to compel assent. And while it posted observers to the work sites (two to each section, working ten days on and four off), protocol kept them at a distance; if a moose crossing, say, was not built according to design, there was nothing the JFWAT man could do but call it to the attention of the state or federal surveillance officer, who would speak to the Alyeska supervisor, who, in turn, after consulting Fairbanks or Anchorage, might or might not direct the contractor responsible to make a change—and finally an order of some kind would reach the men on the job. By then, however, the crews would have moved on and in a practical sense there was nothing to be done; which was, perhaps, the purpose of this elaborate chain of command.

Under the body of environmental legislation that has come into being since 1970—having to do with air and water quality and therefore with permissible discharges of smoke and sewage —the new Environmental Protection Agency has become the fastest growing and possibly biggest of all the federal regulatory agencies. Several aspects of the pipeline fell within its purview: the work camps generated carbon monoxide from heating and electrical plants, sewage which had to be treated, and so too the permanent facilities at Prudhoe Bay and Valdez; at Valdez also to be monitored were the ambitious plans to prevent or control discharges from the tall stacks of its power plant and from the tankers' ballast tanks. Accordingly, at the start of construction the EPA established a pipeline office of its own and dispatched a sanitary engineering Ph.D., Gene Dickason, to Anchorage to man it. Since he was on his own, with a secretary, and had similar responsibilities for all of Alaska, there was not much he could do in a practical way but look on from a distance and process the required permits as Alyeska applied for them; and accept Alyeska's reports on spilled oil, another duty, and the means taken to clean it up.

Alaska has a lot of government for the size of its population. Under state laws paralleling the national ones, environmental monitoring and permit granting were distributed among several divisions of two different departments of the state government, the Department of Environmental Conservation and the Department of Natural Resources. (The latter also administered the original oil leases at Prudhoe Bay and retained control over the plans by which the field was developed.) In addition, the Department of Labor was responsible for occupational safety on the project—essentially as a collector of statistics furnished by Alyeska—and for persuading the contractors and unions to honor the principle of Alaska Hire (see Chapter 7) by giving preference to residents. The state legislature, in creating the SPCO, had intended that it take charge of the Alaska Hire program but had left the DEC and DNR at liberty to develop their own programs of pipeline surveillance, if they chose to—and they did. This created an interesting situation. Chuck Champion let it be known that he could find no slot in his organization for Alaska Hire, and, at the same time, could see no point in another set of monitors for sewage and oil spills when his FSOs had eyes and noses of their own. The effect if not the intent was to enhance his visibility within Alaska and at the same time let Alyeska's managers know, if they had not noticed, that the State Pipeline Coordinator was someone to reckon with. Whatever it may have accomplished, this conflict within the state government, over which Governor Hammond coolly presided from Juneau, provided a recurring counterpoint to the heavy drama of pipeline construction and surveillance. Before the final note sounded, however, all concerned would have more on their minds than bureaucratic prerogatives.

In 1972, while waiting for Congress and the courts to decide the fate of the pipeline, the state legislature created the Alaska Pipeline Commission and charged it with drawing up rules for the operation of pipelines and approving their rates; two years later, as work was at last starting on the haul road, the governor appointed three commissioners to perform these tasks. As a matter of law, the Constitution, and common sense, all of their duties were already in the hands of other, national agencies, in particular the Interstate Commerce Commission. Except, perhaps, as a gesture of independence, it is not clear, therefore, just what function the legislators expected the commissioners to have, and the gentlemen in question were at a loss as to how they were to occupy themselves. During construction, they solved this problem by attempting to ferret out reasons for the pipeline's mounting costs, which would determine the charge for running oil through it and directly affect the state's revenues (see Chapter 11). Once the work was completed, the bills were in, and the tariffs submitted to the ICC, the commissioners would be on hand (along with nearly every other branch of the state government) to argue for lower rates. Meanwhile, they remained on the sidelines and endeavored to keep their eyes open.

Five federal bodies played parts in planning the pipeline and were intermittently interested once construction began. The Bureau of Land Management, for instance, an agency of the Department of the Interior, administers all federal land not otherwise assigned. It therefore served as Alyeska's landlord, within the framework of the Right-of-Way Agreement, for tracts required for storing equipment, mining gravel, or disposing of spoil from the pipeline ditch. Both the Corps of Engineers and the Coast Guard have responsibilities for structures that affect waterways, and at Valdez the Coast Guard had the special function of overseeing Alyeska's tanker traffic control system and providing the land-based navigational aids it would need in order to work. Since the pipeline route crosses Army and Air Force bases near Fairbanks, the Department of Defense had its own quite specific interests, expressed in a separate contract attached to the Right-of-Way Agreement; and, more seriously, the responsibility for protecting the pipeline and tankers from sabotage and attack, matters so touchy that little has been said of them in public. Once the pipeline was completed, it would be up to the Department of Transportation to certify that it was fit for public use.

In addition to all these interested parties, there was Congress: the Pipeline Act had invited the Secretary of the Interior to report periodically, and while the Senate chose to wait and see, the House did its own investigating. Thus, at moments of crisis, accountants from the Comptroller General's office (an arm of Congress) or investigators from the staffs of various subcommittees turned up looking for trouble; or the subcommittee chairmen themselves arrived to hold hearings and take testimony. With the apparatus of secrecy, accusation, and leaked findings, their reports made good copy for newspapers and television and served to remind the public that something was happening in Alaska and the House knew it (see Chapter 10).

All this was cumbersome, but less so in practice than it sounds. Where some part of the project came under the authority of several agencies, as it often did, they endeavored to co-ordinate their review so as to produce a single decision rather than a succession of contradictory permits. The ballast-water-treatment plant at Valdez, for example, affected water quality and was therefore subject to the EPA; and to the state Department of Environmental Conservation by the same reasoning. The Corps of Engineers was interested in what the plant might do to the harbor, the state Department of Natural Resources in the effect on tidelands (a "resource," productive of fish, crab, mollusks). And there were, of course, the two pipeline offices. The five agencies arranged to hold joint public hearings (as some of them were required to) and to issue identical permits. The work went forward and the laws were satisfied.

Alyeska's relations with the two pipeline offices effectively began in late May of 1974 with the submission of its general plans and preliminary design, as the Right-of-Way Agreement required. (Work on the haul road was already well along, of course, but as a separate entity, under the authority of the state highway department and the SPCO.) Review of the plans took only about a month—formulation and refining of the principles on which they were based had already occupied nearly five years, and in any case the serious questions would not have their final answers until the ground was opened and pipeline work begun. Thereafter, and again in accordance with the Right-of-Way Agreement, the over-all plan was broken down into "construction segments" (see page 278), each of which became the subject of a detailed application. Review of these applications by the two surveillance staffs, consultation with JFWAT and other advisory groups, and inspection of the sites affected was carried out within a 90-day time limit, though if revisions were suggested or further information requested for an NTPA, the schedule did not start till Alyeska delivered the new material. By working extraordinary hours through the fall and winter, the two staffs managed to process enough permits so that pipeline work could begin, as intended, in the spring of 1975. By the end of the 1975 season the APO and SPCO between them had turned out something over 700 Notices to Proceed, and this phase of their work was essentially complete, though some large questions remained.

Under Alyeska's contracting system, as it developed, every phase of construction was subject to a specific work order. The NTPs it received from the two surveillance offices were evidently incorporated by implication in its work orders (as the Stipulations from which they derived provided the legal framework for its general contracts) but seem never to have been systematically correlated. Alyeska's fundamental responsibility in the project was thus a divided one: on the one hand, to make sure the work conformed to the work orders under which it was done; but at the same time to harmonize the work with the revelant NTPs. For both purposes, Alyeska was required by the Right-of-Way Agreement to develop for approval by the APO plans for quality control and quality assurance. The first meant a staff of inspectors observing the work as performed and verifying that it matched the specifications. Quality assurance involved a second group of inspectors overseeing the QCs and spot-checking their work. In this scheme the state and federal field officers constituted a third layer of inspection, watching the work itself, the quality-control procedures, and the quality assurance. If all had worked as intended, therefore, it does not seem that they would have had much to do, and that was one of the things Andy Rollins meant by his hope that the APO need never come to public attention. The reality was a little different. Government proposes; men dispose.

When Alyeska, in the spring of 1975, submitted its plans for quality control and quality assurance, both Rollins and Champion were publicly skeptical, and the argument continued through the summer. At the end of May, Alyeska took over most management responsibilities from Bechtel but left its former CMC in charge of quality control. In mid-August, Alyeska offered a revised plan and a week later it was approved by the Authorized Officer. The most important change was one we touched on earlier (Chapter 6): the QCs (like the state and federal fieldmen) were given authority to stop the work if it did not comform to plan. It was a frail authority: if exercised, it would place the lone QC in the position of telling his Aleyska supervisor, the various contractors' managers, and perhaps hundreds of disgruntled men on the job site that they were doing the work wrong, that they must stop and start over. As one would expect, that did not happen very often. When it did, the luckless QC found himself the object of persuasions ranging from loss of job to a broken arm to a stick of dynamite lying suggestively on the driver's seat of his car.

It was apparent, then, that surveillance would require more than "some people that were just hired to make sure that Alyeska didn't muddy the waters and things of that nature," as one FSO described the first learning year of the project. The need was for engineers qualified by training and Alaskan construction experience to judge the entire spectrum of pipeline technology, with the authority to impose that judgment on the Alyeska managers and the physical presence, like a six-foot city cop, if need be, to persuade the work crews before they turned into mobs. What these multiqualified engineers could do was to spot-check the QCs and QAs. It was dull work, undramatizable even in institutional press releases.

Most of the matters that occupied the two surveillance teams were not in fact highly technical. In designing the system, Alyeska's engineers had demanded and gotten precise standards. The aboveground pipe, for example, was to be a minimum of two feet off the ground so that circulating air would carry off heat radiated through the insulation and keep it from thawing the underlying permafrost—the purpose of building aboveground in the first place; where the line crossed moose migration routes identified by JFWAT, the minimum was eight feet. There were similarly precise standards for the depth the VSMs were to be set in the ground, for the depth of below-ground pipe, for the depth of pipe below stream bed in underground stream crossings, for the slopes to which the right-of-way and gravel sites were to be regraded; for the size of gravel or crushed stone to be used in bedding, padding, and filling the below-ground pipe; and so on. Whether these overly precise standards were functional—whether, for instance, pipe set seven feet off the ground instead of eight would keep moose from crossing—was in many cases not known and probably would not be till years later, when the system was in operation; erosion problems, however—a poorly contoured cut through a frozen hill, say—would show up immediately and might even threaten the work pad and the pipe before either had been completed. In any case, the standards had been set and where the work did not conform, the surveillance officers were obliged to take note and cause something to be done about it.

A more complicated and no less recurring question was beyond the control of Alyeska's designers and the two surveillance teams: the construction mode—whether the pipe could be laid below ground or would have to be raised on VSMs. Again there were precise standards. When ditching revealed a "negative geotechnical surprise"—more moisture and muck in the frozen ground than the standard allowed—there were two choices: shift the route of the pipe to one side or other of the intended alignment so as to avoid a pocket of permafrost that, when thawed, would not provide stable support for the pipe; or abandon the buried mode and run the pipe aboveground. Both changes required the approval of the pipeline offices and cost money, more for remoding, which added the cost of VSMs, support beams, insulation, and their installation to that of lost time and work. And since the practical effects of mixed soils, frozen gravel and muck left by old glaciers, were often problematical, there was room for argument; an argument generally won by the APO, playing it safe and insisting that in doubtful cases the pipe go aboveground.

Monitoring the welding of the pipe was the most difficult aspect of the job and, given the risk to the entire system in a single faulty weld, the most sensitive. Judging the quality of a weld by eye and interpreting the X rays required by the Stipulations are tricky matters at best, about which skilled inspectors can reach very different conclusions. There was not much the surveillance officers could do but take the word of the QCs and QAs that their programmed procedures were being followed—and spot-check their work in person, as occasion presented. The checking was necessarily spottier than in most other aspects of the job.

All through the project, as a kind of *basso-continuo* to the melody of construction and surveillance, there were oil spills, fulfilling the doleful predictions of the environmentalists. On an icy, steep-graded curve of the underbuilt and poorly maintained Elliot Highway north of Fairbanks, a tank truck rolled, emptying a thousand gallons of heating oil destined for a camp beyond the Yukon. Or at Galbraith Lake the pipe feeding oil to the camp heating plant, laid in the frozen gravel pad on which the camp was built, came unthreaded and leaked all winter, seventy or a hundred thousand gallons' worth, no way of estimating. Or in the cold the rubber hose broke on a fuel truck feeding machines along the line and its pumps spewed diesel at a hundred gallons a minute. The surveillance men were onlookers to these incidents, not directly part of their job, not much they could do about them. The Stipulations required Alyeska to report the spills to the relevant authorities—the pipeline offices, but also the EPA and DEC, and the Coast Guard where the oil might enter a navigable watercourse—and take steps to clean them up.

In the course of construction there were thousands of such incidents, representing hundreds of thousands of gallons of spilled fuel. On average, according to Alyeska's reports to the DEC, there were 16 spills a day all along the line, 85 gallons each, totaling nearly 1,400 gallons; fewer in summer, of course, more in the winter months when steel and rubber crumbled in the cold, and freezing ground heaved buried fuel pipes apart. When a spill occurred, the immediate need was to keep the oil from entering a watercourse, and in winter the cold helped—the oil thickened, flowed like lava, and did not soak deep into the ground. With the abundance of heavy construction equipment along the line, it was usually possible to haul in a nearby bulldozer and shape an earthwork dike to contain the oil; then it was a matter of picking it up and removing it—pumping the oil from a big spill into a tank, spreading an absorbent which could then be loaded onto a dump truck and carted away for burning. (Hay was the usual absorbent: most of the small crop raised in the Matanuska-Susitna Valley north of Anchorage, where during the Depression glowing New Deal propaganda about rich soil and long summer days had persuaded a few bankrupt farmers from the Lower 48 to homestead.) By these crude means it seems that contamination of water sources was usually averted; the months-long leak at Galbraith was an exception—the oil trickled along the natural drainage from the camp and entered the lake, and in the spring there was a visible slick across the surface. The JFWAT biologists observed, wrote reports, pondered the effects on the life of the streams and the oil-patched earth, and considered the possibilities for rehabilitation if there was anything anyone could think of to do. As for the surveillance officers, they had always more pressing concerns.

As of early May 1975, after less than six weeks of pipeline work, the federal field representatives had performed something over 6,000 spot checks and reported 1,131 instances in which the work did not conform to the Stipulations, according to a study by the Comptroller General produced at the behest of one of the interested House subcommittees. About 18 per cent of the work, in other words, was below standard—or at any rate failed to match the specifications set forth in the NTPs.* The officer on duty at any one time was responsible for an entire section of the pipeline, roughly 150 miles, and work would be going on simultaneously at many points along the route. How much of the work he could see is therefore problematical and there is no way of knowing whether this record was representative of the whole job —the performance of its quality-control department is not a kind of information Alyeska has ever chosen to include in its voluminous public relations program. The repeated changes in its organization and procedures give some indication that this disquieting early sampling was indeed accurate: at the end of May, as we noted, Bechtel's responsibilities were restricted to quality control, and in August the program was revised—enough, at least, to secure the AO's O.K.; finally in March 1976, as the new season was beginning and the welding on the pipeline was turning into a first-rate scandal (see Chapter 10), Bechtel was removed altogether and Aleyska assigned its own men to quality control.

It is not clear what practical effects, if any, these changes had. So far as welding was concerned, it appears that questions of quality continued to arise at about the same rate or perhaps more so. In the fall of 1975, according to Chuck Champion's deputy Cy Price, the weld reject rate was running at about 20 per cent, and although that estimate can be matched with Alyeska's welding production figures released at the same time (see page 236), the actual number of rejects for 1975, as they came out later, was a good deal higher. As of midsummer 1976, when welding had become a very sensitive issue indeed, I was told that rejects were running between 30 and 40 per cent, at times as high as 80 per cent, and again, these estimates appear to be conservative. Even the least of these figures is extraordinary by the standards of pipeline work elsewhere in the United States. Ad-

---

* Despite a lot of effort, I did not persuade the APO to produce comparable figures for the volume of work performed by its field representatives during the remainder of the project nor for the amount of nonconforming construction that these inspections identified. My *impression* is that nonconformance continued at about this level for the project as a whole and was often a good deal higher in such areas as welding and radiography (see Chapter 10).

mittedly, no other pipeline has been subject to such intensive inspection, and all concerned in this process had become hypersensitive to the scrutiny of Congress and the public. The rejects ranged from minor defects easily corrected—"cosmetic" was the term applied, they might affect looks and the images on the X-ray film but have no significance for the functioning of the pipe, though who can predict with certainty?—to others so serious and so obvious as to require cutting out a length of pipe and replacing it with a "pup." The available estimates make no distinction between these two extremes.

Where a surveillance officer did encounter work of whatever kind that was substandard, the Stipulations gave him the power to order that it be stopped. It was a power rarely exercised. Against it was the double purpose of the authority under which he acted—his job was to ensure conformity to the Agreement, but *not to delay the job.* Less obviously, if he followed the work closely and kept on reasonable speaking terms with the Alyeska managers and their contractors, the job should not stray so far from specification that only the drastic medicine of a stop-work order would suffice to set it right again—having to issue such an order, therefore, reflected on his competence for the assignment. In line with Chuck Champion's aggressive style, the state men were expected to use their powers fully and did in fact several times issue stop-work orders. For example, stream crossings had to be completed during the few spring weeks when, according to JFWAT, the work would not impede migrating fish or so silt the stream bed that their roe could not hatch—if the work ran late because of a delayed thaw or for whatever reason, the FSO had no choice but to stop it even if that meant waiting till the following spring. Similarly, if the work extended beyond freeze-up in the fall so that the gravel intended for bedding the buried pipe froze solid, the work had to be stopped—when the oil moved, the gravel would thaw, its moisture flow out, leaving unstable gaps under the pipe; or so the guess of the engineers who wrote the specifications. The federal officers were more cautious. Although it must have occurred, I was unable to find an instance in which an AOFR initiated a stop-work order. Mr. Rollins had a word for those of his delegates who impeded the work—they were "not project-oriented" and did not stay long on the job.

Nevertheless, the power remained a threat and at least provided some support for the embattled QCs.

In practice, by the time a surveillance officer noticed substandard work, it was already finished and too late to stop. There were two things to be done. He could circulate a nonconformance report (NCR), taking note of the discrepancy and requiring correction as a condition for eventual certification of the pipeline—regrading and seeding an overly steep cut, for instance. Or Alyeska could change its plans to reflect what its contractors had actually done, initiating a design-change request (DCR) which would go through the same process of review and approval as the NTPA it replaced. By the end of the 1975 construction season, Alyeska had submitted nearly 700 DCRs or the equivalent—almost as many as its original NTPAs. Chuck Champion figured that "80 per cent" (often quoted) of the pipeline was redesigned in the course of construction. This coincidence of numbers was, apparently, what he meant.

The pilot, slender, blond, and young, gentled the controls and the helicopter slowed to hover speed and tilted in a wide, counterclockwise turn. Beside me, Bill Thompson, the state field surveillance officer in charge of the final section of the pipeline, leaned and pointed through the Plexiglas screen, talking across the syncopated humming of the rotor: down there a couple hundred feet was the thing we had flown up to see, Thompson Pass and the screaming thousand-foot drop the pipe would have to make to get through to the terminal at Valdez. The helicopter, crisply blue and white, with jet pods on either side of the cabin for speed, looked and smelled new, on call, with its pilot, to the fieldmen at Sheep Creek Camp, paid for, like everything else connected with surveillance, by Alyeska's owners; a lovely machine, with the precise, ungainly beauty of an insect imagined by an engineer, quiet enough at this speed for us to talk without headphones.

My son and I had driven over from Anchorage the previous afternoon, camped on the way, and met Bill in midmorning at a place called Tiekel Cache, a log roadhouse on the Richardson Highway about sixty miles north of Valdez, left over from the stage-coach time. In 1943 when he came north to run construction for the Corps of Engineers, the road was still the sandy track laid down during World War I, linked to the port town by a narrow bench carved in the wall of Keystone canyon and known as the Goat Trail—that steep, rocky, narrow. Since then, he had worked, he figured, maybe 2,000 jobs as a civil engineer in the supervision and inspection division of the Corps; settled in Anchorage and raised a son; retired; and on May 1, 1975, with the breaking of ground in Section 1 of the pipeline, came back to work for the state pipeline office. A short, thick, dogged man, he looks his age with the hard hat set square on his head, shading his eyes, younger when he takes it off; younger still in the Kodachrome identification picture sealed in the plastic badge pinned to his collar.

All day we followed him in his yellow Alyeska supervisor's truck down the work pad, stopping often to look, talk, listen; it was the way he worked, not much different for having visitors from Outside. He had been up in the morning at five-thirty, worked after breakfast several hours in his office at Sheep Creek, and would be back there again for three or four hours till bed-time—he did not care much for the camp movies, he said, and there was always the paper work, a daily log of what he had seen and done, weekly reports, "write-ups" of nonconformance, DCRs, an evening conference with the Alyeska managers. In between, while the workday lasted, he drove up and down the line, 25,000 miles on the truck since a year ago May, talking with the supervisors, QCs, foremen, keeping his eyes open. We stopped to look at cathodic protection wires sprouting from buried pipe (see page 195), again to study the crossing of the north-flowing, glacier-fed Tsina River, which had cost three weeks—water filled the pipe ditch, preventing welding, and half a dozen pumps could not keep it dry; a problem expensively solved by boring holes in the sodden gravel containment dike and filling it with 30,000 pounds of grout; complicated by a narrow six-week "construction window" in June and July, between migration and spawning of grayling and Dolly Varden trout, when the work had to be finished or wait till next year.

"We've had some minor problems," Bill was saying, "where you work something out with maybe the supervisors—but it's not my business to go down and tell every Tom, Dick, and Harry what he's supposed to do. *My liaison is supposed to be with the Alyeska representative.* Nine times out of ten he'll come to me and say, Bill, will you get with the contractor and see what his problem's supposed to be and try to work it out with him? And you come back the next day and you find they did it some other way. After the damage is done, there isn't a whole lot you can do except write it up and say, You've got to restore it. . . ."

All this was official and correct—the FSOs were supposed to observe, approve or disapprove, not volunteer—but aggressively applied. The project, I noticed, was always "we" to him except when it violated the Stipulations or the supervisors refused to take advice; then it was "they." There had been an Alyeska man a month back, "He was paranoid and he objected to anybody's suggestion other than what he thought about; the contractor couldn't even get along with him, work with him. It got to the point where we was just locking horns daily. I finally went to his boss, wrote it up, documented it over a three-week period, and I said, Now the situation's got to be changed. They shipped him out, sent him to Fairbanks." By now, late June of 1976, the final year of construction, that was how such problems were solved, at least when they concerned the state surveillance men. Alyeska was not always so compliant. The first year of pipeline work culminated in what seems to have been a determined effort, abetted by the rivalries of other branches of the state government with an interest in the pipeline, to replace Chuck Champion with a new and more amenable official. In November 1975, Chuck responded with an astute threat to resign; went out of sight for several days; flew to Juneau to confer with the governor; and came back with a statement of complete support and the assurance that so far as the state was concerned his authority in matters connected with the pipeline would be final and, if questions arose, they would be settled directly with the governor. Thereafter, when the SPC or anyone on his staff expressed an opinion, Alyeska found it prudent to listen.

In the afternoon, skipping lunch, we worked our way down the right-of-way toward Thompson Pass, the immediate object of Bill Thompson's concern and my curiosity. (The names, it seems, are coincidence.) The final thirty miles of pipeline, from the pass to the terminal on the south side of Port Valdez, presented probably the most difficult work on the entire route and had been saved till last. On a map it looks not too bad, and the sketchy geophysical evidence reported by the early surveyors had done nothing to disabuse the pipeline designers back in Houston; even now, the Alyeska managers in Anchorage made light of the difficulties, but they were multiplying and, as we studied them on the spot, looked nearly insoluble. Climbing through the pass, the route skirts fingers of two small, retreating glaciers. By road the high point here is about 2,700 feet but treeless in this latitude, the grassy heath shaped in hummocks that suggest corresponding lenses of ice, buried deep; on either side the mountains rise to five and six thousand feet, glacier-capped, and in winter the pass fills with a mass of wet, heavy snow almost too great to estimate (see Chapter 5). The obvious route through the pass was already occupied by the state highway. It was necessary therefore to climb above it, following the shelving rock around the glaciers, then drop precipitously beyond Sheep Creek. Here through the five miles of Keystone Canyon was another difficulty: the Lowe River, rising at the south end of the pass, cuts a narrow track between 2,000-foot, waterfall-mantled cliffs before it spreads through many shallow channels in a gravel-choked delta to empty finally at the head of the Valdez fjord; there was just room in the canyon for the boiling river flow, with the old Goat Trail on one side, the present highway on the other, cut into the rock, and one short, primitive tunnel at the narrowest point. Here again there was nothing to do with the pipe but cross the river, climb straight up to the top of the cliffs, and follow high benches on the south bank of the river to the terminal.

There was a further difficulty. The whole area is as scenic as it sounds, more so, perhaps, than any other in Alaska that is immediately accessible by road. At Mount McKinley, for instance, you can drive to within thirty miles of the north slope where, because of local weather, the mountain is concealed by cloud through most of the summer and the distance between represents a couple of days of arduous hiking and climbing, enlivened by the abundant grizzly population. The few miles of glaciers and jagged peaks north of Valdez are by contrast close enough to get into with little effort and highly visible from the road—an area that sooner or later, most Alaskans, most tourists, will visit. In the circumstance, several state agencies were concerned in what the pipe would do to the look of the place. To mitigate its effects, they decreed that, whatever the exact route, the right-of-way and work pad be kept to a minimal width, with a single access road from the highway, and that grading and cutting of the slopes be done within strict and narrow limits.

At the foot of the slope just north of the Worthington Glacier, we parked our car again and got in with Bill Thompson—he was going to try to get up the unfinished work pad to show us the drop down to the pass from above. It was a four-door crew carrier with a box behind where he carried half a yard of loose gravel, for traction, good, he thought for a 30 or 40 per cent grade in the double-low gear of his four-wheel drive, maybe 50 per cent. The percentage is a ratio of rise to run—a 12 per cent grade, about the limit on most highways, means a rise of twelve feet in a lateral distance of a hundred; a 100 per cent grade climbs a hundred feet in the same distance and is therefore equivalent to a 45-degree angle—climbable on foot, though from top or bottom it looks vertical. Bill shifted down and we started up. The bulldozers were still working to clear boulders from the work pad and not much gravel had been spread—scarce down here, they were having to quarry rock, crush and screen it to size, and even with half a dozen machines processing fifty tons an hour, it was not enough. The truck swayed from side to side, clawing its way among the cobbles and boulders, the whole body flexing on the frame. Halfway up, the engine groaned to a stop. We climbed out, lifted the hood—a new pair of batteries, the mechanic had not bothered to strap them down and they had fallen over, breaking the connection; water and acid dripped from one and in the thin sheet metal of the rocker panel had burned a hole the size of a quarter. We cautiously righted the batteries, slammed the hood, and walked. The truck had reached its limit.

The top of the slope was as far as they'd cleared—beyond, another wild drop beside a little pond dammed by glacial débris, another climb above the pass to the work site. We went back to the truck, drifted it down in reverse, and went around by the road through the pass, got out. We were down from the top of the pass, and the slope above us was thickly grown with small pines cleared in a 75-foot swath where the pipe was to go. Bulldozers had been up, making the cut, 50 feet deep by now, but still had not reached the bedrock predicted by the surveyors, essential for stability in this zone of maximum earthquake danger: the mountain was mantled in thick layers of glacial till, a puddinglike mix of rounded cobbles and damp muck the red-brown color of humus. In places the grade was reckoned 124 per cent, an angle just short of 60 degrees. At the top, a thousand feet up, they had rigged a winch to a big Cat to haul men and machines up and down so they could work. While we watched, from just below the crest where a group of surveyors was setting stakes, a black bear started and raced in a straight line across the slope, parting the shrubby trees and brush, and vanished back toward Thompson Pass. The men at the top shouted, pointed, waved, and we watched his track out of sight, shading our eyes. You hear about the speed of a charging bear but hope never to have to face it. This one, scared, looked to have tumbled through two hundred yards of dense brush in about ten seconds. It is healthier, obviously, to see him going than coming.

It was past five by now and the work was shutting down for the day. Bill stayed a little longer, chatting with one of the foremen, then headed for his truck and we followed him back to Sheep Creek Camp. He had gotten us one of the rooms reserved for staff from the state pipeline office and Cy Price in Anchorage had provided state visitors' badges to get us in the camp. It was a kind the security man at the gate had not seen before and he hunted through a reference book of permissible badges till he found it, new that spring and not much used; another point Chuck Champion had succeeded in making.

The meat in the mess hall that night was rib steak, pan fried and well done, filling the plate, more if you wanted to go through the line again. There were a thousand men in the camp at the time, three hundred more expected by midsummer—not the biggest camp on the line but one of the big ones; every afternoon now the new hires arrived from Anchorage in a gaudily painted sight-seeing bus Alyeska had chartered from somewhere for the season, with a trailer truckload of baggage bringing up the rear.

After dinner, Bill called in the helicopter and we set out on our meandering flight toward Valdez. For part of the distance it was still the only way to see what was happening; on the south side of Thompson Pass, where the route climbed up and over Keystone Canyon, even the Cats were stumped—to get them up to where they had to work, they had to be lifted, dangling, on steel cables hung from a big sky crane, sixty-five flights so far, and the men followed day by day in passenger helicopters.

It was ten o'clock by now, a blue haze beginning to fill the golden light with a look of late afternoon, and evening coming; hot all day inside the truck with the windows closed against the boiling gray dust from the work pad, better now. The helicopter completed another banking swing above the cut down the side of the pass, then with a small adjustment of the angle of the rotors skimmed the mile across to the opposite face like a puck on ice and climbed what would be the pipeline route up the wall of Keystone Canyon. I climbed it with my eyes, rucksack tugging at shoulders, sweating, panting, hands black with the pitch from the stubby pine trunks, the gray-brown outcrops of rock crumbling when you grabbed and pulled; only in imagination—the climb from this side is unreachable, cut off by the Lowe River, too wild to cross without a huge detour upstream. Once up, once you have got there, it looks good—the thick-sewn matchstick trees give way to brush and sedge and choosing your route, as you can from a helicopter, in imagination, you have an easy walk along the top and a not too strenuous descent to the terminal where the pipeline ends. There are in fact peaks, topped with vestigial glaciers, carved and drained by creeks dropping down their sides to Port Valdez, a shallow Alpine lake half-stifled by gravel mining, but from here, above, the whole mass along the south rim of the fjord has the look of a single form, a roughly rectangular solid twenty miles long with a deep bowl-like hollow in the middle, above the half-finished structures of the terminal. Liscum Slide the engineers have named this form, though it is not on the maps—perhaps ten thousand years ago, when the ice sheet melted and earth rebounded, lifting the mountain, half of it slumped and slid into the depths of the fjord; you wonder if it might happen again, but now, in this petty span of human time between two ages of ice, it is not high on the scale of likelihood.

The terminal is down there at the edge of the fjord, like a schematic drawing: rows of oil tanks, like buttons on a shirt, on benches carved in the slope, the square metallic shapes of power plant with its stacks, ballast-water-treatment facility, off to one side the packed structures of the construction camp, which will go when the work is done. The helicopter hovered another minute along the edge of the slope, then turned and slithered back across the shelving air to Sheep Creek Camp. The blue light was turning gray when we set down on the pad. Evening, down here in the shadow of Thompson Pass, had come.

In the morning a little after five-thirty Bill knocked on the door of our room, as promised. I was still struggling to wake. I washed and shaved in haste while my son dozed and we went along to the dining hall in time to join Bill for breakfast. By seven he had waved us a brisk farewell at the camp gate and retreated to his office for that morning's paper work.

We drove on in to Valdez where George James, the SFSO at the terminal, was expecting us; and learned more from him about the design of the tanker port, its problems, in three hours of looking, talking, and driving around the site than I had in three previous days of Alyeska's tightly guided tours. He was a good man to learn from. Bears again, roaming the slopes. For the purpose of keeping them out, among other things, Alyeska had thrown a chain link fence around the whole terminal—and discovered it had enclosed several that lived there, two boars and two sows so far, uncounted cubs and yearlings, getting restive; a few weeks back, in the lot outside the gate where workers from town left their cars, to be bused into camp, one got between a man and his girl and their car, caught and mauled them—they fought, first one, then the other, driving it off, but the girl, flayed and flensed, alive, was still in the hospital.

In the course of the day, in and out of the terminal, I made repeated efforts to find George James's federal counterpart, a man named Arne Echola, also expecting us, and finally hunted out the trailer camp beside the Valdez airport where he lived, but he eluded us. He had moved his wife down from Anchorage, someone said, bought a boat, sure, that's where he'd be, it was Friday afternoon after all—out on the boat; but we had learned enough from George by now so that missing him seemed no great loss.

I had better luck the next day with the other federal field rep in the area, Duane Carson, based at Glennallen Camp near the north end of Section 1. He met us at the gate and our state visitors' badges were still valid to get us in. (I had had much talk about this with Bob Gastrock and the question had apparently been answered by Mr. Rollins—the AO did not think it proper to supply passes that would allow us to enter Alyeska's camps and right-of-way, observe his fieldmen at their work; why *of course* you can *talk* to them, meet them outside somewhere, a roadhouse, sure, ask all the questions you want. Cy Price's badges had anyway soothed that particular irritation.)

In climate and in most other ways the region through which this part of the line is built is a transitional zone: a rolling, upland plain dotted with lakes, shaped by the Copper River, the route followed northeast by the Glenn Highway to its link with the Alaska Highway and the Yukon border. The region is enclosed by four ranges of mountains: the Wrangells, starting thirty miles east in a cluster of great glacier-covered peaks, Drum, Sanford, Zanetti, Mount Wrangell itself; north, west, and south but out of sight, the Alaska Range, the Talkeetnas, the Chugaches. Down the middle runs the Richardson Highway and the pipeline—and the Gulkana River, a lively tributary of the Copper which Matthew and I had hopes of canoeing ourselves in the next couple of days, redeeming the Kenai disaster, and therefore studied with care as we cruised the pipeline right-of-way beside it in Duane Carson's yellow Alyeska crew truck.

Transitional: the zone lies near the southern limit of permafrost, where it is discontinuous and often mucky, barely frozen through the Alaskan summers at an average temperature of 31° F. and sheathed in a thick active layer. The pipeline here had had more than its share of surprises and consequent changes of mode or alignment; there were several short stretches where the JFWAT people had insisted the pipe be laid underground in unstable permafrost and refrigerated—for the sake of crossings by the remnant bands of a small and shrinking caribou herd (see page 185). The work was nearly finished by now. What remained was the athletic dance of girdling the aboveground pipe with its sheets of steel-clad insulation, tightening down the bracing rods on the anchor supports, grading off and reseeding the slopes of the work pad, gravel sites, stream crossings; in places a rim of the spindly new grass was already sprouting in the slashed earth, acid green among the earth tones of spruce and aspen and heath.

It seemed we'd hit Duane Carson on a bad day, in a bad week. Gray and slight, mild-mannered, he moved in an uncomfortable half crouch, one shoulder slumped—back bothering him again. That was why, perhaps, he seemed oddly remote, compared with the two state men we'd been traveling with, from the work and those doing it, he knew the Alyeska manager, that was who he worked with, but kept a passively correct observer's distance from the contractors and their crews. He was in fact informative about the varied engineering complications in his section but took some drawing out; like, I thought later, the portly Scot who taught me Dante at college, a superb linguist who switched at will among half a dozen Italian dialects and was also deeply learned in the literature—but never shared his knowledge unless you could evoke it by a carefully formulated question or a simple mistake, when he might talk for an hour. As we worked our way down the right-of-way from Hogan Hill, the limit of Section 1, to the camp, I tried to get Duane talking about how he worked, how closely he followed the job, and what, in a practical way, he could do when something didn't come out as intended.

"So far as you're concerned," I asked, "what can you do when some of these things don't meet specifications—like the lateral supports not being twenty-four inches off the ground? You've got to go in and report on it to somebody, don't you?"

"This is normally a function of QC," he said, "Quality Control, and then they've got their Quality Assurance people. I've got direct contact with either them or the Alyeska project manager. Most of this stuff, if it's like that, it really doesn't bother me too much, it's their pipeline, and normally they will come in with a design change that goes in the office in town which will authorize that change."

"Would you get in on that in some way? The office would have to consult you?"

"Normally Alyeska automatically does this themselves—before I have to say anything. I don't have to force it, *it's done.*"

"Have you had cases where they're not going through that procedure? Or you spot something?"

"If I spot something I'll normally call it to the attention of the QA—and if I can't get anything done then it's the Field Memo route that I have, I can write, tell them, you do this. It's kind of a shaky point to direct the contractor what to do. You gotta always remember that this is private property—privately owned, privately financed. . . ."

"But being built under regulations set by Congress, finally," I said with some impatience. The right-of-way being here mostly in sight of the highway, with frequent access roads, Alyeska had posted it with signs explaining the private property idea and warning against trespass. This was valid enough in the sense that it was a leasehold acquired by contract with the government, but it was also in the first place public land and I was the public; that was *my* land they were telling me to stay off, and in any case just where the leasehold began and ended would remain indefinite until the job was finished and the surveyors had staked and mapped their final boundaries.

"Well, *right,*" he corrected himself, "—under Stipulations which say that they will adhere to their drawings and specs. But to change something, they go into a Design Change Request. I keep an eye on what the problem was but when the DCR comes out, approved by Alyeska's office plus my office, then I don't have any more problems. It's taken care of itself. . . ."

The big question at the center of all these lesser ones about the conduct of pipeline surveillance seems simple enough: What exactly did this large expenditure of human energy, emotion, and intelligence (not to mention money) accomplish? Or: What did Alyeska do or not do in consequence that it would not otherwise have done? It was one I asked myself and everyone connected with the pipeline repeatedly and in many forms. It did not turn out that there was any very good answer. Or there were too many answers, they were so narrow in scope, so technical, as to be nearly impossible to recount or explain. This is not to say that the multitudinous concerns of the pipeline watch were trivial. The building of the pipeline, like the making of a work of art, was the synthesis of a million distinct acts directed toward a final whole, each the product of a corresponding choice, and it would be a fool's judgment to say of any one that it is not vital to the end result. But observing this process made, as I said, dull work.

Chuck Champion, when I demanded instances, made a speech about compelling Alyeska (or its contractor, Fluor) not to cut down and remove the curtain of trees along Pine Point, the narrow spit of land coming out from the Valdez terminal to which two of the tanker berths were to be anchored. Possibly he was addressing the stereotypical writer-environmentalist he assumed must be inside my particular skin. Outside Alaska, the matter of the peregrine falcons, an endangered species, has been often cited by environmentalists and the Department of the Interior to show that Mr. Rollins and the APO were indeed doing their job. At some point after Alyeska put forward its plan for Pump Station 2, on the Arctic plain north of the Brooks Range and about sixty miles south of Prudhoe Bay, it was objected that it would lie uncomfortably close to a falcon nesting site on bluffs above the Sagavanirktok River, the goal of a two-thousand-mile migration that in this case has its start, probably, somewhere in the mountains of California. On the supposition that building and operating the facility would inhibit the birds' nesting and breeding and thereby contribute to their extinction, Pump 2 was in May 1975 replanned about three miles north, and seasonal restrictions were placed on a nearby airstrip. The federal surveillance people in Anchorage were diffident about this apparently clear-cut instance of their effectiveness. It was uncertain whose idea it had been, but quite possibly Alyeska's. One might infer self-censorship, one of the premises of a surveillance system that could not cover literally everything, but it seems that Alyeska had other reasons for the change, among them complications in routing the pipe along the unpredictable Sag. In any case, although the outer shell of Pump 2 was to be built immediately, the pumps would not be installed nor the station operating until sometime in the indefinite future when enough oil was producing to lift the line's capacity to 2 million b/d.

Champion's pines and Rollins's birds are symbols, the tools of rhetoric, the stuff of poetry. It is dangerous to disregard the power of symbols—the pipeline itself is a symbol—but foolhardy to confuse symbol with substance, the real world. The purpose of surveillance was to control that reality. Had it?

Throughout the latter part of 1974, while the haul road was building and the hundreds of NTPAs were making their way through the two pipeline offices, Chuck Champion and his staff welding experts, with their Cook Inlet pipeline experience, were engaged in a bitter debate with Alyeska, the APO, and everyone else over the welding rods to be used. In electric-arc welding, the rod is the length of metal that serves as the electrode and by melting fills up the beveled joint at the end of the pipe; it looks something like a sparkler. Alyeska had specified its rods to the same standard as the pipe itself: that the metal should balance malleability and yield strength down to a temperature of −20° F. The calculation was that although the winter temperatures in most parts of Alaska go far lower than that, the pipe metal and the welding material would not—in operation, they would be warmed by the oil inside, and the line would never be shut down so long for cleaning or other maintenance or for repair as to drop below −20°. (That left, of course, at least two winters for the welded pipe to survive at minimum temperatures, but Alyeska had already tested a sample loop of aboveground pipe at Barrow, and it had passed.) Chuck and his men rejected this line of reasoning. "Minus twenty," Cy Price scornfully remarked, "is a temperature which has no significance in Alaska—minus twenty is somewhere between a warm summer day and what the yield temperature gets in winter!" The same argument applied, of course, to the pipe itself, but there was nothing to be done about that; it had been ordered and long since delivered. The SPCO therefore concentrated its attack on the rods.

The SPCO insisted that the rods be such as to retain their metallurgical properties down to −50° F. and let it be known that it would go to court to enforce its judgment: "I said," Cy Price concluded, "that I *would not allow* them to start construction until the welding criteria was upgraded." The threat of a suit, reinforcing the powers defined in the Right-of-Way Agreement, was too great a risk, and in January 1975, after months of resistance, Alyeska yielded, in time to order the new rods, two different types specially developed to the state's specifications, for the start of pipe work in the spring. This was probably the most important and far-reaching change imposed by the surveillance process. It was also, I believe, a distinctly double-edged achievement. It meant that chemically the rod metal would be quite different from that of the pipe with which it was to be fused, and since the quality of the welding was to be the greatest and most controversial difficulty in the entire project, it seems reasonable to look for at least part of the explanation in this difference; and that was one of the points the welders made in their own defense, when the quality of their work became matter for public concern. It seems likely also that the character of this debate and its outcome were the immediate reason why Alyeska, later in 1975, made a serious effort, as it seems to have, to dispose of the State Pipeline Coordinator by political means.

Alyeska had already tried this approach to good effect in another connection: sewage. The story is exemplary of the tensions among those state agencies with some legal authority over the pipeline project—and of Alyeska's resourcefulness in using them for its own short-term purposes.

Under state law, the Department of Environmental Conservation was in charge of several matters concerned with water quality: among them, monitoring oil spills, approving sewage-plant designs, inspecting the plants once in operation, checking the discharge to make sure it met the legal standard. The Right-of-Way agreements required Alyeska to respect such laws but were vague as to what the state could do to enforce compliance —in particular, whether Alyeska was obliged to bear the cost of a water-quality inspection program on top of the general pipeline surveillance. The head of the DEC at the time was Max Brewer who for twenty years had run the Navy's Arctic research station at Barrow, quite possibly the most knowledgeable man in Alaska about the Arctic environment—and also, one would have thought, the obvious first choice for State Pipeline Coordinator. Mr. Brewer concluded that in order to do his job where the pipeline was concerned he would need a staff comparable to that of the SPCO. Alyeska intimated that it would go along, though without specifying just how far, and on that assurance Mr. Brewer in July 1974 hired a young sanitary engineer, Gil Zemansky, and sent him about the business of inspecting the sewage-treatment plants in the pipeline camps then under construction. Eventually, he intended a staff of twenty-one who, at Alaskan civil service rates of at least $28,000 a year, would add up to an annual budget of about half a million dollars; to get around the vagueness of the Right-of-Way agreements, this cost was to be approved and billed quarterly through the SPCO. Chuck Champion did not like this arrangement any better than Alyeska, though not, of course, for reasons of cost. Five of the proposed positions were moved to the SPCO, but Mr. Brewer continued to insist on a staff of some kind, even if reduced. In November, Jay Hammond's new administration came in, Republicans with a difference, and Max Brewer was out of a job, but the argument continued.

Meanwhile, Gil Zemansky, singlehandedly policing all the pipeline camps (and the permanent Arco and BP camps at Prudhoe Bay) from a base in the DEC Fairbanks office, was having troubles of his own. Alyeska had not gotten around to submitting plans for its sewage-treatment plants or securing the DEC permits required by state law. The plants, as built, were undersized for the rapidly expanding camp populations, manned by untrained operators, and the effluent, to use the polite term, in many cases came out unchanged or, if anything, concentrated; where the settling ponds into which it flowed were excavated in permafrost, as they most often were, it could not percolate through the ground and went on filling till a thick gush of raw sewage overflowed the dikes and entered the nearest streams or rivers. Lab tests of samples confirmed the evidence of eye and nose.

When no improvements followed from his recommendations, the young inspector began filing notices of violation and sending back alarmed reports to Juneau. Alyeska responded by making difficulties about feeding him and putting him up in the camps or, north of the Yukon where it controlled the transportation, getting him there in the first place; Zemansky, a licensed pilot, rented a plane (the state had none to spare) and did his own flying, but Alyeska, while it could not prevent him from landing at the camp airstrips, refused to service the plane—apart from gasoline and oil, once the cold weather comes there is no way to keep a plane functioning unless it is kept warm while on the ground.

In November, the new commissioner of the DEC, Ernst Mueller, confirmed his predecessor's vague standing instructions and told the inspector to keep up the work. In January, while the camps were largely closed down by winter, he was sufficiently impressed, after prolonged persuasion, to order Alyeska not to reopen them until the various sewage violations were corrected. At this point, it seems, Alyeska turned its powers of persuasion from the gnat that had been stinging it to the commissioner and the governor himself, and in February Mr. Zemansky found himself transferred from Fairbanks to Juneau and, soon thereafter, fired. In the meantime (after asking and getting permission from his boss), he had given straightforward answers to questions about his job put by a young reporter from the Anchorage *Daily News,* and the result had been headlines about Alyeska's numerous violations of the law. Those headlines, which so far as the relevant Alaskan statutes are concerned were plain truth, were evidently the stick that broke the gnat's back. Shortly thereafter, although little had been done to meet Gil Zemansky's objections, the DEC issued the permits that allowed Alyeska to operate its sewage plants, and the law, if not the reality, was accommodated.

When I approached Gil Zemansky a little more than a year later, in his office at the University of Alaska in Fairbanks where he had retreated into research and teaching, I expected to meet a shaggy-headed zealot; that was the impression formed from people at Alyeska, the SPCO, and the Anchorage *Times* (its publisher, Robert Atwood, owned a tourist camp at Deadhorse whose system Gil had inspected and found wanting). What I found was an extraordinarily clean-cut Annapolis man of six or seven years back, by way of mechanical and nuclear engineering, who had put in his duty time, resigned, taken a master's in sanitary engineering, and come to Alaska about the time work on the haul road was getting started; with that air of square-shouldered, innocent rectitude that, once having gone the course, they seem never to lose. He was, in short, the perfect, foreordained victim to be fed through the rollers of two petty state bureaucracies and the cutting wheels of Alyeska's notions of cost effectiveness.

A year later, he was able to quote *verbatim*, with proper citations, the statutes applicable to what had been his job and was now his case—he was suing the state, with assistance from the employees' association, on grounds of wrongful discharge, and from my own experience of the law's delays it seemed probable that he would win; and, more to the point, *was in the right*. Meanwhile, he was capable of making sewage both intelligible and interesting; and of punctuating his discourse with apt quotations from Ibsen. Alyeska had chosen the wrong enemy: in so doing, it had drawn his attention to more general questions, such as welding, the motives of pipeline offices, the arithmetic of profit, and the strategies of oil.

That is, however, in the long run. In the short, Alyeska obtained its little victories. After reluctantly paying the bills for the DEC surveillance during Gil Zemansky's tenure, it whittled away at the staff requirements, offering to fund sixteen monitors, then nine, then four, with corresponding adjustments in budget. It produced research to show that twenty-six Alaskan villages were pumping into the streams ten times as much untreated sewage as its construction camps; and declined finally to pay anything for differential enforcement of the law. It was an effective point. The DEC had the same responsibility for the villages as for the camps, but with a difference in where to send the bill. In the villages the "honey bucket" is proverbial. For most of the year, because of frost, a privy does not function. You accumulate the slops in a pail and periodically empty it out the front door. When break-up comes and this frozen coating thaws and flows toward the streams where you collect your water and net your fish, diseases are endemic, among which hepatitis is the mildest. So far as the camps were concerned, the DEC continued its protests and Alyeska for its part hired a sanitary engineer of its own, but things went on much as before, and when the camps were filled the treatment plants failed and the sewage overflowed, in places entering the wells from which camp water was drawn, though apparently it caused no traceable human diseases and left no lasting effects on the water. Since consequences the law was intended to prevent did not in fact follow, we might conclude, therefore, that the law was unreasonable and properly ignored; that was evidently Alyeska's conclusion, with the further

motive that strict adherence would cost permanent money while
the camps were temporary. For Americans, however, that argument is weak. Legalism seems to be ingrained in our character:
if often we expect too much of law—legal solutions to problems
that are human and permanent—it is *our* law, after all, of our
making, with our approval. As is often said, the cure for bad law
is not to flout it but to enforce and change it.

In matching the construction of the pipeline and its camps to
approved design and common sense, execution to intention, the
law, it seems, was as clumsy and spasmodic as in most other
areas of human behavior. Much the same was true on the strictly
environmental side of the surveillance program—the preservation of the fish, land animals, and birds, with their habitats—
which in the pipeline debate had seemed a more pressing issue
than the technical and legal ones; with the difference that the
JFWAT biologists had no power, except reason and persuasion,
to impose their views.

"Our charter here, as we interpret it, " Jim Hemming told me
in the cramped JFWAT office on F Street in Anchorage, "is to
maintain as many options for the future as possible: not just big-
game animals or fish but right down to sculpins and ground
squirrels." Jim, a biologist, had been doing research on caribou
on the Arctic Slope since 1965. "The critters," he diffidently
called the animals that had occupied so much of his life; beneath
the necessities of science and quantifiable data, there was affec-
tion. "As an advisory group," he continued, "our clout is limited
to being able to come in with good, solid professional justifications for what we ask for. We have to come in and sell and sell
and sell and sell. We're much more vulnerable in a sense because
you can lose credibility over one item and they shoot you down
on ten others."

These were the poles between which the JFWAT men had to
operate. On the one hand, "options": we simply do not know
enough about the interconnections of the living creatures within
an "environment" to distinguish one harmless event from another
that in the course of years may produce disastrous and irre-
versible change—and therefore, caution. But on the other hand
the engineers who designed the system wanted certainty, though
often enough, when it came to translating their designs into the

realities of construction, engineering proved to be no more exact or certain a science than biology. Thus, as we have seen, where aboveground pipe crossed moose trails and the engineers demanded and got an eight-foot minimum standard, the pipe as built varied as often as not by four or five feet. Or when a fish-stream crossing would entail two days' siltation and the biologists argued that the silting would do some harm but could not in honesty say how much, the builders, in the absence of certainty, were free to ignore their advice. Or where in place of deep burial below the stream the engineers proposed shallow burial and control of the stream flow by bulldozed "river training structures"—lower initial construction cost against higher continuing maintenance—it could not be said with certainty just what difference the change would make to the life of the stream, and the common sense arguments about long-term cost were outside the biologists' presumed competence. Similarly the sulfuric acid generated by burning sulfurous Prudhoe Bay oil in the power plants at Valdez and the pump stations: it was known that SO_2 in sufficient concentration is deadly to the lichens which are the foundation of the tundra ecology and therefore of the whole chain of life supported by tundra—but not exactly *what* concentration was enough or whether gases from the power plants would reach it.

Where a precise standard could be established and was disregarded in construction, the JFWAT advisers had the same recourse as the other surveillance groups: document the variants and demand that the finished product be made to conform to its specifications. As their complaints multiplied, Alyeska began to take them seriously and in the second year of pipeline work reinforced its quality-control program with biologists—and, more obviously than in the case of the camp sewage plants, improvements followed, corrections were agreed on. Thus, for instance, it was possible to make the moose crossings usable by removing the gravel work pad beneath the pipe; in only one case was it necessary actually to rebuild. But there are limits to what rehabilitation can accomplish, even in theory; and nothing that touches life in its myriad forms can be wholly predictable.

"We're talking about damage that's been done to the terrestrial habitat, to tundra," said Carl Yanagawa, the black-bearded state biologist who as JFWAT's assistant supervisor had shared the team's administrative chores from the beginning. "We don't know of any way to restore or to lessen or compensate for the damage that's been done to the tundra all the way from Galbraith to Prudhoe Bay. We don't know if it's major right now —we know it's an impact we're going to be living with forever."

The uncertainty remains; it will be years before we can know the outcome, if we ever do. Yet looking back at the project through the day-by-day irritations and frustrations of his work, Jim Hemming near the end of construction was able to take a moderately hopeful view of what they had accomplished:

"I think," he concluded, "considering the fact that it's a unique project and also a unique approach for industry, being involved with these types of control—I think it's worked. We'd all say that it could be done a lot better and a hell of a lot less painfully. Industry has a long way to go. But this is a start. . . ."

It is a conclusion, I think, that can stand for the whole surveillance effort.

This is the place to consider what it cost for Jim Hemming and all the others to hold Alyeska to the standards under which the pipeline was built. All of these costs were borne, as we have noted, by Alyeska and therefore formed part of the total bill for construction. That bill, in turn, will ultimately be paid by whoever buys the oil—which is to say, by all of us. Apart from one's natural curiosity about what things cost, therefore, we have an immediate interest in knowing whether we got good value for our money.

Before the start of construction, the Department of the Interior spent more than $12 million on evaluation of alternative pipeline routes and of the TAPS proposal and on the mammoth undertaking represented by the Environmental Impact Statement. From the start of construction to the end of the 1975 fiscal year (June 30), it took about $15 million to operate Mr. Rollins's APO, and the budget continued at the same rate of $12 million per year. To the end of construction in June 1977, therefore, the total came to $34.4 million, and allowing for continued monitoring of cleanup and operation to at least the end of the year, it looks as if the final amount would be around $40 million. The state effort came to about one tenth of this total. These figures cover direct expenses billed by the two pipeline offices but not what Alyeska provided in kind to the several sets of surveillance officers while they were on the line—room and board, office space, ground transportation, planes in and out of camp. In all probability, then, the entire cost of surveillance would be contained in the round sum of $50 million.

That is, of course, a great deal of money, as you and I judge such things. Measured against the $7.7 billion that the construction estimates had reached by mid-1976, it is less than 1 per cent and does not seem so grand; and it shrinks further as the costs trickle in and continue to swell. Without falling victim to the bureaucrats' inability to distinguish tens of millions from hundreds of millions of dollars, we might reasonably conclude, therefore, that the cost was not disproportionate to the job done and that, if the achievements were decidedly mixed, we at least got as much as we paid for.

There is another sense in which surveillance cost money: by substituting legal for engineering judgments; by imposing design and construction requirements which were not in fact functional and therefore wasted money and effort. Sometime in 1975, when the official pipeline estimate was still under $7 billion, an Alyeska public relations man (Larry Carpenter, it seems, who had been a luminary of Fairbanks television) put a value of $2 billion on this difference of opinion. Some months later, Ed Patton had refined this cost upward to the precise-sounding figure of $3.7 billion, nearly half the then estimate for construction. Whatever the number, it is ideal for rhetorical purposes: it rests on so many disputable judgments that it can be argued to infinity without conclusion. One can guess in a general way what they mean: the amount of at least debatable aboveground pipe, costing double the buried pipe (from Alyeska's standpoint it was often better to submit and spend the extra money than to sacrifice the construction schedule to the time it would take in doubtful cases to defend below-ground pipe and secure the necessary permits); the pump stations and port structures built for earthquake magnitudes that in reasonable probability will not occur in the comparatively short working life, twenty or twenty-five years, of the pipeline system; the river crossings that anticipate floods unlikely to occur more than once in a century. The Alyeska people would not, and perhaps could not, substantiate their complaints with evidence or supply the estimates to go with them. Without that, the numbers are vain; figments. One is prepared to listen, but is left guessing.

Late in 1975, at the end of the first year of pipeline work, Congress sent the accountants of the Comptroller General's office to Alaska to find out why the project was behind schedule. Their report, released on February 17, 1976, was hopeful about the schedule but raised questions about the quality of the work being done. By then, however, schedule had become secondary to quality and the questions of quality centered on welding. In July, therefore, President Ford dispatched an executive group to find out what was happening, and at the same time John Dingell, chairman of the House Subcommittee on Energy and Power and an old doubter, held hearings and sent his own staff to do a parallel investigation. The results, which reached selected members of the press over Labor Day, were alarming. As hard-to-answer questions about the welding multiplied, the pipe began to sound like a poorly bottled jar of tomatoes about to explode on a shelf in the root cellar. Alyeska, it seemed, had little control over its quality and no assurance. In the camaraderie of surveillance, what for Alyeska were merely errors of managerial judgment might turn out to be violations of the law by the Alaska Pipeline Office. The Authorized Officer was reported to have accepted "favors" from an Alyeska contractor—a trout-fishing weekend in Idaho back in the Corps of Engineers days.

In the midst of this ruction in the henhouse, Mr. Rollins abandoned his frankness and became as nervous as Alyeska about his staff's opinions, when publicly expressed. For Andy Rollins there was after all no magic cape. It was not possible to remain invisible.

Chapter 9

IMPACTS

> Well, heck! The impact of this pipeline is terrific!
> —Robert B. Atwood

He sat straight-backed as a schoolteacher in the padded leather swivel chair, leaning his forearms on the oversize desk: Robert Atwood, the publisher and editor of the Anchorage *Daily Times*, as it was then still called. (Three months later he dropped the *daily* from the masthead as superfluous—everyone by now knew that, barring Thanksgiving, Christmas, and New Year's, the paper came out every day of the year, the only one in Alaska of such frequency.) Tall, white-haired, and close-trimmed, countryman's steel-rimmed glasses; a handsome, ruddy, rectangular face seamed in the cheeks and around the unsmiling eyes with shrewdness and his sixty-odd years. He came north in the mid-30s, bought the improvident *Times* in the villagelike City of Anchorage, married the sister of Elmer Rasmuson, who was founding the National Bank of Alaska; settled, stayed. The paper grew with its town and Alaska, imperceptibly. In the 50s, convinced that statehood was prerequisite to serious growth, Mr. Atwood and a few others formed a committee to promote it, with himself as chairman; bought oil leases cheap, on the Kenai Peninsula and persuaded oil companies to prospect them, the Department of the Interior to open the Kenai Moose Range to drilling. The Swanson River strike followed (see Chapter 1), others, then statehood, Alaska's solvency being thus attested. The newspaper and the banks prospered, with the state, as predicted. With the profitable sale of his Kenai tracts to oil companies, it was said, Mr. Atwood could at last hold his head up among his in-laws. The bank grew, with the coming of the pipeline in the 70s, to half a billion in assets, ten offices in Anchorage and twenty-one branches scattered in an arc around the Gulf of Alaska from Metlakatla in the farthest Southeast to Adak in the Aleutians, reaching Fairbanks in central Alaska; on its board, still chaired by Elmer Rasmuson, sat directors ranging from Jesse Carr of the Teamsters to Emil Notti of the AFN and Locke Jacobs who, with the oil strike, had risen from clerk in an Anchorage Army-Navy store to broker of oil leases, a millionaire several times over with a hand now in most of the land deals that control Alaska's oil and its other minerals. The *Times*, with its pages of profitable small ads and fat Sunday editions, became the newspaper of record for Alaska, filling out its columns with handouts from oil companies and the government, from publications in the Lower 48, but rarely originating news.

Besides his own part in forty years of Alaskan history, his paper's coetaneous files, and a huge acquaintance within the small world of Alaskan power and influence, the publisher was said to possess a remarkable personal library of Alaskan materials; several people in the first days of my arrival in Anchorage had told me Atwood was someone worth talking to, that he could, if he chose, be helpful, save time, though he might have designs of his own on his Alaskan stuff. Or perhaps the proffered name was only another symptom of what was beginning to seem an Alaskan disease—not many know anything in particular or will express themselves, and they pass you along, relieved and apologetic, to someone else who is sure to have something to say. In any case, I telephoned, made an appointment. Since I was still at the beginning, still trying to get my bearings, formulate the necessary questions about Alaska and its pipeline, there was no limit to what I needed to know, what might in the end turn out useful or enlightening.

The interview started badly. As I produced the tape recorder from my briefcase, Mr. Atwood demanded to know why I used that thing and I politely explained, mentioned previous books and magazine pieces for which I had found it helpful, but he was unimpressed, thinking, perhaps, of his reporters spending his money on fancy machines, taking up time on tapes, exactly transcribed statement and opinion; the paper does not have a reputation for generous pay, and its reporters tend to spin off into public relations or, after a time, go back whence they came, Outside.

The questions continued and I kept answering. He was interviewing me. I shook loose with a question of my own about "development" and its effects, the impacts, as they are called, of oil and pipeline, and for a time he talked. These were issues fought out and settled, so far as Alaskans are concerned, in the whether or not and why of the pipeline debates, but now they were beginning again in the question of what to do about the Alaskan public interest lands Congress called for in the Native claims settlement (see Chapter 4); recurring themes in *Times* editorials and its presentation of the news. About the answers Mr. Atwood had no doubts.

"The thing we hear most of the time around Alaska right now," he was saying, "especially with the present state administration, is negative—*negative,* all the time negative. Impact seems to have a connotation of something undesirable. Up until now, impact wasn't necessarily bad; it could be something good. Everybody in the state's all the time trying to *minimize the impact.* Well, heck! The impact of this pipeline is terrific. You look at the payrolls and what's happened to our economy and what's happened to people and their families and their bank accounts, their spiritual—themselves, their ceilings that have been lifted on them and their ambitions, it's terrific, the impact. And it's all over Alaska. . . ."

He paused, sat back in the desk chair, face dark in the glare of light from the broad window behind him. Beyond a flight down, was 4th Avenue—not the 4th Avenue of the all-night bars and liquor stores, pawn shops and flophouses, drunken Natives and their debased women, but a modern avenue of expensive shops and restaurants, new office buildings, separated from that older Anchorage by a scant six blocks and as many decades of time, with the city hall midway between. The view out the window was dominated by the creamy tower of the Anchorage Westward Hotel, built since the earthquake—in defiance of earthquakes and the subarctic weather—where from the Petroleum Club at the top on a clear day you look north up the Knik Arm of Cook Inlet to the shining, cloudlike mass of Mount McKinley, floating on the horizon a hundred miles off. He had grown up, as I did fifteen years later, on the North Shore of Chicago, but nothing remained of the nasal drawl of upperclass Chicago: the voice had the sharp twang of metal guitar strings, pitched high by age.

He went on, developing another side of his homily, Jay Hammond's new-style Republicans, inaugurated at the beginning of the year: "We got this negative attitude from our state administration. It's new, it's different, and I think it's temporary—and it's not going to endure. I don't think it represents the majority of Alaskans and I think it's going to be out in the next election—if not sooner, because I think Jay Hammond's going to want to be re-elected and he wants to move over and do something, and that's his problem, with all the no-growth negative people he's surrounded himself with. . . ."

He paused, modulated to another idea, resumed—maybe there was some use to be made of this writer after all: "The other—opposite—extreme—and I'm thinking of an attitude for a book: I think your book would have much more interesting—more public acceptance and a more permanent spot in the literature of this nation—if you took a positive view with the spirit and uplift to it that comes when you look at this huge piece of real estate of Alaska, that's been here forlorn, forgotten, beyond the perimeter of public interest for generations, ever since its purchase from Russia—misunderstood, run by Congress when Congress didn't know what to do with it, so they did nothing. . . ."

I listened politely, caught at the thread of history and asked about the economic forces, salmon canners, loggers, miners, that had resisted statehood, but he responded in further generalities. A couple of editors came in behind me, called for a meeting of some kind, and my time was up. I packed the tape recorder away, thanked him, and left.

I stopped back a week later intending one more try at getting the man to talk about the things he knew. In the meantime I had flown up to Prudhoe Bay, then to Fairbanks and on to the Yukon, where the last segment of the haul-road bridge was fitting into place the day I got there, finally over the coastal mountains and glaciers, in thick snow and rain, to the lightless airfield at Valdez for an introductory look at the terminal—beginning, I thought, to get at least a sense of the job ahead and the kinds of things I would have to know before I could write of this place and what was happening to it and in it; and along the way had picked up a streaming dose of someone's Alaskan flu and was feeling miserable and—worse, my ears blocked with phlegm by too many take-offs and landings in unpressurized small planes—half deaf.

The talk this time was shorter and sharper; in the interim, evidently, Mr. Atwood had given thought and concluded I would serve no useful turn after all. Who's your publisher, he demanded, then, who's your editor, and I told him. Then: what's your *angle,* he wanted to know, what's your *slant;* and I tried to tell him that since I was barely started I could hardly say what I thought of the project, that until I had learned as much as I could know about it I would not know what it meant because its meaning would grow from, and be founded on, facts and I would not be able to master them if I started with preconceptions, an angle, about what they were—but the idea did not penetrate; it was not that kind of mind. And finally, rumbling in my stuffed ears: after forty years in the country and flying up and down the pipeline right-of-way in a low-level helicopter and knowing the top men who were building this thing, *he* did not feel ready to write a book about it, and how did I think I could come up here on a trip and then go back and write some superficial little—That was enough. "I don't *do* superficial work," I said. "I've got years ahead of me on this and I *don't know* where I'll come out or what I'll write. But however it turns out, it's going to be what I have seen and thought and felt, what I know for myself—not what anybody else tells me to think." I said good-by and left.

The trouble, it occurred to me later, was perhaps that the old man had ambitions of his own. Regularly, in addition to the editorials, there appeared in the *Times* under his name and photograph little reminiscences of Alaska and Anchorage, in the manner of the history lessons with which Colonel Robert R. McCormick, the isolationist Roosevelt-hating publisher of the Chicago *Tribune,* liked to favor his readers in our respective boyhoods. With due allowance for the difference of time and place, then, he had become the Colonel McCormick of Anchorage. I had by then met or heard of a dozen people who claimed to be writing something about Alaska and its pipeline, but if Robert Atwood was among them it did not seem that he would be serious competition.

In the months and then years that followed, as I pursued my investigations up and down the pipeline and in most of the nation's's oil towns (I missed Tulsa and Oklahoma City), I had often to ponder those two ideas: *impact; angle*—a writer's view of what he would write and therefore of what he would see, feel, think, know.

Impact in that time was the Alaskan code word for whatever one disliked in one's daily life. If the Anchorage traffic was heavy on your way to work and the streets torn up, it was an impact. If the whores in the Fairbanks bars were thicker on the stools than customers, impact. Crowded schools in Anchorage and the kids coming back by night, apparently, to set fires and break windows; months of waiting for a Fairbanks telephone installation, then finding that half a day at a time you get busy signals because there are not enough circuits, or standing in line to get to one of the two functioning pay phones (there are a few others on the streets with their wires cut and their guts ripped out). The expression derives, I suppose, from the environmental-impact statements that preceded the pipeline and means the social effects anticipated from it, a metaphor built on a metaphor—not, one would think, a very good tool for getting at realities. Since we continue to hear of impacts wherever anyone proposes to hunt for oil or gas or dig for coal, it is worth considering how and whether the Alaskan pipeline affected the society in which it was built—and if for the worse, as the word implies; *impact,* a blow, a striking onto or into.

Atwood's angle is a newspaperman's short cut: with luck and the right instincts, it enables him to exclude irrelevant particulars while getting quickly to the gist of what has actually happened. Since most of the writers prowling my territory were newspapermen of one stripe or other, I found myself often explaining that that was not what I was nor what I was doing. It was necessary, therefore, to think carefully about just what that was. It was not the instances to flesh out a predetermined reality that I wanted, no quick revelation and confirmation, tomorrow's stale wonder; an example, it is said, is not a proof. It was the totality I had to grasp of many men's experience, excluding nothing, and when I knew enough, if I could know enough, the connections of all those particulars and their pattern would become apparent, and that pattern would be truth, and whole, and in it, part of it, all those human motives for human acts that the newspapermen do not often like to think about—not merely what the actors said had moved them or thought had moved them but the motives behind those motives. It would be well then to remain skeptical, reserving judgment, accepting nothing uncritically; and patient. With patience, one would find the pattern.

So that opening round with the old battler of Anchorage had not been wasted. Whether he meant to or not, he had made me aware of a combination I would have to watch out for.

When the pipeline debates ended and the environmental arguments were exhausted, two kinds of impacts remained in the public mind: on the one hand, disruption of the small and fragile Alaskan society by the coming inflow of industry, transient strangers, money; on the other, the prospect of a booming prosperity that would put right the things that had been wrong with that society, that had made it the nation's ward since the beginning of the American presence, conferring at last the sovereign, self-sufficient autonomy promised by statehood. For Alaskans and their Outside observers during the debates and since, which of these prospects seemed likely to prevail depended on values, motives; one's angle. It was the pessimistic view, naturally, Atwood's "negative," that was most on people's minds—they *assumed* the pipeline and its oil meant prosperity, that there would be money to be made for whoever wanted it. Their forebodings were, at any rate, much talked of and were duly reported, making good and properly accusatory copy. These reports, certified by print, were enlarged in talk, fresh instances found—example making proof—and were in turn reported; and so on, a circle, self-fulfilling. The social "impact" of the pipeline is in the first place the story of how that was reported, in Alaska and else-where; it is necessary to distinguish the reality from its news-paper version. For reasons that are only partly due to the care-lessness and shortcuts of newspapers and their writers, the two do not make a very good fit.

In the spring of 1974, as Alyeska pushed its haul road north from the Yukon and attempted to organize for the more difficult task of laying pipe the following year, the Fairbanks North Star Borough appropriated the judicious sum of $13,450 to establish what was called an Impact Information Center, reporting to the Borough's mayor and guided by an Impact Advisory Committee. The Borough is a 7,500-square mile administrative entity sur-rounding the city of Fairbanks and, in the Alaskan system, corre-sponds roughly to a county elsewhere. As a strategic rail, road, and air terminal, Fairbanks would serve as the main supply cen-ter for the whole pipeline; as the only area of considerable popu-lation directly on the route, it was also the home base and hiring center for many of the pipeline workers to come, relieving Alyeska of the need for building several work camps—and the recently signed Project Agreement had insisted that those who lived within reach of the work sites be allowed to live at home and not be obliged to go into a camp. The Center's immediate purpose was to combat the voluminous rumors already circu-lating, expressing the citizens' anxieties about the effects of the pipeline. To do this, it was necessary to match up what could be learned of Alyeska's activities and plans with reliable informa-tion about local costs—housing, food—and the ebb and flow of population, employment, wages, police activity. The volunteer committee included at the start Sam Kito, then vice-president of Doyon, the Native regional corporation headquartered in Fair-banks; a clergyman; the presiding officer of the Borough Assem-bly (i.e., the legislative council); and Larry Carpenter, a local television newsman lately hired to do Alyeska public relations, the only committee member who remained in place to the end of construction. The Impact Information Center moved into a rented storefront a few doors down from Larry Carpenter's pub-lic relations office over a 2nd Avenue bar, with a full-time direc-tor, an information officer (Mim Dixon, a University of Alaska Ph.D., through most of the period), a secretary, and began turn-

ing out typewritten reports every couple of weeks for the benefit
of the Borough planners and anyone else who was interested.
The Borough's original budget for these activities was continued,
helped out with "impact funds" from the state.

The Center's work was complicated by the fact that, in a statistical sense, most of the information it needed in order to assess
the social effects of the pipeline was not known with any certainty. The cost of living, for instance: everyone knew it was
going up—the real question was how much and just what this
had to do with the building of the pipeline. To answer this question it is necessary first to know the costs of the myriad items
people spend their money on at a particular time and in a particular place in order to live and to relate these costs to industry
and trade, employment, wages, population, family size, and so on,
then to keep continuing track of all these matters. If this sounds
elementary, it is because the job is regularly done for us in most
American cities and the nation as a whole by an agency of the
Department of Labor, the Bureau of Labor Statistics—for the
purpose the country is divided into what are called standard
metropolitan statistical areas that take in most of the population.
In Alaska, however, only the Anchorage area was considered
worth studying in this way, and despite some prodding by Mike
Gravel, one of the Alaskan senators, the Department of Labor
declined to spend the several hundred thousand dollars a year it
estimated would be needed to find out what was happening in
Fairbanks. Since neither the Borough nor the state government
was prepared to pay it either, the Impact Center was reduced to
sending Dr. Dixon around periodically to the supermarkets to
check the prices on the shelves and study the classified ads in the
News-Miner to learn how rents were progressing.

Similarly with the population: apart from the 1970 census, it
was not known, in the course of construction, just how many
people were actually living in the city and the surrounding Borough or—if their numbers were rising, as they seemed to be—
what mix of single people and couples with children the increase
consisted of. There were at least three separate estimates, all
significantly different: one, by the Internal Revenue Service from
income tax returns, as a basis for federal revenue sharing; another, by the state, for purposes of its own; and a selective sur-

vey by a University of Alaska research body which touched incidentally on the size of the population. The lack of such
information places obvious difficulties in the way of effective
local government. If you do not know what your population is or
how it is changing, you cannot readily provide those basic services—police and fire protection, schools, for instance—that
require comparatively long-term planning. Moreover, without
those vital facts, it is hard to know what to make of such information as you do have, as we shall see. For example, if the police
report twice as many robberies this year as last, is that because
the people are somehow more prone to robbing and being
robbed, or only that there are twice as many of them? Or, more
generally, we might add: *does* crime inevitably increase with
population, by a kind of natural law?

In any case, although what we know of Fairbanks during the
pipeline years is sketchy and incomplete, it is the best we have.
Despite much talk and grumbling opinion, duly reported, and
elaborate research planned at the Anchorage branch of the University of Alaska (but not carried out), no comparable efforts
were made to *find out* what was happening, either in Anchorage,
in Valdez, or anywhere else in Alaska.

With allowance for the difference in scale, the problems of the
state government were like those in Fairbanks but, spread over
the vast acreage of Alaska, diffuse and undramatic, and we heard
less about them in the South 48. Some of these we have touched
on earlier: the vacancies in the thin-stretched roster of state
troopers, corresponding with the number hired into the Alyeska
security system (those who made the jump felt excused by past
skimping on pay and lack of advancement); the decrepit highways, by implication worn out by pipeline traffic. All through the
building of the pipeline, the cost of state government increased,
and with it, presumably, the demand for its services. The resulting deficits in the state budget were made good by borrowing—
despite the years of delay, the oil *would come* and the state was
considered a prime risk—and by ingenious new taxes, principally
on the oil companies; the goose and its golden egg. The common
element, that is, is money, and not just money in the abstract but
particular money: *who shall pay.*

In the summer of 1975, as Alyeska boomed into its first season of pipe work, these half-formed anxieties began to appear as specific and alarming images in widely circulated reports on what the project was doing to Alaska and its people. In an article by Winthrop Griffith called "Blood, Toil, Tears and Oil," published in the *New York Times Magazine* for July 27, 1975, the images clustered around two broad themes. On the one hand, it seemed, Alaskan life was being debauched by endemic alcoholism and suicide, by imported whores and corrupted Native girls, broken families and abandoned children; in Fairbanks, there was no "decent rental housing for less than $600 a month" and the divorce rate in the first year of construction had risen by 25 per cent. Mr. Griffith's conclusions were embellished with photographs—a girl strutting a Fairbanks sidewalk in a thigh-top miniskirt, two pugs in broad-brimmed hats who may be Okies, drinking from paper cups at an afternoon barbecue—and barstool anecdotes, but, apart from the references to rent and the divorce rate, were supported by no evidence. At the same time, the article argued that the pipeline was not only disruptive but dangerous: "The casualty figures for pipeline related workers," Mr. Griffith insinuated, "may well turn out to be higher than for any other major construction project in the nation in modern times." Although he had nothing concrete to say about pipeline accident rates (other than to dismiss Alyeska's figures), he did cite two anonymous guesses about deaths: one that "500 men will die on this pipeline job"; another, a Bechtel computer study estimating that 273 of *its own* employees would be killed before the job was finished (Bechtel at the time, still the CMC, had about 1,500 men in Alaska, so that, in terms of death rates, if this prediction were true, working on the pipeline for Bechtel would be about 7½ *times* as dangerous as serving in World War II). Mr. Griffith had one final point to make: he did not care much for the Arctic landscape, though all he could find to say of it was that it was "eerie."

In all of these views, the writer was in a sense simply confirming what one would expect—so much so, indeed, that one need hardly travel to Alaska and spend time on interviews to find out. If you concentrate several thousand men on the loose in one small place and supply them with cash, enough of them will behave in ways not likely to improve their own lives or the life around them. And heavy construction of any kind is dangerous for those who do it; crossed with the considerable hazards of Alaskan weather, it would, one would think, be exponentially so. But these are suppositions, working hypotheses, aids to investigation if candidly and openly used; in a word, angles (that word again), not conclusions. And it is precisely a writer's job to examine his suppositions (and everyone else's), determine whether they are or are not supported by evidence, and *not to say what he does not know.* But Mr. Griffith, in this case, had no such scruples.

Readers in other parts of the country may object to what appears to be one more instance of Eastern Seaboard provincialism: taking seriously anything written in a long-ago Sunday supplement—even if the sponsoring paper is the New York *Times.* I do so for two reasons. First, the piece was an early and sensational version of the pipeline experience, and as such it altered the ways in which that experience was perceived, and thereby the experience itself; ripples spreading outward from a stone tossed in a pond. (Alyeska, by the way, responded promptly and at length in a letter signed by its president Ed Patton, giving the unsensational statistics of its industrial safety department, but the *Times* did not print the letter, shortened, until September 7th and allowed its writer equal space for a flippant self-defense; the reply, in short, had no effect.) Second and personally, since the piece was among the first I read about pipeline work before going to Alaska to see for myself, it occasioned months of effort to find out what basis, if any, there was in fact for its assertions. We shall turn presently to consideration of the facts Mr. Griffith led me to.

In September 1975, about the time I was getting my first look at Alaska, the Los Angeles *Times* sent a team of reporters north to do some looking of its own. The series of articles that resulted was published in November and within days was syndicated around the country in papers as far off as Orlando, Florida. The Los Angeles *Times* also ambitiously assembled the pieces in the sort of reprint pamphlet in which it is customary to submit one's offerings to the Pulitzer Advisory Board. (The series won no prizes in any of the Pulitzer categories.)

The Los Angeles writers were more persuasive in their rhetoric than Mr. Griffith had been but provided little more in the

way of supporting evidence. The gist was covered in the opening
sentence of the installment that ran November 25th under a
front-page banner headline: "Lawlessness, a helpless government and the stranglehold of a single Teamsters Union chief"
(i.e., our old friend Jesse Carr). In the matter of "lawlessness," the reporters found informants to tell them that "missing" small tools amounted to "$200,000 a month at each of
the 20 camps," which, if you do the arithmetic, comes to nearly
$50 million a year, say $2,500 per man in wrenches and screw-
drivers; that 200 of Alyeska's trucks were similarly "missing"; and
that one trucker had gotten away with "millions of dollars worth
of prime lumber and building materials" from a single camp (it
was said to be the trucker himself who made this particular
claim). They also produced statistics to show astonishing increases in various categories of crime. In Fairbanks, for instance,
in the fiscal year that ended in June 1975, murders were up 50
per cent, prostitution arrests no less than 700 per cent. There
was, however, no mention of the actual numbers of cases, which
were less impressive. The murder percentage meant three cases
against two the previous year; arrests for prostitution had risen
from two to sixteen, a significant increase, certainly, but not
quite the epidemic suggested by the percentage.

Apart from these few statistics (and yet another guess at the
Fairbanks population)—and some lively references to alcoholism
and big-time gambling in Valdez—the reporting team uncovered
three interesting facts. For one, it seemed that twenty or so of
the warehousemen working the Alyeska equipment dump and
supply center in Fairbanks, Teamsters all, were ex-cons that
Alyeska had been unable to get bonded; which sounds indeed
like setting the fox to guard the henhouse. Again, two gentlemen
with Italian names, said to have connections with Tucson
Mafiosi, had settled in Anchorage and gone into business (a bar
and a jewelry store, respectively). Finally, like others who managed to visit the work sites unattended, the reporters found men
standing idle or sleeping in the buses that had brought them out
from camp. The reporters attributed the blame for this situation
about evenly to Alyeska's careless management, to feeble and
undermanned government, and to the reckless self-interest and
intimidating power of Jesse Carr and his Teamsters.

The series, naturally, was greeted in Alaska by protests from sources as different as Mr. Atwood's *Times*, its struggling rival, the comparatively liberal Anchorage *Daily News* (the state's only morning paper), and Governor Hammond and his attorney general, Avrum Gross, who were unhappy at the gloomy pronouncements in which they were quoted; and of course by Ed Patton and Jesse Carr. In editorials and further reportage the Los Angeles *Times* repeated and elaborated its charges until the end of the project, and except in Alaska, they were generally accepted as fact. By January, the Department of Justice was estimating pipeline thievery at a billion dollars, a sum that would have involved getting away with the work camps themselves and the construction equipment and trucks—and would still have left room for the small tools and all 800 miles of the pipe. (Later, the syndicated columnist Jack Anderson claimed the pipe too was vanishing, indeterminate miles at a time, ten miles altogether, though no one could say where it was going or what it might be wanted for.) Both Ed Patton and Avrum Gross offered to prosecute any charges brought forward, but the reporters were silent. Those like the trucker with his "millions" in lumber who might be charged were also sources and therefore protected, so the theory goes, by the First Amendment.

As with the Fairbanks crime statistics—indeed, *any* crime statistics—I found that it is well to be cautious in drawing conclusions from what one observed along the line. Those pipeliners, for instance, standing around, sleeping in the buses—on the face of it, an obvious case of malingering abetted by careless or incompetent supervision, but the reality was not often so simple. For one thing, in the scheduling of all the different jobs involved in laying pipe, the different trades do not do each other's work; which is both self-evident and written into every labor agreement, so that while one team is doing its job others are likely to be waiting for it to finish—welders for the helpers cleaning the pipe ends, tapers for welders. Further, if the weather changed—rain, snow, rising winds, falling temperature—after the crews were taken to the job site, some or all of them could not work but, by contract, would be guaranteed a minimum pay and might as well, therefore, wait it out, a four-hour half shift if the hitch came in the morning, eight hours if later. Or if opening the pipe ditch revealed an unexpected soil condition that changed the design—or worse, a doubtful case that would have to be argued out between Alyeska's design engineers back in Anchorage and the two pipeline offices—the sequence of jobs would be broken or, often, stopped, with the same result as from bad weather: men and equipment idle and in the wrong place for the work they were scheduled to do. It is, in short, unwise to judge from what one has seen unless one is sure of all the circumstances. Nevertheless, the Los Angeles reporters and those who followed them at the least raised some of the fundamental questions—about the competence of Alyeska's management and the quality of the labor it managed, the conjunction of both with the state and federal monitoring programs, with government in general and with the union's leadership; and about the interacting goals and interests of all of these bodies and their mutual effects. These are questions serious and complicated enough to reserve for the separate consideration they deserve, in the next two chapters.

In the meantime, it is necessary to get straight, so far as we can, on the facts from which conclusions may be drawn. In a place and time where it is not possible to obtain reliable information on such elementary matters as the resident population, this is not easily done.

If your first introduction to Alaska comes by way of the falling-down drunks on the sidewalks outside the 4th Avenue bars in Anchorage—or 2nd Avenue in Fairbanks—you would figure the place had more than its share; and that there must be some truth after all in the old American clichés about Indians and alcohol. And if that introduction came anywhere between the spring of 1974 and the summer of 1977, it would seem obvious that the condition had something to do with the pipeline. The real questions, however—how many alcoholics are there and what exactly is their connection with the pipeline?—cannot be answered by casual observation.

For a start, it is less clear than it might seem just what alcoholism is, and until that question is answered you cannot begin to say who has the condition or how many there are of them in a particular place. Most definitions are so subjective and indefinite as to be difficult to interpret or apply in specific cases. Is an alcoholic simply one who drinks a lot, either habitually or on an occasional bender? Or is alcoholism a "disease," as is often said—or perhaps several diseases under a common name? But if so, it is not one that can be as reliably diagnosed and treated as smallpox or pneumonia, say, or most mental illnesses. The difficulties are apparent in the most recent definition adopted by the National Council on Alcoholism:

> A complex, progressive disease in which use of alcohol interferes with health, social and economic functioning. Untreated, alcoholism results in physical incapacity, permanent mental damage and/or premature death.

In considering how widely this description applied to Alaskans, it is reasonable to ask first of all how much liquor they consume—it is agreed, at least, that an alcoholic drinks. The answer, it seems, is that the reputation is deserved. For 1974, what is known in the trade as the "apparent consumption" of distilled spirits—total sales divided by population—was 3.3 gallons per capita. This in turn was about 70 per cent more than the national average and substantially higher than in affluent, suburbanized states known for their drinking, such as Connecticut, Illinois, or California. Indeed, the only places in the country where the apparent consumption is higher are New Hampshire and Nevada (about double) and Washington, D.C. (nearly triple)—all with comparatively small populations, disproportionate numbers of tourists, and low liquor taxes. (When the Washington population and liquor sales are combined with those of Virginia and Maryland, the per capita consumption is not much above the national average—the suburbanites buy a lot of their drink in the capital because it's cheaper.) Leaving these three aside, it appears that Alaska is indeed the nation's drinkingest state. Moreover, similar ratios hold whether we add in the equivalent in wine and beer or allow only for the drinking-age population—those over fifteen, eighteen, or twenty-one, depending on how you care to define it. Finally, while the sales of spirits have been on the rise everywhere since the mid-6os (by about 50 per cent), Alaskan consumption has also being going up—and at a rapider rate.

Curiously, however, the proportion between Alaskan consumption and the national average was slightly lower in 1974 than it had been ten years earlier.

All of these estimates, to repeat, are for Alaska as a whole, based on the entire population and total sales of spirits within the state. In Fairbanks, which made such an impression on the reporters, sales of liquor nearly doubled between 1970 and 1976, but it is hard to say just what this means in terms of individual consumption because of uncertainties about the size of the population (see below). With conservative estimates of the combined civilian and military population, these totals come out to about 3.5 gallons of liquor per capita in 1970, somewhat more than that year's Alaskan average and close to 6 gallons for 1976. These figures require qualification, however: as the hiring center for nearly all pipeline labor (and for the Prudhoe Bay field as well), Fairbanks was also the source for most of the liquor going into the camps, which as a matter of observation was substantial despite Alyeska's no-liquor rule—but just how much the men in the camps accounted for, there is no way of estimating. If we distribute the Fairbanks liquor supply among the Fairbanks population plus 20,000 or more workers in the camps, we come out with an average of about 4 gallons per head—still more than the Alaskan figure and representing a sharper increase since 1970; but a good deal less startling than 6 gallons, particularly since all the pipeline and oil-field workers are adults.

Another common-sense indicator of the prevalence of alcoholism can be found in the causes of death within a population. Perhaps *derangement* is a better word for the condition than *disease*—it is associated with violent death, by accident, murder, or suicide. Thus, unlike every other American state, in Alaska, consistently for many years, accidents have been the commonest cause of death, generally about twice the national average rate. In Anchorage recently it was calculated that 81 per cent of the traffic deaths—accidents, of course—could be attributed to drunkenness, in contrast to the National Council's estimate of 50 per cent for the nation at large; a stranger walking or driving the streets does so in fear and trembling—in any day he is sure to witness several smashups, ranging in seriousness from fender crumplers to head-on collisions in the downtown one-way maze.

In Alaska generally, however, traffic accidents account for only about one sixth of all accidental deaths.

Alaskan suicide rates over the past fifteen years have declined slightly but have remained about 50 per cent higher than the national rates. During the same period, the Alaskan homicide rate also declined while the national one doubled—in 1960 there were proportionately half as many murders nationally as in Alaska, but at the end of the period a few more. (Bearing in mind my own caution about percentages, I note that in 1960, when the Alaskan population was 226,167, the comparatively high suicide rate of 15.5 per hundred thousand meant 35 actual deaths.)

In the 1940s, a researcher named E. M. Jellinek devised a comparatively objective system for estimating the incidence of alcoholism from deaths due to cirrhosis of the liver. With modifications, the Jellinek formula has been generally accepted among students of alcoholism and suggests that nationally something over 3 per cent of the population is seriously affected by drink. It yields unreliable results in Alaska, however, for three reasons: the population is too small; the death records are inaccurate and incomplete; and not enough Alaskan drinkers live long enough to die of cirrhosis—if you wander out into the Alaskan winter, pass out and freeze, your death will be recorded, if at all, as due to "accident," not alcohol. With these differences in mind, a state government study published in 1973 counted those incapacitated by alcohol in Alaska's cities and villages. From this census it appeared that nearly 16,000 people could be classed as alcoholics—more than 9 per cent of the then adult population twenty and over, *three times* the national rate. The study produced other interesting numbers. The incidence for Natives was significantly less than for whites—7.6 per cent against 9.5 per cent. On the other hand, in predominantly Native towns such as Fort Yukon, with a population of perhaps 500, the rate was over 60 per cent —more than half the people would be drunk most of the time.

Within Alaska the alcohol problem is much deplored in the newspapers and by state officials and there are a number of programs for education and treatment, though they suffer from stop-and-go funding. Since the coming of the pipeline two specialized organizations have been formed: one sponsored by the Alaskan contractors' associations, Alyeska, and the unions (Jesse Carr is its vice-president); the other a Native group supported by state seed money and business donations. The first of these aims at identifying alcoholics as they come out of the hiring halls and steering them to Alcoholics Anonymous groups in the camps; out of the many thousands of workers who flowed through the Alyeska system during the pipeline years, fewer than a thousand with some kind of problem seem actually to have been identified, and of these perhaps 200 landed in the AA groups. The Native organization aimed primarily at training counselors and placing them in the villages and at producing educational materials in the various languages—posters, radio and television spots. All of these programs, including those of the state, suffer from a lack of serious information: except in general and subjective terms, those who run them know very little about the real extent of the problem they deal with and therefore (the point for our purposes) have no way of measuring how it is changing or whether their efforts are producing any results.

Although the evidence is sketchy, incomplete, and out of date —I have summarized all there is—it does not seem to offer much basis for the newspaper talk of the pipeline's demoralizing the Alaskans with drink. That is all it was—talk; an "impact" that on the whole did not occur. The truth is that alcohol is a major and long-standing Alaskan problem, more so than anywhere else in the United States; and that if the pipeline has affected it in any way, it has been by making a few people conscious that it exists and impelling them to attempt to do something about it.

A different kind of index of a society's condition is the stability of its marriages, of which some indication is to be found in numbers of divorces. With this in mind and in response to the well-publicized anxieties about how daily life was being affected by the pipeline, the Fairbanks Impact Center began keeping track. Thus, in 1970 there were 351 filings for divorce in the local courts, while by 1974 the total was up to 586, and for 1975 (a projection from incomplete figures) it appeared that it would be over 800—more than double. As is often observed, however, divorces have been on the rise throughout the country during the same period—from a rate of 3.5 per hundred thousand of population in 1970 to 4.6 in 1974, an increase of 31 per cent.

Comparison of what was happening in Fairbanks is difficult because of uncertainties about the size of the population. The reason for this is that the U.S. census, presumably the most reliable source for such statistics, does not distinguish between resident civilians and the substantial military population within the Fairbanks North Star Borough—about 15,000 soldiers and dependents at Fort Wainwright and the Eielson Air Force Base in 1970, perhaps 12,000 by 1976. Estimates of the civilian population are therefore projections from samplings and quite approximate —according to a University of Alaska study in early 1976, about 42,000 against 32,000 in 1970, an increase of 33 per cent. (For 1976, the borough planners used an estimate 10,000 higher, but from other evidence, such as the school population, this appears to be incorrect.) For comparison with the national rates, these estimates yield divorce rates of about 11 per hundred thousand in 1970, 14 for 1974*; if the military population is added, the rates are 7.5 and 11 respectively.

The conclusions to be drawn from these comparisons are like those from our investigation of alcoholism. For one, divorce was far more common in Fairbanks before the pipeline than in the United States as a whole, and it continued to be after work started—about three times so in both 1970 and 1974; it does not seem the best place to live if you mean to stay married. At the same time, while the divorce rate has risen during these years, the increase appears to have been a little slower than in the nation at large—about 27 per cent higher for the Fairbanks civilian population in 1974 than in 1970 compared with 31 per cent nationally; and, for what it's worth, the increase in the divorce rate was significantly lower than the population increase. If the pipeline has had any effect on family life, it has probably been that of spreading around a little more uncommitted income—divorces cost money. One of the fringe benefits added to the 1974

---

* In the absence of a 1974 population estimate, I have used the early 1976 estimate and the 1974 divorce figure; nationally, 1974 is the latest year for which the divorce rates are available as I write. The incomplete Fairbanks divorce total for 1975 looks as if it would be substantially higher than 1974, but there is no estimate of the rapidly changing population to match it with. It should be noted also that the Impact Center's divorce figures are for divorces filed, not those actually granted, and therefore are probably high.

Teamsters wage contract was a kind of legal health insurance, a prepaid legal trust created by employer contributions; using a law firm housed in the union's new office complex in Anchorage, members were thus able to get all their legal work, including divorces, at nominal cost, or none at all. When I tried to find out if divorce work has increased as a result, Gil Johnson, the firm's head partner, thought not but was unable to produce any figures; the Teamsters, as we noticed earlier, are not very good at numbers.

Money keeps coming up in these considerations. Inflation was the impact most feared before the project started, most complained of while it continued, and was much written of by the reporters; Winthrop Griffith's pipeliner who in the spring of 1975 could find no "decent" rental housing for under $600 a month was a fairly mild instance. At the time in Fairbanks there were about twenty rentals on the market, ranging from a furnished "efficiency" at $150 a month to a three-bedroom unfurnished apartment at $650 (plus the considerable cost of heat and light) and a dozen mobile homes and cabins—it depends, perhaps, on what you mean by "decent"; both in Fairbanks and in smaller Valdez there were cases of landlords finding ways of removing long-standing tenants in order to raise their rents by one or two hundred dollars a month. In downtown Anchorage at the time the going rate for a run-of-the-mill hotel room was $35 a night, $5 or $10 more in Fairbanks, though in both places there were rooms to be had for $20 or $25 at lesser hostelries away from the center of town.

Food prices were also stiff, though with less effect than housing in all its forms. Milk in Anchorage or Fairbanks, tanked in fresh or "partially reconstituted," was about double what it would be in the Lower 48, and the limited local production from the Matanuska Valley north of Anchorage cost more. In general, Alaska produces few of the essentials—food, manufactured goods, even building materials—and what it does produce costs more, by reason of the steep local wages (see Chapter 7) among other causes. Recently when a German firm, Prinz Bräu, set up Alaska's first brewery, in Anchorage, the beer sold for 50¢ more a six-pack than national brands shipped up from Seattle; the famous salmon and King crab are too valuable an export for much local consumption—I can buy both more readily in Philadelphia and at no higher price than in Anchorage.

Examples might be multiplied, but tediously—there is no doubt but that a considerable inflation ran concurrent with the pipeline. As with the other "impacts" we have considered, the real questions are how great it was, how it compared with what was happening elsewhere, what connection it had with the pipeline project.

As we know all too well, inflation has been a fact of life everywhere in the country throughout this period. Between 1970 and mid-1976, the Consumer Price Index, a weighted average of basic living costs in selected small, medium, and large cities around the country, went up by nearly 50 per cent; gas and electricity, fuel and other utilities, rose even faster, by nearly 75 per cent. As we also know, or should know, the most important single factor in these increases has been the sevenfold rise in the price of imported oil, affecting all other costs—not only directly, in the cost of fuel, but indirectly, in the manufacture, processing, and transportation of all the goods we consume; from 1974 to 1976, following the OAPEC embargo and the first big jump in oil prices, the inflation rate averaged about 8.5 per cent per year, nearly double the rate for the preceding four years since 1970. If the cost of living has not risen at the same rate as imported oil, it is because through most of this time we have been producing more oil domestically, at controlled and comparatively low prices, than we have had to import. Since virtually everything that Alaskans use must be brought there from somewhere else, we would expect them to be more affected by these circumstances than other Americans, and in fact by 1976 the inflation rate in Anchorage had reached 12.5 per cent.

Some indication of what these dry percentages mean for Alaskan costs and how they differed from those elsewhere in the country is found in the cost-of-living allowance granted federal employees assigned there: in 1976, 22.5 per cent extra in Anchorage, 25 per cent everywhere else. It seems that this allowance was skimpy. The same people who compile the Consumer Price Index also periodically translate it into hypothetical budgets for average families—husband, wife, a brace of kids—living at varying levels of comfort. Between 1970 and 1975, the intermediate budget for Anchorage was consistently 30 to 36 per cent above the U.S. urban average, and both rose by about the same amount, nearly half. In the absence of a similar study for Fairbanks, the educated local guess was that costs were about 15 per cent higher than in Anchorage: that is, if it cost over $21,000 in 1975 for an Anchorage family to keep itself, the corresponding budget in Fairbanks would be about $24,500.

Whether high prices are a hardship or not depends, of course, on what you can afford. (The reader will pardon, I hope, so obvious an observation; it was one rarely made by reporters exclaiming over the price of steak in Anchorage or of apartments in Fairbanks.) From the same University of Alaska study that produced the Fairbanks population estimates mentioned earlier, it appears that immediately before the pipeline started, an average Fairbanks family was indeed having trouble paying its bills; from 1970 on, local businessmen had invested heavily in anticipation and, while they waited for the work to begin, suffered a recession by which some were ruined. From 1974 on, though, median family income stayed several comfortable jumps ahead of living costs, and by 1975, when the family budget had reached $24,500, its income was around $30,000 and in 1976, $35,000; for the comparatively small number of Fairbanks families whose income came directly from the pipeline, the median in that year was over $40,000. (The corresponding national figures—median income for a family of four—were little more than half the Fairbanks estimates: $15,885 in 1975, $17,315 in 1976.) For most people in town, then, the cost of living had become like the weather—something to talk about while warm and dry indoors by the fire, not one of life's serious problems.

There are indications that this prosperity was widely and generally distributed. By the end of 1975, while the rest of the nation wallowed deeper into recession and rising prices, unemployment in Alaska, for the first time ever, dipped below the national average—in fact if no doubt briefly, the boom had lifted the state from the depression that, with occasional respite, had been its lot for decades. In the state as a whole by the end of 1975, personal income as distinct from hypothetical family income (that is, the total wages and other income divided by the total population) was about $9,500—more than half again above the national average and $2,000 to $3,000 beyond such affluent states as Connecticut, New York, Maryland, Illinois, and California; only the District of Columbia, at over $7,700, came close. Between the deflated miniboom of 1970 and the boom year of 1976, assessed property values in the Fairbanks North Star Borough (rated at their market value under state law) rose from $229 million to more than $1 billion, and by 1977, when the pipeline was finished, were projected to reach $1.3 billion, a more than fivefold increase; and tax rates were reduced (almost) in proportion. While a substantial part of this gain must have been due to inflation and another large part to new construction—including the valuation of the pipeline itself, ninety miles or so of which is within the Borough—it appears that the real wealth of every Fairbanks householder had been largely increased; just the kind of pipeline "impact" tirelessly proclaimed by Robert Atwood's Anchorage *Times*.

Moreover, if housing and other costs did not rise in proportion to property values or income, the early outcry was perhaps part of the reason. In the summer of 1975, the governor appointed rent review boards in Fairbanks, Anchorage, and Valdez, their modest costs met by what were called impact funds provided by the state (a notion we shall return to): a tenant hit by an increase could bring a complaint and the board, on reviewing the landlord's costs, was empowered to reduce the rent. Thereafter, although rents remained high by the standards of cities elsewhere—remember, suitable land, a basic factor, is scarce in Alaska and therefore expensive, and developing it is costly for reasons ranging from construction wage scales to the presence of permafrost—they also fluctuated quite widely, as did the actual demand for housing. There is at least some evidence that the threat of a hearing, as well as the expansion of the market by new building, put restraints on cost increases, and that similar restraints were at work on other basic costs, such as food.

Although the extent of Alaska's inflation and its general prosperity during pipeline construction are fairly well defined, the role of Alyeska and the oil companies in these conditions is less clear. The company politically bought or leased what it could of its trucks and heavy construction equipment from Alaskan agencies, but very few of its capital needs—pipe, bridges, pumps, buildings, and so on—were or could be procured within the state (or even in the United States). As for the glamorous pipeline wages, the biggest part of the total pipeline cost (see Chapter 11), the entire pipeline crew in the peak construction months represented no more than 12 or 13 per cent of the Alaskan work force; and in fact, as we noted earlier (see Chapter 7), despite well-intentioned maneuverings by the state government not many established Alaskans formed part of that crew at any time, and perhaps two thirds of the enormous wages were not even spent in Alaska but went straight into banks Outside, where those who earned them still lived and meant to return.

If, then, we look for the direct and immediate beneficiaries of the pipeline, we find them not among Alaska's ordinary citizens but in their state and local governments. Thus, a very substantial part of the cost of the pipeline (and of the oil wells, camps, and other facilities at Prudhoe Bay) was added to the tax rolls of the various local governments through whose jurisdiction it passed, improvements to private property in the eyes of the assessors; not to mention the enhanced value of all property lying anywhere near the line. Similarly, if much of the huge Alyeska payroll was actually spent elsewhere, all of it was subject to the state income tax, which, at a maximum rate of 14.5 per cent, has the distinction of being the highest in the nation. Indeed, early in 1976, while Alaskans and others were still debating the pipeline's "impact," one economist, Robert Richards of Elmer Rasmuson's National Bank of Alaska, was arguing that the state's inflation and its lively spurt in population (by then up about one third since the 1970 census) could both be accounted for by the growth of state and local government: 15,000 new employees since 1964 plus twice as many dependents plus 22,000 immigrants to provide the services paid for by their handsome salaries, which were in turn provided by the bonanza of the 1969 Prudhoe Bay lease sale. Although the arithmetic works out, I am not sure the argument can be sustained. Nevertheless, looked at attentively, Alaskan government turns out to be a very curious affair. It is, for a start, the state's biggest business.

When the $900-odd million poured in from the lease sale, the Alaskan legislators were like the proverbial lottery winner, the money burning holes in their pockets and a thousand ideas for spending it dancing in their heads: distribute it among the citizens, establish a permanent investment trust that would make it possible to pay dividends instead of collect taxes, move the capital from its Southeast backwater to a kind of subarctic Alaskan Brasilia to be built between Anchorage and Fairbanks and accessible to most of the population. None of these things has happened, though they are still talked of and moving the capital has been approved by a referendum (no one has thought just how this would be paid for except, vaguely, from future oil royalties). Instead, as with the lucky lottery winnings, the cash has gone in dribs and drabs. The direct road from Anchorage to Fairbanks was at last completed. Schools have been built (and staffed) in outlying areas where formerly the population was considered too small to support secondary education. Useful social programs—in alcoholism, for instance, at $3 million a year—were enacted and funded, and in general the size and responsibilities of the government were expanded. By the end of 1975, the oil bonus had all been spent.

Ever since the Prudhoe Bay strike, the state's budgets have been drafted as if the oil revenues were already in hand, rising from $154 million in the 1970 fiscal year to nearly $700 million in 1977. This has been done, as the politicians have piously explained, without raising taxes. Instead, as the state's expenditures have run from $100 million to $200 million per year more than its revenues, the difference has been made up from the shrinking bonus money or from borrowing. (The banks also look forward to the oil royalties and, as I said, consider the state a prime risk.) When the bonus money ran out, the legislators devised ingenious schemes for compelling the oil companies to balance their budget, among them a tax on oil reserves to be deducted from future taxes on production—in effect, an interest-free loan to the state.

What this has meant is that in recent years the per capita state budget has been running at better than $2,000 per year. In contrast, a large state like New York, reputed a spendthrift, gets by on about $850 per year, and in California the cost of government (as in fact it is) is around $750. Alaska's per capita debt is likewise formidable—about $2,000 per capita, two and a half times that of New York. Besides expenditures which it directly controls, Alaska benefits from federal programs and a substantial military budget that together amount to nearly twice the state budget. Collectively, all these sources of government employment soak up about one third of the labor force: currently, around 15,000 jobs with the state, 19,000 in the various federal agencies, perhaps 13,000 in local government; in addition, Alaska's strategic value brings in about 25,000 soldiers and other military personnel. The state in particular is a good boss: its unionized employees, with few compunctions about striking, typically earn $30,000 per year, and perhaps 200 get over $40,000, with four- and five-week vacations and retirement at age fifty-five. Until Alyeska came on the scene, no other Alaskan industry could compete, in money or in number of jobs (since then, heavy construction generally has kept up, in competition with Alyeska). But that will end, has ended. Government remains.

There are, of course, practical reasons why government in Alaska should cost more than in other states (more than double the New York per capita budget, six times what a poor state like Arkansas spends). Vic Fischer, the University of Alaska economist, and Chancy Croft, the energetic and well-informed president of the state Senate, both gave me similar explanations, and they are sensible. For one thing, coming to self-government at this late point in the nation's social history, the Alaskan population is simply too small to be provided efficiently with the kinds of government services Americans now generally expect (or that those who govern think they expect or should expect). Further, Alaska's large distances and the many areas that most of the year can be reached only by air, an inherently expensive means of transportation, add substantially to all costs, including those of government. Finally, as we have seen, everything in Alaska costs more than it does elsewhere, and its government is subject to the same inflationary forces as the entire economy. Nevertheless, these reasonable explanations disregard the most obvious fact about government in Alaska: its large number of employees and their high pay and benefits. In relation to the small Alaskan population, the 15,000 state employees represent a bureaucracy *three times* the size of the civil service that runs the federal government; if we add in those who serve the federal and local governments in Alaska, there are proportionately *eight times* as many of them as in our vast federal bureaucracy.

Just why Alaska has so much and such costly government is not, of course, the question that concerns us. What does emerge from these considerations is that the pipeline's economic "impact" was both less and less destructive than it was predicted or was portrayed at the time; that the inflation that occurred had more to do with the facts of Alaskan geography and climate and with factors beyond the state's borders (as historically has always been the case) than with Alyeska; and that the size and nature of Alaskan government are a more powerful influence on Alaskan society than any other. But Alyeska remained a convenient whipping boy, and inflation was the whip.

Other kinds of pipeline "impacts" were anticipated and were widely reported: accidents and deaths in large numbers; pilferage and miscellaneous thievery on a huge scale. Since both safety and security were within Alyeska's direct responsibility to tolerate or discourage, they would seem to be clear-cut instances of "impact," if they occurred.

A substantial body of federal and state law and regulation requires all employers to keep and report precise and detailed records of injuries to their employees so as to ensure that they are properly compensated. What are called lost-time accidents are the basis for such reporting—injuries in some sense caused by the job that are serious enough to require medical treatment and result in at least a day's absence from work. (A mangled finger would be covered by this definition but also a heart attack —if it could be shown that it was induced by the work.) From these reports the tireless Bureau of Labor Statistics computes an immense number of different accident rates—by state, by industries and their subdivisions, by time of year, and so on. Against these standards it should be, therefore, quite easy to determine just how safe the Alaska pipeline work was. For Alyeska, however, there are complicating factors. The BLS incidence rates are calculated per hundred employees working 40 hours a week, 50 weeks a year—2,000 hours altogether; for Alyeska, with its extraordinary work weeks and constant turnover (and for other employers in like circumstances), it was agreed that the standard would be hours worked rather than numbers employed, even though a typical pipeliner might put in 3,000 or more hours in a year. At the same time, most Alyeska employees would be living in Alyeska's camps—it would have responsibility for some injuries in off hours (not all) that would not be reportable in more normal situations.

There were a couple of other things about the Alyeska safety program that aroused skepticism. For one, its guiding principle was the chilly concept of "cost-effectiveness." The volunteer firefighters in the camps, mentioned earlier (Chapter 7) and deplored by the Teamsters Union for reasons of its own, are characteristic of this approach; but although there were a number of serious winter fires in the northern camps, it seems that, in fact, no injuries or deaths resulted. Apart from running the fire-control program, the main work of the Alyeska Safety Department was that of education—instructing new hires in cold-weather survival, producing and circulating videotaped spots concerned with such things as the wearing of hard hats on the work sites and staying clear of grizzlies—and keeping the records required by the laws. The working end of the company's medical program, run by a separate department, consisted of two physician's assistants and a first-aid room in each camp. As with the Army in Vietnam, the theory was that an injury requiring more than first aid would be handled by air evacuation, plane or helicopter, to a hospital in Fairbanks or Anchorage; that is, that the cost of providing and equipping a doctor in each camp would not be repaid by any corresponding benefit. (The same approach prevailed in the oil camps at Prudhoe Bay—not, of course, Alyeska's responsibility—until the spring of 1976, when, after grumbling from its employees, Arco and BP both agreed to station doctors at their base camps.)

The pursuit of cost-effectiveness likewise led Alyeska to practice self-insurance—probably the first large company and project to adopt this system, which has since become fairly common in American industry. That is, instead of paying premiums to an insurance company to settle any claims against it for injury or death, Alyeska negotiated and paid its claims itself (and for a comparatively modest fee hired a subsidiary of the Insurance Company of North America to do the actual paper work). The theory is simple enough: if you can keep down the number and the seriousness of the injuries, you save at least the considerable profit margin built into an insurance company's premiums, and you come out ahead. The self-insurer has, of course, a strong motive for being tough in settling its claims as well as careful about safety.

When all of these qualifications have been entered, these doubts pursued, it nevertheless appears that Alyeska's safety record for the life of the project was pretty good. The injury rate for construction in general is the highest of any U.S. industry, running in recent years from about 18 to 20 per hundred workers employed (*or* per 200,000 man-hours)—that is, if a company employs a hundred men, something will happen to about 20 of them in the course of a year. Further, reflecting the fact that much of the work is done in remote and difficult country, often in extreme cold, the Alaskan rate for construction is a good deal higher than the national one, anywhere from 25 to 30. In contrast, by the end of January 1976, when Alyeska had expended about 76.5 million man-hours since the start of construction, it had incurred 5,350 reportable injuries, which works out to an incidence rate of 14—about half the Alaskan rate and significantly below that for all U.S. construction. The incidence was in fact low enough to arouse official skepticism, was investigated, and seems to be valid. Moreover, virtually none of the accidents reported had anything directly to do with cold—frostbite, trench foot, and so on—and, contrary to expectation, it does not seem that other features of the Alaskan winter—the twilight, ice fogs, white-outs—were serious factors. It looks, then, as if the Alyeska safety program was indeed "effective" in its practical results— and in its cost as well. Again as of the end of January 1976, Alyeska calculated its loss ratio at about 29 per cent—that is, its claims paid out amounted to that percentage of what its insurance premiums would have been. In commercial insurance, on the other hand, the actuarial assumption behind the rates for heavy construction is a break-even point of about 70 per cent— claims must be held below that point for the insurer to make a profit on its premiums. Hence, it appears that Alyeska's actual costs for self-insurance were coming out to about one third of its expected premiums.

Before construction began, authorities such as the Alaska Department of Labor predicted from past experience that there would be one or more deaths per million man-hours worked—

that is, about 80 by January 1976, perhaps another 50 by the
completion of the project, both estimates depending, of course,
on Alyeska's own highly elastic estimates of how much labor it
would take to build the pipe. In fact, the official total at that
point was 21, and by the summer of 1977, when construction was
effectively complete, it was only 25. It appears that these figures
represent strict compliance with the relevant laws and no more—
those deaths for which Alyeska could be compelled to pay compensation. About the same time, from an Alyeska source outside
its Safety Department, I obtained a circumstantial list of every
death that had occurred in connection with the project since
May 1974, and the total corresponded with a similar record kept
by the Alaska Department of Labor. It was 36.* The differences
illuminate Alyeska's views of cost-effectiveness. The unofficial list
included two men killed in a plane crash at Fairbanks after an
airport worker topped off the fuel tank with diesel fuel instead of
aviation gasoline; several road accidents (truckers on public
highways, hauling supplies but not employed by Alyeska or its
contractors); a heart attack in one of the camps; two drug over-
doses in October 1975, at nearby camps below Fairbanks. There
were also, in addition, a couple of alcohol deaths—dts and/or
freezing. Nevertheless, both the official list and the unofficial one
(evidently compiled to answer impressionistic reporting by the
likes of Winthrop Griffith with complete and provable facts)
make the same basic point: in comparison with the dire predictions and the experience of most heavy construction, Alyeska's
record was quite respectable. The line was a place to keep head
up and eyes open, all right, but working there was very far from
a death sentence.

* It should be noted that the Alaska State Police kept their own records
of their activities along the pipeline, which were summarized in annual reports. These give yet another total for pipeline deaths, considerably higher
than the others and approaching the pessimistic early predictions: 29 in
1974, 31 in 1975, 15 in 1976; 75 altogether. Without knowing the circumstances of each case (they are not in the reports), there is no way of
judging. It appears that some of these represent natural causes not attrib-
utable to Alyeska or its contractors, while others, south of the Yukon where
the right-of-way runs beside public roads and existing communities, are coincidental events, accidental or otherwise, having no connection with the
pipeline; but whether there are enough such discrepancies to account for
the great difference in the totals, I do not know.

From early 1976 on, therefore, although I continued to scrutinize the Anchorage *Times* obituaries and take note of the hortatory signs posted around the camps (at the entrance to Pump Station 8, for instance, in February: "532 days without a lost-time accident"—the same camp that at start-up in 1977 suffered the most spectacular smashup of the entire project; see Chapter 10), I no longer followed Alyeska's accident and death rates with the same close attention. Safety, it seemed, was not going to be one of Alyeska's big problems, and there was no story in it; another "impact" that, on the whole, did not happen.

In the question of the massive thievery reported on the pipeline, we find the same broad pattern: sensational accusations, largely unsupported; evidence that, while intellectually not very satisfactory, points in the opposite direction.

For the camps, the right-of-way and the haul road, law and order were provided by the system described earlier: private security forces contracted to Alyeska, Nana north of the Yukon, Wackenhut south, their senior members deputized so as to be able to make formal arrests, both groups reporting to and managed by Alyeska's own security department; assisted by a small number of state troopers, either resident in the towns between the Yukon and Valdez or assigned to some of the northern camps. The two security groups reported their police activities, including thefts, to the troopers as they occurred and also compiled their own summaries, listing all such incidents. The state police kept their own records day by day, including both incidents reported by the security guards and others in which they were apparently called in directly, as by a camp manager. These three sources of information—from Nana, Wackenhut, and the local troopers—constituted the annual pipeline reports of the Alaska State Police referred to in the preceding footnote, and they are, as I said, unsatisfactory: different in form, contradictory to each other and in themselves. By analyzing this raw material—that is all it is, and hardly that—I have arrived at my own estimates of pipeline thievery, which differ from those of the state police and are rather higher than the numbers tossed off by Alyeska's executives.

When these estimates are combined, it appears that the total of goods reported stolen in 1975 was something over $2 million,

about $300,000 less the following year*—nearly $4 million altogether. (The values are evidently those given by whoever reported the loss and are inconsistent and probably too high.)
Most of this sum is represented by trucks, buses, and crew carriers, virtually all of which were recovered; driven to other
camps, parked in the wrong places, "borrowed" for an evening at
a local roadhouse—that's what it sounds like. The rest is about
evenly divided between small tools taken from the workshops
and personal possessions—cameras, stereos, Arctic clothing—
lifted from bunk rooms or lockers. The net loss, by my calculations, was over $400,000 in 1975, under that in 1976; $792,670 altogether.

Although nearly invisible among the gigantic costs of the entire project, this is not a paltry amount. Is it *a lot*—in Alaska, in
the oil business, in 1970s America? It is said that thievery in
American oil fields runs to $10 million a year (for what it is
worth—the figure turned up in a self-serving speech by an
Alyeska vice-president). In 1975, when I estimated Alyeska's
losses at $419,707, after-hours fires and other vandalism in the
Anchorage schools were reckoned at $600,000. In any case,
whether to the loss or the publicity, Alyeska reacted. It assigned
undercover men to circulate through the camps, pursue rumors,
bring charges; and it persuaded the unions to acquiesce in its
right to search the men's outgoing baggage, with the further
right for anyone who objected to sort his duffel first, behind
closed doors. As a result, two arrests were made, with one conviction (by July 1976, when I discussed the program with Bob
Koslick, Alyeska's number-two security man); for 1976 Nana reported stuff worth $5,036.91 as "recovered in baggage," and
though Wackenhut did not use that category, it presumably got
back about the same; small change. Despite the reports of
Alyeska trucks and buses turning up as far off as Mexico—
yellow, 80-passenger school buses—it appears that no more than
twelve vehicles were unaccounted for, and these were wrecks
along the right-of-way, eventually deposited within the chain

---

* I have excluded from my calculations X-ray film stolen in February
1976 and valued by Alyeska at $500,000. This pertains to a separate ques-
tion which we shall consider in the following chapter.

link fences of the Fairbanks storage yard, where I saw them, or
were silently dismantled as sources of spare parts. The machines
on this job had a short life.

The available records do not support the notion that Alyeska's
stores were largely and systematically pillaged in the course of
construction. There remains the question of whether the records
themselves really show what was happening, and it was vociferously raised. The records of the two security organizations and
the state police, though inexact and not to be relied on in detail,
seem to me at least broadly indicative. Beyond this, however, is
the much broader question of Alyeska's own records: unless you
know how many wrenches or truck tires you are supposed to
have on hand, you cannot determine whether any of them are
missing and perhaps stolen. Alyeska's requisitioning, purchasing,
and inventory systems were several times revised but remained
chaotic to the end, matters to which we shall return in another
context (see Chapter 11). Apart from the main supply depot at
Fairbanks, every camp had its own considerable stock of parts
which for big items—tires, engines, transmissions, scraper blades
—meant several acres of loose storage around the repair shops;
in winter, needed items were easily "lost" under the snow,
smashed by prowling fork lifts, damaged by the weather, and
were reordered; the controls for small tools and parts seem also
to have been loose. It was therefore possible for substantial
amounts of parts and equipment to slip through the cracks in
Alyeska's supply systems, though whether this would in fact constitute a large sum in the project costs or greatly change the theft
records, I think there is no way of determining.

The final question concerns the fifteen or twenty convicted
men, Teamsters all, concentrated in the Alyeska warehouses at
Fairbanks—convictions including several murders and an attempted murder, burglary, narcotics traffic, all documented.
These were men that Doug Jones, a long-time Anchorage police
sergeant recently assigned to a joint city-state drug unit, had
"worked with on a firsthand, stick-'em-up-type basis—and I've
arrested some of them personally for murder and assault to com-
mit murder. . . ." Although the police I talked to were naturally
curious about the situation, they were also quick to point out
that their investigations had produced no proof that any crime

had occurred, no supportable charges; that the men in question had mostly belonged to the union before their involvement with the law—they were hired as Teamsters, not as cons; and that in the present state of American justice there was no legal way for Alyeska (or the layers of subcontractors through whom the warehouses were actually managed) to refuse to hire a man on the strength of his criminal record or otherwise take cognizance of that record. It may seem a grossly sentimental, if not insane, view of civil rights that works to place a man in a ready situation for repeating past crimes, but that is the view that now prevails, many times re-enforced by our courts.

Through most of 1976, though it no longer had much to say about pipeline thievery, the Los Angles *Times* kept up a defensive drumfire of charges concerning Alaskan crime in general: murders, robberies, and other forms of violence; big-time gambling, prostitution, and drug dealing promoted by "organized crime" moving in from Outside. The state's governor and attorney general, officials in Anchorage and Fairbanks, voiced similar complaints. I therefore spent a good deal of my time in Alaska hunting up men in the law business who could tell me something specific about what was going on: working cops, such as Doug Jones, in Anchorage and Fairbanks; Colonel Pat Wellington, the director of the state police; Chuck Dulinsky, the FBI man in Anchorage, with a special interest, under federal law, in interstate crime and drugs (some categories of offenses that have been softened or eliminated under state law remain serious under federal statutes). All were a good deal more cautious than the newspaper talk and political speeches and said much the same things: yes, it did seem that more crimes were being committed, some of these new people pulled in by the pipeline, not like our old-time Alaskans, made you a little careful about locking your house up, but it didn't look much worse than comparable places elsewhere —crime was on the rise all over the country. So far as big-timers from other states were concerned, some had indeed come up for a look, early in the project, but had decided Alaska was too small to bother with; and the local hoods had things pretty well sewed up in the drug and gambling line. The girls who had added Anchorage to the circuit seemed to be free-lancing.

From the statistics of the municipal police departments in Anchorage and Fairbanks, and from state records for other areas, it appears that the numbers of crimes have indeed risen steeply. In Anchorage in 1975, there were, for instance, twice as many murders (ten against five) as in 1971, nearly twice as many robberies and burglaries. The total for all categories of serious crime* rose by 63 per cent against a population increase of 41 per cent—faster. Things were livelier in Fairbanks—twice as many assaults, burglaries, auto thefts, four times as many robberies and larcenies. The total number of crimes was up 262 per cent compared with a population increase of 31 per cent. In both cities, only rape showed no marked rise. State figures during the five years for those areas without their own police rose a little faster than in Anchorage, far more slowly than in Fairbanks, with the several forms of violence—murder, negligent manslaughter, rape, assault—out of all proportion to the small population.

Like all other statistics, these have limited meaning by themselves—without comparison, we cannot say whether they represent a lot of crime or little. The usual means of comparison are the numerous crime indexes compiled by the FBI from the annual reports of the nation's state and local police agencies: totals for each category of crime related to population and expressed in a rate per hundred thousand. This rate, or index, can be derived for the whole country, for regions, states, or cities, for urban and rural areas, and so on; and for all crimes or for particular kinds of crime. Thus, for the United States as a whole, between 1971 and 1974 (the latest year available as I write), the index for serious crimes rose from 4,140 per hundred thousand to 4,821, about 6.5 times the rate of population growth; statistically, one might say that in 1974 about 5 per cent of all Americans were touched by one of the serious crimes. In Anchorage, on the other hand, the serious crime rate was 2,713 in 1971, 3,141 in 1975, and was also lower in both years than the component national indexes,

---

* The standard categories are: homicide; forcible rape; robbery (by force or threat); aggravated assault (causing injury); burglary; larceny (theft, without violence, of more than $50); vehicle theft. The totals are swelled by the nonviolent crimes of larceny, burglary, and vehicle theft, in that order; most stolen vehicles are in fact recovered.

except for auto theft and, despite a decline, rape. Anchorage
likewise came out ahead in comparison with the crime rates for
American cities of similar size and for Alaska as a whole. In
other words, it appears that by most standards, Anchorage was
and remained a fairly safe place to live—except for the risk of
having one's car stolen or being raped.

Things were different in Fairbanks. There, the serious crime
rate rose from 2,741 in 1971—about the same as Anchorage and
low by other relevant indexes, including Alaska at large—to
7,550 in 1975, which looks alarming. There are qualifications,
however. These rates are based on the conservative population
estimates used elsewhere in this chapter and are therefore probably high. Moreover, the main reasons for the difference were
large increases in larceny and auto theft. In other words, a midnight stroll along 2nd Avenue was still a long way from the
Dodge City experience purveyed by the newspapers—or from
the reputedly tough districts where I have lived, at one time or
other, in Chicago, New York, Paris, and London. Nevertheless,
from the evidence, it looks as if the reporters had scored a moderate hit. Fairbanks became a less inviting place to be than it
had been.

Alaska's wide-open reputation had a few real consequences. In
the fall of 1975, a fellow from Kansas bought a 32-foot house
trailer, hired a bodyguard, persuaded three goodhearted tarts
that prostitution was legal in Alaska (as in Nevada) and would
make their fortune at $100 a trick, and set out for Valdez via the
Alaska Highway. A few days before Christmas, the caravan
paused at Glennallen, 130 miles north of its destination, and the
girls were posted to distribute handbills; and the crew was at
once arrested, tried, and convicted. More seriously, in July 1976,
after weeks of evidence, a federal grand jury in San Francisco
charged a Valdez gambling and prostitution scheme to nine persons: a couple of minor Alaskan politicos with Teamster connections who years back had resigned office under shadow, an elderly retired bootlegger; girls and spear carriers. The Los Angeles
Times greeted these vindications with many column inches of
color, and one takes a grand jury's conclusions with respect. By
February 1977, however, after the usual delays, the case has run

its course: two dismissals for want of evidence, jury acquittals for the rest. There *was* some smoke down in Valdez, but no proof of much fire.

After this extended review, what we come to, I think, is this: that nearly everything said or published about the social effects of the pipeline and the conditions of its work was uninformed, fanciful, or self-serving. One of the purposes served, it seems, was the obtaining of "impact money"—federal funds channeled through the state and eventually to areas supposed to have been harmed by pipeline activities. Thus, for instance, official complaints about what Alyeska trucking was doing to the roads—particularly the asphalt Richardson Highway from Valdez to Fairbanks and the gravel Elliot Highway beyond, both carrying thousands of heavy loads from the two double-joint operations—were loud and continuous. Both roads were in fact lightly built and, while not bad in winter, regularly broke up in a springtime acne of potholes about mid-June when the underlying roadbed and frozen subsoil finally began to heave and thaw; by the spring of 1976, not much remained in the way of road surface through the Thompson Pass above Valdez and on several stretches of the highway farther north. What had happened, however, was simply that the state had cut road maintenance to a minimum—so much so, I was told by a man in the business, that most of the small Alaskan contractors that normally subsisted on this work were by 1976 effectively out of business. The Teamsters shutdown in February 1975 (see Chapter 7), which for four days had the effect of a general strike, was aimed essentially at the state, in protest at the dangerous condition of the Elliot Highway. The state also canceled or postponed a number of projects: among them, a new highway, partly built, connecting the fishing town of Cordova, below Valdez on the east side of Prince William Sound, with the Richardson Highway; and new construction on the outskirts of Anchorage that might have stilled the complaints about the slow and heavy traffic in and out of the downtown center. In any case, Alyeska was a convenient villain in all of these problems.

The mounting deficits in the state's budgets were reason enough for this policy, perhaps. The practical result was that by late spring of 1976 Senator Gravel had succeeded in managing $70 million worth of help through Congress, and the road crews went promptly to work; by the fall of 1977 on the Richardson Highway, they were nearly finished, not merely restoring the worn-out surface but in many places, widening, straightening, and rerouting the old road, bypassing the tunnel through Keystone Canyon, near Valdez, with bridges. The solace Alaska got for its troubles was in effect a brand-new highway.

The well-publicized grievances about the crime rate led to a similarly happy result: a $425,000 grant to investigate "organized crime"—though law-enforcement professionals insisted there was no such thing in the accepted sense of the expression, and one obvious reason why things were jumping in Fairbanks was simply that, despite the growing population and the huge increase in tax receipts, *not one* uniformed officer had been added to the force since 1973 and only a single administrative person. (The budget had tripled, due to hefty raises for the chief and his subordinates, more modest ones for patrolmen.) After the Los Angeles *Times* took them at their word, the governor and attorney general talked a little more softly. It was feared that the tourists, more than 200,000 of whom stream north every summer to enrich Alaska by perhaps $100 million, would stay away, taking their money to some less dangerous climate instead.

Again in Fairbanks, the planners had forecast a drastic rise in the school enrollment, meaning crowded classrooms, double sessions, more teachers, new schools to be built, but also more money from the state, in Alaska as elsewhere. None of this happened. Ignoring the common experience of northern construction and Alyeska's own programs for its work camps, the planners had assumed that those drawn by the pipeline would be normal families—wives, children, and what-not—whereas in fact most were either unmarried men or, as we have seen, had left their dependents at home; an unintended side benefit, perhaps, of the tough talk about the scarcity and cost of housing. The modest state allocations for the Impact Center itself, the Rent Review Board, and a few other such programs did, however, come through.

From all of which I venture a general law: *No level of government will spend what it can persuade a higher level to give.* The evidence is to be found not only in Alaska in its pipeline days but in every town council, state legislature, in Congress itself. *Who shall pay*: it is over such questions that the most passionate arguments are invariably raised.

Everything we have considered here so far has been concerned with the social history of the pipeline, with the ways in which that reality was perceived and reported—and with the differences between the two. The newspapers and their writers, the necessities of circulation, were, however, only half of the equation. The other half was Alyeska itself and the information —public relations—people through whom it communicated with the press. In the world we live in, those who would report, record, the workings of corporations and governments depend on such people, and vice versa; symbiosis, the biologists call it— each lives by the other. The discrepancy between reality and report began in this peculiar relationship.

Back in the early days of TAPS, an oil-company committee determined the pipeline's public relations policies, as other committees determined other aspects, and this oversight continued to the end. The intent at the outset was that the project should be as different from all others in this respect as it clearly would be in most other ways: free access, an uninhibited flow of accurate information; so far as the budget was concerned, a minimum of opinion-shaping advertising. This policy came from the top of the oil companies, either from their chairmen and presidents or from information vice-presidents who, in the present state of the business, work intimately with them, as at Sohio. That was the intent; but it worked out rather differently.

The people entrusted with this policy were mostly Alaskans: Robert Miller, a state government information man; Beverly Wilson (Ward after she married), who had taught English and journalism at the high school in Ketchikan; and John Ratterman, a newsman from the oil country who toward the end of construction succeeded to the management of the department. Others in Anchorage came and went, as also at Valdez. The two men at Fairbanks were chosen for their intimacy with the prickly local scene: Larry Carpenter, a television newsman, and Dean Wariner, who had been managing editor of the *News-Miner;* in the nature of things, their hiring by Alyeska left gaps in the local coverage that were never properly filled. In addition to its own staff, Alyeska hired a Washington PR firm during the first round of congressional hearings in 1970 and kept it on; later another in New York (with branches in Seattle and San Francisco) served mainly as a processor and distributor of press releases originating in Anchorage.

What this complicated organization produced was a steady flow of releases (about one a week through most of 1975 and 1976) reporting routine construction progress and occasional sensitive incidents such as serious oil spills; and a color-illustrated, slick-paper tabloid quarterly made up of free-lance feature stories on various aspects of the project, generally well and interestingly presented and sometimes quite informative. In the more serious matter of keeping the news lines open in both directions—answering questions with reliable information, channeling access to the work itself and to those doing it—the organization functioned less effectively.

After some early fumbling, Alyeska's policies became increasingly restrictive. Thus, visits to the work sites were allowed only in the company of an Alyeska representative (usually Beverly Ward) at times and places of its choosing, and only on condition that one refrain from talking to the men; to make sure this condition was adhered to and the time limited for direct observation, overnight stays in the camps were prohibited, though there were privileged exceptions. Further, from an early stage, interviews with senior Alyeska managers were permitted only by fairly elaborate prior arrangement and were usually monitored by a public relations person; in the spring of 1976, when things heated up over the quality of Alyeska's welding, its stolen and falsified weld X rays, and its quality control in general (see Chapter 10), its executive vice-president Peter DeMay (one of the borrowed Exxon men who did *not* go back when construction finished) ruled that *no one* should be interviewed without his first approving. Alyeska adduced reasons for these rules: the huge number of requests for press tours (perhaps fifty newsmen a week actually turned up on average, more in summer), the scarce and expensive bedspace in the camps, the need for protecting visitors from harm and itself from liability, its executives from misquotation or misrepresentation; in general, the need not to interfere with a difficult and tightly scheduled job. The arguments were reasonable enough on the surface, though none, in my observation, was entirely true. It might also be said that, as part of an industry notable for secretiveness, Alyeska was no more self-protective than most of the big oil companies now are. With all allowances made, however, it seems inescapable that the chief practical use of these policies was to restrict and control information about the project to what favored the company's view of itself. This was a long way from the original aim of the public relations committee. Perhaps the committeemen had not thought through the consequence of being candid about a project that had been surrounded by controversy from the start and would not get less so, no matter how honorable their intentions.

There was a more fundamental problem: simply that too often the Alyeska public relations people had no answers to straightforward questions of fact and could not or would not be bothered to dig them out; and if you played by their rules, there was no one else in the company to turn to. A small characteristic instance. In June 1976 I went over to Valdez for a look at progress on the terminal and it was agreed that I would be escorted through, after a delay of several days. I used the time hiking and climbing in the vicinity and camped on a hummocky shelf of tundra off the road near the top of Thompson Pass. Between there and Valdez the pass descends steeply and narrows into Keystone Canyon, sheer rock face on one side, roaring, impassable river on the other, the road carried on a ledge cut into the rock, and passing at one point through a fifty-yard tunnel, narrow, old, cribbed with square-cut timbers; a literal bottleneck, no other way in or out of town by road. On the morning I packed up and started down the pass, minutes before I got there, a trucker hauling an oversized Alyeska housing module had misjudged, run his load into the cribbing and jammed, knocking loose a latticework of timbers. While I watched, a Cat arrived, chained up, and tried to drag the module out but could not budge it; no telling how long it would take to clear and repair the tunnel (five days, finally, it turned out). I changed plans, headed back north, and after a hundred and thirty miles, at the first working pay phone, put in a guess-what-happened call to John Ratterman in Anchorage: had they had much of this before, I wanted to know, with the wide loads and all the traffic up from Valdez, knocking into the tunnel? *Never heard of it* was his answer, and it was all the answer I got. Later I learned that, though this was the worst, there'd been dozens of such accidents, and in consequence Alyeska had only recently put up the money and materials to replace the most battered timbers with steel beams, which would now have to be redone. Characteristic, as I said: one went on asking, politely, but wasted time; if answers were important one learned to look elsewhere.

"One of the things that defies credibility: they say that they don't want you to go to the construction field because you might get injured, and *that's* why they always have someone with you." This is Tom Snapp talking in the second-floor office on 2nd Avenue in Fairbanks where he gets out a lively tabloid called the *All Alaska Weekly;* an unlikely Southerner who came up here about seven years ago, with scrupulous, old-fashioned notions of objectivity in reporting—he has refrained from running stories (about some of the reasons for Alyeska's mounting costs, for instance) that he believed were accurate but whose facts the company would neither confirm nor deny. He gets himself another Coke from the machine—it is summer now, July of 1976, and warm—lights another cigarette and continues: "In a pipeline camp when you're just going from barracks, like a hotel, how you going to get *injured?* But they won't even let me go around the camp. Instead of going off with them, I tell them, you just go off with the others and I'll just wander around the camp and talk, but they won't even let me. They cain't sell that to me, because it doesn't make sense: because they're afraid that by my questioning I'll dig up things that are unfavorable. . . ."

So far as the public was concerned, Alyeska's restrictiveness produced two opposite results in the reporting of the pipeline, neither satisfactory. Tom described one of them:

"They came through, big newspapers and magazines, and all over, Alyeska was taking them on these tours, which I call Cook's tours, and then they would furnish some background data, and their stories *all took the same line*—the same stories, over and over. Big over-all picture of construction, describe the men and the machinery, put in the environmental thing, *get that in.* Invariably they'd come down here to 2nd Avenue where we had a problem with prostitution, increased activity, and they would describe the street scene and the bar scene in Fairbanks. I felt that it was overdone. . . ."

The other side of this situation was that if you could not get what you needed from Alyeska or what they told you excited skepticism, and if you were serious about your job, you found other sources: disaffected Alyeska types, friends in the camps, gossipy barroom acquaintances—Laborers, welders, roughnecks, Teamsters—in town on their R and R, sources with their own distortions, imbalances, incompleteness; and you found your own way into the forbidden places, the camps and work sites. A large corporation engaged in an essentially public enterprise such as the pipeline does not have a choice between revealing its activities and selectively concealing them; rather, its choice is between allowing itself to be fully and accurately reported or being reported with the kinds of hit-or-miss distortion that warped the stories of massive thievery and slack productivity. It does not have any serious secrets that can be kept. It is an axiom of public relations that to be useful you must be trusted, and to be trusted you must mostly know and tell what you know, without evasion. The corollary is that if you have a bad story, you must let it out straight, with all its known facts; if you try to stop it, slant, improve, evade, or conceal it, it will only get worse. At critical points, Alyeska behaved as if this law had been repealed or did not apply to itself. Or perhaps the error, which is fundamental, originated not with the public relations department but with Alyeska's management. When things got tough, it was Peter DeMay who thought he could keep a hand on who said what to whom. The combative general denials, which, as I have remarked elsewhere, were rarely supported by evidence, were consistent in tone and sound like Ed Patton himself.

Since none of the major American newspapers evidently felt that a full-time reporter in Alaska would earn his keep, all depended on Alyeska's handouts and guided tours, supplemented by occasional hit-and-run investigations such as that of the Los Angeles *Times;* and, for day-by-day continuity, on the wire services and such odds and ends of local reportage as they picked up. Through most of the project the Associated Press was represented in Anchorage by a young woman named Tad Bardamus whose special relationship with Alyeska made possible extended visits to the camps and work sites, with Dean Wariner as her guide; the "privileged exception" mentioned earlier. From these visits came reassuring feature stories about goodhearted pipeliners. When unavoidable news did occur, "Mostly how the AP get their deals," Tom Snapp explained, "is that one of the papers will come out with something and they will do a story on top of it. So one of their sources was Alyeska. So then they'll go and talk to Alyeska and Alyeska will give them—they'll tone down the story, the Associated Press will dispute another story and they will quote and they won't do sufficient investigation to determine who is right and wrong: a kind of a surface deal." Within Alaska's journalistic community, the deal was known as the "sweetheart arrangement." Toward the end of construction, Miss Bardamus was reassigned to London. Coincidentally, Dean Wariner quit his Alyeska job and moved there too.

There were Alaskan exceptions. In Anchorage where the slender *Daily News* limped behind Robert Atwood's *Times* in circulation, advertising, and the volume of news it could afford to carry, its publisher Katherine Fanning, like Mr. Atwood a well-connected Chicago émigrée, but of a generation later, sought strength in a mildly liberal editorial line and serious investigative reporting—and in the spring of 1976 was rewarded with a Pulitzer prize and other honors for a series on the Teamsters run the previous fall. (Honor but no financial improvement; as an economy, Mrs. Fanning had contracted the paper's printing and business services to the *Times* and, when the deficits got worse, brought suit for damages.) Till near the end of construction the *Daily News* was able to assign two reporters, Sally Jones and Rosemary Shinohara, to the pipeline and the oil business generally, and between them they several times produced original work—on the sewage problems in the camps, for instance (the *News* published the accusations while the *Times* ran the denials), and on the mismatch of the pipeline's capacity with that of such West Coast refineries as were capable of processing Prudhoe crude; thus reopening the fundamental first question of whether the market could absorb what the oil companies intended for it and, if not, what they really had in mind.

There was one other exception: Richard Fineberg, a gaunt, intense, mustachioed man who, for a time covering oil and gas for the *News-Miner,* scored one of the few genuine beats that actually originated with a newsman (a major oil leak at the Galbraith Lake camp, of whose full extent Alyeska had not seen fit to inform the reporters or the public). Later, free-lancing from Fairbanks, he was able to write more broadly of the ways in which the project's intentions had outrun performance; and to reflect on his colleagues' failure, for the most part, to keep up with either.

Work elsewhere was not altogether dross. Although the Los Angeles *Times* kept whacking away at the dead horse of its failed Pulitzer, it also uncovered several minor but indicative Alyeska missteps (among them the money thrown away on the anticorrosion pipe coating that did not stick—see page 175; for a daily paper, I suppose, a research effort incommensurate with its rewards). When the welding troubles at length came to a head, the *Wall Street Journal* managed at least to raise some of the right questions, with vigor. Nearer my own home base, on the other hand, the increasingly parochial New York *Times,* although it pursued the butterfly of "energy alternatives"—windmills, say, for every household in the land—and sometimes posted a man to Alaska for a quick look, mostly made do with the bits and pieces of pipeline news that came over the wires.

Even when the reporting of the pipeline was good according to the norms of journalism—energetic, accurate, pertinent, detailed—it suffered serious problems of emphasis: some things of minor practical significance in the long run were written into major issues—because they made good stories, perhaps, because they were accessible—while others of far greater import went untouched; we have considered instances of both kinds here, but others, more serious, are ahead. The reasons are obvious. The mere physical extent of the pipeline and of Alaska itself is more than one man or many in a lifetime can grasp and contain; by the time you had gotten from A to B, you had filed your piece and gone home. This was the advantage of the serious Alaskans —Tom Snapp, Richard Fineberg, Sally Jones, until she was budg-

eted off the *News:* that they could stay around long enough to
be edited by the passage of time; with the disadvantage that
they remained at a far remove from all those Alaskan decisions
which, as always, were being made in other places. The technical
extent of the job was even more formidable: all the environmental disciplines, the many branches of engineering, complicated by their mix of mathematical exactitude, unverified data,
and contradictory opinion; the manifold political, social, and
finally philosophical connections by which the project was linked
to the national life. Alyeska's policies about information exacerbated these already formidable difficulties: obstructor of
firsthand observation, stimulator of skepticism, consumer of
scarce time in the checking of even the most trivial facts; the
wonder of the pipeline, perhaps, is that anything more than
Alyeska press releases was printed. These difficulties—and the
time-bound inability to make the long connections—are inherent
in the trade, that is what history and the writers of books are for;
but to know all is not necessarily to forgive.

At the end of June, when the Keystone tunnel was open again,
I went back to Valdez, by prearrangement. My son Matthew had
flown up from school to see Alaska with me but would not be allowed along on the tour, *absolutely not;* it was an arrangement I
had neglected to think of in one of the phone calls to Anchorage.
While we waited to start, in the rectangular Alyeska PR building
overlooking the small-boat harbor across Port Valdez from the
oil terminal, I chatted with the author of this ruling, the new
man in charge, Ron Mierzejewski, and wrote the name carefully
in my notebook, with the distinction of being born in Southeast
and recently brought over from state government work in Juneau
(his predecessor had quit, gone into the grocery business, locally); blond, round-faced, talkative. *A party of journalists*—
there would be a party of journalists making the tour today—
and I explained once more that that was not what I was nor my
interest. He tried weather, the formidable snows this past winter
and how·they had exceeded the averages, impeding construction,
and quoted statistics—erroneously, the same kinds of errors that
had nullified the original terminal design.

Our guide that day was Andrea Harrison, one of Alyeska's

local resources: a twenty-year-old third-generation, born Valde-
sian, whose grandfather had taken over the old Army base, Fort
Liscum, across the Port, and in the 30s turned it into a family
fishing camp and cannery; whose grandmother Oma, by complex
arrangements with Locke Jacobs and the Arco landsmen, had
sold the quarter section as the site of the work camp from which
the terminal was built, a tiny bit of Alaska that was indisputably
private property. (The Alyeska buses taking workers over from
the town were identified by the name her grandfather had given
the tract, Dayville; from his own name.) A nice girl, connected
with the local history, and interesting; but I had made the
rounds with her before and there was not much more to learn.

The journalists were the daughter of a Sohio official and her
bearded husband, up on vacation, who talked about their baby
son and things back in Cleveland; and a fellow who taught, he
said, at a place "near" the University of London, the sort of silly-
ass Englishman who seems to have become the Old Country's
national export, on his way home from something at the University of Alaska. We drove around the terminal—those were the oil
tanks, Andrea pointed, and that was the tall stack (newly
erected) of the power plant; the Engishman leaped out to take
"pickies," as he called them, with his little camera. On the
sharply layered fragments of the graywacke with which the terminal roads were littered Andrea slashed one of the Blazer's tires
and we limped to a repair shop for a replacement; and were
back in Valdez in an hour and a half.

We retrieved my son, dozing over a book in the cab of my
truck, and went on to lunch at the Sheffield, down the road from
the Alyeska PR building, visually superb though with its ups and
downs in food and service. The Englishman glared across the
table. I did not have much to say. A pipeliner occupied the one
functioning pay phone, sipping from a bottle of beer, talking to a
girl, and I went back and forth, anxious about unofficial appoint-
ments over at the terminal to occupy the rest of the day.

Eventually I got through.

Chapter 10

MURPHY'S LAW

Whatever can *go wrong* will *go wrong.*—Old Saying

At the end of November 1975, according to Alyeska's arithmetic, the project as a whole was 40.5 per cent complete, compared with a goal of 43 per cent. Of the system's three chief components, the Valdez terminal was by the same calculation about eight weeks behind schedule while the pump stations were ten weeks behind; the pipeline itself, on the other hand, which during the summer had lagged even farther, had now caught up and was thought to be six weeks ahead—the pipeline had reached a point that, according to plan and given the virtual shutdown of outside work necessitated by the deep Alaskan winter, had not been expected before sometime in the spring of 1976, depending on just when that came. There would be some further progress through the winter: at the pump stations and at Valdez the structural work was far enough along so that inside finishing could continue comparatively unhindered by weather, and even on the pipeline some jobs were pushed on into December and resumed in January after a brief break, though the main effort was to get the heavy equipment into the camp shops and repair the ravages of the first year of pipe work. All considered, although the project had fallen a little short and the time made up had been at the expense of a considerable expansion of the work force, things looked not too bad at the end of this first season and should get better. The first year had been a learning year, the Alyeska managers told each other and anyone who asked, they were past that now, no longer short of equipment, supplies, beds; and Frank Moolin's revision of the *Pipeline Department Control Manual*, designed to impel greater productivity by the contractors and their men, was in the offing.

Like most engineering calculations, Alyeska's genial percentages told both more and less than they meant; they are best not taken at their face value. They were based on estimates of the numbers of man-hours needed to perform the multiplicity of tasks of which the project consisted—that was the sense in which jobs as different as laying pipe and building a pump station, welding and ditchdigging, were comparable—and these estimates, in turn, were the basis for all other Alyeska budgets. Behind the estimates were assumptions about the productivity of the labor coming out of the hiring halls (see Chapter 7) and the possibility of controlling and improving it; and about the comparability of similar tasks performed under varying conditions, among which weather was only the most obvious variable. So far as laying pipe was concerned and whether designedly or not, the toughest jobs had waited till the end.

Further, in order to achieve its impressive percentage of completion, the Pipeline Department had taken serious chances in several places along the route, running late in the season to make up time lost during the good weather. It was to be a funny winter, cold and snowy in many parts of Alaska early on (in the minus 20s in Fairbanks, with ice fog, by mid-November, a lot cooler than on the Arctic coast), turning comparatively mild as it progressed. As it turned out, these risks were unfortunately taken and much that was expensively and hastily done in November and December was undone by the coming of the Alaskan spring. Again that urgency, about which we have wondered before: the Alyeska managers and the oil men behind them talked calmly enough about their enterprise, rationally enough, but the recurring quality in their decisions and actions is a desperation to get the job over with and get out. The motive of reputation—fulfilling the promised schedule, even though it left no room for accident, error, or engineering second thoughts—should not be underestimated: for the men, it meant future career, enlarged or shut down, according to performance; for all the contractors, many of them oil-company dependents to begin with, the same, in terms of future contracts. Nor should the national interest be disregarded, though Prudhoe crude was no longer the trump card it might have been any time up to November 1973.

The burr under the saddle, though, was money. Alyeska and the oil men had run longer schedules through their computers, but the fixed costs were always against them: "indirect" costs they were called (as distinct from the "direct" costs of laying pipe, setting VSMs)—the camps, equipment, administrators—"that gotta be paid whether the work goes on or not," as Frank Moolin put it when I stopped by his office at Fort Wainwright later that winter to talk. "The most effective way to reduce cost is to get the pipeline done and dismiss our major contractors, close up our camps, get rid of our construction equipment." It was not of course Frank's money or Alyeska's but the oil companies', a lot even for a giant like Exxon, all they were worth for the smaller partners, and borrowed in a rising money market, in which the Rockefellers' Chase Manhattan Bank was often a principal, at interest rates of 10 and 12 per cent; borrowed to near the limit of what the market could supply. And in the oil towns where such things are pondered, Houston and Dallas, New York, Los Angeles, probably the visible cash, the seven or eight billion and the interest, was not the final motive but the golden hoard of oil, and the money laid out, the time, in infuriating expectation of attaining it. Time, as they say, is money—and capital, life; everything.

This motive shivered down to Alyeska's managers and out through its interlocking contractors, and at the nerve ends the various members reacted, twitched, jumped, and moved.

Pipe work began on March 27, 1975, with a 1,400-foot string of pipe to go under the Tonsina River halfway between Valdez and Glennallen. The joints were welded together, the ditch excavated below the stream bed, the pipe taped and pulled across, the massive concrete saddles lowered to hold it down while the ditch was filled around it with gravel—and the saddles slipped off, the empty pipe popped to the surface, damping the jubilation of beginning. It was finally in place and buried three and a half weeks later, but it was a year and a half before anyone discovered that the pipe had been crushed in the process. By then, in the midst of more general anxieties, it was impractical to repair—excavate the pipe, lift off the saddles, cut it out, replace it—and it was bypassed with a new string placed with new weights in a new cut a short distance upstream. The Tonsina was thus the first begun of the 800-odd stream crossings, as well as the first pipe laid; it was also the last of the crossings to be completed. In the time it covered and in its errors and accidents it was the whole project in a 1,400-foot synopsis.

Farther south at Valdez a more serious difficulty had already revealed itself, though just how serious was not yet known. A chief premise of the terminal site, countering the anxieties aroused by the 1964 earthquake, was that the eighteen giant oil tanks required by start-up and the three smaller tanks for tanker ballast water would rest on bedrock only six feet below the sloping surface of glacial till. When the Cats set their blades to work in 1974, they discovered that the rock foundation was a good deal farther down. Eventually as much as sixty feet of muck and gravel had to be cut away, moved, disposed of, swept under various rugs: about 15 million cubic yards, nearly a quarter as much as the useful gravel mined and spread for the entire 800 miles of pipeline work-pad, haul-road, and pump-station foundations. By late 1975, when four million yards of till had been shifted and half the tanks were nearing completion in the thickening snows, Alyeska sought permission to dispose of some of its overburden by dumping it south of Valdez in Prince William Sound. An outcry followed over possible harm to commercial fishing; the city of Cordova, farther down the sound, offered to accept the stuff as fill if Alyeska would bear all costs of barging and spreading, at $11 a yard for a million or more cubic yards. A less expensive solution was found in a system known as reinforced earth: packing the till below the base of the East Tank Farm* and hold-

* The 18 oil tanks planned for the pipeline's start-up volume of 1.2 million b/d are grouped on two level rock platforms carved from the mountain: 14 at the east end of the terminal, known as the East Tank Farm, 4 more higher up and to the west, the West Tank Farm, where there is room for 14 more when and if Prudhoe production and the line's capacity require them. The first 18 tanks, holding 510,000 barrels each, represent about a week's supply at the initial rate. The additional 14 tanks would bring the capacity up to around 2.2 million b/d, perhaps 500,000 more than the present Prudhoe reserves are considered capable of producing.

ing it in place with panels of cast concrete interlocking like pieces in a jigsaw puzzle. (The panels were fabricated in Seattle and barged up during the winter, Alaska being no place for quick-curing concrete.) Hasty and slipshod surveying was probably sufficient explanation for this error, as those on the scene assumed (see page 200); but in the pipeline debates the bedrock argument was prominent among those favoring Valdez as the terminal site.

Finding the bedrock was only one of the reasons why work at the terminal ran slow. When finally uncovered it proved to be a kind of shale, or phyllite, with greenstone and graywacke mixed in: a black, fine-grained substance sedimented in thin layers, breaking up in knife-sharp slivers—phyllite, "leaflike" in its Greek origin. The rock, appropriately, looked like the kind geologists associate with oil but was worthless for this or any of the structural purposes for which builders use crushed stone in volume (while Fluor, Alyeska's terminal contractor, was thinking up ways of disposing of its rock problem, it was buying crushed stone in quantity to mix with cement and trucking it over from Valdez, where there was an inexhaustible supply of suitable material carried down the Lowe River from the mountains); good, in short, for nothing except slashing tires on the temporary gravel roads around the terminal site. This rock continued unpredictable, a danger as well as a nuisance. In September 1975, as the builders cut back into the rock behind the two tank farms, a dynamite blast loosed a torrential slide of fragments amounting to thousands of cubic yards. Although it apparently did no serious damage, the slide was a stern reminder of just how unstable the rock was and of what it might do if, for instance, another earthquake came. Two costly precautions followed from this accident. First, the size of the cutback was greatly increased, both to reduce the gradient and to enlarge the clear area behind the tanks, a job not finally completed till after the first oil entered the line nearly two years later; for a time, in an effort to fit the extra volume to the slipping schedule, the blasting crews worked nights as well as days, and Alyeska ads in the Anchorage *Times* warned planes away and told them to keep their radios off. Secondly, it was necessary to hold the slope in place with rock bolts: that is, hundreds of holes were drilled horizontally 60 to 100 feet into the cliff, using rigs turned on their sides and hoisted into position by 90-ton cranes, and rods were inserted to hold steel plates screwed tight against the outer surface (expensive machines—the estimate for this part of the job was about $4 million); and, as a final step, since water in this wettest of all the Alaskas is a major force—dropping from the skies as rain or snow, freezing and thawing, pressuring upward from below ground—much of the porous, fissured rock slope was sealed with asphalt or gunnite, as were eventually most other exposed surfaces within the terminal.

The rock bolts solved one problem but brought another into view—perhaps that is in the nature of engineering and why, as we have discovered, it provides so fertile a field for public debate, whose only answer is experiment: try it out and see if it works. Stitched together as it now was with steel rods, the rock probably could no longer break loose in the myriad fragments that would form a slide but only in one juggernaut mass that would carry much of the cliff face down with it, and whatever stood in its way; perhaps the kind of mountain-demolishing catastrophe that produced the nearby geological curiosity known as the Liscum Slide (see page 308). In fact, in September 1976, a year after the first accident, something like this did happen, over behind the West Tank Farm, when an overcharge of dynamite sent a huge chunk of rock sailing across the open space and into the last of the nearly completed tanks, crumpling dozens of the heavy plates—8×30 feet, nine or ten tons each—from which it had been welded together. Poor blasting technique was the verdict in this case. Men were fired, replaced, the techniques changed, more closely watched, and it did not happen again; but rebuilding the tank took almost a year.

Whether measured in volume or the cost of dealing with it, the snow at Valdez was a problem not inferior to the rock and till. Removing it—clearing it, loading it on trucks, dumping it in the Port—was worth $14 million a year. Nine feet of snow fell in one two-week period in the 1975–76 winter, burying the crated equipment and supplies stacked around the terminal's temporary dock. By then perhaps half the oil tanks had their walls up, their roofs in place, and what remained was inside finishing. The roofs were designed as flattened cones, on the theory that once the oil was in its heat would melt the snow, the melt would run off and, if it did not refreeze within the dikes built to contain possible oil spills, be carried off in the terminal's sewers; in the meantime, till the oil came, the job was to be done by men with snowblowers hoisted the 62 feet up to the tank roofs—guardrails were provided for their protection. Nine feet of the wet, heavy, clogging snow was about all the tank roofs could support, however, and when it came, the 10-horsepower snowblowers simply tunneled in and buried themselves. Giant 225-ton cranes were brought in to lift rakelike devices and scrape the snow off, and incidentally the guardrails as well; and were the big item in the cost of snow.

Earlier in that same difficult winter, Alyeska had risked working through November at both ends of the line. At a crossing of the Tsina River, which rises near the Worthington Glacier and boils north down from Thompson Pass to empty into the Copper, the contractor had five hundred feet to go. All that remained was to fill in the gravel around the welded pipe, but the gravel was frozen into chunks and would not pack—when it thawed, void spaces would be left below and around the pipe, and under the pressure of oil and the shifting fill, the pipe in theory could break. After days of arguing about it, the state surveillance man, Bill Thompson, stopped the work; in the spring, the fill and bedding were removed, the pipe and its anticorrosive tape were found to be damaged, and the work was redone.

On the Arctic plain about the same time, near the Happy Valley Camp where the line crosses back and forth among the shallow channels of the Sagavanirktok River, Alyeska faced a similar dilemma. Early in the pipeline debates, scientists of the Navy's Arctic laboratory at Barrow had protested the whole northern section of the route, preferring one farther west—the Sag River flood plain presented opposite problems, the force of deep-frozen ice in winter, its roaring release when spring came in June; the route was chosen chiefly because it was a few miles shorter, therefore, in theory, cheaper (see Chapter 5). The engineering answer was to coat the pipe with cement, applied in the field just before lowering-in and backfilling, to give it enough weight to withstand the lift of ice and flood. This was done along the Sag in November, hardly the best time for curing cement, but the gravel collected at the site was frozen in a heap and, short of dynamiting, could not be used; the pipe was lowered into the ditch and left for spring. In June, when the Sag broke through, the half-cured concrete was gone, pulverized, swept north and deposited in the river's alluvial fan, and 1,700 feet of pipe by Alyeska's count had been floated and dropped on the gravel bank; when I saw it a few days later from the cab of Bill Granger's truck, it was wrinkled as a cast snakeskin, or like a beer can that has been crushed in a fist, stamped on, and tossed aside. "It's so simple," Frank Moolin told me with his smiling frankness, "you have to think, well, why in the hell didn't we think about it at the time? The thing should have been backfilled. . . ." But in the temperatures that went with November work beyond the Brooks Range, there was no way till spring of transforming the fill to the plastic condition in which it could be used.

That particular gamble was not made good till November of 1976, when the smashed pipe was replaced, recoated, and covered in; as if to demonstrate that pipeliners could indeed do their job as intended, regardless of weather. By then Frank Moolin, in personal charge of the northern section of the line, had more serious and immediate concerns.

Getting the oil moving as planned in the summer of 1977 rested on one basic premise: that all the pipe would be in place and the line essentially complete by the end of the 1976 season; after that there would still be an elaborate sequence of steps to get through before start-up—pressure-testing the pipe (and the tanks at the pump stations and terminal), training the crews that would operate the pumps, valves, and system controls and be responsible for dozens of contingent and subordinate procedures, such as leak detection—but most of these could be squeezed into the spring, between break-up and full summer; provided the line was complete. Achieving that goal came down finally to what could be done in two short sections of the route, perhaps five miles altogether, more than five hundred miles apart: Atigun Pass in the north, near the Brooks Range divide; and Thompson Pass together with Keystone Mountain (above the canyon) across the valley of the Lowe River.

The problems were similar in both areas: steep grades and extreme weather, signifying the likelihood of avalanches; complicated by permafrost at both sites and the all but unmeasurable snows of Thompson Pass. The avalanche dangers ruled out the normal solution for pipe in permafrost, running it aboveground on vertical supports. In both places the plans were similar: clear the unstable surface soil, cut the pipe route into the bedrock, box it protectively in concrete. At Atigun the work was carried through about as planned, but slowly. In the Thompson-Keystone area it was hindered in the first place by the extraordinary grades (see Chapter 8) that made it impossible to get men, equipment, or the pipe itself up to the work sites by any normal means. That barrier was crossed with daily helicopters for the men, a Sikorski Skycrane to lift the Cats and other heavy equipment, dangling from cables, to the top of Keystone Mountain; later, winches to let the earth movers down the slopes and pull them up again, a system of cables and towers, like a ski lift, to fly the 80-foot double-joint pipe up from the valley. At that point it was discovered that the Thompson Pass plan had been based on the same kind of careless and self-serving early studies as were made at the terminal. Bedrock was not a few feet below the surface as predicted but fifty or sixty feet down through a cushion of earth and gravel. As we noted earlier (see Chapter 8), the natural route here for the pipe was not over the mountains but through the pass and canyon, beside the Lowe River—but that obvious route was occupied by the state highway. As early as 1974, however, the state had planned to move the highway, straighten and widen it (if it could persuade Congress to provide the money), and when it did, the old right-of-way would be available for the pipeline. It was suggested that Alyeska consider that solution, but the company ignored it until the machines were in place and excavation revealed the full seriousness of the difficulty in its chosen route.

Alyeska executives and engineers, state surveillance men and officials from the several interested departments of government —the area is both grandly beautiful and highly visible from the highway, therefore an accessible symbol of government responsibility and effectiveness—converged on the pass, walked and climbed the route, talked. Alyeska did some halfhearted engineering studies of the alternative; the various departments gave thought to the limiting conditions (width of right-of-way, numbers of access roads, restoration) that would be imposed if it were adopted. In the end, Alyeska's decision, apparently made by Peter DeMay, was to follow its original route, for which all the necessary state and federal Notices to Proceed, though based on faulty information, had already been granted. Redesign and a new round of permissions, it was said, would consume more time, hence cost, than might be saved by easier conditions for construction, an argument often repeated in the summer of 1976 as the project approached its deadline and quite possibly true, but it was also entirely consistent with the attitudes that had prevailed since the pipeline was first thought of: that except under extreme compulsion Alyeska and its owners *would not* be swayed from their own determinations by any outside arguments, however reasonable. The state, in effect, shrugged its shoulders and let the matter go.

At Atigun and in the Thompson-Keystone area, excavation and deep-cut rock work (started on Keystone Mountain in late summer of 1975) continued into fall of 1976, altogether perhaps a million cubic yards to be blasted and removed, half of it on Keystone itself. The pipe joints were eased into the trench, supported by blocks of Styrofoam that left room for the welders to work under the pipe. The last weld was finished, on the southward slope of Thompson Pass, on October 20th, under five feet of snow which laborers worked to clear with shovels, dubious conditions for outdoor welding, which is as sensitive to wind and damp as to cold; welding at Atigun, in colder temperatures but otherwise better weather, was completed a few days later. At both places, backfilling went on into December. The Atigun pipe inside its eight-foot-wide concrete coffin was wrapped in an insulating cocoon of Styrofoam twenty-one inches thick before the fill went down. At Thompson Pass the ditch was filled with soil cement, a low-strength mix with earth, more stable than the usual gravel, covered over with a thin, smooth-graded layer of soil where in time grasses would root and hold. On December 7th, Alyeska was able to announce that its deadline had been met and all pipe was in place; except for a few remaining welds to be made in the spring, connecting long strings of pipe still to be tested, the work was essentially complete.

All through its struggle in the two passes, Alyeska was engaged in final testing of the line as it was completed. For the tanks at the terminal and the pump stations, the system used was called hydrotesting, an engineer's term meaning that the tanks were pumped full of water which was allowed to stand for twenty-four hours while monitors looked for leaks through the welds holding the plates together. The pipeline itself was tested hydrostatically: filled with water under pressure and, again, held that way for twenty-four hours to see if the welds leaked. Short of pumping oil through the line, this was the most persuasive of all the tests, inspections, and calculations performed on the system's welding (and equally, for the pipe steel itself), practical, not theoretical. The pipe was specified to have a maximum operating pressure of 1180* psi (pounds per square inch); the test pressure was supposed to be 125 per cent of the operating pressure, or 1475 psi, which in turn was figured as 90 per cent of the "design pressure," 1638 psi, the point at which the pipe would burst. (Alyeska's contracts with is Japanese mills called for testing each 30- or 40-foot joint at 132 per cent of the maximum operating pressure—the pipe began as flat steel plates which were formed into a tube and were held in that shape by a weld from end to end; that weld was the point of the mill test.) These are tiresome numbers but important, and (apparently) extreme: the *actual* operating pressure of the pipe when moving oil would vary greatly with the topography—perhaps 300 psi over level ground, even less going downhill, with the oil moving fast and the pipe, probably, never entirely full; approaching the maximum only on long uphill grades.

For testing, the pipeline was divided into 80 segments ranging in length from a few hundred feet to 23 miles. Near the northern end in the fall of 1975 there were experiments, using water pumped from the Sagavanirktok River and heated (it would, in theory, be warm enough not to freeze till the test was finished

---

* This is the value published by Alyeska and used consistently in reference to hydrostatic testing, and the pressures that follow derive from it. It was actually the maximum operating pressure for pipe of .462-inch thickness, whereas the heavier .562-inch pipe would presumably stand greater pressure, and the pipe's endurance might also be affected by chemical differences apparently specified for various conditions; but since the different grades of pipe were intermixed, the standard for the hydrostatic tests was that of the weakest.

and it could be drained), but serious hydrostatic testing did not begin till mid-June in the following year. Difficulties appeared, particularly in the elevated pipe south of Fairbanks, where the permafrost it was meant to circumvent is generally thin and not much below freezing: under the weight of the water in the pipe, pilings sank in the frost as much as a foot or more and the lateral pipe supports with them, and leftover oak cribbing was stuck in to fill the gap; fifty pairs of supports sank in this way and another hundred were marked for maintenance crews to watch once the line started operations.

The reason adduced for this alarming side effect was that in the VSMs in question Alyeska had not gotten around to installing the heat-exchange tubes (see Chapter 5), meant to keep borderline permafrost from thawing, when it filled the pipe with water; management, once again, the ability to cause things to happen in a reasonable and intended sequence. (The reasoning was that of the State Pipeline Coordinator's Office; on this problem Alyeska kept officially mum and, practically speaking, it went unreported.) As a matter of fact, from early in the project the federal monitors in the Alaska Pipeline Office had been agitating for Alyeska to perform systematic field tests to demonstrate that the elevated pipe design would actually work in the varied conditions for which it was intended, but Alyeska did not get around to furnishing satisfactory data till the spring of 1976, when more than half the VSMs were already in place. (There had been early tests of elevated pipe, empty, but with a concern different from the one revealed by the hydrostatic testing—how the steel itself would stand extreme winter cold, whether the pressure of the frozen ground would eject the VSMs like toothpaste squeezed from a tube.) In the meantime, the VSMs had been a source of continual difficulty, from the beginning.

As we noted earlier (Chapter 6), the sheer volume of drilling required for VSMs to support more than half the line aboveground was enormous—greater (and in much larger diameter) than all the oil-well drilling needed to get the Prudhoe field into production; and portable drilling rigs that would do the job in the many varieties of frozen soil simply did not exist. Exxon, considering that it knew a thing or two about making holes, stepped in and led a group of Texas drilling companies through the de-

velopment of suitable rigs and felt sober satisfaction when it
winnowed from these experiments six different designs that it
judged acceptable—or so the company's drilling manager, J. V.
Langston, when I talked to him in Houston in the spring of 1976,
though he did not mention the difficulties that developed when
the machines were at last delivered to the field. In general, they
did not work very well. A drill that made a lovely hole in frozen
till might break its bit when, as happened frequently, it hit the
rounded cobbles of an ancient stream bed in the next hole along;
other rigs were so clumsy to dismantle and reassemble that they
could produce no more than two or three holes in a day out of
the 76,000 that would finally be needed and were, in a practical
sense, not portable at all; or relied for control on a tangle of hydraulic lines which worked well enough in summer but froze or
crumbled when the weather turned cold.* On the other hand,
once the rigs were rebuilt in the field so that they could actually
be used and the crews had learned to work with them, they went
so fast that the gangs setting VSMs could not keep up and often
lagged miles behind; management again. The difficulty was that
when the holes were left open for days at a time, they filled with
surface water and soil, which froze; or the walls of the hole
thawed and sloughed into the bottom. Either way, it was often
necessary to bring the rigs back and redrill; or start fresh. These
problems were solved after a fashion, though with much waste
effort—by topping the completed hole with an inflatable ball,
like a cork in a bottle; or by casing it in (as in most oil wells)
with pipe of sufficient diameter to hold the walls in place.

In many places along the route, the design of the VSMs called
for "corrugating" the pipe at the bottom to increase its holding
strength and keep it from sinking into the permafrost: that is, a
device was lowered into the pipe to blast holes at intervals (by
the force of air, water, or explosive) so that the metal, flanged

---

*These complaints came from Alyeska's contractors, perhaps in justifica-
tion of their low productivity and consequently rising costs, and they provided one of the project's recurring themes. Most also demanded to master
the drilling techniques with practice holes off the pipeline right-of-way, but
Alyeska (or its owners) refused to bear that modest expense, preferring on-
the-job training. To judge from the increased productivity that followed in
the drilling operations, it seems that this approach worked, though whether
it in fact saved any money is doubtful.

outward around the hole, would lock into the surrounding soil; a technique like that used in an oil well, where the production pipe is perforated so that the oil can enter and be drawn to the surface. Rather often, however, the corrugation process simply broke the VSM pipe or split it up the side, and the pipe had to be discarded and replaced; the specifications for the steel, some of it produced in Canada, most in Japan, were changed but the problem persisted, unpredictably.

A good many other difficulties came up to defeat the engineers' foresight and in general make their managers avoid aboveground construction where they could, and not only because it was expensive even when it worked as intended; but these are indicative.

In July 1976, a month after Alyeska discovered that some of its VSMs were sinking, an incident occurred below Thompson Pass that was abundantly reported and excited the attention of Representative Dingell and his colleagues on the House Subcommittee on Energy and Power. In the course of applying pressure to a water-filled 480-foot string of pipe, the technician in charge of the hydrostatic test managed to puff it up like a sausage skin and rip a seven-foot gash in the half-inch steel of one of the joints. "Human error" was Alyeska's all-purpose explanation of the accident—meaning by implication that its design and all the calculations on which it was based, back to the original order for the pipe, had not failed. What actually happened was that a pressure gauge stuck and showed only 187 psi whereas, mathematically, it would apparently have taken nearly ten times that pressure, 1560 psi, to cause that particular rent. Further, since it was the pipeline welding that was chiefly on the public mind at that point, Alyeska pointed out that it was the steel itself, not the welded joints or the mill-produced lengthwise welded seam, that had failed—the welds, that is, had met the fundamental standard of being stronger than the metal they joined. And finally, it was noted, the actual operating pressure of the pipe at that point would be only about 300 psi, far short of that which had caused the break. If the reader will turn back to those tiresome numbers that describe the strength of the pipe, however, he may conclude that bursting the pipe under a pressure of 1560 psi is less reassuring than it sounds. While it is one third above the specified maximum operating pressure, it is not in fact so much greater than that called for in the hydrostatic test; and it is rather less than the "design pressure" that the pipe was supposedly built to endure.

One other incident marred the tranquillity of the testing process before it was finished but went unannounced and, so far as the public at large was concerned, unnoticed; again its meaning is ambiguous, both disturbing and reassuring. At the end of September, in an elevated test segment beside Hess Creek, north of Fairbanks and not far from the Yukon, the crew pumped water in from the south end. They had not, however, taken note of a "displacing pig" inside the test segment (a metal cylinder inserted to clean the pipe, and for other purposes), nor of a one-way check valve built into the pipe north of that point—designed to stay open so long as oil moved in the right direction, slam shut if it reversed, as it would in a line break farther north when pumping pressure would cease and the oil in the pipe would run back down the slope to the Yukon and out the break; a feature meant to limit the oil spilled in such an accident. In any case, the valve worked as it was supposed to. The pig, pushed backward by the water column, reached the valve, the clapper banged shut. The water kept coming. The pig banged away at the valve inside the pipe. Under the continuing pressure, the pipe stretched like a piece of rubber hose and shimmied back and forth on its Teflon-coated support beams, bounced up and down with the intense vibration. Before the crew discovered what it was doing, the pipe had actually leaped high enough to top the VSMs and had fallen on the work pad beside them, wriggling like a snake. This behavior was in fact about what the designers had intended as a defense against earthquake—that in such an event the pipe should move freely on its supports, though vibrations great enough and of such duration as to jump it over the VSMs—actually about *twice* the force generated by an earthquake measured at the maximum 8.5 on the Richter scale—were not in their calculations for this comparatively low-risk zone; but the design had never, so far as I know, been put to a practical test. The result of this mishap was that several pairs of VSMs were knocked flat, a few hundred feet of insulated pipe was elongated, peppered with dents and otherwise knocked

about, and the whole segment had to be replaced, a job that took
two months. Neither the welds nor the pipe itself, however, had
broken. The earthquake-protection system, then, had been put to
a real if unintended test, and it had passed; but in a way that
once more raises questions about the management of the project,
from the vice-presidents in Anchorage on down to the straw
bosses on the work sites. Human error.

The hydrotesting of the tankage at the terminal and pump stations was uneventful—the welding held, no leaks occurred. Nor
was there, so far as I know, any further excitement on the pipeline itself. With heated water the hydrostatic testing continued
into late November, when the streams from which water had
been drawn were sealed to the bottom by ice. The pace was
slow, twenty or thirty miles a week along the whole line, not
much faster than the original laying of the pipe, possibly because
Alyeska was finding more serious uses for the available supply of
man-hours. The testing of the remaining 160 miles of pipe resumed on April 1st and was completed on May 25, 1977, in time
for the first input of oil.

The difficulties I have sketched all had elements of random
error but were also both managerial and inherent—in the work
itself, in the environment in which it was done—in about equal
measure. All were immeasurably exacerbated by another,
touched repeatedly in passing, which showed itself early in 1975,
in the first year of pipe work, and was not finally resolved (if it
was) until a few days before start-up in 1977: the welding of the
pipe. It became the fundamental problem from which all others
derived—and necessarily so, perhaps. Most elements in the
pump stations and the terminal—piping, valves, tanks, power,
communications, the pumps themselves—were designed in multiples like the veins, arteries, and nerves in a living body ("re-
dundancy" is the engineering term), so that, within limits, any
one might cease but the system as a whole would remain intact,
safe, and functional; even the main oilbearing pipe could, as we
have seen, withstand considerable abuse without failure. Not so
with the hundred thousand or so welds that held that pipe together: if one broke, the system stopped, with more or less
disastrous consequences—not, probably, the apocalypse imagined

by the environmentalists but serious nonetheless. And therefore pipeline welding became the focus of many of the anxieties that the project had aroused.

Actually, the welding problem involved two separate questions: the objective quality of the welding in actual use; and the validity of the means by which that quality could be verified in advance. That complicating and almost metaphysical distinction was generally blurred in the accounts and was ignored by those congressmen who found welding by itself tangible enough to be politically useful. But it is important to bear in mind.

We looked earlier at the temperament of pipeline welders and the ways in which it connects with the work they do. It is time to consider the work itself.

Several preparatory steps precede the actual welding together of two pipe joints. The ends are beveled mechanically so that, when joined, they form a 60-degree angle, closed at the inside edge, widening to the outside. The beveled ends are then sandblasted clean and heated (usually by the gas jets of a spider torch placed inside the pipe), and the new joint is lifted by a side-boom tractor, aligned with the pipe to which it is to be welded, and held in that position by an internal clamp. Welding then consists of filling up the beveled angle between the two pieces of pipe with welding rod melted and fused with the pipe metal by the welder's torch; it takes fifteen or sixteen pounds of rod to complete each weld. The first layer is normally applied by four welders, one to each quadrant of the pipe, and their work constitutes what is called the root pass. Five or six more passes, depending on the thickness of the pipe, are required to complete the weld, each normally laid down by teams of two welders; each pass is made in quadrants, one fourth the circumference of the pipe, corresponding to the length of the welding rods, and perhaps fifteen welders may work on any one weld, each marking his own part of the work so that it can be identified. Since welding is quite sensitive to temperature, in cool weather each pass would normally be made under the cover of a canvas welder's "tent," a kind of awning unrolled around the weld to keep off wind.

Each step in this process is critical: if one is imperfectly done, the rest may fail and, finally, the entire welded joint. Hence in the quality-control program Alyeska developed, an inspector was to be present throughout the process to monitor, approve, and certify that each—beveling, cleaning, heating, alignment, each pass and the completed weld—was acceptable; quality-assurance inspectors, in turn, were to monitor the work of the QCs, on a spot-check basis. The final check was the radiographic examination of all welds required by the Stipulations incorporated in the Right-of-Way Agreement (100 per cent radiography in contrast to the usual practice on U.S. pipelines of verifying about 10 per cent of the welds by this means). The completed weld was wrapped with photographic film, irradiated by X rays or gamma rays, the film developed and inspected. If the film revealed a defect in any of the welding passes, it was to be repaired or, in extreme cases, cut out and replaced, and the repair would then be refilmed. The radiographic process, subject to its own program of quality control, thus provided a permanent and comparatively objective verification of the quality of each weld, and one that could be analyzed and if need be re-examined at leisure. The QC/QA program as a whole, if it worked as intended, would in theory provide something approaching absolute assurance of the integrity of every weld on the line. At the same time, since every part of the program, including radiography, depended on visual assessment by the inspectors and on their ability to relate what they saw to the physical reality of the welds and their practical function, it left unlimited room for interpretation.

As we have noticed elsewhere, from early in construction there were serious disputes among the four parties concerned—Alyeska and Bechtel, its construction management contractor, the state and federal surveillance offices, each with a rather different interest in the outcome—about welding, about the quality-control program and, quite broadly, how that and the project as a whole should be managed. We have seen also how some of these disputes were resolved. The state welding experts succeeded in specifying a welding rod that in their view would suit the extreme temperatures of the Alaskan weather—but that was chemically different from the pipe steel on which it would be used and, perhaps, not entirely compatible; it was a rod, anyway, with which few of the welders had any experience and for which, as this became apparent, they required special training;

and as the project advanced was often and loudly resented by them. As first submitted, Alyeska's QC/QA program was critically received by both surveillance agencies and was not formally approved, much revised, until more than half the first season of pipeline work was already spent, but the big change demanded, the QCs' authority to stop nonconforming work, was a sword of paper and remained so till the end; and as we shall see, the program's failings were already apparent and would become increasingly visible. Bechtel, an international firm with troubles enough in managing other projects (among them a controversial and much-delayed nuclear power plant in Midland, Michigan), was fired as CMC about June 1, 1975, while the quality-control program was still being argued out, but remained for almost a year as QC supervisor. Some of the issues between Alyeska and its chief contractor can be guessed but they will not be fully known until the resultant lawsuits come to trial, if they do, and at least some of the documents become public; whether Bechtel's diminishing role made any practical difference in the way the job was run is another question altogether.

On March 26, 1975, *one day* before the official start of pipeline work on the Tonsina, Alyeska signed contracts with the two radiography firms that would be responsible for X-raying the welds: Ketchbaw Industries south of the Yukon, the Exam Company to the north. From near the start, there was more work than they could keep up with. The early rejection rate for welds was extraordinary, 50 or 60 per cent, in some places and at some periods nearly 100 per cent. The reasons for rejection might be anything from a major and obvious flaw to a minute discrepancy in one quadrant of one welding pass, of no practical significance, probably, for the functioning of the pipeline. Nevertheless, all flaws had to be repaired and, under the sweeping requirements of the Stipulations, all repairs had to be X-rayed. Since the same X-ray crews were doing new welds as well as repairs, they were in effect being called on to grind out as much as twice the volume of work for which they had been contracted, and the hours they put in showed it—up to twenty hours a day and still not keeping up. As a result, the welders got farther and farther ahead of the radiographers—miles, days—and the QC staff failed to keep pace with either the complex sequence of inspections required in welding or with the quality control of the radiography itself. The radiography companies complained, asked authorization for more and larger crews. Alyeska, which by contract controlled the number of men assigned to a particular task in relation to its budgeted man-hours, refused; a characteristic decision.

Naturally, the work suffered. Welders fell into errors of technique and repeated them over and over, unchecked by the final authority of the X-ray film. New or repaired welds were buried or covered with insulation before they could be filmed because no QC was present to insist on that final verification. Some of the film itself was improperly exposed or developed, unusable. Nagged by the layers of supervisors stretching back to Fairbanks and Anchorage, the radiographers found a way of making good, catching up. If you had no film of a weld or the film was bad or showed a trivial defect, you could replace it with an acceptable duplicate film of a satisfactory weld—all that was needed was to supply the proper identifying number of the weld and its documentation. It was one of those necessary discoveries, like fire in man's earliest history, that spread so rapidly up and down the line that it seemed to have many origins, all simultaneous and none in particular that was traceable.

The first public wind of these short cuts was a law suit filed on September 10, 1975, by Peter Kelley, one of the radiographers. Sometime in June, his supervisor ordered him to change numbers on his X-ray films. When he objected, he was fired; filed grievances with his union and the Alaska Department of Labor; and in September brought suit under the Alaska Hire Law against his employer Ketchbaw—not that he had been given an improper or possibly illegal order but that as an Alaskan resident, living in Fairbanks, he had been discharged and replaced by a nonresident; a lawyer's notion, one would guess, of a provable, easy case.

Later, Alyeska would insist that when Kelley filed his suit it already knew that something was amiss with welding and radiography, that back in July its QA inspectors, carrying out their prescribed functions, had uncovered 39 "deficient" films that had to be reshot in Section 2 (*not* Kelley's section) and 7 welds so seriously flawed that they were cut out and replaced; and in August had ordered a complete review of X-ray films in Section 3, where Peter Kelley had worked until discharged. These reassuring assertions did not come, however, until a year afterward, in September 1976, when the ripple of anxiety started by the ex-radiographer's charges had swelled from a minor labor dispute to a major scandal that threatened to delay or stop the entire project; and were unsupported by evidence. Alyeska's audit of the Section 3 films—to find out whether every weld and repair had indeed been X-rayed, whether the films matched their welds, were technically acceptable and showed any uncorrected welding defects—did not actually start until September 11th, the day *after* Kelley filed his suit.

On October 8th and again on November 13th, Frank Moolin held combative press conferences in Fairbanks to tell the Alaskan reporters that the Section 3 audit had been completed, that a few falsified films had indeed turned up (20 was the number in October, 132 in November), that the problem was really comparatively minor, his department had it well in hand; and so on. But whatever information he may have had on those two dates, it was, as it turned out, very far from final.

On December 8th, a Ketchbaw supervisor was found dead in his apartment in Fairbanks; it would seem that he had been sent to Alaska to find out what was really going on and to defend his company by producing evidence that fiddling the X rays was not confined to one section of the project or to Ketchbaw's jurisdiction but was commonplace all along the line. A Fairbanks coroner's jury took its time looking into this death and did not get around to a verdict until February 20, 1976; suicide, it concluded, by ingestion of cyanide.

Meanwhile, on the last day of December, Alyeska canceled Ketchbaw's contract and announced that change on January 8th, the day it signed a new contract with the pair of California radiography companies that would replace Ketchbaw on the southern half of the line, Peabody Testing and Bill Miller X-Ray, Inc. (conveniently—they were already working together on the pump stations and the terminal). On February 5, 1976, Alyeska brought suit against Ketchbaw for violation of contract, claiming minimum damages of $10,000, the total to be determined by trial. On February 22nd, a Sunday, 358 X-ray films from Section

2, all of them falsified or otherwise unsatisfactory, vanished from Delta Camp; Wackenhut, the Alyeska security force south of the Yukon, reported this loss to the state police on February 24th as a theft and gave it a value of $500,000, by far the biggest single item so reported during the life of the project. There was a bustle of investigation—lie-detector tests, state police searches of trucks crossing the Yukon bridge and at other checkpoints—but it recovered no film, produced no culprits, and was officially dropped less than a month later, on March 13th; Chuck Dulinsky's FBI office in Anchorage was called on for help but declined to take a hand—the crime, if it was that, had not been shown to have interstate connections to which any federal statute would apply. It was suggested at one point that the film had simply gotten into a camp garbage can and been burned with the other trash; which, all considered, sounds not impossible, though if it happened—358 strips of film more than twelve feet long, an awkward, bulky bundle of some weight—it could hardly have been accidental.

On February 23rd, the day before the film's disappearance was reported to the police, Ketchbaw brought a suit of its own against Alyeska but also against the state and federal pipeline offices, alleging that all were implicated in the falsification of its X rays—by causing or allowing welded pipe to be buried before it could be filmed, by failing to identify unacceptable or duplicate X rays; and that all these practices were even more prevalent in the Exam operation north of the Yukon. In compensation, Ketchbaw claimed $40 million in damages. Its brief included confidential Alyeska reports which it had somehow ferreted out, showing incidentally that Alyeska had for some time been concerned about both the quality of its radiography and the effectiveness of its quality-control program. These documents, however, were dated in late September 1975, about two weeks *after* Peter Kelley brought his suit.

At some point in these proceedings, Alyeska concluded that it would be necessary to audit its records of all the field welds, 30,800 of them, completed during 1975. Oddly, just when this investigation began is not one of the items included in the "Weld/Radiography Chronology" the company published in September 1976, but to judge from the kinds of misinformation on which Frank Moolin was still relying in his November press conference, it must have been later—quite possibly much later. My concern for this date is not niggling. The sheer volume of work required and Alyeska's perennial shortage of technicians qualified to evaluate the films made time a fundamental factor in the effectiveness with which the investigation could be carried out; and when completed it raised, I should say, about as many questions as it claimed to answer.

Let us consider the magnitude of this problem. Out along the pipeline are the 30,800 welds, identified by location and code numbers in various documents, among them the reports supposedly made by the QCs for every weld; back in the vaults in Fairbanks (or presently gathered there from the pipeline camps) are, say, 40,000 pieces of X-ray film, maybe a lot more than that since in theory both the original welds and any repairs, sometimes several on a single joint, were X-rayed and the *average* repair rate for the two full years of pipe work was something like 30 per cent. Now, all these films supposedly carry the same identifying numbers as the welds and their accompanying inspection reports, but you know that some of these numbers have been doctored: how do you figure out which? Alyeska's solution was something it called radiographic fingerprinting. It worked like this: every piece of pipe is held together by a weld from end to end, and when two joints are lined up for welding the ends of these longitudinal seams fall in different places, probably never in quite the same relationship on any two welds; further, in making the girth weld that joins two pieces of pipe, the welder starts his final pass from somewhere near the top and leaves a little lump of metal at that point, called a "top button," which is visible on the X-ray film along with the ends of the two longitudinal welds, and the chances are pretty good that the relationship of all three of these reference points will never be exactly repeated. All you have to do, then, is measure these distances carefully on all your films—from top button to the two longitudinal seams— feed the measurements into a computer and let the computer tell you which ones are identical. Ingenious, what? Once you know which films are duplicates despite their different code numbers, you can do a similar analysis of the QC reports and figure out which welds are not documented or are covered by reports that have been altered or are otherwise fishy, and the welds so identified (or not documented at all) are probably ones that are in some way defective or anyway deserve direct inspection. Finally (as Alyeska implied it did), you can subject all the remaining films not singled out by these means to further examination, and if any of them show possible defects, you can, again, take a further look at the actual welds.

This all sounds rather impressive, but there is one very large gap in the procedure. Unless you go back and take the same measurements for *every weld* and compare these with their X rays, it does not seem that you can have any certainty that they match. That was not a step, so far as I know, that anyone suggested, certainly not Alyeska and not in public, and for good reason. Most of the welds were already buried, in some cases under stream beds, or, if aboveground, were at least swathed in insulation; digging them up again, where that was in any practical sense possible, would risk damaging the pipe, would further tear up the right-of-way and its surroundings, and would be very expensive. Indeed, we might well conclude that the purpose of Alyeska's exhaustive attention to the X-ray films was precisely to avoid this possibility. That was how it worked out, anyway. And there were quite enough comparatively small questions without opening up that big one.

Toward the end of April 1976, Alyeska produced a report on its investigation of the weld X rays that came to be known as its White Book (copies circulated in a flexible white fiberboard binder). The materials it comprised carried several dates about that time; it was formally transmitted to the two surveillance offices on April 26th. The first page carried a minatory notice in all caps:

THIS BOOK AND ITS ENTIRE CONTENTS ARE THE PROPERTY OF THE OWNERS OF THE TRANS ALASKA PIPELINE SYSTEM, AND REMAIN SO WHEREVER FOUND, AND ARE CONFIDENTIAL AND PROPRIE-TARY TO THOSE OWNERS, CONSISTING OF COM-MERCIAL INFORMATION OF A CONFIDENTIAL AND PROPRIETARY NATURE.

As a student of the language of law, and of oil-company law-yers in particular, I would say that the warning is about as strong as it could well be made. The oil men, that is, considered what was inside the White Book secret and wanted it to remain so. In fact, however, the enterprising reporters of the Anchorage *Daily News* had a copy of their own within days, prompting Alyeska (apparently) to release a summary on May 14th, though never officially the entire document; at the end of June, Chuck Champion, the State Pipeline Coordinator, concluded that under the state's freedom-of-information statute he was obliged to pro-vide copies to any citizens who asked. So much for secrets.

After all this, it is not much of a book: a collection of reports by several hands, contradictory, self-serving, inconclusive. Its most useful parts are two one-page tables that enumerate, by sections of the pipeline, the discrepancies in radiography and welding turned up by the audit of the X-ray film. Except that they add up to the same total, the two tables do not appear to be consistent with each other or within themselves, but they are the best we have in this thorny matter, and they seem to have defined the terms under which Alyeska and the state and federal monitors acted till the completion of the project. The more intel-ligible of the two summary tables (to me, at least) divides the problems into four categories. For one, 37 welds were found to contain defects so serious as to require cutting out a length of pipe and replacing it with a pup. Missing or duplicated radio-graphs added up to 895—that many welds would have to be X-rayed, according to the Stipulations, to determine if they were acceptable, though presumably they would not be. The films also showed 1,911 welds that would need various minor repairs but would probably not have to be cut out; and 1,112 where the film was ambiguous—the welds might or might not be defective, and whether they were would be determined by visual examination. The total of all categories was 3,955, of which nearly half oc-curred in Section 5/6, the northernmost part of the line. In the newspaper accounts, it was invariably said that Alyeska had ad-mitted to 3,955 bad welds, but it had done nothing of the kind: what it admitted was 37 welds that were clearly unac-ceptable, 1,911 that called for limited, sometimes trivial, repair, while the rest either needed new X rays or could not be adequately judged from the films alone; and, be it remembered, we are talking always about images of welds, not the welds themselves, which so far as the White Book was concerned had not yet been examined. In a practical sense, however, the distinction is academic. In every case, under the ground rules by which the pipeline was built, it would be necessary to uncover the welds, X-ray most of them, and repair those that film and visual inspection showed defective; finally, re-X-ray the repairs. Alyeska set to.

The White Book loosed a chain of unanticipated questions. They were formulated by the several interested parties—reporters, the surveillance offices in Alaska, federal officials in Washington, congressmen—but were prompted by common sense. For instance. If 13 per cent of the welds performed in the field in 1975 were questionable, what about those now being done, in 1976? or the double-joint "factory" welds in the shops at Valdez and Fairbanks, before pipe work started? or, for that matter, the longitudinal welds made at the mills in Japan? And the Alyeska audit—could its findings be verified by an independent auditor? was its bookkeeping on radiography and welding such that it *could* be audited? The "fingerprinting" process on which Alyeska relied to distinguish one film from another depended on exacting measurements of images on the X-ray negatives, but were these measurements accurate enough for the certainties demanded of them—or *could* they be? The audit had assumed that no films were exchanged between sections, but why was that impossible? And finally, given the rings of authority with which the project was surrounded, how could radiographers conspire without the connivance of the QC/QA inspectors, of government surveillance men and executives on up through the ranks of Alyeska and the various departments of government? While Alyeska endeavored to keep its attention on the repairs and the laying of new pipe, these were the questions the interested parties asked, and for which they sought answers.

In the search for answers both the *Wall Street Journal* and the New York *Times*, for the first time in months, sent reporters to Alaska in early June for on-the-spot investigation and gave considerable space to their reports. On June 21st, John Dingell's House Subcommittee on Energy and Power held a public hearing in Washington, with Peter Kelley as its star attraction (and

Ed Patton in the supporting cast). Three weeks later, President Ford dispatched an investigative team from the Department of Transportation, which by law would eventually have to certify that the pipeline, a common carrier, was fit for public service. Simultaneously, Congressman Dingell posted subcommittee staffmen on a parallel quest. Andrew Rollins, the Department of the Interior's Authorized Officer in charge of the Alaska Pipeline Office, found these congressional attentions in particular distasteful; since the Ketchbaw charges in its $40-million suit, the APO had become not only highly visible but, to a degree, a public adversary, like Alyeska itself. When the two investigators insisted on interviewing his staff, Mr. Rollins wanted an APO lawyer present; the investigators refused; then compromised by agreeing that the staff members could make a written record of the questions and answers. (To find out if the AO was in fact perusing these reports and using them to discipline his people against embarrassments, they included loaded questions—"Is it true that part of your job is to provide Mr. Rollins with liquor and women when he visits the work camps?"—and obtained a red-faced, firsthand denial from the AO himself; whatever his faults, those were not among them.)

By these and similar tactics, like Alyeska with the reporters, the APO managed to limit what the investigators could find out: the subcommittee report, which in good Washington fashion found its way to the New York *Times* in early September over the Labor Day weekend, several days before its official release, was strong on disquieting gossip and random observation from several days of flying up and down the line, but short on substance, like most of the newspaper stories that had preceded it and for the same kinds of reasons. To that extent, the old soldier had won the battle, but the war was far from over.

The stew continued to bubble. In May, bypassing the APO, the Department of the Interior had ordered Alyeska to stop tapping, burying, and insulating the pipe until it could provide foolproof assurance that the prescribed weld-inspection process, including radiography, would really be completed while the welds were still accessible. This work was allowed to resume after four days, but eventually, in July, Alyeska acceded to a further order that required it to identify every weld with a unique and permanent code number visible on the X-ray film (something it had not yet thought to do) and to furnish the APO with a photographic print of every film; Alyeska glumly referred to these procedures as "redundant markings and radiographs."

The APO took a couple of defensive steps of its own. It instructed its field representatives and contractual consultants to undertake something more than a spot check of the welding inspection process. And in May it commissioned Arthur Andersen & Company, the Chicago-based international accounting firm, to examine the methods by which Alyeska had performed its audit of the 1975 radiographs and welds. The Andersen report, completed two months later in mid-July, was carefully inconclusive: the records were not, it suggested, auditable; there was no way of determining from them whether the findings in the White Book were accurate or not. Alyeska revised its methods, but in January 1977, after a further examination, the Andersen accountants once more declined to certify them.

In July, anticipating the doubtful findings of the first Andersen report, Ed Patton produced a sweeping defense of Alyeska's radiographic audit and its welding in general, concluding: "I remain as confident as ever that the welds on this project are equal to the highest quality ever produced." That *equal to*—curious, one sees the mind working—why not say straight out that they *are* the best welds ever?—but no, watch out for superlatives, always exceptions, disputable; so, "equal to," judiciously, having it both ways. Looked at thus, it is an exact instance of what is known in rhetoric as equivocation, seeming to say what it does not quite assert, raising doubts where it means to soothe. In September the company produced and distributed an elaborate packet of materials explaining its welding problems and defending the means by which it sought to resolve them, and much of this was straightforward, informative, and, I think, accurate; but also with disturbing lacunae like the one noted earlier, the omission of the date on which the investigation was begun (page 394). The practical responses to the problem by those concerned, as distinct from what was said, were similarly equivocal.

Back at the beginning of June 1975 Bechtel had been dismissed as CMC but stayed on as supervisor of Quality Control. Then sometime in March of the following year it was relieved of that role as well. Both changes sound decisive, radical, and from them one would expect clear-cut changes in procedures to follow, but I am not at all sure that this was the case. It appears that the practical effect of that first change in Bechtel's contract was simply that the Bechtel managers were taken over as a group by Alyeska, for want, presumably, of others to fill those slots, and it is not clear, if it mattered, just who signed their paychecks after that. Bechtel's dismissal as *supervisor* of Quality Control is likewise equivocal: we do not know from this that it no longer performed the work, only that it was now supervised by Alyeska, whatever that may mean (as supposedly it always had been, by Alyeska's Quality Assurance inspectors); in fact, from my own observation in the summer of 1976, it seemed that pretty much the same QCs were still doing the same jobs and, whichever payroll they were on, were known as Bechtel men. So far as practical effects are concerned, the congressional investigators in their whirlwind tour in July 1976 found the same anomalies that had been complained of all along: the new radiographers (if they *were* new—again, one wonders) still shorthanded and working miles behind the welding firing line; the welding itself (and other critical jobs, such as river crossings) being performed with no QC or surveillance officer in attendance or even on the job site, and contrary to Alyeska's official standards—the pipe ends uncleaned, no protective welding tents; the same reports of supervisory indifference and physical retaliation that all through 1975 made the QCs' stop-work authority meaningless and did so still. It was, however, a quick tour; these were, perhaps, mere instances, not proof.

These violations of procedures notwithstanding, there was at least, one supposed, the APO and its technical staff, provided by Mechanics Research, Inc., to make sure things were done right. Had not the APO assured us that, starting with the 1976 season, its own people would be looking over the shoulder of every QC and verifying his inspection of the welds? Well, no, it seems that impression was mistaken. Near the end of a report on the welding question completed in late June 1977, a few days before start-up, the APO described, with exact qualification, what it required of MRI: "100 percent inspection (not just spot check-ing) of certain selected welding crews and radiographers by sec-tions." *Nota bene:* this does not say "100 percent inspection" of welders and radiographers; it says that *some* welders and *some* radiographers in *some* sections of the pipeline were to be in-spected "100 percent" (whatever that may mean). It is, in short, another striking example of equivocation, worthy to stand beside Mr. Patton's own in any textbook of rhetoric.

On December 10th, still worrying his rat, Mr. Dingell con-vened a further subcommittee hearing, this time closed to the public so as to protect the several former radiographers who testified. The subject was the double-joint welds done under con-trolled conditions by automatic welding machines and not so far seriously challenged; and the functioning of the APO. What was said of the federal surveillance team and its boss can only be in-ferred, but the radiographers insisted that there had been high-jinks with the shop welds and X rays, just as with those done in the field; and (as the welders had been saying for some time, defensively) that they had found unreported defects in the few inches of the longitudinal welds that showed in their circum-ferential X rays.

In consequence, Alyeska performed an audit of its double-joint radiography, completed in February 1977, which confirmed the subcommittee testimony: about 600 welds were added to those to be dug up, examined, X-rayed, and repaired, and apparently this was done. The same study reassuringly showed that defects in 2,318 of the double-joint welds had been caught and repaired before leaving the shops in Valdez and Fairbanks—5.5 per cent of the 42,000 such welds performed, not far from the industry standard for pipeline welding in the field. Alyeska also glanced at two other nagging questions, defects in the longitudinal welds and the swapping of faked radiographs back and forth among the five pipeline sections; and reached a statistical conclusion that both were unlikely. (The longitudinal welds had been through a series of inspections and tests, back in the Japanese mills, comparable to that required of the girth welds.) On the other hand, what seems to have been a random sample from the in-ventory of 1975 radiographs showed that perhaps *20 per cent* of the measurements from which the weld "fingerprinting" derived and on which the identification of welds to be examined or repaired was based was erroneous. That finding, if correct, would, I should say, invalidate Alyeska's entire repair program, by this time nearly complete; but that was not a question that anyone at this point—it is now sometime in February 1977—was prepared to pursue. It should be added that all these bits of information came from Alyeska and were incorporated without comment in the APO welding report referred to earlier; apart from its two unsuccessful efforts at verifying Alyeska's audit, it does not appear that the federal agency availed itself of its contractual right (under the Right-of-Way Agreement) to examine the pipeline company's records.

As for the APO: following Mr. Dingell's hearing in December, mysterious paragraphs began appearing in newspapers (see page 323) suggesting that Mr. Rollins had accepted favors, at some time in the past, from contractors he was now monitoring as Authorized Officer; a proper Washington dénouement. In February, as Mr. Carter's new administration took hold, Andy Rollins was silently unseated and retired to Dallas. His deputy Jack Turner was left in indeterminate charge and in August, after a six months' trial period, succeeded him in the office. About the time Mr. Rollins was becoming a nonperson, his state counterpart, Chuck Champion, completed his final report, took a long vacation, and came back to resign—with honor, and not carried home on his shield; and went back to the oil business where he had started, in a job with one of the companies contracted to explore Pet 4, the extensive but still unproved Naval petroleum reserve west of Prudhoe Bay. His deputy Cy Price remained to conclude the activities of the SPCO.

While these Byzantine maneuverings transpired, Alyeska went on doggedly opening bell holes around the 3,955 "questionable" welds (or, eventually, 4,800, counting the 600-odd double-joint welds and 200 welds made in 1976 that further auditing showed to be improperly X-rayed or documented), inspecting, repairing, re-X-raying. At the beginning of September, the company offered evidence why the Department of Transportation should exempt 612 of the welds from this further examination—trivial defects according to the X rays, environmental damage at frozen stream crossings that outweighed the welding risk; but went on digging.

The Department took its time over this evidence and four months later, at the end of November, consented to overlook three of the 612 welds; but by then the repair program was effectively complete. The final joint was welded on May 30, 1977, and on June 16th the Department of Transportation certified that so far as welding was concerned the pipeline was in a state to receive oil. A considerable number of formalities remained to be gotten through, governmental and practical, but the pipeline was now officially almost to the point of start-up; ready to receive the long-awaited oil.

From what we have seen, it is obvious that Alyeska's welding/radiography fiasco was expensive. How expensive? In announcing a new cost estimate of $7.7 billion for the project (June 1976, the last made public during construction), Alyeska attributed substantial parts of the $700 million increase calculated since late 1975 to the repair program and to additional materials and equipment, much of which must have been due to the repairs; but as usual did not specify just how much. Earlier in the 1976 season the company let out an estimate of $55 million for the repair program and this was several times repeated without change; so far as I know, no actual final figure was ever given. Since the estimate was made before most of the repair work had been done, when big items like the audit and repair of the 600 double-joint welds were not yet known to be necessary—and since every one of Alyeska's published estimates, from the beginning, had been notably elastic—it seems likely that the real cost was a great deal more. For one thing, the $55 million apparently did not include the cost of the audit of 1975 X rays, which in May 1976 Alyeska estimated at $4.5 million. Further, if Alyeska's $500,000 valuation of the 358 X rays that vanished in February had any meaning, then the cost of the additional X rays required by the repair program would be around $6.5 million, at least; which, considering all the high-priced men and machines involved in the actual repairs, does not leave much in the pot for excavating, inspecting, welding, taping, insulating, backfilling. State investigators (see Chapter 11) did estimates of their own which came out to $61.7 million, $80 million, or $100 million, depending on various assumptions about costs, and none of these is unreasonable.

Whatever the exact figure, these are the *direct* costs of repair: they make no allowance for the original cost of all the work that was now to be undone—money, in effect, flushed down 4,800 holes in the Alaskan landscape; nor for the probably incalculable disruption of work and dilution of managerial attention at a critical point in the project, when the project as a whole was no more than half complete and, in addition, all the finishing work that would prepare it for its scheduled start-up in 1977 was still ahead. In August, for instance, Frank Moolin was pulled from his job as pipeline project manager and sent to the Galbraith Lake camp in the Brooks Range to take charge of Section 5/6, where new work lagged behind the other sections and the volume of repair work was far greater. Thereafter, the day-by-day management of pipeline construction was in the hands of an assistant in Fairbanks, K. E. Eliason; and, evidently, of the Anchorage vice-president Peter DeMay, whom we have met before in his role as builder of barricades between the public and the project.

Whatever the final bill turns out to be, if it can indeed be calculated, Alyeska's effort to make good on its welding and X-ray promises represented a sizable public charge—in the sense that like every other pipeline cost it will ultimately be repaid out of our collective pocket. We may naturally wonder, therefore, just what we got for our money. To put it exactly: are we to understand that without these desperate efforts the pipeline would have failed, with all the disastrous environmental and economic effects that such failure implies? And: was the pipeline work any more effectively performed in the project's concluding months than in 1975, before the X-ray and welding troubles had become a national scandal? Surprisingly, in view of the immense expenditure of human effort, emotion, and money, the answers to these questions are less obvious than one might wish.

In the final year it was said (and repeated by Chuck Champion in his last report) that Alyeska had greatly expanded its radiography crews and quality-control staff, and that the QCs were zealously and needlessly rejecting marginal welds that were not

in any functional sense deficient; and, contradictorily, that the
reject rate had been cut in half compared with 1975. On the
other hand, it was still possible for unprepared visitors (such as
the congressional staffmen) to search a work site and find neither QC nor X-ray crew, and there were places that summer
where X-ray crews were dismissed in the face of mounting
backlogs of work. And if the reject rate in fact dropped by 50
per cent, it would mean that 20 per cent of the field welds performed were still not passing inspection, compared with an average of 40 per cent the previous year (in addition to all those
identified after the fact)—an improvement, of course, but not so
hot in a business in which 5 per cent is considered a no more
than tolerable limit. As we noted in passing, after all the attention, an Alyeska audit of its 1976 welds still found about 200 possibly defective joints that had slipped past its QCs—and we
know nothing of the methods or completeness of this further review, except that, like the questions about the "fingerprinting"
measurements, it came too late in the game to reopen any
difficult questions.

Against this doubtful performance, both Alyeska and the APO
submitted to independent laboratories what were apparently actual welds identified as questionable in radiographs and cut out
from the line; and the welds were subjected to the full series of
destructive tests and certified as functionally sound. From these
physical tests, Alyeska, while grimly pursuing its repair program,
insisted that most of its repairs were of no practical significance.
The two surveillance groups were in broad agreement.

> A vast majority of the 3955 weld discrepancies identified by
> Alyeska's Quality Assurance [Chuck Champion wrote in his
> final report] were of a cosmetic nature and definitely not haz-
> ardous to pipeline integrity. National publicity . . .
> prompted . . . the expenditure of millions of dollars to make
> unnecessary repairs.

Jack Turner, when I talked to him a day or two after he'd been
confirmed as Mr. Rollins's successor, said much the same thing,
more emphatically:

The significance that ultimately came out of the welding thing . . . was that from an integrity standpoint we've got an excellent quality material, the pipe, we had a good welding procedure, a very ductile system, and there was no anomaly that was ever checked, including cracks . . . , that would ever cause the line to fail. . . . In the real world, many of the dollars were spent to take care of something that had no significance. . . .

It is necessary to allow for the self-interest at work in both views: the state's in limiting all pipeline costs and for the same reason—revenue—doing nothing that would seriously impede the start of the oil flow; the APO's similar interest in getting the job done, with the perhaps stronger goad not of money but of defending its own performance against congressional and journalistic antagonisms. Nevertheless, with these allowances made, I think we are obliged to agree: the year or more expended on the welding repair program probably did make some difference, but not much—I am not at all sure that any disasters were probable if nothing had been done; or that they were averted. The welding as such was not the real issue.

We come back to the radiographs and that tenuous distinction between image and reality. The image in this case is a fuzzy, light-colored band half an inch wide, striated and textured in gray that suggests shape and depth in its two dimensions, standing out from the solid black of a strip of photographic negative; this is the weld as it looks to the radiographic technician examining it on a light box in a darkened trailer shielded somewhere along the right-of-way from the glare and noise of the pipeline. Maybe somewhere along this twelve-foot band of textured light is a dark spot a sixteenth of an inch long, which you and I will notice if it is pointed out; possibly it means a defect in the weld itself—incomplete fusion of the metal, a void bubble of gas, a bit of foreign matter covered over by welding material—and possibly that is a weak point through which under pressure the whole weld might come unzipped; or possibly not.

It is not much to go by—not, one would say, an absolute standard on which to base absolute decisions, let alone moral judgments. In fact by the time Alyeska's welds and their X rays became matters of national concern, there were newer methods of evaluating welding, more objective, less dependent on the subjective interpretation of a thousand different technicians, and some of these had figured in the acceptance of the longitudinal pipe welds back in the Japanese mills. They were not mentioned or allowed for in the law under which the pipeline was built and the surveillance offices operated. The law permitted little interpretation and no exceptions; a clumsy, sweeping, scythelike instrument.

It is often said, by officials of government and large corporations, that some great public work, such as the pipeline, is being done to "rigid" standards—as if that rigidity should assure the rest of us, with our fragmentary awareness and technical ignorance, our newly awakened environmental nervousness, that the completed work will necessarily be harmless, benign. Well, whatever is rigid will break if enough force is applied, and that is what seems to have happened to Alyeska's welding and radiography standards. Perhaps then, if such matters must indeed be governed by exact law, we had better learn to accommodate enforcement as we do in most other areas, to the human realities: the gap between intention and capacity that appears in every human undertaking. We can learn from the willow, that scrubby, contemptible plant, rarely attaining the dignity of a tree, that grows in thickets along most northern streams: when break-up comes and the ice goes out, it scours the banks, smashing the willows flat, stripping them to a tangle of broken stubs; but the willows survive and grow where no tree can, and their roots secure the banks which are the basis for all streamside life.

Alyeska, as we have seen repeatedly, had some rigidities of its own, from its original choice of route on the strength of information that was often ludicrously sketchy to its persistence in running the pipe down the steepest walls of Thompson Pass in the face of a far simpler alternative. The same set of mind, it seems to me, is behind the constant failures in the scheduling of work that we have noticed in the course of the project—failure, that is, in the fundamental requirement for the management of a job of this kind. This failure links to another no less serious: the managers' consistent inability to estimate or control the quantities of labor that the job would need (and hence their costs, a matter we shall look at closely in the following chapter). Thus, for a start, the peak work force soared from an early guess of about 5,000 men to about 6,000 on the haul road in 1974 and 22,000 or 23,000 in the two years of pipeline work, and Frank Moolin's seemingly radical productivity reform in the final year had virtually no effect on the size of the work force or the man-hours expended. (One might of course argue that without that plan things would have been worse, given the huge quantities of labor soaked up by the weld/radiography troubles.)

Alyeska's estimates, in turn, rested on assumptions about productivity in the various phases of the work that were never realized; there is evidence, in fact, that although these assumptions were several times revised downward (and the revisions reflected in changing estimates of total cost), productivity showed a continuous decline from start to end, and the final level, measured against the company's own standards, was on the order of 50 or 60 per cent—that is, it was taking nearly twice the man-hours anticipated to complete particular tasks and the project as a whole. Assuming that Alyeska's estimates of man-hours were valid in the first place—an admittedly large assumption—what this means is simply that the company was paying for a lot of work it didn't get. Part of the reason for this is to be found in the unions: the untrained and inexperienced workers processed through the hiring halls, coupled with the extraordinary turnover on the job; the excessively narrow jurisdictional limits among the crafts (that was the reality behind all those stories of bus drivers sitting idle in their buses for ten hours a day), reinforced by what were in fact jurisdictional strikes, contrary to the Project Labor Agreement; the broadly interpreted rules governing travel time and requiring full payment for part shifts or part days, which, if perhaps valid elsewhere, made less sense here. The other part of the reason lies with Alyeska's management, in about equal measure, I should say: the ineffectual scheduling that brought crews to work sites where they could not work—or did work that had later to be redone, of which the welding/radiography mess was merely a pervasive and highly visible instance; the haphazard estimating that caused some work sites and jobs to be oversupplied with men while others (notably the QCs and radiographers) were often shorthanded and were being cut—a problem made worse by Alyeska's efforts to control the work assignments in detail from Fairbanks or Anchorage. Low productivity, then, was the natural product of the two bureaucracies, Alyeska's and the unions', working against each other. It is not necessary to cast moral stones at the men or to assume that many were there merely to draw pay and avoid work. A good deal has been gossiped to this effect in newspapers and magazines, more recently in books, but that was not what I saw; it is the view of some who signed on in order to write about the work, not to do it.

There is another side to the radiography problem: the motives of the men who falsified the X rays and of the QCs and construction supervisors who connived with or compelled them to do so. In the circumstances—the vanished negatives, the suspicious death, the violent threats—the motive looks like that old devil money: the oil companies' urgency for profit (and dread of loss through delay), passed down the line, the contractors' hopes for bonuses* and future contracts, and therefore the packets of dirty bills in envelopes, changing hands. I can only say that no one who looked into this obvious explanation found any evidence that it was so. It seems that managerial ineptitude—the basics, scheduling and estimating—was enough to account for all the troubles. There is one more point. In view of the elaborate documentation that Alyeska was required to produce in order to secure the state and federal Notices to Proceed for each phase of the project, it would seem both sensible and efficient for those plans to have formed the substance of its contracts and work orders. That was not the case: the two processes, approval and contracting, ran separate and parallel, unconnected. From what we have seen of the effects, we risk one more conclusion: Alyeska did not take very seriously the conditions imposed on it —quality control and 100 per cent radiography, environmental sensitivity and all the rest. Another oil company rigidity.

For many years now assumptions about American "efficiency" have been part of the stock in trade of popular comment on

---

* The details of Alyeska contracts have remained "proprietary"—secret— but provided for bonuses to the section contractors for completing work ahead of schedule, and probably to at least some of the subcontractors as well.

events in the United States, reflexively repeated to explain or justify every large program from space exploration to urban re-
newal: efficiency—the ability to define more or less distant goals
and the means of reaching them, then to organize and control
the available resources so as to attain the intended goals exactly,
with a minimum of waste effort; maybe in the aftermath of national self-disesteem following from the war in Vietnam we hear
this cant a little less. It is as if since World War II we had
inherited the view the Germans once had of themselves, by
which they were likewise understood in other countries. For my
part, considering the number and variety of failed social programs that have been promoted and allowed to run their course
over the past twenty years on the same premise, I can never hear
such talk without a shudder. If there is such a thing as a national
character, efficiency in that sense is not one of our traits: our
successes have been those of improvised teamwork inspired by a
situation of desperate urgency and immediacy, generally perceived as a crisis. TAPS is precisely a case in point.

The Trans-Alaska Pipeline act and the Right-of-Way Agreement that followed made a couple of assumptions about American efficiency: they defined the expeditious delivery of the oil to
market, with minimal effects on the environment, as a national
goal; and they prescribed the means to that end and created a
system for enforcing these determinations by continuous monitoring of the actual work. Efficiency, however, was not what
came out of this logical arrangement—as we have seen. The ability of the surveillance staffs and their Alyeska counterparts to
control the work or even to keep detailed track of it was quite
limited; the mutual resistance of monitors and managers was
among the project's most prominent recurring themes. And yet
out of this muddle, the basic goal of getting the oil moving was
met, and with environmental effects that were at least limited if
not minimal, though in the nature of things it will be years before we can say exactly what they were. It was not the efficiency
intended by Congress and promised by the oil companies that
made up the difference: it was inspired improvisation.

At the beginning of July 1976, when W. J. Darch moved over
from BP to take Ed Patton's place as president of Alyeska, the

project's troubles were near their climax—congressional and presidential inquiries ten days later, serious talk of a year's delay. Dr. Darch—the doctorate, thirty years back, in chemical engineering from what was then University College, Cardiff, where he had played international-class rugby—had been, like Ed Patton, a builder of refineries in difficult places, though in greater numbers and variety: Norway (like Patton) but also northern Canada, Australia, several parts of Africa and the Middle East; when I met him a few days after his arrival in Anchorage, he had a slight cold picked up a week earlier moving in and out of air conditioning and the 120° heat at Abu Dhabi, where he had been responsible for completing the biggest-ever liquefied natural gas plant. From the man and the circumstances, it looked to those of us with an interest in Alyeska and its motives as if the first team was coming in, Ed Patton in his new job as chairman had been politely fired, other drastic changes in management would follow, and Bill Darch had been handed the knife. The Anchorage oil men whose job was to keep an eye on Alyeska, however, insisted that the change was long planned, the timing coincidental, and that the new president had been chosen to run the start-up and operation of the pipeline as his predecessor had been chosen for construction; the two jobs are so different, of course, that the same team is unlikely to be good at both. That was how it turned out; we speculators about motives, in this case, were wrong. "The health of any company is set by its people," Bill Darch had answered when I asked how he expected to do his job. "In the operating phase, I would hope to know everyone in the organization by first name." That was possible. It would be a small and stable group running the line, not the transient mass that had built it.

From the day he took office, the new president evidently had his hands full with preparations for start-up. These were of two basic kinds which we may distinguish as regulatory and functional: submission of thousands of documents to the two surveillance organizations for approval (and/or to the Department of Transportation's Office of Pipeline Safety, whose certificate was also required before the oil could flow); and testing the pumps and other machinery, hiring and training the men who would run them. The repair and rehabilitation program—the welds and

radiographs, of course, but also things like regrading and seeding
the stripped earth along the right-of-way and around materials
sites—ran simultaneously with the completion of construction
and was, strictly speaking, the job of the construction group, but
Dr. Darch apparently took a hand, particularly in rehabilitation,
and was responsible for securing the formal approval of the federal and state agencies attesting that the work had in fact been
satisfactorily done.

For a start, the Stipulations called for complete operating
manuals to be submitted for approval 180 days before start-up:
instruction books describing every job on the line and telling
those who would fill these jobs, in programmed steps, how to
proceed, in normal operations and emergencies, in accordance
with the various regulations; since at this stage there seems still
to have been hope that the line could start as early as March
1977, that set the deadline at the end of August. The first step,
then, was to compile a complete list of the regulations to be met,
from the applicable laws and the official rulings of the many
agencies concerned but also from Alyeska's voluminous corre-
spondence on particular points. What came out of this study was
ten fat Xeroxed volumes covering the system's general operations, the handling of tankers at the terminal, surveillance and
monitoring, quality assurance, safety, environmental protection;
with four volumes devoted to plans for dealing with oil spills—
section by section along the line (for the purpose it was divided
into twelve sections determined by the locations of the twelve
pump stations), at the terminal and in Port Valdez, and in
Prince William Sound. All this material, which was prepared for
review, comment, and approval by the APO, was comparatively
general and descriptive. In addition, there was an unspecified
but apparently large number of detailed manuals telling, in more
technical terms, how the whole system was supposed to work—
the maintenance of the pipeline itself, with its many valves, the
complex of operations represented by the pump stations and the
terminal; these were *not* submitted for review but were closely
held within Alyeska, presumably for the sake of security and not
without reason—if you know how the system works, you know
also how to stop it. The whole package added up to a sizable
bookcase of material, comparable in sheer mass to the original

environmental studies and the project description under which the pipeline had been built.

Meanwhile, it was necessary to deal with—identify, correct, secure documented approval for—the thousands of cases that had been subject to nonconformance reports at the time of construction or had been uncovered since, an effort not inferior to that required by the welds and X rays. There were those VSMs that had settled in the course of hydrostatic testing, others that appeared to have jacked up out of the permafrost, though Alyeska argued that in most cases it was the work pad that had settled beneath the elevated pipe; at animal crossings where the elevated pipe had been set too low, it would be necessary to excavate the work pad in order to meet the established standard. The weld repairs had the incidental benefit of opening up improperly taped buried pipe for approval or correction. As spring returned, two thousand men were occupied in strictly environmental matters—clearing the right-of-way of the general detritus of construction, regrading eroded stream crossings, revegetating —and a day a week of Bill Darch's time, out of the office and in the field, in the personal style he had predicted, very different from his predecessor's remoteness. "Revegetation" is an engineer's term for planting grass: four different mixes of seed based on annual rye (there are not many native grasses), mixed with fertilizer, to be sprayed on bared slopes by machines called hydroseeders; in theory the grass would hold the soil long enough for slower natural growth to take hold. Where this worked—where the machines arrived on schedule and spring rain and flood did not either fail to appear or come so abundantly as to wash out the seed—the grass luxuriated; and was eaten to the roots by a surging population of voles and other small herbivores. In the spring of 1977 Alyeska refined this idea with a million and a half willow cuttings, to be sprouted and planted in places where the grass had failed (the state surveillance group, wary of costs and doubtful of results, had reluctantly assented to federal insistence). When the cuttings died, it looked as if start-up would be postponed for a year, but the APO agreed to let the oil flow on condition that Alyeska try again, in the spring of 1978.

Department heads and other operational executives were hired about the same time as Dr. Darch or, in several instances, rather sooner—he apparently had little or no say in picking them, in contrast to Patton's considerable freedom in recruiting his original construction staff. Assembling the technicians who would be responsible for the actual work in the pump stations and at the terminal went on through the winter, perhaps 1,500 jobs altogether. A serious effort was made to fill about half these slots with men of considerable experience in pipeline operations in the Lower 48 (twenty or twenty-five years, I was told, was typical) and they were then partnered with less experienced workers on a buddy system. In line with the principle that had prevailed throughout construction—and with the usual oil business practice—the number of employees directly working for Alyeska was deliberately limited. Security remained in the hands of Wackenhut, but now for the whole line—Nana had lost its contract north of the Yukon; the pious intentions of the pipeline act were made good by throwing the maintenance contract to another Native corporation, Ahtna, centered on the Copper River between Valdez and Fairbanks, which for the purpose formed a joint venture with one of the out-of-work Alaskan highway contractors. (Since the new company was nonunion, there was much grumbling by the construction unions, already uneasy about Alyeska's intentions, followed by limited picketing at the camps still open and, shortly after start-up, a formal strike by the few union workers still on the job.)

By the spring of 1977 the assembled crews were ready to start practicing the lessons they had been studying in their manuals. They began with the easy jobs, following up the hydrostatic testing of the pipe with pigs sent through to flush out ice, leftover water, and whatever dirt and junk remained inside from construction or testing. A similar device, known as a caliper pig, was used to check the roundness of the pipe—and revealed that in a fair number of places the empty buried pipe had been flattened by frost heaves during the winter; in most cases this was apparently not serious—it was possible to seal the pipe, pump it up with air pressure and pop it back into shape like a bicycle tire— but at about ten points it was necessary to dig further bell holes, cut out a length of pipe and weld in a replacement.

In the course of cleaning the pipe, the crews made another curious discovery: bullet holes. One of the things you notice if you drive northwest across the country is that the farther you travel the more frequent are the bullet holes in the road signs until in Alaska you discover that almost no signs or other public markers remain that have not been destroyed by multiple gunshot—unless they happen to be brand-new replacements. This is all the more striking if you happen to start from Pennsylvania, where there are more hunters and more deer and other game taken than in any other state; but it seems that they save their ammunition for the field and for the most part leave the signs alone. At any rate, from the beginning there had been lively speculation in Alaska as to what kind of target the aboveground pipe would make, what gun and load it would take to puncture it, how close you'd have to be, what would happen to anyone who got in the way of a jet of oil spurting at several hundred psi. The wounds discovered in preparation for start-up provided only partial answers: they had penetrated the pipe insulation, dented the pipe, but had not broken it. Repair was fairly easy: a half-circumference of pipe six or seven feet long, welded on top of the dent; replacement of the insulation.

The next step was breaking in the machinery and mastering its operation: power plant, vapor-recovery system, ballast-treatment plant, communications systems, and so on, at the terminal; trial runs at the pump stations, particularly of the pumps themselves and the Rolls-Royce turbines—jet engines in design, running slow—that power them. Although each station is an elaborate installation covering several acres, its heart is a set of three pumps housed to one side of the main pipeline and connected to it by a system of valves, manifolds, and pipes: one pump to move the oil at the limited flow intended in the first year or so of operation, 1.2 million b/d, two running simultaneously for the maximum for which the pipeline was designed, 2 million b/d, when production reaches that point; the third as relief and insurance for the other two. Starting one of the turbines is a ten-minute sequential procedure something like the countdown followed in firing a rocket, with an automatic shutdown if any step is performed incorrectly or out of sequence or one of the system's components functions improperly. To practice this procedure (and switching from one pump to another, cleaning and other mainte-

nance) the pumps and their connecting pipes were closed off
and filled with imported Arctic diesel fuel, turned on and run.
There was one disadvantage. The diesel, circulating, heated up
as the heavier crude would not in its transit, and after eight
hours the pumps and pipes had reached their limit and had to be
shut down. Each pump in the five stations that would go into
service at start-up was run in for 120 hours by this stop-and-go
method—not only for training but as a heavy test of the machinery to reveal, for instance, excessive vibration due to defective
parts, inadequate mounting, misalignment of the piping.

The final stage in the preparations came at the end of May,
with tests of oil-spill contingency plans arranged by the federal
monitors. In each of the three sections in which the line was
divided for operations, the monitors produced a scenario—size
and location of the break—and translated that into programmed
data to be fed into the computers in the control center at Valdez.
It was then up to the technicians there to interpret the data, act
on it in shutting down the line and determining where the break
had occurred, transmit instructions to the oil-spill crews. On the
spot, sheets of black plastic simulated the volume and flow of oil
on the ground—further matter for interpretation, choice, response, depending on gradients, the proximity of streams and
lakes. The crews, apparently, passed their tests.

After the Department of Transportation's certification of the
pipeline on June 16th, one more approval remained: that of the
APO. Since the beginning of April the Pipeline Office had been
pondering two fat volumes compiled and submitted by Alyeska
and known as the *Start-up Presentation/T.A.P.S.*: a rough-and-
ready listing of every discrepancy that had been identified
throughout the system, with description and location, matched
with documents showing both that the flaw had been corrected
and the means by which that had been done. On June 19th, a
Sunday, Jack Turner and his colleagues concluded that the job
was done—or that anyway what remained was no cause for
delay—and informed Alyeska that it could go ahead. The oil entered the line at Prudhoe Bay the next day, at 10 A.M. For the
purpose Alyeska flew a last planeload of newsmen up as the
switches were pressed to open valves, start pumps. A few
workers and supervisors stood around for pictures, pressed ears

to pipe to catch the rumbling, baritone throb of heavy oil moving in its metal vein. There was no ceremony, after nine and a half years the oil men had had their fill of ceremonial events; but it made dull copy.

After all the lessons in humility it had received in the course of those years, Alyeska moved the oil with extreme caution. Nitrogen, cooled to a liquid at —290° F., was pumped into the pipe first to form a barrier between air inside and the inflammable vapors of the oil—if touched by fire, the vapors will burn (or if confined, explode), but only in the presence of oxygen. Behind the nitrogen came a cleaning pig fitted with wheels to clank along the pipe walls so that teams of men walking beside it, watching for leaks, could hear and know where it was; and then the oil. The pumps would be run at low speeds, the pressure of the nitrogen would slow the oil to less than walking speed, around 1 mph on average, faster on the downslopes; the first oil would take about a month to travel the 800 miles to Valdez, compared with four or five days (depending on volume) once full operation was attained.

Within the first few days a couple of incidents enlivened this careful routine. Up on the Arctic plain a bulldozer operator backfilling the gas line that powers the four northernmost pump stations managed to break the pipe; in the south, near Pump Station 10, a fellow tried to drive his dump truck under a piece of elevated pipe with the dump up, and left a dent. Nothing serious, unannounced and barely reported; the oil kept moving. Then late in the afternoon of the Fourth of July, near Pump Station 8 a few miles south of Fairbanks, linewalkers noticed nitrogen gas seeping up through the gravel work pad at a point where the buried pipe makes a sharp bend down and up to resume its elevated mode. A little more serious, though the oil column was still several miles north, and for three days the valves were closed, the pumps stopped, while the cracked pipe was dug up, cut out, replaced. Alyeska inferred that an inexperienced crew had put liquid nitrogen into the line without vaporizing it and the cold had broken the pipe; what was not said was that at or near that point the pipeline builders, like clumsy dentists hammering an inlay into a tooth, had had extreme difficulty forcing the pipe into position for a hookup weld of two strings of pipe connecting the main line with the pump station—the two strings did not quite align.

On July 8th, the day after the oil got moving again, a disaster occurred at that same pump station. What happened, as it was reconstructed in the inquiries that followed, was that as the first of its three pumps received and moved the oil—given the short time since the previous incident it could only just have arrived, the pump could not have been run long—the maintenance team in the pump house noticed a build-up of junk in the strainer on the pump (the pumps here were the only ones on the line so equipped) and passed the word back to the control room, 300 feet away, to shut the pump down, switch on another pump, so they could take the strainer out and clean it. The first pump was stopped, its valves closed, and the firing sequence for the second pump and turbine was begun. At the last stage in this procedure something did not work and as the system was designed the new pump's discharge valve would not open. Rather than leave it at that while they found out what was wrong—it would have meant shutting down the whole pipeline—the man in charge of the control room elected to go back to the first pump, which he knew was working properly despite the build-up of sludge in the strainer. He initiated its start, opening a valve. Meanwhile, however, the crew in the pump house had opened up the pump and removed the strainer. Hence, when that first valve opened, a geyser of oil poured out from the 26-inch opening where the strainer had been, driven by the full pressure within the pipeline, which at that point was about 350 psi. The force was enough to blow the roof off the pump house, smashing lighting fixtures, breaking live wires, and a flash of electricity ignited the oil, nearly 300 barrels of it before the line could be shut down, and it burned for hours. One of the mechanics, Charles Lindsey, was killed; five others were injured, one seriously, and were flown over to the hospital at Fairbanks.

All the procedures, including the removal and cleaning of the strainer, were ones the crews had practiced repeatedly, but in this case there were a couple of things they did not do. The maintenance men, in the first place, were required to get a written work order before starting any job, and if they had, the supervisor in the control room would presumably have known that the pump was open. In addition, the first step in the start-up process was supposed to be for the control man to walk the 300 feet to the pump house and verify that all valves were in the proper position, something that is obvious at a glance; in the haste to avoid stopping the oil flow, he did not bother. In view of these rudimentary mistakes, the investigators' verdict of "human error" seems fair enough. Yet why, one wonders, had the designers provided no means of showing, in the control room, something so obvious as an open valve on a pump to be started? And why had their system been made to depend on the rigid observance of so many human procedures, including the nuisance of the written work order and the control man's physical inspection of the pump house a hundred yards off from his control board? If these peculiarities are indicative, it would seem that this first big accident will not be the last.

While the oil alone would have done considerable damage, it was the explosive burning of about 12,000 gallons of crude that destroyed the pumps and their housing. The fire-suppressant system had been designed on the premise that, whatever happened, the pumps would remain enclosed, an unusual situation in itself, since pipeline pumps in most other parts of the world need no such protection from weather or earthquakes: in a fire the system was to fill the pump house with an inert gas, Halon, which by excluding oxygen would quite immediately damp the fire—a useless provision if the roof blew off or a wall caved in; or, one would think, someone merely left a door open. At some other pump stations the Halon system was backed up by another spraying a blanket of foam, which probably would have done the job at Pump 8 but apparently was not provided.

The disaster did not in fact stop the line, but this time it took longer to get it started again. Replacing the pumps, turbines, and other machinery would take up to a year, at a cost of perhaps $50 million. In the meantime, it was possible to bypass Pump Station 8 (the mainline pipe was undamaged) and bring on Pump 9, forty miles south, planned for use later in the year when the system reached its first level of operation, 1.2 million b/d. Because of its location, however, the new pump station made it impossible to move the oil any faster than 800,000 b/d, a difference that for the oil companies collectively represents a loss of gross income amounting to about $2 billion—and for BP, the wiping out of the profit incentive for raising production that was written into its agreement with Sohio. The changes and the accompanying investigations took up ten days. Alyeska succeeded in restarting the line on July 18th. The one bright spot in this gloomy affair was that the pipeline control system, at least, had worked as intended. The system had transmitted the signs of trouble (a sudden, radical drop in line pressure) to the control center at Valdez, and the technicians there had interpreted them correctly, just as in the drills, and responded by closing the valves all along the line and shutting down the pumps; it took them the minimum time to do this, four minutes, as predicted.

The troubles were not over. On the very day the line was restarted, at Arco's Flow Station 1 at Prudhoe Bay (a plant which gathers the oil from all the Arco wells, processes it, and transmits it to the pipeline—like a pump station but on a much bigger scale), workers managed to leak 40 or 50 barrels of crude from one of the pumps, an error that sounds much like that at Pump 8; but since the flow station is outside the pipeline system, the accident had no effect on the movement of oil. The following day, twenty-three miles south of Prudhoe, a man driving a bulldozer knocked a two-inch vent off the top of a valve and a fountain of high-pressure oil spouted a hundred or more feet. (Cautiously, Alyeska had left the pipe ditch open at this point in case start-up showed the need for further repairs; the luckless driver had been backfilling the ditch.) Once more the system responded, closing valves on both sides of the break, but the volume of inflammable crude made it suicidal to bring in trucks or any other internal-combustion machines—there was nothing for it but to wait—four hours—till the fountains subsided and a man could get near enough to hammer a wooden plug into the vent; by then at least a thousand barrels of oil, maybe twice that, had sprayed across the landscape. The repair took eight hours. The oil was moving again on July 20th.

Sometime in the night of that same day one Larry D. Wertz—a young man living in the woods a few miles north of Fairbanks, working a little, on and off, at trapping, a mining claim—undertook to do something about the pipeline. With a couple of no less addled pals, he attached explosives to sixty feet or so of aboveground pipe and its attendant VSMs. They set off their charges and got out. The VSMs were bent, two or three sections of insulation were torn up, but the oil-filled pipe, though dented, was essentially unharmed. Astonishingly, it was apparently not till July 25th,* five days later, that anyone noticed what had happened or reported it to the state police; the troopers collected the trio over the next two days. On the first page of the massive operating plan it submitted to the APO, Alyeska had declared that "a daily ground and helicopter oil-spill monitoring effort" was among its intentions, at least for the first several months of operations. That was evidently not how it worked out; once again there was a large and crucial difference between intention and reality. Nothing but his own incompetence had prevented Larry Wertz from putting the pipeline out of business. One of the things we have got for the billions that built the aboveground pipe is a sitting target 425 miles long—and for more dangerous weapons, clearly, than guns and bullets.†

The litany continues. The oil by now was getting close to Valdez, about eighty miles north. On the day Alyeska reported this latest incident to the police, a crew at Pump Station 12, putting nitrogen into the line in anticipation of the oil, pumped it instead

* The second date is not certain but there was indeed a gap of several days between attempt and discovery, which Alyeska lamely tried to cover by saying that someone *had* noticed the ripped-up insulation but thought it meant only some new kind of repair. Ha! In general, Alyeska was extremely reticent about letting out details of how the bombing was done or why it failed, fearful of encouraging someone to try again and succeed; I shall emulate the company in this respect.

† This modest prediction was fulfilled several months after I wrote it. Around February 15, 1978, in the same area a few miles southeast of Fairbanks and accessible to public roads, an explosive blew a two-inch hole in the elevated pipe. From what one can learn of the circumstances, the operation was sophisticated as well as successful: the break was discovered not by Alyeska's control system or its inspectors but by a private pilot who chanced by in a small plane, by which time it was estimated that 8,000 barrels of oil had been spilled. This volume represents the contents of about eight miles of pipe and means that the leak went undetected for at least a quarter of an hour; suggesting that whoever was responsible knew enough to deactivate the automatic valves in the area or possibly the whole communications system. After its slow start, Alyeska managed to patch the hole and get the oil moving again within twenty-four hours. So far (April 1978), neither the perpetrators nor their methods have been uncovered.

into a 50,000-barrel surge tank (a safety feature meant to relieve the line of dangerous pressure, as for instance from the sudden closing of a downstream valve). Nothing happened to the pipe, but the tank puffed up like a roasted marshmallow (and took a month to rebuild). Two days later, as the troopers were bringing in Wertz's two chums and the oil column reached Thompson Pass, Alyeska hopefully announced that it would be in Valdez later the same day. A Department of Transportation inspector, however, keeping ahead of the oil, found a possibly defective weld at the point where the pipe enters the terminal and ordered that it be cut out and replaced. The oil was allowed to keep moving its last twenty miles, but slowly, while this work was hastily completed. Simultaneously at Pump Station 10 a crew had spilled a significant volume of the diesel used in breaking in the pumps, but the spill resulted only in one more humiliation for Alyeska and a mess to be cleaned up.

At one in the morning on Friday, July 29th, the oil at last reached its destination and across the Port in Valdez someone heard the news and turned on the fire whistle by way of welcome. The first tanker to load tied up two days later, on schedule, the leeway allowing the time needed to get enough crude into the storage tanks: appropriately, the 120,000 dwt *Arco Juneau*, one of five the company had ordered in the first go-ahead days of discovery. One more stoppage occurred the next day, for seven hours: mysteriously, at Pump 1 near Prudhoe Bay and at recently activated Pump 9, 550 miles south, the fire-control systems both went off, filling the pump houses with Halon (and in the north, a blanket of fire-suppressing foam); investigation uncovered no reasonable cause, and this odd coincidence was pronounced an accident, this time almost harmless—the one-time cost of running the fire system and replenishing its chemicals is only about $45,000, and there are not many other things on the line that can be bought for that. Thereafter, Alyeska men talked glumly of the unpredictable sensitivities designed into their system ("We've built in so many safety features," Bill Darch remarked, "that if a *fly* buzzes past these damn things they literally go out"); the fire system was one of the things they had in mind.

In any case, by now there was enough oil in the tanks to complete the loading of the *Juneau* despite the interruption, and late in the afternoon of August 1st the ship sailed for Arco's beautiful new refinery at Cherry Point, 1,315 miles to the south. Another tanker, Sohio's *Invincible,* had already taken her place at the loading arms. The night before, Alyeska had arranged a modest feed for the locals, shark meat and fried salmon, with kegs of beer, set out along the dusty main street of Valdez, across from the Pipeline Club and the town's one other bar. Predictably, whoever ordered the beer had underestimated. It ran out in an hour.

A few days later, I stopped by Bill Darch's office in Anchorage to chat about the start-up procedures—all that planning had been his intense responsibility, and there was probably no one else in the company so well qualified to talk about them. Alyeska since the day the oil first flowed had abandoned its "Cook's tours" in its redoubled anxieties about security, and if you played by their rules there was not much to do but talk—though as before there was nothing of importance on the line that a determined man could be prevented from seeing or finding out for himself. Meanwhile—I had decided against sailing south with the *Juneau,* it seemed it would be crammed full of reporters—I was waiting around for a chance at a tanker out of Valdez, all the careful, long-planned maritime schedules tangled by the urgency to move the oil out and the stop-and-go functioning of the line.

Seated behind his desk, back to the big windows looking across the flat Anchorage hinterland to the mountains, Dr. Darch was as quick and articulate as when I'd first met him a little over a year earlier, with the urbane and quite personal courtesy that much travel often confers on his countrymen. He seemed tired now, almost languid—and heavier, puffy in the face and around the eyes. Understandably, of course: the sheer stultifying mass of the start-up plans and approvals, the days out on the line seeing to the environmental rehabilitation work with its very small rewards in growing grass and sprouted willow; the training of the crews, standing by to react to and monitor the oil-spill tests; and finally the succession of accidents and shutdowns in their apparently limitless variety, when he had rushed to the airport and a plane so as to be present and see for himself. He did not seem disturbed by all this, anxious, but puzzled, yes, very much so: what more, exactly, could a man think of to do that had not been done?

"We talk about 'human error,'" he was saying, his voice giving the phrase a small, dismissive emphasis. "But what concerns *me* is, *how do I stop it happening?*" We had been talking particularly of the breakdown at Pump 8, the what—in some detail and accurately, I think—and now the why, and he worried around the question, finding no answers. "Consider a car going downhill," he continued slowly, shaping the analogy, "—with a sharp right-hand bend at the bottom. You've given a guy a car with a steering wheel, and with brakes, and you say to him, O.K., this is what you do when you come downhill, let's practice it, do it again—do it again. Then you turn him loose and he just doesn't put the brakes on or steer. *That's the problem.* How do you stop that?"

For me also it was the essential question. Maybe, I suggested, it was the manuals, they were asking too much of the manuals and the men who were supposed to follow them. He made a gesture of pulling a booklet from a hip pocket, opening it in his left hand, with his right turning dials, pressing buttons—the start-up of the pumps—in their ordered sequence; along the way from Cardiff to Anchorage he had done a lot of these jobs himself, that was how he did them. No, the manuals were O.K., they had gone over them, the National Transportation Safety Board had reviewed them again. (Later I heard they had objected, not that what was in the instruction books was inadequate but that the men left them protectively in office drawers; they were experienced men, they *knew* how to do the goddamn job and did not need to read it in a book.) Maybe so, I agreed, but I thought his British education might have given him an undue notion of what a contemporary American could take in through print—or of the discipline needed to translate an intricate written procedure into exacting action; the two, literacy in the full sense and discipline, *self*-discipline, are indissolubly connected.

It was not an answer but it was, I think, a start toward one, and perhaps he went on thinking about it. For now my time was about up and for him the multiple problems went on, still pressing. We talked a little more, courteously: American literacy or what passes for it among the members of the television generation, as by now they mostly are. It was, I said, a matter in which I take an immediate and practical interest: if they can't read what I write, I'm out of business.

On the morning of August 5th the *Juneau* entered Puget Sound with its 824,803 barrels of Prudhoe crude and was greeted by environmentalists in boats displaying antagonistic slogans (WE DON'T WANT IT); the little boats, of course, were powered by engines fueled by gasoline or diesel. Two weeks later the first gasoline and other products were delivered to Arco dealers around Seattle. Meanwhile, however, the incidents continued. At Pump Station 9, with a scary likeness to the Pump 8 disaster, a crew let a flash of oil out of a pump, only 30 barrels this time, but it was 23 minutes before the fire system went on. (After the accidents at pumps 1 and 9 someone had ordered that all the fire systems be set on manual, not automatic, and it was that long before anyone thought to act.) A few days later, a Mobil tanker delivering the first Prudhoe load to Los Angeles suffered an electrical fire in its engine room, surrounded by tanks of fuel, was towed to a shipyard, and went on burning for two hours; but the oil did not catch. Things were quieter after that. There was no further trouble until September 12th, when Alyeska discovered a leak of diesel fuel, 3,500 gallons, at Pump Station 3 on the northern downslope of the Brooks Range. This time there was no talk of "accident" or "human error." Someone had opened a valve and left it running.

Toward the end of August, with the line running again, ships loading, I flew back to Valdez to board the *Manhattan,* glorious of memory for its Arctic passage but now, after a year and a half of lay up and not much traffic before, a zombie among tankers.

Chapter 11

THE PRICE OF THE PIPE

> This contract works altogether different from any other con-
> tract I've ever been involved in. This contract is a cost-plus-
> negotiated-fixed-fee.
>
> —W. O. Thompson

"The original nine-hundred-million-dollar estimate," Ed Patton
was saying, "was what I call a horseback—what the industry
knew about constructing a major-diameter pipeline—and there
hadn't been any built yet, so that itself was an extrapolation—
they just arbitrarily assumed if they doubled that they'd have
this thing covered. . . . And then we made an annual update,
based on what changes there had been in the ground rules—and
it just kept marching up."

We were sitting at one end of a long, sumptuously polished
table, teak or walnut, in the windowless conference room adjacent to his office, sipping the ritual cups of coffee. Midafternoon
by now, we had been talking for nearly an hour: outside, the
February twilight would be wrapping Anchorage in its worn
gray velvet, but in here the light from the fluorescent panels in
the ceiling was discreetly cool and blue, neither bright nor dim,
and never changed.

I had spent the previous week with an Air Cavalry troop pa-
trolling a stretch of woods along the Tanana southeast of Fairbanks, in borrowed Arctic gear and carrying an oversized and
much-admired pack containing my own notions of the necessities
for survival; it had been minus forty or colder most of the time

and in the mornings, above the spindly pines frosted thick with glittering rime, helicopters fluttered toward the command post, dropping smoke bombs and firing noisy blanks. (In Anchorage in that odd winter, when I got back, the temperatures were not much below freezing, the snow thin and by day slushy on the sidewalks, and "We're having a heat wave," I remarked to the cabdriver on the way from the airport to 4th Avenue, for the small laugh. "What's this about everyone going to Hawaii? It's here!" A hundred and fifty miles down the Inlet one of the volcanoes, Mount St. Augustine, had erupted, covering the town and its thin snow with a fractional inch of gray pummice.) The joint exercise—the one American brigade of Arctic troops, another with ski training in mountains, Air Force units, Rangers flown in from Fort Benning, a scattering of sailors, Coast Guardsmen, Canadians—had been organized by the United States Readiness Command, USREDCOM in the lingo, a small group in Tampa whose job is to plan for military emergencies and, if they come, co-ordinate the several services. This year's theme had been the defense of pipeline and pump stations—represented by miles of snow fence, some disused buildings on an artillery range—against renegade semiguerrillas, simulated by the Rangers. "Finest soldiers in the world," they tell you, in the spirit of every proud unit, and were furious when a team of light-traveling Eskimo Scouts from Nome, officially National Guard, caught a Ranger platoon in their tents, in daylight, sleeping off a night of towing 200-pound sledgeloads of supplies and equipment on unfamiliar skis. Nevertheless, it seemed to me—in the confused intelligence, the alternating furor and numb inactivity of something like battle, and against its carefully chosen and programmed attackers—the exercise had succeeded: no vital part of the line had been captured or blown up.

By now, eight months along in my private pipeline project, I had reached a state about like that of an industrious college sophomore, less green than grass, conscious at least of volumes of ignorance and beginning to be able to ask some of the questions I would have to try to answer. The Alyeska public relations people in this case had seen fit not to monitor the interview, figuring, maybe, that their president could look out for himself. After six years of standing on the pivot point between the

owners' committees and the several branches of government, all
the other organized and discordant forces on both sides of the
line, answering the questions, repeating, insisting, Mr. Patton
had disciplined his face to what seemed like impassivity, and
behind it, emotions, thought. ("You have to have a guy that the
eight owner companies reach agreement on," he observed, when
I asked, getting started, how it felt, "needs to be in Purgatory,
anyway.") A broad rather flat face, florid under the silvery hair:
in a crowd, with the glasses, you might guess a school superin-
tendent in a well-off suburban district, except for the obdurate,
unmoving eyes.

At close range he talked quietly, reflectively, and as far as I
could see without reticence; with none of the bluff bonhomie
that had rubbed Fairbanks sensitivities in speeches when he first
came to Alaska, nor the dogged combativeness that ran through
the style and argument of the Alyeska press releases and seemed
to be personal and his own. So, affably, we had worked through
from the beginning, as he had experienced it: the jobs he had
done before, for Exxon, and the qualities the owners may have
seen there that persuaded the oil men to pick him for this one;
the way he had set about it, assembling his managers; the evolution of the project's cost estimates, with some of the reasons that,
as they went on rising, would be enlarged and elaborated—the
accelerating inflation since its first conception, the changing, arbitrary and sometimes conflicting design standards imposed by
those outside who could have no more than vicarious responsibility for the end result. Ten days earlier the company had announced its latest estimate, $7 billion, the only remaining uncer-
tainty the workers' performance in the final year—Moolin's
productivity plan; no allowance for the welding/radiography
scandal or the other troubles that were ahead.

It was cost that had begun to interest me. At the simplest, one
is merely curious to know what anyone, any organization, must
do to spend $7 billion. Except perhaps in the dying days of the
Weimar Republic, when marks were reckoned in trillions, it is a
task beyond human imagining. If you could enlarge your needs
and fancies to $1,000 a day, it would take you 19,178 years to
spend $7 billion, starting on the day of your birth. Left in the
local savings bank at current rates of interest, the sum would

earn about $1.25 million a day. If you were condemned to
dispose of it in the course of a biblical lifetime of threescore and
ten, it would be a labor of Sisyphus—$273,972 a day. None of
these comparisons is much use; the resultant numbers are all too
big to take in. I had been hoping the president would give me
some easy particulars to make the budgets intelligible; how
many cases of paper napkins or gallons of pop the project consumed; how much was pipeline, terminal, pump stations; how
the costs divided, at least, between labor and everything else.
These were not terms in which Mr. Patton would talk. The
undifferentiated cost lay like a lump of clay on the table. We
talked around it a little longer, but for this set of questions, it
seemed, I had come to the wrong place for answers.

There are reasons why oil men are generally closemouthed
about their companies' costs and most other details of their
finances. They are, I think, the same as for the proliferation of
corporate structures within and under the large oil companies
that operate across the whole spectrum of the business, from oil
wells at one end to gasoline pumps at the other. Most of these
entities look on paper like conventional corporations—officers,
stock, income, expenses and, presumably, profits, taxes, dividends—but have no existence apart from the parent company
which owns their stock and appoints their officers. Thus, for instance, the president of the Exxon Pipeline Company generally
occupies a quite modest slot within the hierarchy of Exxon,
U.S.A. which is in turn only one among the host of national subsidiaries of the Exxon Company, the old Rockefeller flagship still
incorporated in New Jersey and headquartered in New York.
The parent company's annual reports and other public documents go into some detail about its total finances but reveal
nothing, except in casual and general terms, about the operations
of Exxon Pipeline or any of its other subsidiaries. This arrangement is defensive: it limits the parent company's liabilities in
each operating area, including its liability for federal, state, and
other taxes; it also makes it difficult to discover costs and profits
at any stage in the flow of oil from wellhead to consumer, information that would be useful to a competitor in setting its own
corporate strategies, though for Alyeska, with its eight owners,

such knowledge would be rather widely spread, worth little effort to conceal.

In the case of a pipeline there is a particular reason for keeping mum about the details of its costs. Like a railroad or a truck line, it is a common carrier. Hence, by long-standing American law and practice, its rates are regulated like those of a public utility—what it charges for its services, which controls what the public pays for the oil it carries as well as the producers' profits, is subject to the approval of the Interstate Commerce Commission (or of its state counterparts for oil and oil products produced, transported and sold within a state rather than across state lines).* These rates are set so as to enable the pipeline owner to recover his entire construction cost over the life of the pipeline. In addition, under a 1941 court decision, he is allowed a reasonable annual profit on his basic investment. While the purpose of this ruling was to hold the profit to no more than 7 per cent, in practice it has meant that the rates are almost invariably set so as to earn that maximum percentage.

There are a couple of things to notice about this practice. For one, it assumes that the pipelines so regulated are truly distinct entities, dependent for their survival, like any business, on balancing income and expense so as to earn profit. This is true only in a rather limited and special sense; oil supply and demand are so volatile, changing sometimes from hour to hour, that Exxon Pipeline Company, say, will of necessity accept oil from whatever producers can keep its lines full and will in turn deliver it at the other end to any refiner with spare capacity to take up; the controlling reality from start to finish of this process is the one we have noticed before—once found and tapped, *the oil must flow*. In broader terms, however, the fact is that the pipeline company is rarely a freestanding structure supported by its profits: it is a subsidiary—it *subserves* the oil company that owns it. (Alyeska's nominal owner-operators are all subsidiaries of this kind; three of them—Mobil, Phillips, Union—are subsub-

---

* Since the price of American-produced oil has also been regulated, after a fashion, for many years, the effects of pipeline costs have been muted, but the relationship is nevertheless real and immediate, as we learn from the way these two theoretically separate issues interacted in the case of the Trans-Alaska Pipeline.

sidiaries, *Alaskan* corporations organized for the purpose.) In the reality of oil, that owner company can no more do without the services its pipelines provide (or turn to other means of transportation) than the public can. To stay in business, it must move its oil whether its pipelines are profitable or not. What the ICC does or can do about regulating pipeline rates is limited by these two necessities.

The other interesting point about this regulatory process is that if pipeline rates (or tariffs, to give them their proper name) are set so as to repay investment with profit, then the more a pipeline costs, the higher its tariff will be—and also the profit it will earn. The only real limit is the price of oil from which the tariff must come, whether that is set by the market, by politicians and government economists, or by a combination of the two, the present American system.

It is perhaps for this reason that the oil companies, when they have done so at all, have talked about their pipeline investments in undigestible lump sums. That is the form in which the ICC generally receives the evidence supporting a proposed tariff for a new pipeline, and in the past it has been normal for the commissioners to take the oil companies at their word. It would, then, be impolitic for the oil men to analyze such total costs; to do so would only raise questions, provoke argument.

Despite the unprecedented size and cost, the disparate style and interests of its owners, all this was as generally true for Alyeska as for other oil-company pipelines. It was not to be expected, therefore, that Mr. Patton, his managers, or the oil men behind them would have much to say about where the money went. As I brooded on this afterward, however, it seemed to me there was a perhaps simpler explanation, and that was simply that *he did not know.* And I remembered fragments of that conversation: groping, I had wondered how much of the $7 billion was labor, how the budget divided up among the pipeline sections, surely that information was fresh in computer printouts; and so on. Oh sure, the president had answered, somebody would know, but it wasn't worth the time or expense of running it off.

That is a puzzling admission from a company president. Styles differ, but in my observation if you are successfully engaged in any enterprise it is not a kind of question about which you can be ignorant or indifferent, whether you are the current Henry Ford or the owner of the neighborhood drugstore: you know what your stock is and where, and what it costs and how it's moving; your appetite for these minutiae has no limits. Maybe this is another of the ways in which oil is unlike other businesses, though here too there are exceptions, corporate heads who know their trade as intimately as the corner druggist knows his.

By the beginning of February 1976, when I had my talk with Mr. Patton, the state of Alaska was showing a keen interest in Alyeska's costs. With its hopes of building a new capital and other large ambitions, its by now habitual deficit budgets, the state was concerned for its cut of the oil money and just when that would come. To find out, it had made projections of the rate at which the Prudhoe field would be produced, of total pipeline costs and the resultant tariffs, of the cost of delivering the oil by tanker to whatever markets it was destined for, and the price it would fetch at the refineries once it got there; since each of these factors is variable, the computers had generated dozens of different estimates of state income year by year, depending on the assumptions used. All had one thing in common: the more the pipeline finally cost, the less the state would get. Its 12½ per cent royalty is based on what is called the wellhead price of the oil—the selling price at the refineries less all the costs of getting it there, of which the pipeline tariff would be by far the biggest; and the tariff would be determined mainly by the cost of building the pipeline. From the beginning the state thus had a general interest in keeping that cost down, discouraging avoidable expense. Now, with all the talk of waste, thievery, and other forms of inefficiency in the management of the project, it had a more specific interest: to demonstrate that a part of that final bill had been needlessly incurred and should not, therefore, figure in the tariff—an unprecedented argument in the decorous proceedings of the ICC but one that the state, an interested party, seemed to be preparing to make. To do so, of course, it would have to find out a great deal about just how Alyeska was spending its money.

Perhaps, then, there was more than gentlemanly indifference to the inability of Alyeska's managers to talk about their costs: a *willful* ignorance. What you don't know you can't tell; and it cannot hurt you.

Our own interest in finding out how the money was spent and whether wisely or not is only a little less immediate than the state's. The price of the pipe does not directly affect our personal incomes. It *does* affect what we pay for oil and therefore the cost of everything else that depends on oil. We too are interested parties.

Back in the days when the oil companies were making their basic decisions about what to do with their discoveries at Prudhoe Bay, U.S.-produced oil of comparable quality was selling in Los Angeles, the nearest big market, for around $3 a barrel. In other words, the 9.6 billion barrels they cautiously figured they could produce from the field would bring in about $30 billion. This was the bankroll out of which all their costs would have to be paid—early exploration, lease bonuses, field development, transportation, taxes; and of course profit. The early studies from which the pipeline system was decided—Ed Patton's horseback estimates, apparently—yielded a combined pipeline tariff and tanker cost of about 60¢ per barrel, say $6 billion over the life of the field: enough, that is, to cover pipeline construction and interest, operation, and, even if all the oil was to be sold within the United States, an entire fleet of American-built tankers, as required by national law, to haul it to American ports. A third or half that amount would take care of developing the field—the hundreds of wells to be drilled, the gathering lines, flow stations, headquarters camps—and recovering the $900 million in lease bonuses and the millions in exploration costs. When these and all other expenses are taken out, including the state's royalty cut, you are still left with a very large slice of the original pie, both proportionately and in money—$18 or $20 billion, say. Moreover, disposing of the crude and profiting from it is only the beginning. At the prices prevailing in the early 70s, the oil refined and delivered to the gas stations would be worth altogether close to $150 billion as it came out of the pumps.

If you were an oil man contemplating those early projections of expense, income, and profit, you would, I think, figure there was no way you could lose. You would be like a man with a million dollars in the bank and the plans for a hundred-thousand-dollar house in his hands: you might haggle over details, you might make a lot of little mistakes in the course of construction and even some big ones, but none of this would be serious enough to make a great dent in your bank account; or so you would think. The oil men differed from the house builder in one important respect: while most of their costs increased drastically and the pipeline itself to eight times the original estimate, their capital also was on the rise, though not quite so fast. The same force that was inflating their costs along with other American costs—the price of imported oil—was multiplying the value of the Prudhoe crude, eventually by about four and a half times. At the end, although the percentages were less rosy than they had been, the actual money to be realized looked a great deal better. In addition, however these things turned out exactly, there was always, from start to finish, the 26 trillion cubic feet of gas in the reservoir to sweeten the percentages, to be sold outright to utilities that, with their own assurances about costs and profits, would provide their own transportation, with no trouble and little expense to the oil companies—gravy, as historically natural gas seems always to have been to the oil business; but not just yet, not indeed until such time as gas should be allowed to rise to something approaching the equivalent cost of oil and coal.

None of these fortunate projections, of course, could have any meaning so long as the crude stayed in the ground. Moreover, the oil business is what the economists call capital-intensive, meaning that it requires large volumes of money in order to make money—money that these days is of necessity mostly borrowed and goes on costing whatever the borrower does with it. Time, once again, was money: to recover their early costs, cover their current ones, generate the capital to hunt new Prudhoe Bays, the owners' most urgent necessity was to get the oil moving as quickly as possible. Despite numerous gestures toward cost-effectiveness, the actual cost of doing so was secondary to this fundamental motive.

It is a motive that accounts for much that is otherwise odd in the way Alyeska managed the job, with decisive effects on what it cost. The basic system of construction contracting, for instance. Our millionaire with his house plans would most likely take them to a contractor or perhaps a string of subcontractors—carpenters, masons, plumbers, and so on—and get bids on the work. Once these were agreed on, they would form the substance of contracts subject to mutual conditions—in essence that the contractors would do the work according to plan and the owner pay the specified price, with penalties on both sides if either failed to perform. If the job were a complicated one, the contract would probably provide means of dealing with changes in costs or specifications; if time were important, it might set a deadline for completing the job, with a bonus for coming in early and penalties for delay. All this is so usual in construction, whether the job is a spare room added to a house or a multimillion-dollar office building, that I remind the reader of the principles only because Alyeska's approach was entirely different.

Alyeska's arrangements with its contractors were among the closely held secrets, but you could not be around the job for long without picking up at least the general idea. Bill Thompson, for instance, the state field surveillance officer for Section 1 of the pipeline: before that and retirement he had put in more than thirty years in Alaska with the Corps of Engineers, a lot of that time as a supervisor and inspector of contractors' work. "The way this contract works," Bill said when I asked about Alyeska's system, "is altogether different from any other contract I've ever been involved in and I've been involved in probably a couple of thousand altogether, with the Corps. This contract is a cost-plus-negotiated-fixed-fee—what the fixed fee is I don't know, ten per cent, twelve per cent, or what. It does not stipulate any cost whatsoever." This was the arrangement for the execution contractors, those with general responsibility for the five sections of the pipeline. That responsibility was not the usual one of performing work according to a detailed plan at a stipulated price. Rather, Alyeska undertook centralized control of each contractor's performance, specifying each phase in work orders that apparently set schedules and budgeted the crews and man-hours to be used in meeting them; and supplied the camps and food, the equipment and maintenance, it calculated would sustain those specifications. In return, the contractor had only to submit invoices showing the man-hours expended under the Alyeska work orders and he was reimbursed for his labor costs. His profit on the job was a percentage on top of the cost of labor—10 per cent for execution contractors, as it appears from other evidence; plus a somewhat smaller percentage for management and other overhead. The profit, while not enormous, was clear.

The contractors' situation was thus very much like that of Alyeska's oil-company owners: costs and profit assured, they could not lose. It is evident also that, so controlled and protected, they had no strong motive for practicing economy. Alyeska attempted to make up this weakness by close attention to its contractors' billings. By the summer of 1976 in Bill Thompson's Section 1, for instance, seven or eight Alyeska auditors were on duty, not only examining the records of the execution contractor, Morrison-Knudsen, but going around to the men on the crews to ask if they'd really done the work and worked the hours the contractor claimed. There were also a couple of modest incentives. In theory the contractor redid at his own expense any work judged substandard by Alyeska's Quality Control men or the government surveillance officers; but it does not seem that the auditors were well equipped to distinguish new work from work that was being redone. And at least toward the end of the project there were bonuses for contractors that met or exceeded their schedules, though I do not know that any was ever earned or paid.

Defensively and late, Alyeska advanced reasons for these makeshifts: essentially, that its plans were so continually changed by government directives that no contractor could prudently rely on them in preparing a fixed bid. Even without that limitation, however, it appears that the plans were still incomplete when the time came to let contracts and start construction: they were not in a biddable state. Once again, schedule, not cost, was the primary motive.

That point was underlined by Alyeska's arrangements about construction equipment. It will be recalled that originally the execution contractors and their subcontractors brought with them about one third of the 15,000 pieces of equipment required, while Alyeska early in the project determined to buy or lease the rest, provide all maintenance, and centralize the assignment and scheduling of the equipment as it did the entire labor force; the final total of machines on the job—everything from light pickups and portable heaters and generators to giant cranes—was about 18,000, valued in the Alyeska books at $800 million,* of which half seems to have been purchase price or leases, the rest upkeep, repairs, and a substantial inventory of spare parts. Although there seemed to be compelling reasons, as we noticed at the time, this was a critical decision for the way the job would be run and what it would cost. Except for a small number of managers, a major contractor's work force comes and goes; his chief asset is the equipment itself, whether bought or leased. Since this is also a sizable fixed cost, his ability to work at a profit depends on how constantly he can manage to utilize his machines, and he thus has the strongest possible motive for scheduling them efficiently and keeping them in good working order. It is evident that Alyeska's attempt at central control precluded this important incentive. In theory Alyeska's representatives in the work camps had at least some authority to make the kinds of day-by-day adjustments in the scheduling of crews and equipment that are essential to efficient management of construction. In practice, partly because any change imposed by new conditions in the field might affect compliance with the Stipulations and require the approval of the two surveillance offices, it seems that such decisions were most often passed up through the supervisory chain and were made in Fairbanks. What this cumbersome process frequently meant was several days of delay during which crews and equipment either sat idle or continued with work that would later have to be redone.

There were recurring efforts at cutting costs, often centered on camp maintenance and food: "indirect costs" in the Alyeska jargon (as opposed to those that contributed directly to progress on the line) and therefore a continuing annoyance. What it sounds like is that one of the oil-company panjandrums would visit a camp, notice that the men were emptying freezers full of ice cream by way of after-dinner sustenance, and complain, and

---

* Alyeska evidently wound up owning whatever machines the contractors had brought with them. I am not sure just when this transfer took place, but at the end of the job Alyeska contracted with Morrison-Knudsen to renovate and auction off all that were not totally worn out. The three years were tough on equipment, but continuing inflation gave some advantage to fixing the machines up instead of junking then.

then, for a time, the ice cream stopped; or read a cheery story in the papers about the welders' lunchtime steaks, the evening cookouts and lobster tails, with similar results. These petty, stop-and-go economies remind us of two peculiarities of the pipeline project that we have noticed repeatedly in other connections: its multiple layers of management—the oil-company committees, Alyeska, Bechtel (on and off), down to the execution contractors and their numerous subcontractors doing the actual work—with their nearly infinite capacity for conflicting and contradictory directives; and the divergent interests of the oil companies themselves. An anonymous international giant like Exxon would almost necessarily take a different view of what was costly and how to deal with it than would a comparatively small and bustling newcomer such as Amerada Hess.

Alyeska's contracts with its three caterers were apparently like the others in providing reimbursement for actual costs, plus an agreed sum for every meal served, to cover overhead and profit; purchasing of food and other supplies was under Alyeska's central control. Here again, the system seems designed to eliminate any incentive the contractors might have had for efficient use of labor and supplies; their costs and profits were guaranteed. Alyeska's attempts at controlling the catering costs, stirred up by complaints from the oil-company owners and their auditors, centered on the meal counts for which it was billed. Often this derived simply from an inventory of occupied beds in the camp, multiplied by three. When the auditors tried to verify these figures—for example, by counting heads as they came through the food lines or checking the bed counts against time sheets showing how many were actually working in the camp—they rarely succeeded; or they discovered curious anomalies, as that a lot of the men apparently never ate the breakfasts that were being paid for, or when hours for dinner and breakfast coincided (to accommodate those, such as oilers for heavy equipment and some camp maintenance people, who worked at night), the company was being billed for both—or maybe some of the hungry ones were eating two meals at once.

Alyeska made repeated gestures of reform—more auditors, more checking and rechecking of meal counts—but left the system itself untouched. Since these palliatives went unpublicized, it is hard to tell for whose benefit they were introduced, unless to placate the owners' management committees. In any case, the anomalies persisted: the bus driver, for instance, provided day after day with sixty sack lunches for a crew of twenty and, when he complained of the waste, told to mind his own business. (He solved the problem by donating the food to a Fairbanks rescue mission to feed derelicts.) Evidently someone had realized that since food and the labor of preparing it were provided and Alyeska had no way of matching meal counts with the numbers of people actually employed on any given day, the extra lunches were pure profit; perhaps also the system was so clumsy and unresponsive that it was easier to leave the order wrong than to try to change it; safer too—to correct it would be to admit an error in the first place. The particular case of the bus driver and his spare lunches is one that got into the open where it could be checked (by Richard Fineberg, who went to the rescue mission and asked). The contracting system, however, was such that it must surely have generated many similar instances.

As an old-time construction stiff, Bill Thompson was scandalized by the disproportion that seemed to be built into the system: on the one hand, the rigid pursuit of small controls and economies—the auditors' head counts and spot interviews with men coming off the work sites, their futile efforts to reconcile inventory records of nuts and bolts with the stock actually in the warehouses and camps (inventory control and purchasing, another set of problems never solved); on the other hand, except by a kind of head-on collision, will hitting at reality, the inability to anticipate and deal with difficulties such as those encountered at Thompson Pass or in the preparation of the Valdez terminal site, which clearly had major effects on the total cost of the project. "My experience in the last year," he said, summing it up—this by now was the end of June 1976, the project was rounding the final turn toward completion—"my experience is that they've been, you might say, nickel-and-dime conscious but the dollars didn't mean a damn thing. They've harped on little incidental things and made big issues out of saving money here and there—and then turned right around and something that was a million dollars, that was *insignificant*, that was a *happening*, that was a *fact of life*, forget it!" We stood on the first bench up from the road, near the bottom of the descent from Thompson Pass. Above us on the terrible grade the earth movers churned at the deep cut, thirty feet down by now, it looked, but they still had not reached the predicted bedrock in which to anchor the pipe.

He seemed to have arrived at pretty much the same conclusion I was reaching. The problems had to do not with the contractors and their men, how they carried out the work actually set before them, but with the stiff hand of Alyeska and its management, holding them tight. "There's all kinds of equipment setting idle," Bill continued, "day in and day out. They took all the responsibility and initiative away from the contractor, didn't matter if he did it ass backwards, he got paid for it and made his overhead. The jobs that I've worked on, even on cost-plus—but the contractor had an initiative in those jobs; if you saw that he was padding the payroll, the equipment, you kept track of that and that was deducted. This job it isn't, the equipment belongs to Alyeska, they *give* it to the contractor, and as far as even trying to maintain and keep that equipment running, the contractor is strictly dependent on Alyeska for every damn little part and nickel-and-dime part and grease and what have you. . . . On a normal construction job, that contractor who's working for hard money, he's going to have that stuff there when he needs it. . . ."

Hard money: that was the unadmitted reality on this job. The changes in procedures that came down from Fairbanks and Anchorage—the attentive auditors, Frank Moolin's tough talk about productivity—naturally had little effect. And they were devised by the same people, in the same set of mind, as those who had invented the system in the first place.

Even in this brief overview it seems obvious that saving money was not the prime purpose of Alyeska's contract system but rather completing the project as rapidly as possible. It was in fact the most economical or "cost-effective" strategy to adopt: defensively, by completing the work and bringing the fixed costs to a stop; aggressively, by producing the field and turning the oil into income without further delay. The state on the other hand would argue—with passion, much circumstantial evidence, and its power of subpoena—that Alyeska's managerial style was not purposeful at all but incompetent and that a big part of the cost incurred was needless and should not figure in the pipeline tariff or be subtracted from its own oil royalties; an argument of some complexity which we shall take up presently, in its place.

Our immediate interest in Alyeska's contracts is much simpler. It was not a very good system for predicting costs, for controlling them while they were happening, or for sorting them out afterward. We do not know, that is, exactly where the money went because Alyeska itself does not know. From odds and ends of information, from inadvertent admissions, we can, however, form some pretty good guesses.

The only one of Alyeska's published estimates to include any sort of breakdown was one in October 1974, the first, Ed Patton thought, to reflect the actual conditions under which the project would operate. The breakdown does not in fact add up, but the total at that point was given with some exactitude as \$5.982 billion. Of this, \$2.636 billion was attributed to labor, with another half billion for contingencies, most of which must have been to the same account. The pipe itself, nearly all delivered and paid for, came out to \$150 million,* while the cost of the supports for the aboveground pipe was given as \$500 million; both figures check with other sources. On the other hand, the estimate allowed \$1.5 billion for construction equipment, nearly four times what Frank Moolin thought his gear was worth and double the (probably) inflated figure Alyeska published at the end of the job when it was preparing to sell it off; and it included nearly \$2 billion for work by Alyeska, Bechtel and Fluor, as distinct from the execution contractors, that was apparently distributed among all the other accounts.

At the other end, in a speech to the Town Hall of California in May 1977, Ed Patton argued that most of Alyeska's costs either

---

* It will be recalled (Chapter 3) that the original cost of the pipe was \$100 million. Alyeska by 1974, I think, had bought or at least ordered some additional pipe to cover rerouting that had added eleven miles to the line's original length (and would go on ordering to make good the pipe chewed up in the welding repair program). The main difference, however, must be the failed experiment in epoxy-coating the pipe as a rust preventative (Chapter 5); plus, perhaps, storage costs since 1969 added to the capital cost by the same peculiar accounting that doubed the book value of construction equipment.

were due to forces beyond managerial control or were imposed by the Stipulations. The biggest was $3.2 billion, attributed to inflation during the project's four-year delay. In the absence of any explanation, I did my own arithmetic, based on Alyeska's 1972 estimate of $3 billion and the U.S. inflation rates prevailing from 1972 to 1976; assuming that the 1972 estimate was meaningful, the extra for inflation would have been about $1 billion. Some of the other costs given in this speech seem, however, to have some basis in reality.

My guesses derive from these two extremes and from a lot of bits and pieces in between.

For the pipeline itself, it appears that the total cost of construction falls within a lump sum of $3.5 billion. This was the figure tossed off by Frank Moolin at project's end when, with several associates, he formed a construction consulting firm in Anchorage for the purpose of sharing his managerial experience with late-comers—such as a new pipeline to carry Prudhoe gas, which by then was somewhat more tangible than a daydream in the committee rooms of Congress and the Canadian Parliament. ("It's a tough act to follow," he had answered, noncommittally, when at one point I wondered what he thought of turning his hand to next; Frank Moolin & Associates, Inc., was to be the new act, the capital provided by an ambitious Fairbanks construction company.) Of this total, the biggest item that can be identified (from Ed Patton's speech) is the $1.1 billion cost of the 425 miles of aboveground pipe. This is plausible and may well be too little: at $13,000 per set, every 60 feet, the materials alone—VSMs, horizontal supports, and so on—come to about half a billion, which was also the figure given in Alyeska's October 1974 estimate; if the drilling of the nearly 80,000 holes for VSMs were at all proportionate to the drilling cost at Prudhoe Bay (where the oil companies were spending $1.5–$3 million per well), Alyeska may well have put a billion dollars into drilling alone. As it is, the $600 million we can infer for drilling, installation, and insulation (not, I think, including the welding of the pipe itself) seems comparatively modest.

The pipeline camps were another sizable cost that we can single out. Initial construction apparently amounted to about $10 million apiece, averaging the seven northern camps cheaply run

up in 1970; as with the construction equipment, camp maintenance seems to have been added into the capital account—
rebuilding of heating and fuel systems after the several big leaks
and fires, three complete renovations of the sewage plants Gil
Zemansky found so noxious—and effectively doubled the original
investment. On this basis, I think $400 million probable for the
twenty pipeline camps. (At one point, when there was a prospect of selling the camps for use in a proposed trans-Alaska gas
line, Alyeska was talking $750 million; this seems so high that it
must have been either a pushcart bargaining figure or a result of
confused accounting that lumped in the operation of the camps,
perhaps even their catering, with the original construction
costs.)

Finally, on the basis of Alyeska's 1974 estimate, the pipe itself
came to $150 million; and the welding repairs may have
amounted to as much as $100 million.

These pipeline costs add up to $1.75 billion, exactly half
Moolin's total of $3.5 billion. That is, all other pipeline construc-
tion—site preparation, and work pad, ditching, taping, welding,
backfilling—would have to come out of the remaining $1.75 bil-
lion, and this seems not unreasonable. Welding alone—in the
field, in the double-joint shops at Valdez and Fairbanks—probably represents about half of the balance.

In the course of construction, the estimates for the Valdez terminal more than doubled, to $1 billion by the summer of 1976
and a perhaps final $1.1 billion given in an Alyeska public rela-
tions piece in April 1977. The construction camp there went on
building, slowly, almost to the end of the project; from references to the size of the work force, 110 men spread over 22
months, I have estimated the cost at $36 million. The state,
which attributed the cost of terminal site preparation to careless
early planning, gave that a value of $183 million. In his speech of
complaint, Ed Patton valued at $175 million the vapor recovery
system built into the terminal power plant—his complaint was
that this was the price of meeting the air pollution standard of
the Stipulations, though the same system also produced the inert
gases pumped into the storage tanks to insulate the crude against
dangerous contact with oxygen; in the tank design adopted, an
essential safety feature, I would say, by any reasonable standard.

In the same speech Mr. Patton gave the cost of the ballast-water-treatment plant as $140 million. Adding up these identifiable items, we are left with about half the total cost, a little less than $600 million, to cover everything else; which seems a reasonable proportion, as with the pipeline, though one would like to know more.

The twelve pump stations constitute the third big element in the complete system, for which in July 1976 Alyeska was quoting a total bill of $800 million. The rebuilding cost circulated after the disaster at Pump Station 8, $50 million, sounds like the rule-of-thumb estimate for the whole job—say $600 million for actual construction. Each pump station was supported by its own temporary work camp, smaller than those for the pipeline, which I estimate at $7 million apiece. With a guess of $5 million each for maintenance, we have a camp total of $144 million, leaving only $56 million of the $800 million total for odds and ends.

Adding up the three large divisions of the pipeline system, we get the following:

pipeline	$3.5 billion
terminal	1.1 billion
pump stations	.8 billion
	$5.4 billion

In Alyeska's organization, several operations that contributed to all three divisions of the project were run like subsidiaries, their costs kept separate. Assembling the construction equipment and keeping it up seem to have been handled in this way, at a cost of $800 billion, as we noted earlier. The same approach was evidently used for "Logistics"—the department responsible for collecting the necessary equipment and supplies in the Lower 48 (with Seattle as the chief staging area), delivering them to Alaska, and distributing them among the camps. For another of its public relations efforts, Alyeska rounded out these costs—barge, plane, and truck transportation—to $1.5 billion. When the 361-mile haul road north of the Yukon was completed in the fall of 1974 (to the more costly standard of a state secondary road rather than the temporary work road originally planned), its cost was put at $180 million; since Alyeska valued it at $300 million in later references, I infer that it spent a further $120 million on the rebuilding and general maintenance that were needed in the two remaining years of the project.

These departmental charges total $2.6 billion. When we add them to the costs of pipeline, terminal, and pump stations, we get a grand total of $8 billion. This was the working figure the state used in forecasting its oil revenues and it was generally confirmed in the evidence the oil companies submitted later to the ICC in defense of their pipeline tariffs (see below). It compares with the $7.7 billion given in Alyeska's latest published estimate in July 1976. Evidently between then and start-up almost a year later the total crept upward by another $300 million, much of it in weld repairs.

Such as it is, this total, $8 billion, is about as accurate as we are going to have for the construction of the pipeline system—essentially, I think, what Alyeska paid out against the billings of its contractors and suppliers. It is not, however, the entire cost. For example, in its October 1974 estimate, Alyeska allowed $1.776 billion for work it performed directly and for management services by Bechtel (for the pipeline) and Fluor (for the terminal and pump stations). This total cannot be reconciled with the other breakdowns given in that estimate. Unlike the service departments, it appears that most of these general-management costs were distributed among the project's three large divisions. There may, however, be a sizable part, perhaps hundreds of millions of dollars, that should be allowed for design performed by the oil companies or Alyeska and for the salaries and other costs of the company's Anchorage headquarters staff, from Ed Patton on down.

From 1974 on, as parts of the system (such as the work camps) were completed to the point of assessable value, they became subject to state and borough real estate taxes. As percentages, these were pinpricks in the total cost, 2 per cent altogether, but they were payable in every fiscal year and added up: $211 million between 1974 and 1976. I am not sure whether this particular set of taxes figured in the construction total or not; it seems more likely that they would have been deferred as charges against pipeline operating cost, where they would have had some effect on income subject to state and federal income taxes.

By far the biggest cost of building the pipeline was never mentioned in the arguments that preceded construction and continued off and on to the end: interest; and it was touched on only with great delicacy in the new debates set off in 1977 by the ICC tariff-setting procedures. As you know if you have ever borrowed money to buy a house, the interest you pay over the life of, say, a twenty-year mortgage will be more or less equal to the amount borrowed, depending, of course, on the actual interest rate; the interest paid over the life of the mortgage, in other words, effectively doubles the cost of owning the house. The pipeline, like anyone's surburban quarter acre of the Good Life, is a very handsomely mortgaged piece of property.

As we have noticed elsewhere, one of Sohio's attractions for BP was that it provided a convenient base for borrowing in the U.S. money market. The two companies organized subsidiaries whose function was to issue bonds or notes—IOUs on the profits from the oil: Sohio-BP Trans-Alaska Pipeline Finance, Inc., and BP North American Finance Corp. Between them in the course of construction—through these paper corporations, their pipeline subsidiaries, or in their own right—Sohio and BP maneuvered eight major offerings, amounting to $4.415 billion; including the biggest ever, completed in November 1975, $1.75 billion repayable over eighteen to twenty-two years at the stiff rate of 10⅝ per cent interest per year. Where notes or bonds were the means, they were sold through coalitions of brokers led by Morgan Stanley & Company; direct borrowing were spread through a comparable network of major banks among which David Rockefeller's Chase Manhattan took the leading role. As Sohio's chief executive Charles Spahr often and plaintively remarked, his company had borrowed, one way or other, something more than its entire net worth to hold up its end of the pipeline and supply BP with the funds to develop its half of the Prudhoe field.

The other partners followed a similar strategy. By start-up, their collective pipeline debt (according to their ICC submissions) amounted to $8.03448 billion, the entire cost of construction. Not all the terms were as long or the interest rates as steep as in the Sohio-BP 1975 record-breaker, but if the total debt were paid off over 25 years at interest of 8.5 per cent, the cost of using the money for that time adds up to more than $9 billion, just like a mortgaged house; as I said. The terms used in my calculation are, I think, moderate and representative.

The oil companies' money strategies raise a number of abstruse and difficult questions, among them the capacity of private industry, aided by the entire national financial structure, to support a project of the size of Prudhoe Bay and its pipeline; wherever energy is the issue, these questions are going to become bigger and harder, less soluble. My immediate concern, however, is again much simpler. The real cost of the pipeline was never $1 billion or $3 billion or $7.7 billion but double that: finally around $16 billion, all of which must ultimately be paid in the pipeline tariff and therefore in the price of oil, energy, and everything else. Naïvely or deceivedly we all—Congress, the reporters, the public—kept our eyes on the mere construction cost and ignored the interest. In so doing, we missed half the story.

The interest has two similar but distinct effects on the cost of the pipeline, under the precedents of the ICC. That which was paid between start and finish of construction, nearly $2 billion, is termed "capitalized" by accountants and is added onto its capital cost: the real construction cost, then, is not $8 billion but nearly $10 billion. As we shall see, that makes an important difference in the pipeline tariff: a portion of the $10 billion must be paid off in each barrel of oil pumped through the line, but the total also forms the base on which the pipeline's allowed 7 per cent annual profit is calculated; the extra $2 billion in amortized interest is thus worth $140 million a year in extra profit and, over the life of the pipeline, more than pays for itself. The remaining interest, close to $8 billion, is likewise spread over barrels of oil but has no direct effect on profit.

There was one other large but neglected factor in the price of the pipe: tankers. To move the oil through the line to Valdez is nothing, by itself; what counts, for profit and national utility, is the means of delivering it to its markets. In the present state of the American merchant marine—few ships, small and obsolete—that was a big order. Practically speaking, it meant building a fleet of 30 supertankers and having them in service by start-up, with more to come as the flow of oil increased. And what that means is another large chunk added to the cost of the system, at least $1.5 billion* in outright construction, and as much again in interest.

The enormous difference between the cost of the tankers required for the Alaskan trade and the cost of the pipeline—say $3.5 billion altogether against $19 billion—is worth attending to: it means on the face of it that you can move the oil by tanker the 2,000-odd miles from Valdez to Los Angeles for about *one sixth* what it costs you to take it the 800 pipeline miles from Prudhoe to Valdez (in fact, rather better than that, as we shall see when we consider tanker tariffs in the following chapter). The cost of this pipeline was extraordinary and unprecedented. Nevertheless, the essential difference is true: where there is a choice— or, as in Alaska, the possibility of combining tankers with pipeline so as to minimize the pipeline distance the oil must be transported—that choice is self-evident. For the volume of oil to be moved, the initial cost of the tankers is decisively lower, and since the recovery of that capital is by far the biggest item in the day-by-day operating charges, the ships will continue to cost less as long as they operate. Moreover, unlike a pipeline, the tankers are not confined to a fixed right-of-way but can be steered wherever the seas float them, and if for a time they lack oil or oil products to earn their keep, they can as readily switch to hauling other bulk cargo, such as chemicals or, recently, grain, with no more conversion than a thorough washing of their tanks. From this it follows that the four-year debate over TAPS and its alternatives—the several suggested Canadian pipeline routes, each a minimum of 3,000 miles in length—was essentially fatuous and futile and would have been more obviously so at the time if any of the parties had based their arguments on anything resembling

* In 1973 Rogers Morton, still Secretary of the Interior, suggested $1.7 billion for the tankers, the only time, so far as I know, the subject was alluded to in the public debates; this rule-of-thumb estimate evidently came from the oil companies and was repeated by them without change or qualification in other contexts. We shall look again, in Chapter 12, at the arithmetic of tankers, but it is a reasonable and perhaps generous figure. It would provide enough tonnage to carry the oil on a continuous, efficient schedule, to Los Angeles, with time out for North Pacific winter storms and maintenance; and some leeway, I think, for longer voyages, as to Japan. Most of the necessary ships were contracted for early and were therefore not much affected by inflation.

the real and final costs. Or so it would be if the fundamental premise is correct—that the system will in fact deliver the oil to markets where it can be used. That question, as we shall see in our conclusion, is still open.

When we have ascertained the entire cost of the transportation system, the pipeline with its facilities plus the tankers at the other end, our knowledge is still incomplete. The field itself must of course be developed, produced, and that too costs money which, while it has no direct bearing on the pipeline tariffs, will ultimately be repaid from the same capacious pot: the sale of the crude.

To take first things first, development involves drilling enough wells to produce the field efficiently, once its size is reliably estimated, without extracting the oil so fast as to leave an excessive amount irrecoverable in the ground as in the bad old days in Pennsylvania and Texas: the principle of "maximum efficient recovery," or MER (see Chapter 3). As originally planned,* the pipeline was to be run at about 800,000 b/d in the first months following start-up, rising to 1.2 million b/d by the end of 1977 and 1.5 or 1.6 million b/d within about one more year. That rate represents the MER for the main Prudhoe reservoir in the geological formation named Sadlerochit, estimated as being capable of producing 9.6 billion barrels of oil in its lifetime. Raising production to the pipeline's maximum capacity, 2 million b/d when all twelve pump stations are working, depends on what the oil companies decide to do about the several other oil-bearing formations in the area (see page 24), a question about which they have so far refused to be particular. This schedule probably reflects calculations of the MER over the life of the field and possibly estimates of growth in the markets for which the oil is destined, but the main consideration, I believe, was simply an orderly progression in the use of men and equipment for drilling the wells, the ordering and installation of pumps and other heavy machinery needed to process the oil as produced and move it south.

* The Prudhoe development schedule was presumably adjusted to the reduced pipeline capacity resulting from the destruction of Pump Station 8. Probably, therefore, the peak levels of production will not be reached until about a year after the dates set in the plans I have outlined here.

The planning of the field called for about a hundred producible wells by start-up, fifty more by the end of the year or early in 1978, and around two hundred altogether a year after that. Since it appears that a typical Prudhoe well will produce at about 10,000 b/d, it will be seen that these totals allow considerable leeway: for shutting down wells for maintenance or repair, perhaps also differences in the productivity of wells at various locations in the field—like every other natural form, the Sadlerochit reservoir is not uniform but varies considerably in the thickness and accessibility of its oil-bearing zone. By early August 1977 BP and Arco, the two companies charged with developing the field on behalf of all its owners, had essentially met the first stage in their schedule. BP at that point had 43 wells ready to go—drilled, completed with their production strings, perforated, topped with Christmas-tree valves, plugged in to gathering lines; Arco's total was 54. In addition, the two operators were well along in the next phase of development, with 38 more wells drilled but not yet ready to produce.

Drilling the wells is only part of it. The crude comes out of the ground with several other things mixed in: salt water derived from the ancient sea in which the oil was formed, on which now, imprisoned in the pores of its layered rock, it in effect floats, and as the oil is drawn off, the water level rises and comes through the pipe in increasing volume; grit from the original sands and sediments in which oil and water were enclosed; the various airy molecules which collectively are called natural gas, dissolved in the crude like carbon dioxide in soda pop—at Prudhoe, about 700 cubic feet in every barrel, of which, as it happens, 12 per cent is in fact CO_2, the carbonating gas. All of this stuff must be removed and disposed of before the crude is delivered to the pipeline. The refiners who pay the pipeline tariffs on volume, so much per barrel, dislike being charged for transporting useless brine and sediment; both, besides, will corrode or otherwise damage their expensive equipment. The gas would of course be doubly dangerous at every stage in the oil's movement: it represents pressure, again like a carbonated drink in a bottle; compressed and exposed to air and a spark, it will burn, explode.

To clean up the crude for sale, the two operators built large processing plants: essentially a sequence of vessels connected by

pipes and valves in which by stages the water and gas are drawn
off by heat and pressure, with huge turbine-powered pumps to
drive the oil through and on out to the pipeline. To protect these
systems (and incidentally the men who serve them) against the
Arctic winter, the machinery is enclosed in windowless, prefabricated, steel-walled structures set eight feet off the ground on
cement pilings, for the sake of the permafrost: gray-green
anonymous buildings that in functional looks and size suggest
airship hangars or the rocket facilities at Cape Canaveral; it was
getting these immense structures to Prudhoe Bay, preassembled
in Seattle and mounted on 400-foot-long ocean-going barges,
that constituted the chief urgency and drama in the successive
summer barge lifts organized by the oil companies. The plants to
do these jobs are called gathering centres by BP in its western
half of the field, flow stations by Arco: national differences in
terms (and in details of design) but similar in function. Four
such facilities were ready for start-up, each linked through a web
of pipes to from three to six drill sites with their clusters of wells
(again, differences in approaches to design). Each plant was
rated at a capacity of about 360,000 b/d, so that two more would
be needed within a couple of years, when the field reached its
projected maximum production of 1.6 million b/d; two more
were planned, in the event the other reservoirs in the area turned
out producible or new ones were found.

By law through the Division of Minerals and Energy of its Department of Natural Resources, the state has a considerable
voice in the design and operation of the field: in particular, the
disposal of the water and gas cleaned from the oil. Since the
whole field and all its facilities are never more than a few miles
from the sea, that would seem the obvious place to get rid of the
salt water, returning it to its source, though at a remove of hundreds of millions of years, but that was not how the state ruled.
Rather, it was to be pumped back into the ground. There was an
environmental argument for this—the effect of concentrated
brine, mixed with other dissolved minerals, on the life nourished
in the mild waters of the Arctic shoreland was at least uncertain.
The productivity of the field was the more functional reason:
forced back into the rock below the oil zone, the water would
help keep up the pressure needed to lift the crude to the surface,

a secondary recovery technique widely used to enhance the production from old fields. What it meant immediately at Prudhoe Bay was more wells to drill, more cost—by start-up, three on the BP side, two on Arco's.

Gas in the past when it came as a by-product of oil production was often regarded as a dangerous nuisance of comparatively small value; the safest and simplest thing to do with it was simply to burn it off (as is still done in some fields in the Middle East, for the same reasons). The state, however, unwilling to put a torch to future royalties, decreed that the gas be pressured back into the reservoir until such time as it could be produced. In the life of the field, this would yield benefits like those of the water-injection system, but at considerably greater cost: to do it would require a plant as elaborate as the flow stations and bigger, to clean and compress the gas for reinjection, more wells to be drilled (seven functioning at start-up, three more nearly completed); raising the gas to the pressure necessary for reinjection, about 4500 psi, consumes up to 10 per cent of the gas to fuel the compressors. In the sharing of responsibilities among the partners, it was agreed that Arco, which with Exxon would own most of the Prudhoe gas when it was finally produced, would take on the gas-conditioning plant and wells, while BP would build and operate the central power plant for the field. Although the owners were in no hurry to begin producing gas, this arrangement put some limits on how long they could wait: the more oil they produced, the more gas they would have to dispose of—more wells, more expense, more gas consumed to run the turbines in the plant. That was perhaps one of the reasons why they decided to use the gas rather freely in the meantime as a source of power—for the four pipeline pump stations north of the Brooks Range divide (where it was carried by a small-diameter line), for the flow stations/gathering centres at Prudhoe Bay.

In sniffing out the costs of these facilities, I found the oil-company department heads for drilling and construction both better informed and less secretive than their Alyeska counterparts, ready enough to talk, at least in general terms, about what the work was costing. Drilling was always expensive at Prudhoe Bay, about four times the Texas standard: uncertainties about how to deal with the deep layer of permafrost, which, if thawed from the heat of drilling, often turned out to be cobble-size rocks loosely packed in sand and muck, sloughing into the hole and plugging it; the inherent expense of transporting men and equipment to a remote and unsupplied place, maintaining them in its harsh and unfamiliar climate. The development wells drilled in the field's go-ahead early years came in at about $1.8 million each. Through the years of court action and congressional debate, BP kept on working at a quite moderate pace. Arco, with its partner Exxon prompting over its shoulder, grew discouraged and in 1973 pulled out its crews. By the time Congress had removed the last barriers to development, the OAPEC embargo had transpired, oil prices had begun their dizzy climb, and with them most of the costs connected with oil: pipe alone went up by 51 per cent in one year, drilling by 35 per cent, then 28 per cent. When Arco resumed, on a more urgent schedule, its wells were costing around $3 million apiece or a bit better; in addition, a number of those completed early had been damaged in the interim—drilling mud froze in the permafrost, crushing the production pipe—and had to be reworked. When all of these expensive difficulties are allowed for, it looks as if the average cost for a Prudhoe well was about $2.7 million, or nearly $300 million for the oil wells and the gas and water injection wells needed by start-up; and bringing the field to its maximum production of 1.6 million b/d would more than double that figure—say $650 million altogether.

According to one of the men immediately responsible for the design of the Arco half of the field, the two flow stations ready at start-up were worth $400 million each. Arco had made a seriously false start back in 1970 by ordering and having delivered to Prudhoe Bay the components for two different and smaller designs; they were stored on the ground, disassembled until late in 1975, when parts and machines were salvaged and incorporated in the replanned facilities that were finally built. If this error meant extra cost and if that cost was included in my friend's round-number estimate—neither premise is certain—it seems likely that BP would have gotten its gathering centres for somewhat less. With these allowances, I think $1.5 billion is probable for the four such plants needed at start-up, and another

$1 billion—inflation continues—for the two more that would match the field's MER of 1.6 b/d.

From another Arco source, it appears that the construction budget for the rest of its installations was around $900 million: the central gas-compressor plant would have been the big item, probably more than half; and the headquarters camp with its two adjacent topping plants built side by side, an airfield, temporary work camps, many miles of thick-laid gravel road, the gravel pads from which the wells were drilled, docks; possibly the substantial sums spent on early exploration. Assuming that BP's expenses in its half of the field were on the same scale, it looks as if the total cost of bringing the field on stream was around $3.5 billion, while raising it to maximum production would cost another $1.5 billion for flow stations/gathering centres and additional wells. If, finally, we add in the roughly $1 billion that went for lease bonuses, we reach the final totals: $4.5 billion to start-up, $6 billion to production of 1.6 b/d. These broad totals were in fact several times confirmed to me by Arco and Exxon executives who I think knew and had no special motives for concealment. I believe they are reasonable and, if anything, on the conservative side; and they allow nothing for interest on borrowed money, which, as with the pipeline construction work, would probably double these totals.

It should be noted that while BP and Arco were executing the work of development (with abundant advice from their partners) and keeping the bills paid, they were only partly responsible for these large expenditures. All costs were in fact shared among the owners in accordance with the amount of recoverable oil that could be shown to lie within their leases. During construction, working percentages for these shares were set by provisional agreement. It was not until April 1, 1977, that the final divvy was settled.

The main reason for this delay was that every well drilled in the nine years since discovery revealed a little more about the extent and thickness of the reservoir, the amount of oil actually producible from any particular tract; and provided fresh matter for interpretation and argument by oil-company geologists and lawyers. In addition, there seems to have been a fair amount of swapping and consolidation of leases. What resulted from this process was two separate agreements whose purpose was to define the Prudhoe Bay Unit—that is, to combine all the oil-bearing real estate into a single entity—and to specify the manner in which it was to be developed and operated and its owners were to share its cost and revenues.*

The first of these is known as the Unit Agreement and was made between the state and the owners. Its purposes were to establish which tracts or parts of tracts were to be included in the unit and the approximate boundaries of the underlying oil and gas reservoirs; to provide assurances that the owners and their successors would pay their royalties and the various taxes; and to specify the state's interest in matters—the location and spacing of present and future oil and gas wells and other facilities—which would affect the field's production and therefore the state's income. The second, the Unit Operating Agreement signed on the same date, was among the owners alone and was concerned with everything in the operation of the field that would affect its costs and income: in particular, the exact ownership of the oil and gas reserves.

The two agreements together are instructive in the risks as well as the rewards of oil. Of the 303 state-owned tracts actually leased in the vicinity of Prudhoe Bay, more than half turned out to be outside the area where any oil had been found—half a billion dollars in bonuses, in other words, had been thrown on the table and produced no return. Another fifty tracts were thought to have oil of some kind and might eventually make money but were excluded from the unit—they represented those oil-bearing formations which the leading partners were still pondering and had not yet decided were worth producing. A little less than a third of the leased land actually fell within the unit. Of this acreage—246,406 acres by my count from the maps and lease records—Sohio was still the biggest owner, with about 35 per cent, followed by Arco and Exxon with a little over 21 per cent each; Socal, Phillips, and Mobil seemed to be sitting fairly pretty with from 5 to 8 per cent of the land; and the rest was distributed among seven minor owners.

* The unit boundaries and the limits of the Sadlerochit reservoir, as defined in the two agreements, are shown in the Prudhoe Bay map on page 104, along with other important features of the field's development completed when production began in the summer of 1977.

Actual ownership of the oil, as it was argued out from the geological evidence of what was under the tracts, is different enough to be worth recording here in tabular form.

Prudehoe Bay Unit
Lease and Oil Ownership

Owner	Leased Acreage	Share of Oil
Sohio	35.549%	53.1552049%
Arco	21.348%	20.2744718%
Exxon	21.348%	20.2744718%
Mobil	7.581%	2.0938593%
Phillips	5.753%	2.0449947%
Socal	5.129%	0.8433430%
Minor holdings*	3.292%	1.3136545%

* Amerada Hess, Getty, Louisiana Land and Exploration, Marathon, Placid; and five heirs of the redoubtable Hunt brothers.

As the table shows, although Sohio came away with the lion's portion of the oil, all three of the leaders had placed their leasing bets rather successfully (and their share in the oil closely matches their financial contributions to the building of the pipeline); on the other hand, despite the high hopes that had boosted the bidding back in the big lease sale in 1969, everyone else's land turned out appreciably less valuable. The seven decimal places in the percentages are less derisory than they may look. At $14 a barrel, one thousandth of one per cent of the Prudhoe crude (0.001 per cent, or 0.00001 in decimal form) is worth well over a million dollars; each digit in the seventh decimal place means around $13,400 when the oil is sold at the refinery.

The division of the gas reserves, also spelled out in the Unit Operating Agreement, makes a shorter but no less interesting list.

Prudhoe Bay Unit
Gas Ownership

Owner	Share of Gas
Sohio	14.8183707%
Arco	42.1240052%
Exxon	42.1240052%
Mobil	0.2727500%
Phillips	0.2584336%
Socal	0.4024353%

Here the disproportions are even more striking—between land and production share, oil and gas. In the asymmetrical dome shape of the reservoir, the gas is by its nature concentrated around the center and top, and it is evidently in and over that smaller area of the unit that Arco and Exxon had most of their joint leaseholdings. Their coolness toward developing the gas is therefore all the more striking: it underlines once more the practical effect of the long-standing and arbitrarily low controlled price of natural gas—with their enormous share, about 22 trillion cubic feet, Arco and Exxon would have the strongest possible motive for producing it, with as much urgency as the oil, if it were worth producing. There is one other interesting point about this sharing of the gas. It will be recalled that all the Prudhoe crude comes from the ground with a substantial charge of gas, and the estimate of the gas reserves, 26 trillion cubic feet, was in fact predicated on extracting it and, eventually, selling it. The ten owners at the end of the oil list would thus have on their hands around 88 billion cubic feet of gas as a by-product of their oil, less than 1 per cent of the reserves but a goodly quantity all the same. That they no longer appear on the list of gas owners suggests other, private agreements at which we can only guess: bartering, perhaps, their share in the gas for some or all of their charges for developing the oil field.

Throughout the development of the Prudhoe field and the building of the pipeline, the two successive state governors, William Egan and Jay Hammond, vied with the legislature in devising new taxes on the future oil wealth. Their purpose was to provide for the continued expansion of state government and its services while making good the resultant deficits in the budget; and they were politically canny in that they would accomplish these ends without raising the already stiff Alaskan personal income and sales taxes. Most voters apparently found no fault with this arrangement, but it naturally gave the oil men chilblains and heartburn.

The first move in this game of wits came on the heels of the Trans-Alaska Pipeline act, in a pair of tax bills that went into effect on January 1, 1974. The first of these doubled the existing gross production tax (often incorrectly referred to as a severance tax—see page 18): like the standard royalty specified in the original oil leases, a percentage of the wellhead price of the oil—the selling price at the refineries less all costs of transportation. The precise form of the tax was a little delicate. Although its object was to get at the Prudhoe oil, it was necessary to do so in the form of general law—the tax would apply equally to the much smaller fields around the Cook Inlet which had sirened the oil men north in the first place, fields that after twenty years of expensive development and controlled low prices were still only marginally profitable if at all and might be knocked out of business altogether if the state took a really hungry bite; or so the oil companies argued, and despite the absence of any detailed accounting it sounds as if there may be some truth in the argument. The solution in any case was a graduated percentage calculated on the average daily production of every well: 5 per cent on the first 300 barrels, 6 per cent on the next 700, 8 per cent on production above 1,000. The lowest percentage would apply to most of the wells in southern Alaska; at Prudhoe Bay, where the average production was expected to be around 10,000 b/d each, the tax rates work out to 7.77 per cent per barrel.* The oil men suggested that what the state had done was simply to raise its royalty rate and that this was an arbitrary and one-sided revision of the contracts represented by their original leases (*confiscatory* was the word used heatedly by some, but only in private). The protests were not loud: the state's much lamented fiscal troubles had indeed been complicated by the pipeline delay, it was necessary to commiserate, be good citizens, as the PR men are wont to put it; and the Prudhoe pay was rich enough to accommodate many mouths at the trough.

One can think of another reason why the oil men might have taken the tax increase tamely. If you were in that spot yourself, you would notice that it was based on the wellhead price and that therefore the more you spent on building the pipeline—and

---

* Tied to the same bill was something the legislators called a conservation tax of one-eighth of a cent a barrel ($0.00125). Since the revenue was to be paid into the state's general fund and no particular use for it was specificed, it is impossible to imagine the purpose of this small annoyance. Nothing connected with Prudhoe Bay is really small, of course; at a production rate of 1.5 million b/d, the conservation tax would bring in $684,375 a year.

hence the higher the resultant tariff—the less you would pay in
tax; and that with the recovery of your construction costs assured by ICC custom, the extra expense would cost you less than
it would the state. If that was what you had in mind, you would
then make a great show of economy—depriving the pipeliners of
their steaks and lobster and ice cream—and if anyone asked, you
would point to these efforts and say what a pity it was that you
had not succeeded in reducing the costs. Yes, sir, a *very* great
pity, but we sure did try, didn't we? It seems that the legislators
anticipated this possibility and included in the same bill a penalty for running up the cost: the 2 per cent property tax on the
pipeline system (and on oil production facilities in general)
which, as we noticed earlier, had produced $211 million by start-
up. Since Alaskan property assessments are based on market
price—essentially in this case the cost of construction—it looks
as if the $5 billion cost at Prudhoe Bay and the $8 billion for the
pipeline would together yield the state about $260 million a year,
more than a third of its current budget; if capitalized interest is
included in the two valuations, it would sweeten the property tax
receipts by $32 million a year.

A year and a half later, when the last of the Prudhoe lease
money was gone and the new taxes were not as yet very productive, the state devised the further expedient noted elsewhere
(Chapter 9): a 2 per cent tax on the value of the oil in the
ground and known, therefore, as a reserves tax. Since that value
was calculated quite moderately—a low estimated selling price
less all costs of development, production, and transportation—it
yielded only about $250 million a year shared among the
Prudhoe owners. Moreover, by its own terms it would expire in
January 1978, when presumably the oil would be flowing and
with it the royalties and the more serious taxes; and the oil companies would be allowed to deduct, over an extended period,
what they had paid in reserves taxes from their other duties to
the state. The oil men grumbled that the new tax was really a
forced, interest-free loan which they would have to provide from
money borrowed at rates of 9 or 10 per cent, but again not
loudly, and they went along; one more mouth at the trough. The
pols came forth from the session with a magic message for their

constituents: a budget balanced in the nick of time and, except for a few oil companies, at no cost to anyone with a vote.

All the moves so far were minor and preliminary but suggestive. Stirred by these little victories, the legislators got to thinking about how much the state might *really* get out of the oil and, being Alaskans, they turned to an outside consultant for answers: one Michael Tanzer, an economist who works out of an apartment in the West Eighties in New York. Dr. Tanzer's 95-page report, "Alaska's Prudhoe Bay Oil: Profitability and Taxation Potential," commissioned in July, was delivered in January 1976 and, as its title suggests, it told the legislators what they wanted to hear. The central tool in this analysis was an arcane concept known in the trade as discounted cash flow—DCF. I spare the reader the details, but the conclusions it yielded were striking: that the oil companies would be profiting at rates of up to 40 per cent a year on their Prudhoe investments; that the farther into the future these profits were projected the less actual value they would have for the owners; and that accordingly they could be taxed at rates of up to 80 per cent and not even feel the bite. If from these conclusions DCF sounds to common sense like Funny Money, that is what I think it is; and its application in this case was not made more effective by the fact that the report was quite sketchy about actual costs—a good deal less complete, detailed, and accurate even than the round-number estimates outlined in the present chapter. After reading the report I sought the opinion of several of Dr. Tanzer's peers, New York consultant economists specializing in oil, but it was like asking one doctor for an opinion of another's practice; all I got for my pains was little coughs of embarrassment. The legislators, of course, were dazzled. The economist was their Bacchus, conferring the gift of Midas, and they had only to touch fingers to typewriters to turn the oil to gold.

This time the oil men reacted and, when they could get their voices back, made vehement protests: disheartened by the confiscations in the Middle East, they had turned north in the assurance of conducting their affairs under stable law and the certainties of the American Constitution; now the Alaskans were talking like pale-skinned Arabs. In the next few months, while a succession of new tax bills worked their way through the committees, aimed at turning the Tanzer hypotheses into law and money, oil-company presidents supported by their most personable PR men flew to Juneau to cajole and threaten: the doctor's arithmetic was faulty and overstated the likely profits, DCF was only one of several methods of forecasting income and making decisions and not the most important; to treat the success at Prudhoe Bay in isolation was anyway an error since the return from one field, even the biggest ever, necessarily made up for failure in many others, that was the risky nature of the oil game. Finally and most seriously: Alaska's present assortment of taxes was already substantially higher than in the other big oil states and would be many times greater if the Tanzer proposals took effect; since there was never enough capital to be made from profits or borrowed, they would have no choice but to pull out and put their exploration money into states that offered a reasonable chance of coming out ahead.

These arguments were abundantly reported in the Anchorage *Times* and found allies in the Native corporations whose hopes for profit centered on discovering oil and other minerals on their lands; and among miscellaneous other Alaskans. By now, in the early summer of 1976, with the pipeline troubles becoming daily more acute and congressmen and stockbrokers predicting a costly year's delay, it seemed that any radical change in the rules might stop the play altogether. The legislators listened, concluded that the time was not ripe for roasting the goose, and contented themselves with plucking a few feathers, in the form of new regulations for the tankers out of Valdez. These too would cost money, but they left the tax laws, for the time, untouched.

In June, in the aftermath of these alarums and excursions, I stopped by the unprepossessing Anchorage law office of Chancy Croft, the president of the state Senate. By office and disposition, he had seemed the most effective of the leaders in the latest round, a big soft man with thinning hair who looks to be cast from the same mold as Robert Benchley but with the humorist's gentle, self-deprecating wit left out; I found him well and articulately informed about the state's affairs, hence one with whom it was possible to argue reasonably even in disagreement—the sort of man, in short, to stir hopes for the prospects of state government, perhaps even for democratic government in general.

We talked through the several tax schemes that had been drafted in the session just finished but had not quite made it into law: a handsome increment in the gross production tax, possibly as much as another 12 per cent; two different ideas for extracting the producers' Prudhoe income before it disappeared into the corporate pots, one known as the properties net proceeds tax, the other concerned with excess profits. The timing; maybe next time around they would bring it off.

Still worrying Alyeska's ever-rising costs, I wondered what he or his senators knew, but it seemed they had nothing special or specific—he put my card in his wallet with a promise to let me know what they dug up—or no more than an intelligent man might generalize from careful observation. He went on, talking about the kinds of things that had made the pipeline expensive, the roads Alyeska had had to build, the police services, communications systems, transportation network—"that whole range of services that normally government provides to any private development—because the Alaska political, social, economic structure was just too small: it was all paid for by the companies." This was surprising: it was one of several things Alyeska had been saying, would go on saying, in extenuation. Then he brought me up short, coming to an opposite conclusion: "It really has isolated very dramatically the subsidy that government often provides—and I think that's one of the more significant things about the cost of the pipeline, all lumped in there." We went on chatting, circling the questions of money—Alyeska's, the state's, taxes, budgets, deficits—but that was one of the ideas that stuck, and I went away brooding. It is a very odd view of our government, and though we hear it in several variants and some of the civil servants we are saddled with behave as if it were true, it is one I reject. The road in front of my house, the water that comes out my taps, are not "subsidies" dispensed by a benevolent state. *I pay for* the bloody water and roads with my taxes or otherwise, and those who administer them are my servants, appointed or elected by the one hundred millionth voice of my vote; neither they nor the government they serve *owns* any of this, and if they suppose they are *giving* me anything, then I say it is time we chose ourselves a new bunch. If there are exceptions or inequities in this relationship, we had better get on with changing them too. But under law a corporation is a person as I am, paying its due; the scale is different, not the principle.

Chancy Croft's views on taxes and, more broadly, on the state's interest in TAPS and the oil it was built to carry were spawned in several currents of popular ideas. Each of these ideas focuses a fundamental conflict—of values, self-interest—for which we have not yet found any satisfactory resolution; or perhaps there can be none but only a case-by-case application of principles tempered by circumstances. Either way, it is necessary to be clear about the issues of principle underlying the turmoil of particulars.

The most fundamental of these questions concerns sovereignty over what we call natural resources. By definition a natural resource is one that individuals (or corporations) may discover, develop, exploit, but have no part in creating; and is at the same time of vital interest to all the people, whether directly involved in its discovery or not. Historically in the United States we have taken an ever broader view of what we shall regard as natural resources in this double sense and hence of the means by which the people's interest may properly be expressed, exercised, protected: where once we took it for granted that the interest of the private discoverer and developer would naturally serve that of the society at large, we are now much more inclined to compel that service by direct controls; and the Alaskan pipeline represents the most elaborate exercise of such controls that we have yet undertaken. But the practical question remains: what kinds of controls—of what extent, at what level, to what degree—will best benefit the public interest? And if that is the purpose, I do not think it makes much difference whether the resource in question is privately or publicly owned. Land is the fundamental resource from which all others derive, and land in our beginnings was a common possession before it devolved to private property or to administration by local, state, or national government; that is the first meaning of sovereignty of the people.

The essence of sovereignty is power: final control over public decisions within a defined political jurisdiction. And as de

Tocqueville long ago observed, the people's sovereignty arises from and is expressed in the face-to-face decisions of the most purely local forms of government; that seems to me the plain meaning of the expression in the Constitution and in particular the sense of the reservation of powers clause—the powers of county, state, and national government are delegated, limited, conditional, drawn off from the great reservoir of popular sovereignty, and if that sovereignty is not supreme in matters of immediate local effect for which authority has not been granted to larger jurisdictions, then it is meaningless. Yet as we know in our multiform political system, each level of government exists within and is circumscribed by a higher, more general level, and the history of American popular sovereignty since de Tocqueville's time has been one of powers continuously flowing out from local to state to national government. Perhaps inevitably: there are not many local decisions that do not impinge in some degree on other or broader jurisdictions, acutely so in the case of natural resources which are necessarily of national concern. And once again Alaskan oil and its pipeline are the supreme case, both of the inherent conflicts of sovereignty and of our latest attempts at resolving them.

We come back to that Alaskan politician's catch-phrase about the Prudhoe oil—that it belongs to the people of the state; the premise of the maneuvers about the taxes and other forms of control. But if the slogan means anything, it means sovereignty, not ownership, a quite different idea, just as the Natives in their claims had no quarrel with the sovereignty of the United States over Alaska while insisting on their rights to their land. So the rest of us cannot stand mute when Alaskan acts impinge on our interests: when state taxes, by whatever name, cut seriously into national ones or threaten the availability of an essential resource; when local regulations impede the pipeline and the tankers it feeds, adding costs at the gas pumps that we have no voice in deciding. These are not matters that can be determined within a single state sovereignty, without reference to the others it affects: that is the rule of common sense, set forth in the Constitutional clause concerning interstate commerce. The rule, unfortunately, does not answer particular cases—the prohibition of large tankers in Puget Sound, of a pipeline terminal at Long

Beach, acts that complicated the destiny of Prudhoe oil and to which we shall return in conclusion. Indignation on either side is beside the point. The two sovereignties are real, of the state to regulate what happens within its borders, of the nation to possess its resources. Perhaps recognition of that reality was part of the reason why the time was not right in Alaska in 1976 for a drastic change in the rules. But the itch remained.

Clearly it is difficult in the calmest of times, where important natural resources are concerned, to maintain a just balance between the state or local and the national interest. When the resource at issue is oil, the question is complicated by the residue of public bitterness and distrust that has surrounded the oil companies from the beginning, periodically renewed. In Alaska in the years of the pipeline the whole problem was made immeasurably more difficult by coincidence with OAPEC embargo and the succession of drastic increases in the price of crude oil imposed by the OPEC producers. Moreover, since these events came at a time when U.S. production was in decline while demand continued to rise, Americans were faced with a novel and disagreeable predicament, a growing need for imported oil, most of it from the OPEC countries, with the simultaneous threat that it might be shut off; and multiple increases in the cost of the imports and consequently, though less sharply, of other forms of energy—the beginning, in short, of a general inflation which shows no signs of abating, and cannot while its causes remain.

Among the gentlemen in Congress this situation produced flailing arms and noisy accusations, and the oil companies were convenient targets: accessible to American law, too big to miss, made easier by their habit of burying their heads in the sand so far as public information is concerned. Since the price of crude was up, so too, naturally, were the prices of most of the products made from it, therefore the companies' gross income and presumably their profits as well. Someone coined a phrase, and it caught: *obscene* profits. The congressmen waved thunderbolts of special taxes, corporate dismemberment, and convened hearings on multinational corporations (another catch phrase), particularly those that traded in oil, to determine the extent of the obscenity. The hearings warmed over a good deal of the past history of oil but in fact brought to light little of substance about the companies' current operations or their role in the OPEC manipulations. For example, the hearing reports produced figures to show that the major American oil companies involved in the international trade generated what looked like large income (this was hardly news—they are big companies); and that the U.S. income taxes they paid on this income were derisory, ranging from about 1 per cent to 6.5 per cent. Shocking! What the reports did *not* say was that these figures were derived by ignoring the very large taxes, amounting to from 40 to 70 per cent, already paid on this income by the companies in every country in which they do business.

As a matter of common equity we on the whole reject the idea of taxing the same money twice, and this principle is long established in our tax laws. Thus, if a corporation (or any individual, for that matter) earns income in a foreign country to which it pays income taxes, it is allowed to deduct a portion of that tax from the U.S. taxes that would otherwise be due on the same income, the size of the deduction varying with the proportion of foreign income and tax. Apart from equity and law, this is simply common sense: if, say, as a corporation you earn income of $1,000 in Great Britain on which you would pay around $600 in tax, there simply is not enough left for the $480 due under the typical U.S. corporate income tax rate of 48 per cent.

In short, the evidence about oil-company taxes and profits that came out of the congressional hearings was arrant nonsense. Nevertheless, it made good copy and was widely repeated by politicians and by credulous journalists with an ax to grind (for example, by Anthony Sampson, a British newspaperman, in a popular diatribe against the oil giants called *The Seven Sisters*). And it shaped the frame of mind in which Chancy Croft and his colleagues set about seeing how thoroughly they could milk the Prudhoe cow for the benefit of the state, assuming there was no natural limit to the take. There are real questions here that may be reasonably debated: whether, for instance, we will be better served by replacing privately owned corporations with some form of government-created entity—or on the contrary, by relieving them of much of the burden of government regulations under which they now labor; whether we wish to encourage U.S. corporations to do business abroad or actively prevent them; whether, or to what extent, we as a nation can allow any state to exercise final control over national resources. But we cannot formulate such questions or debate them meaningfully on the basis of bogus facts.

Considering the amount of oratory and printer's ink that has been wasted on oil-company income and profits, it is perhaps worth a little more effort at this point to ascertain the real facts. They are a good deal less glamorous than they have been made to appear; indeed, given our dependence on oil and the huge capital needed to produce it, rather discouraging.

Profit, or net income, is what is left when you have added up all revenues from sales and other sources and subtracted all expenses. That is a very simple idea, but it has several different meanings, depending on how you look at it, what you compare it with; and net income itself has two different meanings, before taxes and after taxes. Thus, profit may be understood as a percentage of total revenue, as an annual percentage of return on a corporation's assets, or as a return on the value of its stock in the hands of those who own it. As a return on revenue, it gives a rough measure of the company's effectiveness in selling its products while controlling and limiting its expenses. As a return on assets, it measures efficiency in the use of capital. As a return on stock, or investors' equity, it must compete against all the other uses investors might make of their capital, such as putting it in a bank to draw interest. This is all true of corporate income before the various income taxes have been taken out. What remains, net income after taxes, has two main jobs to do. For one, it is a company's most accessible source of capital for investment aimed at generating future income and profits. It is also the source of dividends on stock and therefore, compared with the value of stockholders' equity, an incentive to future investment competing with other uses investors might make of their private capital.

If these concepts seem to go by too quickly, concrete figures may make them a little clearer.

Thus, from 1973 to 1974 the revenues of the major oil companies* nearly doubled, from about $134 billion to $245 billion, in consequence of increases in crude oil prices decreed by OPEC. Since they are international companies and buying crude is their biggest expense, their costs increased at about the same rate, their income taxes more sharply. As a result, their net income after taxes increased very slightly as a percentage of return on capital, from 8 per cent to 8.9 per cent; but declined as a percentage of total revenues, from 9 per cent to 7 per cent. These were narrow changes. Indeed, in the thirty-odd years since World War II, by most measures of profit, the U.S. oil industry had differed little from other industries involved in international trade, and all experienced the OPEC effects to about the same degree.

Exxon, as the biggest of all the oil companies, its business most widely spread, was both representative of and influential on this balance sheet for the industry as a whole. Thus, Exxon's revenues rose from $28.5 billion in 1973 to $45.8 billion in 1974 (and inched up the following year), but its costs increased at about the same rate, with much the same effects on its income as the same events had on its competitors'. In dividends paid, Exxon did a little better than the rest of the big-oil group, about 7 per cent on the value of its stock compared with 5 per cent. Given the risks, you might think about switching your money from the neighborhood savings and loan to Exxon's stock, but oil-company stocks in general are not very attractive. More seriously, as I read its annual reports, from 1973 to 1975 Exxon's net income after payment of income taxes and dividends amounted to about $5.4 billion—all the money immediately available for capital investment throughout the world. This sounds like a lot but Exxon's 20 per cent share of pipeline construction and the Prudhoe Bay development alone amounted to about $3 billion (including capitalized interest). What was left would be very thinly spread

* The summaries that follow derive from studies by the Energy Economics Division of the Chase Manhattan Bank, based on the annual reports of the major U.S. oil companies active in the international trade, with several comparable European companies such as BP and Royal Dutch Shell, supplemented, as the bank says, from other sources; together they represent the biggest part of the non-Communist oil business. I have made my own readings of the reports of the Prudhoe and TAPS owners.

elsewhere in the United States and over the far greater part of Exxon's business that is international; it does not look like enough. Neither for Exxon nor for its competitors have the profits been so very obscene.

In 1972, hopefully anticipating that TAPS would soon be started, the Alaska legislature created a regulatory body known as the Alaska Pipeline Commission, consisting of three commissioners appointed by the governor with the concurrence of the state Senate. Their duties were essentially two: to certify new oil and gas pipelines as fit for public service; and to approve their tariffs. Since the law that created the APC did not permit it to supersede the rulings of a national agency such as the ICC on an interstate line—in particular, TAPS—and there were no new pipelines being built within the state, the commissioners did not have much to do except issue annual reports affirming that they were looking ahead to a time when they would have duties to perform.

The time came nearly four years later, at the end of December 1975, with a plan for a small-diameter, 69-mile pipeline to carry refined oil products to Anchorage from a place called Nikiski on the Kenai Peninsula: public hearings on the proposed route and the line's estimated cost; further hearings a year later, when the line was finished, to certify it for operation and approve its tariffs. The commission's investigation served as a kind of preseason warm-up for the big game with Alyeska which was about to begin. Officially, the APC had nothing to say about TAPS or its tariffs so far as these concerned the transportation of oil to other states, but it was the final authority over any oil delivered within the state; and in fact, under a plan that had been evolving since the early days of the Prudhoe discoveries, a small refinery organized by a Dallas company was now nearing completion at North Pole, a village adjacent to the pipeline a few miles southeast of Fairbanks. Since the premise of this project was that the refinery would be able to draw off the necessary crude as it came down the line, the pipeline owner companies, as common carriers, were obliged to establish tariffs for transporting the oil this halfway distance. And since these tariffs would apply entirely within the state, they would be subject to the APC, which thus gained a legal basis for full investigation of Alyeska's costs, on which the tariffs would be based, and for considering whether or not the pipeline should be allowed to function. In addition and more seriously, as the state agency with a formal interest in pipeline tariffs, the APC was Alaska's natural advocate in the battle about to begin over Alyeska's general tariffs as they would affect the state's oil royalties.

Accordingly, on November 24, 1976, the APC contracted with Terry Lenzner, a Washington lawyer who had caught the public eye as the energetic assistant chief counsel of the Senate Watergate Committee, to investigate Alyeska's costs. At the time I envied Mr. Lenzner his powers to compel truth: to subpoena the company's documents, to take its executives' testimony under penalties for lying. In fact, however, compulsion was not much more effective than sweet-tempered guile. Alyeska, after promising openhanded co-operation with the APC's special counsel, began censoring or losing the documents; in distant federal courts it submitted arguments showing why documents should not be produced, why its people should not make depositions or not without the presence of its lawyers. Meanwhile, the ICC procedures unrolled toward start-up: formal requests to the oil companies to submit their tariffs, their delivery in Washington a few days before the oil was scheduled to flow, protests filed by interested parties, in particular the state of Alaska speaking through the APC. On June 15, 1977, Mr. Lenzner filed a 270-page "preliminary statement" in the form of a brief to the ICC disputing the bases for the pipeline tariffs; this was a kind of *Reader's Digest* version of his 600-page complete report on the investigation, *The Management, Planning and Construction of the Trans-Alaska Pipeline System*, produced on August 1st—after the ICC had made its initial ruling but part of the record in the continuing dispute. Mr. Lenzner's fee for his eight months' work was $160,000; the cost of the investigation, in salaries, travel and expenses, was another $800,000.*

"There are any number of layers of the problem of costs,"

---

* These are estimates which I have been unable to persuade the Alaska Pipeline Commission to confirm or deny, despite the state's freedom-of-pubic-information statute.

Chancy Croft had remarked back in June when I raised the question of where Alyeska's money had been going. "The first one is the accounting function of making sure that what they say has been spent has actually been spent. Just where is the check for this item? We haven't really got that resolved yet. The second one is that, O.K., you spent the money and you're claiming it's in the pipeline, but wasn't that really for development of the field? So it shouldn't be allocated to the pipeline and thus show up in the tariff. Then the third question is, O.K., you did actually spend the money, you spent it on the pipeline, but was it good business practice to have made that decision or was it bad management?"

Chancy at the time had expressed something approaching irritation (he is a very equable man) at the APC's indifference to these difficult but necessary questions. It is probable that when the commissioners did finally begin their investigation, the agenda they handed Terry Lenzner and his staff was prompted by his senatorial anxieties. At the same time, the questions they pursued are obvious to common sense if there had been no special coaching from Chancy Croft and other interested parties within the state government.

Actually, in the time available, the investigators concentrated on the third of the senator's three concerns, evidently taking on faith the answers to the first two—that the oil companies really had spent Alyeska's $8 billion and had spent it on the pipeline, not on developing the Prudhoe Bay field. With that important qualification, the report that resulted was an impressive piece of work in both its short and its long version, particularly so in view of the veils of obstruction and secretiveness in which the investigation was conducted. It examined, with an abundance of circumstance, most of the weaknesses in the conduct of the project that we have touched on in the present chapter and elsewhere: the multiple layers of managers and contractors and their consequent slow-burning responsiveness to the day-by-day decisions needed to direct the work in the field; the delays in developing and verifying the design of the system (imposed by the owners' unwillingness to provide adequate budgets until the basic question of the right-of-way was settled by Congress and the courts), resulting in costly design revisions and redoing of completed work; the low and declining labor productivity; the general weakness of the quality-control program, with the radiography and welding troubles that followed from it and the hardly less serious but less publicized difficulties with the aboveground pipe and its vertical supports. From his analysis in these and other areas Mr. Lenzner concluded that in the total construction bill of $8 billion, $1.5 billion was money needlessly misspent. In its effect on the pipeline tariff and therefore on the oil royalties, a rather larger sum would be lost to the state over the life of the pipeline, as we shall presently see.

In my own view, $1.5 billion is probably a fair value for costs that need not have been incurred in the building of the pipeline; it may well also be confirmed (or something like it) by the processes of ICC investigation and lawsuits. At the same time, however, it must be borne in mind that the Lenzner report was not an impartial inquiry but a lawyer's brief whose entire purpose was to assemble evidence in support of a predetermined conclusion—that a substantial part of the pipeline cost resulted from managerial errors and should therefore not figure in the base from which the tariff would be calculated. Thus, for instance, in arriving at a value for such errors the report used hourly rates for labor and equipment that look to me excessive— its equipment rate yields a total for "excess equipment costs" of "over $422 million," which is more than the entire original cost of all the equipment used on the project. Further, given its purpose, the report was under no obligation to consider alternatives, as they appeared at the time, to decisions that in retrospect look like errors—if indeed there were alternatives in any practical sense. As we have seen, the real alternative in the big decisions was most often to delay completion, which in terms of the submerged iceberg of interest and other fixed costs—not to mention the effect on the national economy and the financial health of the pipeline's owners—may well have outweighed any short-term saving. Indeed, put thus, the question is so hypothetical as to be unanswerable, which is presumably why it was not raised. Finally, the investigators showed little interest in the principles on which Alyeska's basic contracts were written—cost plus fixed fee —or the way those contracts and the company's rigidly centralized management worked to deprive its contractors of their primary incentive for efficiency and economy in the use of men and equipment. These are, in my view, the most serious questions of all about the building of the pipeline and should be fundamental for any comparable future project. Their effects on the final cost, however, are probably incalculable, and that presumably is why the APC chose not to investigate them.

It will be recalled that for financing and the recovery of costs, Alyeska was organized as if it were building eight small pipelines, each sized in proportion to the investment of its eight oil-company owners—that is the meaning of "undivided interest." Accordingly in May 1977, about a month before start-up, the eight oil companies, through their pipeline subsidiaries, assembled their tariffs and sent them in to the ICC along with more or less elaborate rules and regulations governing the conditions under which they would accept oil for shipment at Prudhoe Bay and deliver it at Valdez. These formal notices, dated between May 31st and June 15th, effective from two to four weeks later, set tariffs of from $6.04 per barrel (Arco) to $6.44 (Amerada Hess). These tariffs work out to about *18 times* the average for all other U.S. pipelines.* They were also $1 or more higher than the figures the state of Alaska had been using in estimating its income from oil. This was a serious difference. An extra dollar on the tariff meant a dollar off the wellhead price, hence 20.3¢ less per barrel† to the state: around $111 million per year once the pipeline was hauling 1.5 million b/d, nearly $2 billion over the life of the oil field. Naturally the proposed tariffs were greeted with cries of protest—from the state, of course, but also from the Native organizations, which under the Native Claims Settlement (see Chapter 4) had an interest of their own to protect, and, in sympathy, from the ICC commissioners and staff and the antitrust zealots at the Department of Justice. Since the oil would not move unless the ICC approved their tariffs—or at any rate *some* tariff—the oil companies made haste to submit financial justification for their charges.

Each of these documents apportioned the pipeline construc-

---

* Based on gross pipeline revenues, total pipeline mileage, and tons of oil transported; all figures for 1973, the latest year available. This is a rough indicator—the pipelines vary greatly in size, age, and cost, hence in their applicable tariffs.

† The standard royalty of 12.5 per cent plus the graduated gross production tax, which will have an average value of 7.8 per cent per barrel on wells producing at a rate of 10,000 b/d.

tion and operating costs a little differently and used different methods in arriving at the proposed tariff. Interest, however, was the main reason for the different tariffs that resulted: the borrowings by which the companies had provided their shares of the construction costs ranged from 77 per cent in the case of Exxon to 100 per cent for Mobil. Since the terms and interest rates for this money varied considerably, each company's tariff reflected differences in two important factors: the valuation of the pipeline to be recovered from each barrel of oil it carried (construction cost plus capitalized interest—i.e., interest paid up to the point of start-up) which was also the basis for the profit allowed in their tariffs; and the amount of interest due each year, to be spread over the volume of oil transported. On the other hand, they were in general agreement on how long the Prudhoe field would remain economically productive—twenty-five years —and therefore the working life of the pipeline; on how much it would cost—$1.049 billion—to dismantle and remove the pipeline once they were through with it, as required by the Stipulations; and on the pipeline's basic operating costs—essentially, the pay for its managers, technicians, and maintenance crews, the cost of fuel to run the pumps, and state and local property taxes.

Two important variables affect the way these costs are translated into a per-barrel tariff. In the first place, although the total valuation of the pipeline represents a fixed factor—one twenty-fifth to be earned back each year for twenty-five years—what it means per barrel in any one year would vary with the actual rate of production; as we have noted elsewhere, the Prudhoe production plan called for starting slow (at less than one million b/d), accelerating to between 1.5 and 1.7 million b/d, to be maintained for several years, then tapering down toward the end, and in practice the actual production could be expected to vary almost from day to day in order to maintain these averages. Again, assuming that the owners would pay off their borrowed money at a steady rate over 25 years, the interest they would be paying, like the interest on any mortgage, would decline continuously throughout that period, from about $724 million in the first year to $29 million in the final year. Thus, if the pipeline tariffs were free to reflect actual expenditures, they would change at least from year to year if not more frequently, generally higher at the start, probably declining toward the end—depending, however, on how the oil production held up and what happened to pipe-line operating costs. The ICC's rate-setting regulatory process is a clumsy and artificial means of dealing with these realities. In practice, the proposed tariffs represented projected average costs for about the first five years of operation and assumed that at that point the arithmetic would be refigured and new tariffs set.

Taking this averaging approach several steps farther, we can arrive at a kind of generalized average of all the costs which shows approximately what the pipeline tariffs will mean in terms of income and profit. These estimates, based on the oil companies' submissions to the ICC, are summarized in the accompanying table. The final average profit, $559.578 million per year, is on the modest side by the ICC standard for pipelines: about 5.7 per cent on a total investment of $9.7336 billion. To yield the full 7 per cent return, averaged over the life of the pipeline, the tariff would have to be set not at $6.21 per barrel but at around $6.86.

Average Annual Return on TAPS Tariff
(millions of dollars)

Gross revenue @ $6.21 per barrel (1)		$2,384.64
Annual costs		
Operations	$410.944	
Depreciation (2)	389.344	
Dismantling, removal	41.96	
Interest (3)	376.604	
	$1,218.852	
Net income before taxes		$1,165.788
Income taxes (4)		606.210
Profit (net income after taxes)		$ 559.578

1. A weighted average of the proposed tariffs multiplied by the average annual production of the field, 384 million barrels (the 9.6 billion barrel estimate of reserves produced over 25 years).

2. Total of the eight owners' valuation of their shares in the pipeline: the $8 billion construction cost plus $1.7336 billion in capitalized interest or $9.7336 billion, spread over 25 years.

3. Total debt attributed to the pipeline of 86.7 per cent of the valuation at an average interest rate of 8.582 per cent for 25 years, a total interest paid of $9.415099 billion (assuming straight-line repayment of 1/25th each year) averaged over 25 years.

4. 52 per cent, the combined estimate of the effect of federal and state tax rates.

Just how long the Prudhoe field remains productive is a critical factor in these calculations. The 25-year life used by the oil companies looks generous from what we know of the field's reserves and the companies' development plans: it works out to an average production rate of 1.05 million b/d, whereas if the average rate were raised to 1.2 million the 9.6 billion barrels of reserves would last only about 22 years; and in fact by start-up the 9.6 billion estimate was too high—thousands, perhaps millions, of barrels of crude had already been consumed by the Arco topping plants. (During pipeline construction a second plant was added, back to back with the original one completed in 1969.) Both these differences would have a sizable effect on the annual depreciation and interest charges; and would push the tariffs up accordingly.

On the other hand, one might argue that 25 years is an unreasonably short estimate for the usefulness of the pipeline: that with the other oil reservoirs already known in the Prudhoe area or likely to be discovered, the system may well go on pumping for 35 to 40 years. Physically, the pipeline as built would support this argument: the pipe itself may well be good for 50 years, most of its other components for 30 to 40 years. Any such extension of the pay-out period would, of course, have a drastic effect on the recovery of the owners' investments and other costs, and on their profit rates; and would bring the pipeline tariffs sharply down while enhancing the state's royalties and tax receipts. This is the reason, I think, why the oil companies have so far refused to make any estimates of the recoverable reserves in the other oil-bearing formations. This is more than gamesmanship in the short-run strategy of the tariff-setting process. Whether the reserves are economically recoverable depends not only on their quantity and quality but on future market prices for crude and the rates of federal and state taxation. If the recent history of oil shows anything, it is that none of these factors is predictable.

Time was the gist of the argument the ICC adopted in criticizing the proposed tariffs: that the charges should be predicated on some longer period of operation; and it set forth the lower tariffs that would result from this change of assumptions, invited the interested parties to submit briefs on the matter and present oral arguments at a public hearing in Washington. The ICC has no authority to set tariffs for a new pipeline; its power is limited to accepting or rejecting what the owner proposes to charge. In the TAPS case, the commissioners got around this limitation with the novel proposal that they could "suspend" the proposed tariffs and then invite the oil companies to resubmit their tariffs at levels that the ICC would accept. This time it was the oil companies' turn to cry foul, in their briefs and at the public hearing held on June 28, 1977: that the commission's action was contrary to law and precedent, that the proposed tariffs violated the long-established profit structure allowed to all other pipelines. So far as the history of pipeline law is concerned, these arguments appear to be true; on the other hand, a suggestion by one of the Exxon lawyers—that if the ICC tariffs turned out intolerably low, the owners could not operate—was dismissed as bluff, and rightly so. The companies had too much cash laid out to stop, in the case of Sohio its entire net worth put up against it borrowings; and by now the crude had already entered the line and was eight days south toward Fairbanks and Valdez. The nine commissioners voted unanimously to suspend the oil companies' proposals and at the same time "authorized" them to submit new tariffs at the lower rates calculated by the ICC staff. A formal written order, No. 36611, dated July 11th, confirmed this decision, briefly answered the legal counterarguments, and set in motion a full investigation of the pipeline tariffs, to be completed by the end of January 1978; at which time the interim tariffs would be reviewed in a new hearing and either confirmed or changed.

The ICC tariffs that thus went into effect work out to a weighted average of $4.84 per barrel. If we substitute this amount for the one used in the table on page 476, the net profit that results is 3.15 per cent per year. We may argue philosophically as to whether or not this or a larger or smaller percentage provides sufficient incentive to finance and build this pipeline or others; whether, given the interconnected functions of the oil business, a pipeline subsidiary is entitled to any fixed rate of profit. What is clear is that the ICC rates and the procedure by which they were reached represented a decided change in the rules of the game as it had heretofore been played. Beyond that, we may properly wonder whether the appointed members of a regulatory agency or the owner-managers who have placed their corporate necks on the block are the best judges of how to get their investment back. That—how that investment shall be recovered—is the immediate issue, with broad implications for the practical functioning of our society. If, for example, the ICC's judgment of the proper tariffs required thirty years to earn back the pipeline costs and the field were actually played out after twenty-five years, the owner companies would be stuck with a sizable bill that would never be repaid—confiscated, in effect, by government action: something like $1.5 billion already spent on construction, and about as much again in interest on borrowed money, whether already spent or committed in future repayment of the loans. That is a lot even for a group of the world's biggest oil companies. The total, say $3 billion, does not come out of thin air or an old mattress but out of net earnings, the capital with which to find and develop new sources of oil.

Terry Lenzner's Alaskan argument—that $1.5 billion of the pipeline's cost should be disallowed and form no basis for figuring the pipeline tariffs—leads to a similar result and raises similar questions. Reducing the valuation of the pipeline in proportion, along with that part of its debt on which interest could be charged to pipeline operation, I calculate a tariff of $5.47 to yield an average return over twenty-five years of about 6 per cent on the revised valuation. This lower tariff, in turn, would be worth about $1.5 billion more to the state in royalties and gross production taxes over twenty-five years compared with what it could expect from an average tariff of $6.21: it would get back what it took off the pipeline cost.* That money has in fact already

---

* In these comparisons I have for convenience used $14 per barrel as the approximate Los Angeles delivered price for Prudhoe crude; and the wellhead price on which Alaska's royalty and production tax are figured is this less the pipeline tariff and the cost of tankerage from Valdez to the refineries (for the latter, see Chapter 12). Since 1973 the controlled price for U.S.-produced oil has been permitted to rise to an average of about $8 per barrel (late 1977). Since *any* tariff reflecting the real TAPS costs would make Prudhoe crude impossible to produce at this price, it was quietly agreed that it should be allowed to sell at the price of imported oil. The standard import price is quoted on a lighter grade of crude; with allowance for the difference in weight, Prudhoe is priced about 50¢ per barrel below this standard.

been spent, however, in construction costs paid mostly with borrowings on which interest will still be due: the loss to oil companies is then not $1.5 billion but more like $3.6 billion (and including that amount, their actual return on investment would be not 6 but 4 per cent with a tariff of $5.47 per barrel). That may indeed be a deserved penalty for careless management. What it would mean in a national sense is a significant withdrawal from the pool of capital available to meet our need for oil and for energy in general. Given what is now known of the capital that will be required to do that job over the next twenty or thirty years, it is not at all clear that we can afford it.

The state's arguments as to what costs should or should not be allowed did not figure overtly in the ICC's decision, though they presumably would in the investigation that would occupy the remainder of 1977. I write these words in advance, with no way of knowing how that examination will come out or what tariffs the ICC will finally agree to accept, on what basis; though at a guess one would expect a compromise, yielding something to both sides. In a sense, the exact conclusion does not matter, at least within broad limits: whatever the questions of principle involved, the Prudhoe pot is still big enough to cover a lot of little errors, and if they mean losses on one side or the other, they are relative losses, not decisive ones.

Meanwhile, in July, before the ICC ruling had been formally promulgated, Mobil, the combative 100 per cent borrower, had taken the commissioners to court, where it was presently joined by Sohio and then Exxon. By October their suit had found its way to the Supreme Court, where on October 20th the justices ruled for precedent and the oil company arguments: that the ICC had exceeded its authority by imposing tariffs without a full investigation and that those originally proposed should be restored; but with the condition that the difference collected be held in escrow and paid back if the full amount was not finally allowed when the commissioners issued their final ruling in January 1978. A straw in the wind, if any were needed: what it means is that the Trans-Alaska Pipeline, which came to birth in lawsuits and courtrooms, is back where it began. To judge from its history so far, it is likely to be there a good long while before all the questions are answered.

By now, any reader who has followed my argument this far is probably tired of numbers. So am I, brother—so am I; particularly when I remember that all I really wanted from Ed Patton and his colleagues was a few valid examples to bring the pipeline costs within light-years visibility of the family budget—something of that sort. And not to be lied to, not to be insulted with vagueness and generality and sloppy bookkeeping. Yet the arithmetic we have worked through is not an abstract problem written on a classroom blackboard. It concerns fundamental realities: how we shall obtain and distribute the energy on which our civilization runs, in what ways the labor of doing so shall be shared among those who benefit from that enterprise; money's mathematical ebb and flow is merely a convenient means of measuring that double apportionment. These are realities that show every sign of growing crueler as the century lurches toward its conclusion. I think we still have a chance to learn them, master them. The alternative to the rigor and the social discipline required for such mastery is to be lulled to death by the blather of reporters, the evasions of oil men, the sentimentalities of politicians and, God help us, Presidents. Or maybe worse than that: that the next generation will no longer have our chance.

Chapter 12

ACROSS THE WIDE PACIFIC

> WARNING. The prudent mariner will not rely solely on any
> single aid to navigation.—National Oceanic and Atmospheric
> Administration, Chart of Prince William Sound

Getting to Valdez by plane begins in a small, square, shabby
waiting room at one end of the ambitious new terminal building
of Anchorage International Airport. Until recently the little
planes that connect with Valdez—and Glennallen, Cordova, Seward, Homer, Kodiak, the Aleutians—all flew from the old commercial airport, Merrill Field, at the edge of downtown Anchorage, but the atmosphere here is the same, secondhand and
provisional: vinyl-upholstered, oversize chrome-framed chairs
along the walls, worn carpet covering the cracked linoleum-tile
floor, cigarette butts and half-emptied paper cups of coffee; re-
pressed anxiety and delay, as in the waiting room of a city clinic.
In one corner a door opens onto the asphalt apron of the runway
where the Valdez plane, if it comes, will pull up like a feeder
bus at a country way station. Back in the spacious main lobby of
the terminal the restaurants and bars and gift shops are thronged
this morning with passengers in transit to expensive destinations,
Los Angeles, Honolulu, Tokyo, over the Pole to Europe, but it is
an older, working Alaska that waits here, enclosed and quiet,
men in the boots and caps and work clothes of their trades,
women in jeans and shoulder-long unbound hair, encumbered by
babies, going to join the men who have gone ahead.

A Polar Airlines girl at a lectern near the door is reading a list of flight numbers and place names: Glennallen will fly; Valdez and the rest will be delayed an hour or more, if they go. It is past the middle of August; a chill sea mist has blown in off Cook Inlet, the beginning of fall, has thickened, settled. At Valdez, I have heard, a big new terminal has replaced the shack where formerly one arrived, but still no lights on the runway, no system for instrument landings, and as before an early flight will have to get across the mountains, scout the changing weather to see if it can get down by eye, and if not, come back. We wait on its report. The chairs are all occupied by others similarly waiting, and I lean my pack against the wall and crouch beside it.

A man next to me is trying to get to Seward: a leak in the lubrication system of a supply tug serving the oil rigs in Cook Inlet, an advanced Swedish design built under license in Seattle; the company has sent him up to try to fix it. We exchange cigarettes and, when I mention Valdez and my hope of boarding a tanker south later in the week, he looks me over and takes a guess that I am a deck officer of some kind—I don't at least look like an engineer; but no, only a passenger, I reply, and leave it at that. Normally, by habit and principle, I travel formally, in suit and tie, but now I am dressed for the ship, and the khaki shirt and pants cover me like a disguise. A week ago, on the flight up from Philadelphia, half-empty till we joined the tourist-filled Anchorage-Tokyo plane in Chicago, I rode with a young Philadelphian hired through his local for two or three weeks of finishing-up work on the pipeline near Valdez—an insulator, he said, an uncle up there on the same job and willing to pay his own way for the chance of a look at Alaska and a little money over; and the work just now around Philadelphia was slack. So when his question naturally followed—"What trade are you in?" and I said "Writing, I'm writing a book"—the information did not call for special comment, it was a trade like any other, a kind of work, and as such intelligible, and we went on to talk about what he might be doing and the camp he was probably headed for, Sheep Creek, one of the better ones, I thought, though big, when I had been there.

The real beginning of my journey had been in Houston a year and a half earlier, with earnest persuasions to Bram Mookhoek,

Exxon's manager of marine operations and the chairman of the Alyeska owners' port and tanker committee. The cheapness and flexibility of tankers for transporting oil had been the compelling reason for choosing the TAPS route and system, and without that connecting link the system itself was meaningless; and as with everything else connected with oil, secondhand experience and mere information are not worth much—until you can see for yourself, you have no way of forming the necessary questions or judging whatever answers they produce. So for all these purposes I was determined, if I could, to be aboard an early ship out of Valdez, and after some weeks of discussion back and forth Bram let me know that his company would not object. Just what ship that might be or when would be a little harder to say, however, for reasons inherent in the pipeline project and its schedule; and in the American merchant marine as it has developed over the past fifty or sixty years.

All along it had been necessary, in order for the system to work, to mesh two quite different planning and construction schedules: on the one hand, the pipeline and its various components; but on the other, the tankers—having enough tonnage available for the Alaska trade to carry the oil wherever it could be sold once it began flowing down the pipe. Put thus, the problem before Bram Mookhoek's committee was complicated by several obvious variables: the completion date for the pipeline and whether that would be delayed, as by the spring of 1976 had begun to seem possible; the levels at which the system would be run, depending on the development of the Prudhoe field as well as the construction of pipeline pump stations, the Valdez terminal; and the old uncertainty as to just where the oil was going—both where, in general, its owners intended it to go and where at any time the necessities of the market would require it. The basic equation is fairly simple. At a production of 1.2 million barrels a day, which the owners meant to reach within the first year, the oil amounted to 167,000 long tons of cargo and would require around 174,000 tons of tanker capacity per day to move it.* But how far? Los Angeles, the center of the West Coast

---

* Tanker capacity, as noted earlier, is measured in long tons (2,240 lbs.), including fuel and supplies, which take up about 4 per cent of a ship's deadweight tonnage (dwt): that is, a tanker rated at 100,000 dwt can ac-

market, is six days' steaming from Valdez, say 15 days round trip with time for loading and unloading. That means a minimum of 15 times the daily tonnage to keep the oil flowing, with no allowance for weather, repairs, and ordinary maintenance. Seattle is closer but by comparison a small market; San Francisco in between, in distance and in its demand for oil. On the other hand, if the flow turned out to be more than the West Coast could absorb—and this from the beginning was the critical question about the whole project—then some of the oil and maybe a lot would have to go a great deal farther and would require the commitment of proportionately more tonnage. Through the Panama Canal to Houston, for instance, where the oil could be pumped through pipelines to the thirsty Midwest, is more than three times as far as Los Angeles: at least six slow weeks round trip. In all this there was a further complication. Whatever the answers to these other questions, the solution would never be one 174,000 dwt tanker a day out of Valdez. Except for Valdez itself (and parts of Puget Sound) there was (and is) no American port deep enough to take a ship of that size when fully loaded. Typically, all around the continent, 70,000 or 80,000 dwt is about the maximum, 60,000 where a passage through the Panama Canal is part of the voyage. The solution, then, would be a mix of many smaller ships. Where big ones were included for their economies in construction and operation, they would bring complications of their own: scheduling enough of the load for Puget Sound to enable them to get into San Francisco or Los Angeles; anchoring off the ports and lightering into smaller tankers capable of going in.

Combining all these factors and the owners' differing notions of their interests, the marine committee produced a series of estimates of the tankers needed to do the job at successive stages of production. The range was considerable, from 21 ships and about 2 million dwt at start-up to twice that many and nearly 4 million dwt at a peak production of 2 million b/d; probably the biggest difference was in the oil men's changing projections of

---

tually carry no more than 96,000 tons of oil. Crude oils vary considerably in weight. The moderately heavy Prudhoe crude runs about 7.2 barrels per long ton. Depending somewhat on its precise design, a tanker of 100,000 dwt should therefore be able to load a maximum of 691,200 barrels.

West Coast demand and how far, therefore, they would have to carry their crude to market. Along the way was one made in April 1973—35 ships, 3.65 million dwt—which became the basis for Rogers Morton's estimate of $1.7 billion (quoted in Chapter 11) for the cost of the pipeline's tanker leg. All of these plans had one thing in common: depending on the precise assumptions, they were at or beyond the tanker tonnage that the American merchant marine could muster.

Under a law on the books since 1920—the Merchant Marine Act, commonly the Jones Act for its principal author—all cargo transported between American ports must be carried in American ships: American-built, American-owned. There is one exception. By a presidential order in 1942, justified as a wartime measure, foreign ships were allowed to supply the Virgin Islands, and that rule has stood. The original law in its protectionist origin aimed at saving an essential core of the American merchant fleet, and that is about all it accomplished. It is often said that the decline of the United States as a seafaring nation has been due to high construction costs and to excessive wages and restrictive work rules exacted by tough unions. So far as tankers are concerned, the evidence on these points is inconclusive; for American and foreign tankers hauling oil comparable distances—from Texas to New York, for instance, compared with the trade from the east Mediterranean to England or Holland—the current charges per barrel per mile are about the same. What does seem clear is that at comparable rates the foreign tankers are generally more profitable—lower in over-all operating costs, supported by government subsidies (in the case of the leading tanker fleets of Europe and Japan) or subject to no more than nominal taxes (the "flag of convenience" nations, Liberia and Panama, whose ships, when their ownership can be traced, seem to be mostly American-owned). In the years since the Jones Act, while the United States remained a leading exporter of crude, American oil companies and shippers thus had little incentive for expanding their fleets—the exports were carried in more profitable foreign bottoms. Now that the situation has been reversed, the same motive prevails. Over the past ten years, while world tanker tonnage has more than trebled to keep pace with the ever greater demand for oil and the economies of giant ships running the long route around the Cape of Good Hope to Europe, U.S. tonnage has remained almost constant, new construction just about keeping up with the replacement of ships written off and scrapped. The ships are generally old and, by current standards and the limitations of American ports, small, their numbers in balance with the needs of the coastal trade; many of them, moreover, are committed not to moving crude oil but to more valuable and profitable products such as gasoline, lighter in weight for the same volume—tankers so designed are less suited to transporting crude.

What all this meant was a new fleet built primarily to carry Prudhoe Bay oil, and this at least was understood in the arguments by which Congress and the public were persuaded to accept the TAPS plan—that the tanker leg would be filled by new-built American ships elaborately equipped with whatever was newest in the way of navigation aids. The oil companies proceeded according to their differing styles and aims. Arco, with its vision of exact balance between oil supply and markets, the supply essentially all from safe American sources, placed its orders early and took delivery on one tanker a year from 1971 to 1975, two of about 70,000 dwt, then three bigger ships, 120,000 dwt each, with two more of 150,000 dwt scheduled for 1979 and 1980 to match its share in the planned increase in oil production. Exxon, the biggest owner of Jones Act tankers among the American oil companies, was less ambitious: in 1969 and 1970 it added three ships of about 76,000 dwt, built to double in transporting petroleum products until whenever the oil flowed; two smaller tankers commissioned five years earlier would also be assigned to Valdez when the time came. That left Exxon short of tonnage to haul its share of the oil. The difference would be made up by a tanker chartered from a private owner, and it was this decision that brought the *Manhattan* out of dead storage in New Jersey where, except for intervals of carrying wheat to Asia and Russia, it had mostly lain since its two-year adventure through the Northwest Passage. Exxon in any case was not strong on the West Coast—its only refinery was a new, small one at Benicia on San Francisco Bay, capable of processing only about a third of the Prudhoe oil it would own (this was the project, smoothly managed in the face of environmental clamor, that got Ed Patton his job as Alyeska's president); but there was evidently money to be made in carrying its own oil, even if it was sold to others.

Sohio's position was awkward: more oil than anyone else and proportionately large demands on its ability to raise money, both for the development of its half of the Prudhoe field and to meet its share in the mounting pipeline costs; and at the same time no West Coast outlets of its own and a perennial shortage of crude to feed its Ohio, Pennsylvania, and Texas refineries. It took delivery of two new ships in 1970 and 1971 (80,000 dwt each), then waited three years before signing contracts for eight bigger tankers (two of 120,000 dwt, six of 165,000 dwt)—but not to be completed until *after* the field reached full production; and with the provision that the contracts could be assigned to other owners and the ships chartered back to Sohio—an arrangement calculated to stretch the company's limited capital until such time as it should begin to come back from the sale of Prudhoe crude.* From the timing of these decisions, it appears that Sohio was gambling on the export of its oil, a possibility foreclosed by Congress; or possibly on a new exception to the Jones Act. Either way, the company would have been able to charter all the ships it needed on the world market, with no need to tie up cash in tankers of its own. As things turned out, it would be seriously short for two or three years, scraping together whatever surplus tonnage it could get from West Coast refiners, Socal, Mobil, Shell, and from private owners.

One other link in the oil brigade went unnoticed and appeared on none of the maritime committee's tanker lists, in none of the newspaper speculation about where the oil was going and by what means. Amerada Hess, a small owner at Prudhoe Bay with a very modest hand in financing the pipeline, does most of its refining at a big installation at St. Croix in the Virgin Islands. In the course of pipeline construction it had nearly doubled the refinery's capacity, to around 700,000 b/d, which makes it, so far as I can determine, the world's biggest. Hence toward the end of

---

* In November 1976 Exxon took over the contracts for the two final Sohio 165,000-tonners, scheduled for delivery in California at the end of 1978. Although Exxon implied that they were needed on its own account, it is evident from the companies' respective supply situations that, by one arrangement or another, Sohio's crude would necessarily be their main cargo.

August, about a month after the first Arco tanker steamed south to Cherry Point, another came quietly into Port Valdez to collect Amerada's share of the oil, flying the Liberian flag. It was a possibility that had been much argued in Alaska. "If our oil goes to Japan," Chuck Champion had said when I asked what he thought of the advantages of exporting the oil, "then foreign-bottom tankers can take it and therefore the safety and environmental standards will have been bypassed totally and we will have any leaky Greek that wants to come into our harbors to load oil." Strong words—but missing the point of the Virgin Islands exception. The *Hercules,* moreover, was no leaky Greek but a respectable 216,000 dwt supertanker built in Japan in 1971 and ultimately owned in New York by the Maritime Overseas Corporation, which was also chartering American-built ships to Sohio to make up its deficit. The Coast Guard inspectors in Alaska finding no fault, the *Hercules* was big enough to borrow a leaf from the Middle East book and carry its 1.5 million barrels of crude profitably the 17,000 miles around Cape Horn to St. Croix. There would in all likelihood be other exceptions, profitable to Amerada Hess and useful to Sohio and BP, perhaps even a revival of the dormant project for a Panama pipeline paralleling the canal. As the *Hercules* sailed south, Amerada was completing negotiations for another big refinery on a nearby Caribbean island, St. Lucia, whose status is that of an independent state "associated" with Great Britain. A lot of the Arctic oil, it seemed, would be getting to its American market (or elsewhere) by way of the tropics.

In July, with the oil flowing, I had talked once more with Captain Mookhoek about my tanker hopes and this time he offered a bunk on the *Philadelphia,* one of Exxon's new 76,000-tonners launched in 1970, sailing from Valdez sometime the third week in August—with the changing pace of start-up the tanker managers were having difficulty keeping to exact schedules, and just which of the West Coast oil ports it would be going to was also a little uncertain, maybe even Panama. Anyway, lucky, I thought; coincidence is luck. When I got off the plane in Anchorage, however, the news in that evening's *Times* was of another shutdown: the spill at Pump Station 9 in circumstances alarmingly similar to those that produced the earlier disaster at Pump 8 (see Chapter

10). This time the line would be closed indefinitely, while federal officials flew in from Washington to look and ponder and Alyeska shuffled its technicians and considered ways of reconnecting the pump-station piping so as to make this particular human error less likely to recur. Meanwhile the tankers already on hand were backed up at anchorage off the Prince William Sound shipping lanes and those en route were being diverted, where possible, to other jobs. Just when they would be moving again or whether any would have a spot for an inquisitive author was anyone's guess.

While I waited to find out, there was a good deal to catch up on in Anchorage. One of the Native corporations had bought the leading hotel, the Westward, and made a deal with the Hilton chain for management and the name; another had built a stylish shopping mall, painted an emphatic yellow, on 4th Avenue across from the city hall. Chuck Champion had quit as State Pipeline Coordinator and gone back to the oil business. Andy Rollins had long since departed from the federal pipeline office, and after months of temporizing, the new administration had finally named his assistant, Jack Turner, to run what remained of the job, a man I would have to get acquainted with. Alyeska was negotiating to sell off its construction equipment and most of the camps were closed, though just which ones was as usual one of those bits of minor fact not easily ascertained; Sheep Creek, for instance, where my Philadelphia insulator friend was working—definitely one of the shut-down camps, I was told, though I knew for certain it could not be. Visits to the pipeline and the remaining camps were more restricted than ever: the federal and state inspectors could not be kept out, of course, but anyone else—. Of course, they said, you can drive along the public highway and look if you like, get hold of a car somehow: can't stop you from doing that now, can we?

After a week of this, when the various officials pronounced themselves satisfied and the pipeline was allowed to restart, I telephoned Captain Fisken, the BP tanker man in charge of Alyeska's port operations at Valdez. Bill Fisken's voice came over with a crisp, offhand assurance that was both North British and maritime. By now there were ten or a dozen of the big ships backed off in the Sound, but maybe, he thought, the *Manhattan,* coming in from refitting in Japan under charter to Exxon, and what would I say to that? Given the uncertainty of the oil flow, it was hard to say just when, perhaps in a week. I said I would take my chances and come on down. There were things too to be done in Valdez while I waited.

The Polar Airlines girl is reading her flight list again. Mine is on, but no promises; they will give it a try and come back if they cannot land; the rest for today are canceled. I have been advising the tug repairman to rent a car and drive to Seward, two or three hours, easy, and scenic; sometimes the weather lingers like this for days and the little planes do not fly. Now with the announcement, he bolts for the lobby ahead of the crowd, hunting a phone. I swing the pack up on one shoulder, gather camera and briefcase under arms, and follow the others, the lucky ones, out onto the misty runway. The copilot is standing beside the steps, stuffing luggage that will fit into the wings, what doesn't, my pack, for instance, into the tail compartment. A Volpar Turbojet he says when I ask but the curious name means nothing to me: silvery aluminum, then the dark cabin with its porthole windows, seven seats a side, tightly fitted as opposing scrums in a rugby match, and I sidle up the narrow aisle and slip into the last empty seat. The copilot comes past, stooping, pushes the cockpit curtain aside and takes his place; leaning, I could touch his shoulder if there were anything I had to say. Someone on the ground folds up the stairway and slams the door, hard.

The plane route to Valdez combines several that have given Alaskan history its particular shape: up the choking mud flats of Captain Cook's Turnagain Arm, then through the pass at the railroad junction of Portage and over a glacier where once prospectors toiled with their gear in packs and before them the Eskimos of this southern coast; and out onto a broad corridor of Prince William Sound, Wells Passage, also British-named, walled off by barren islands, where for a time Baranov's Russians tried building ships and the fishing today—salmon, halibut, herring, the three species of crab—is still plentiful enough to have furnished one of the subsidiary arguments over whether TAPS and its tankers should be allowed. There is not much of all that to be seen just now. The plane circles, tilts, climbs through the directionless dirty-gray wool of cloud, leaps like a fish into sun and levels off, skimming east. Below us a jumble of peaks pokes through, the Chugach Mountains, eight and ten thousand feet, formed by their looks of the same gray-black rock that had troubled the builders at Valdez, their shoulders creased with unmelting snow, the deep valleys between filled with thick cloud as with eons of cosmic dust ground fine by the ebb and flow of the universe: the riven face of a dead planet to which life in its softness has not yet come. Then for a moment the cloud mass opens and I am looking down through the blue light at the broad ice field that is the Columbia Glacier; the glacier's lip, where the great bergs break off like bits of glass from a shattered pane and plunge into Columbia Bay to begin their southward drift through Prince William Sound, stays hidden.

In front of me now the copilot is tuning the radio to Valdez. A voice comes on loud, blurred with static: another opening over the airport, move along and we can get down; reassured, he switches to another channel and the cockpit resounds with rock music from the Valdez station. The plane swings north up Valdez Arm, descending through alternating layers of clear air and cloud and up the narrow basin which is Port Valdez to its head, the gravel flats at the mouth of the Lowe River. Banking sharp against the mountains, the plane drops to the runway and taxis to a short stop in front of the terminal. The cabin door opens and we tumble out.

A few minutes later, waiting for the cab into town, I remember the sky and look up. Closed in again; it does not look as if any more planes will get down just now for a while.

Valdez is the basic Alaskan paradox reduced to a scale that can be taken in with one slow sweep of the eyes: stores and houses indifferently scattered across the new townsite like tin cans in an empty lot; behind them and all around, the sheer rim of mountains, topped with glaciers from which streams fall in silently cascading ribbons of foamy white down walls lightly spray-painted with the heath-colored gray-greens and browns of the native plants. The empty-lot simile only partly fits: the rudimentary architecture is set on a level plain of oil-dark rubble, the scourings of an ancient moraine scraped together and bulldozed flat—a thousand-acre desert unrelieved by the random green of grass and weed and tree, and on dry days when the wind sweeps the gravel of the two main streets, the black dust lifts and swirls. Here and there a builder has made a self-conscious gesture: the square varnished form of a Tyrolean chalet, steep-roofed and wide-eaved, the second-floor balcony fretted with scrollwork. The landscape derides such allusions. This place and its people have not yet attained a style; if the great quake comes again with its seismic wave—*when* they come—their scratchings will be wiped clean as if they had never been. The elements will be left as they were, mountains and glacial streams and the fathomless deeps of the sea arm which is the Port; and the original question, still unanswered and barely asked—what living connection can man have with this land?

Among these failures of imagination, there is one partial success. From the road, the Sheffield House presents the same bleak aspect as the rest; three low wings around a muddy, rutted yard where guests' campers, station wagons, and light trucks are parked in rows. On the other side, however, beside the water, black topsoil had been spread from somewhere and a narrow lawn. The dining room and bar look out on the small-boat harbor sheltered behind a wooded spit, orderly ranks of sport cruisers rocking at their moorings. (The working boats of the fishermen, 500 of them, run mostly from Cordova fifty miles down Prince William Sound.) From the bedroom windows you can study at leisure the layout of the terminal, the comings and goings of the tug-led tankers at its four loading docks. Across the three miles of Port Valdez these structures—gray-green tanks and administrative buildings, the tall stacks of the power plant and ballast treatment facility—seem as remote and unobtrusive as the features on a relief map. Indeed, as long as there has been a town there have been human workings of some kind over there: the barracks of Fort Liscum, the docks and sheds of a mining venture high up the mountains, the fishing camp and cannery of Andrea Harrison's Grandfather Day. Only the terraces behind the two tank-farms, cut deep for fear of quakes and slides, are obvious, and even these emblematic wounds are ambiguous at this distance: carved in the rock, you might guess, not by engineers and machines but by geological ages of weather.

Altogether, if the terminal and its tankers are what interest you, the Sheffield is a pretty good place to start.

It was lunchtime when I got in from the airport, in a Checker limousine with a leaky roof and a family of Japanese tourists. I made for the Sheffield dining room and helped myself to a table by a window where I could watch the boats. The waitress was red-haired, chatty, quick, an improvement on my last time here. There was a trace of British in her speech and I asked, but no, it was her father, she picked it up from him, she thought. Just off the pipeline? I wondered, testing a theory, but no to that also, she was only up two weeks ago from California, taking a break from college. The crowd too had changed and thinned since the pipeline finished: tourists mostly, the Japanese, miscellaneous Americans, what seemed to be a solitary Scandinavian, flown over from Anchorage for a day or two and the fifty-dollar boat ride out to look at the Columbia Glacier; among them a sprinkling of men in pea-cloth shirts and jackets, jeans tucked into rubber boots, wearing round, black-brimmed Baltic caps—the pilots specified by a recent Alaskan maritime law to guide the tankers through Valdez Narrows, a job worth a shipmaster's pay and overtime and the right to draw it twelve months a year, say $80,000; the last and biggest of the big-money pipeline jobs.

At a corner table at one end of the room I recognized a round, pink, insolent face and brought the almost unspellable name and identity back from the summer before: Ron Mierzejewski, the local Alyeska PR man, holding forth to a couple of chums over a long, late lunch. Not much to keep him busy any more and pretty soon nothing, in fact; he would not, it seemed, be among those Alyeska was keeping on, but he was telling it loudly as a hero's progress, flying to New York to discuss a very top-level job in the corporate headquarters of one of the oil companies. At least I would not have to talk to him; he was not one to remember a name or a face. I sipped my beer and went back to watching the tethered boats. I had arrived, it seemed, for the Salmon Derby—a poster in the lobby, others in storefronts around the town, announcing a beer fest at the end of the week and a prize for the biggest and most fish. Silvers now, the kings and the big-head, hump-shouldered reds were past. Below the window, from a floating dock at the near end of the boat harbor, a man dropped a line in and lifted a fish, flopped it on the planks, an eight-pounder by the look of it and gleaming as the hand-hammered, thumb-polished metal, silver all right, a silver salmon; then another, easier than telling. He hefted them into an orange plastic pail of salt water. Too easy. I wondered if they counted toward the prize.

Back in my room I telephoned George James, the state surveillance man at the terminal, who with his knack for making engineering intelligible seemed like an old friend by now, a trustworthy anchor against Alyeska's official vagueness and evasion; running short of work himself, though he was scheduled to stay on to the end of the year, and wondering what he'd do next, maybe have a shot at reviving his Anchorage construction business now the pipeline furor was past. Sure, George said, he'd been expecting me, have a pass for me tomorrow and take me around—but only, he added, the areas where construction was still going on and subject to the State Pipeline Office; the rest, Operations, was fenced off and outside his authority. I tried Bill Fisken again, got an assistant—the captain was over at Berth 1 where a tanker was loading—but the *Manhattan* was still on, should be coming in on Thursday, loaded and sailing a day or two later; they meant to look her over pretty carefully before they went ahead.

On a knoll a hundred yards west of the Sheffield the Coast Guard that summer had finished its new headquarters for the Valdez operation, a square, heavy, cement-walled building that looks built to withstand earthquakes and big waves, maybe even enemy attack. The site is screened from the town by scraggly trees; in front a plot had been spread with black soil that in time would be lawn. Another exception to the Valdez rule, perhaps: from the top of the rise you can see the whole of Port Valdez and in particular the panorama of the terminal across the water; down the slope, just inside the entrance to the boat harbor, there was a new dock with a broad-beamed, black-hulled Coast Guard cruiser tied up. When I went in, asking for someone who could tell me about what they were doing, the seaman behind the desk waved me upstairs and I found myself in the office of the CO, Commander Homer Purdy, a curly-haired, well-fed ebullient man of perhaps thirty-five. His business card, imprinted in one corner with the anchor-trimmed seal of the Coast Guard in red, white, and blue, carries a string of titles that sum up the job: Commanding Officer Marine Safety Office, Captain of the Port, Officer in Charge Marine Inspection; and a slogan evoking its dual purpose—"Protecting man from the sea . . . and the sea from man."

That purpose is embodied in the operating manual for the Vessel Traffic Service designed to regulate shipping at Valdez and in Prince William Sound, issued by the Coast Guard in July 1977: essentially, a system like that used at major airports for controlling the speeds and flight paths of arriving and departing planes; it has evolved from similar systems developed for areas of heavy ship traffic from the English Channel to the California coast. At Hinchinbrook Island—another of Captain Cook's names, the entrance to Prince William Sound sixty miles out from the Valdez Narrows—incoming tankers of more than 20,000 dwt are required to align themselves in a traffic lane 1,000 to 1,500 yards wide, separated from outbound shipping by a distance of about 2,000 yards. For smaller ships and boats—the Cordova fishing fleet—the traffic lanes are optional, and given the difference in size and speed they will for the most part prudently stay out; similar but broader lanes are also specified in the immense reaches of the Gulf of Alaska. At the Narrows, the minimum distance from shore to shore is about 1,700 yards, complicated by the tip of an underwater peak sticking up 800 yards out from the west bank—Middle Rock. "A can opener," Chuck Champion, the State Pipeline Coordinator, had called it when I asked about possible hazards to tankers going in and out of Valdez, and his eloquence warmed: "The most probable catastrophic occurrence I can think of is Middle Rock at the entrance to Valdez Narrows—a can opener sticking up. You're talking about a one-million-barrel oil spill if you rip one of those tankers open." The boats make their way unconcernedly on either side of Middle Rock, and probably the tankers could too, but with allowance for wind, tide, and current, the safe maneuvering room here narrows for three miles to about 900 yards to the east of this shoal, over depths of from 300 to nearly 800 feet. These are still extraordinary dimensions compared with most other tanker ports in the world, but after the anxieties stirred by the pipeline debates the Coast Guard ruled that traffic past Middle Rock should be one way, one ship in or out at a time, at speeds reduced to 12 knots inbound and 6 knots outbound with a load of oil (16 knots—about 18½ mph—is normal sea speed for a modern tanker); in bad weather, the Coast Guard might declare the area closed altogether. All these orderly plans would, however, remain subject to a principle as old as seafaring: that the master, with final responsibility for the safety of his ship, would in necessity (and with notice to Commander Purdy's traffic controllers) maneuver in accordance with immediate conditions as he perceived them.

The assumption behind the Coast Guard's VTS system was, of course, continuous communication between its control center and all big ships in the area; this was the purpose of the new installation I had gone to visit. Coming into the Hinchinbrook Entrance, a tanker was to identify itself, giving its dimensions and draft, its course and its expected time of arrival at the terminal, and to stay on the Coast Guard radio frequency until it docked; it was up to the Coast Guard monitors at their head-sets and radar screens to keep track of all tanker movements (and of any boats that might stray into their lanes), to advise changes of course or speed, and in extreme weather to close the Narrows to traffic. So that this could be done, the regulations required tankers of 20,000 dwt or more to be equipped with two radiotelephones, one of them battery-operated in case of power failure; and, on the same principle, two radar systems; besides the normal navigation equipment, the rules called for the latest model Loran receiver on the ships—an instrument for establishing exact position by co-ordinating two radio beams, an acronym for *lo*ng *ra*nge *na*vigation. The counterparts of this shipboard equipment were the radio and radar systems the operations center had been built to house, served by antennae at Valdez, in the Narrows, and dotted around the islands and promontories of Prince William Sound.

As we neared the end of our chat, Commander Purdy led me along to the darkened room where the new equipment was being installed. No one as yet was fully trained in using it, but he knew enough to leave me properly impressed, as of course he intended. With a twist of a dial the blinking dots of light on the circular radar screen changed to vectors showing directional bearings of any moving object within range, from a rowboat on up, and at the touch of a button a computer supplied its speed. Another adjustment and the contours of the shores came into view in green-lighted relief; again, and the center of view switched from the antenna on the harbor breakwater to another far out in the Sound, then enlarged the scale to zoom in on a narrow patch of water. The commander apologized for not demonstrating more of the system's capabilities; it was an advanced version of one designed originally for San Francisco Bay, and they were still learning what it would do and how to use it.

If the Coast Guard's preparations sound elaborate even in outline, there was some reason for care. With the coming of oil, Valdez would rise from near invisibility as a center of merchant traffic to one of the nation's biggest ports. At 1.2 million b/d, it would be handling something over 60 million dwt of shipping a year, substantially more than Philadelphia, the country's fourth port. When and if production reaches 2 million b/d, the annual tonnage in and out of Valdez will be surpassed only by New York and New Orleans. On the other hand, these large volumes of cargo represent comparatively few ships. The tankers scheduled for the trade by Bram Mookhoek's maritime committee average better than 100,000 dwt each. Thus, at 1.2 million b/d the actual port traffic amounts to about 1½ ships a day, on average: one ship at one of the four docks, loading crude, a process that, with time for deballasting, receiving stores, and shore leave for the crew, should take around twenty-four hours; another arriving to begin deballasting; a third still three or four hundred miles out in the North Pacific. Even at 2 million b/d, when the traffic would amount to fewer than three tankers a day, it does not seem that collisions are a very likely hazard—in normal operations. And perhaps that is the point of the VTS system: those times of storm or pipeline shutdown when the tankers cannot get through the Narrows and back up in the sound, gather in the anchorages, then come in at tight intervals to load the crude; in trying to connect with my tanker I had already gone through a mild sample of the disrupted scheduling that would result.

So far as the tankers themselves were concerned, the new Coast Guard regulations were quite modest: essentially, a few refinements in traditional ship-handling assisted by equipment that could be added without major refitting or expense (and which the new ships already had, as a decision of prudent management or in anticipation of the rule changes). This limited strategy was greeted with dismay by Alaskan politicians and the environmentalists who interested themselves in Alaska and its oil. They had argued for and expected something a good deal more drastic, and in the current adversary style of American law and politics they responded with charges of deception, conspiracy; with laws and lawsuits.

In June 1973, testifying in defense of the TAPS route before the Joint Economic Committee of Congress, Rogers Morton had this to say of the tanker end of the system: "Newly constructed American flag vessels carrying oil from Port Valdez to United States ports will be required to have segregated ballast systems, incorporating a double bottom which will avoid the necessity for discharging oily ballast to the onshore treatment facility." The Secretary was speaking extemporaneously but the qualification was carefully considered: these changes in tanker design would be required of "newly constructed" ships—*not*, therefore, of those then committed to the Valdez trade, most of them already built or at least ordered. Moreover, the implication that "segregated ballast" would be carried in a "double bottom" suggests that he did not understand what he was saying.* Oil men and representatives of the Coast Guard, which would be responsible for setting and enforcing the rules for the tankers carrying

* After discharging its load a tanker pumps sea water into its tanks as ballast to provide stability on the empty return voyage, amounting in good weather to around 12 per cent of its capacity. A film of oil clings to the sides of the tanks and mixes with the ballast: probably no more than 0.5 per cent of the original load (possibly as much as 1.5 per cent) but in fact a good deal of oil in a big tanker, perhaps 3,500 bbl in one of 100,000 dwt. This residue must then be washed out before the ship can be loaded again and until recently has simply been pumped overboard at sea. A newer system involves setting aside some cargo tanks for ballasting so that oil and ballast water will not mingle—this is the meaning of "segregated ballast." At Valdez, tankers not so designed are obliged to discharge their ballast into the ballast-treatment plant described in Chapter 5, and since this takes time it is an incentive for the tanker owners to swich to a segregated system. The double bottom, which we shall consider in a moment, is a quite different idea, and Mr. Morton's linking it with segregated ballast suggests confusion on his part.

Alaskan oil, offered general assurances, but so far as I can find the Secretary's qualified commitment on these points was not repeated by any official with the authority to do anything about it. Nevertheless, those who were already nervous about TAPS and its tankers seized on his remarks as an explicit promise; and were the more disappointed when it turned out to have no part in the reality—the revised tanker regulations issued by the Coast Guard.

Thus the Alaskan state senator Chancy Croft, in a conversation toward the end of pipeline construction: "We were told by the Coast Guard that there were going to be double-bottomed tankers. And then as soon as Congress acted the Coast Guard started backing off." And Eldon Greenberg, one of the lawyers associated with the Center for Law and Social Policy in Washington (see Chapter 4): "Back in 1973 and '72 there were a lot of promises made, promises that sounded like commitments to ensure that the marine transport leg of the pipeline would be something special in terms of environmental protection. The claims that were made were pretty extensive and they just haven't been fulfilled." Both men were in a position to do something about their displeasure: Chancy Croft by sponsoring a bill that would make double bottoms and other radical innovations in tanker design mandatory for ships entering Alaskan waters; Eldon Greenberg by filing suit toward the end of 1975 (on behalf of the Audubon Society, the Wilderness Society, the Sierra Club, and five other environmental groups) in order to compel the Coast Guard, by its regulations, to require the same features of all American tankers—not only those serving Alaska, not only those to be built in the future, but if possible those already in service and the foreign tankers that carry the swelling volume of imported oil to American ports.* At both levels it was a meat-ax solution to an admittedly serious problem, involving retroactive law and delicate international agreements governing the tanker trade; and ultimately the survival of the U.S. fleet. It is true, however, that Mr. Morton and his allies did not deliver on his

---

* Mr. Greenberg has since left the center, and as I write (late 1977) the case is in abeyance, waiting on the issuance of final tanker regulations by the Coast Guard, when, it seems probable, it will be resumed and the issues ultimately decided in court.

offhand promise. Whether in reality they could have done so and whether, if they had, it would have provided any practical benefit are different questions altogether.

In a modern tanker, except for a comparatively small space at the stern occupied by the engine room and storage for fuel and supplies, about five sixths of the hull is taken up by oil tanks: "wing tanks" along either side with pairs of center tanks between, say four such groupings or sixteen cargo tanks altogether in the typical current layout for tankers of moderate size. Since there is nothing between the cargo and the sea but the steel plates of the hull itself, if the ship runs onto a reef and knocks a hole through its bottom, some of the oil will leak out, with disagreeable results—perhaps the entire contents of one tank, which in a tanker of 100,000 dwt might be 40,000 bbl or more. Hence at first thought a double bottom sounds like a sensible precaution: a sealed inner floor to the oil tanks, creating a void space between the oil and the sea (1/15th of the ship's beam is the proportion most often suggested, which in a tanker of 100,000 dwt, with a beam of 150 feet, would place an insulating space of 10 feet between the oil and the ship's bottom). Then, one would suppose, if a rock broke through the outer skin, it could not penetrate the tanks, no oil would be lost, and the marine environment would be safe. On this line of reasoning, since tankers can also be damaged by collision with other ships, it would seem even better to extend the cushion up the tanker's sides and create a complete double hull; six feet is the usual suggested minimum for the distance between inner and outer hull.

Given the apparent reasonableness of these ideas, it is puzzling to find that nearly all tanker men—masters and their managers, designers, Coast Guard and Naval officers—are unconvinced of their benefits; on the other hand, the vehement advocacy (colored with accusations of bad faith and unenlightened self-interest) has come mainly from politicians and environmental lawyers—from men who with few exceptions have never worked on or commanded a tanker or any other ship, never had the responsibility for managing a fleet.

Construction cost by itself does not seem to be a decisive issue in this argument. If designed into the ship at the time of construction, a double bottom and/or a double hull has been estimated to add somewhere between 4 and 10 per cent to the cost for the same dead-weight tonnage. (Any requirement to add these features to existing ships, however, would simply be a death sentence; they would be scrapped.) The tanker men's objections center on the question of whether such designs would in fact enhance the safety of ship and load or have the opposite effect. A tanker's hull is braced by a multitude of steel beams, typically an inch or more thick and two or three to five or six feet wide, laid crosswise and end to end. An inner bottom or hull would then be welded on top of this structural framework and the protective space thus created would not in fact be void but laced with a network of steel connecting the cargo tanks with the ship's outer hull. Hence, any force great enough to penetrate the hull would probably spike one or more beams through the tanks as well. Moreover, if the space between were empty, it would fill with water and in a grounding the ship would settle deeper and harder, doing greater damage to itself (and in normal operation the ship would ride high and top heavy, unstable); if the space were used for ballast, the water, being uncompressible, would act like a hydraulic battering ram in an accident, crushing the oil tanks. These arguments were supported by computer simulations —there are a few tankers with double bottoms or hulls, used for carrying crude, petroleum products, chemicals, but not enough to tell much statistically about their comparative safety; after studying the grounding of a chartered supertanker, the *Metula*, in the Strait of Magellan in August 1974, in which a quarter of the load was spilled, Shell, which owned the oil, concluded that with a double bottom the ship and its entire cargo would have been lost. Indeed, as a result of such studies there have even been serious proposals within the International Maritime Consultative Organization* to make double bottoms and hulls *illegal* for crude-oil tankers. I think we may conclude that the benefits of these innovations are, at the very least, undemonstrated.

Throughout this debate, neither party has paid much attention to a far more serious point, possibly because environmentalists and politicians are not as a rule much concerned about how they spend other people's money: the effect of these proposals on

* IMCO, a United Nations agency that provides a means of setting shipping standards by general agreement among its national members.

tanker dimensions and load capacity and therefore on all operating costs in the movement of oil. On a large tanker of say 100,000 dwt, 1,000 feet long, with a beam of 150 feet and a draft of 50 feet, a ten-foot double bottom would subtract about one fourth of its load capacity, something over 170,000 bbl: with a complete double hull, the ship's capacity would be reduced by about 34 per cent. The cost of transporting oil in such a tanker would rise in proportion—that much smaller load for the same operating cost. And since the movement of oil by tanker is a fact of life that we cannot wish away, the difference would have to be made good by additional tankers—more ships, more traffic in and out of the major oil ports, more risk of accidents, more expense. Conversely, if the tanker were enlarged to accommodate the same capacity within a double bottom or hull, its draft, as we noted earlier, would keep it out of most of the American ports where oil is needed. Either way, the effect would be much the same: significant increases in the cost of moving oil which ultimately would be borne by all of us who depend on oil products. I am not sure we can afford it; not, certainly, without decisive evidence that it is both useful and necessary.

The argument nevertheless continues. While Eldon Greenberg and his colleagues pursued their lawsuit, Chancy Croft laid before the Alaskan Senate a new bill with similar aims. This was a complicated piece of legislation, running to thirty-two pages in its draft form. Its central purpose, however, was fairly simple: to compel shippers carrying oil from Alaska to adopt a number of specific design features for their tankers. It makes an interesting list: (1) double bottom and/or double hull in combination with a segregated ballast system; (2) a gas inerting system (recovering the oxygen-free discharge from the ship's smoke stack and piping it into the tops of the oil tanks so as to drive off oxygen and thereby lessen the possibility of explosion by vapors rising from the oil); (3) lateral thrusters (small propellers placed near the ship's bows, facing out, to provide extra push in turning); (4) controllable pitch propeller (same idea, with blades on the main propeller whose angle can be changed) or astern horsepower equal to 40 per cent of forward horsepower (to provide faster stopping when the engines are reversed); (5) a spare boiler (to assure power in case the main boiler failed—most tankers are run by some form of oil-powered steam turbine). These standards were to apply to tankers of more than 40,000 dwt—as if to encourage the shippers to use their smallest, oldest, and least economical tankers. In the absence of the last three features on the list, the tanker would be required to be escorted by tugs while in Alaskan waters; but that requirement might be waived by the commissioner of the state's Department of Environmental Conservation, who was made responsible for administering the law.

In its purpose, the bill is obviously unconstitutional, as Chancy Croft was free to admit. He therefore adopted an ingenious strategy to accomplish it. The five design elements were not to be directly imposed. Rather, the bill created a "coastal protection fund" of $30 million, to be used in cleaning up oil spills or for other purposes, at the commissioner's discretion; this fund, in turn, was to be made up from fees charged for obtaining an annual "risk avoidance certificate" for each tanker; and the price of this certificate was to be set by the commissioner in proportion to the presence or absence of the five features specified. Whether this arrangement is any more constitutional than the direct means of attaining the same purpose, I am not lawyer enough to judge; but the oil men, after flying to Juneau to protest, took it tamely enough. The new law, by its terms, went into effect on July 1, 1977, about the time of start-up.

The several items on Chancy's list nevertheless deserve a closer look. We have already seen that the double-bottom/hull concept is of at least doubtful value and probably economically impossible for the general tanker trade (and the legislature adopted the Rogers Morton fallacy, specifying that the double bottom/hull be used for segregated ballast). A gas-inerting system is comparatively easy and inexpensive to install and of some value; and has no effect on seamanship. Lateral thrusters and controllable-pitch propellers, on the other hand—from what I can glean from interviews and from reading the literature—may improve stopping and turning at sea speed but have little effect at the low speeds and in the often shallow port waters in which sharp maneuvering is required; and in an emergency a variable-pitch propeller on a big tanker may risk broken blades or a broken shaft. As to the second, or redundant, boiler: most steam-tur-

bine tankers carry a diesel engine to supply electrical power if
the boiler fails (and turn the propeller to get home at sailboat
speed in such an event); but the boiler rarely does fail—and if it
does, a spare is not much use in an emergency, since it takes
twelve hours to fire up.

It does not seem to me that Alaska or the nation got good
value from this particular piece of legislation. The premise itself
is odd: that a state legislature (or a federal judge, in the case of
Eldon Greenberg's lawsuit) is a proper authority for designing a
tanker. The comment is not entirely flippant. The question is not
so technical as to be beyond the reach of informed common
sense—that is what we have been doing here, informing ourselves; and we had better, since we are vitally affected. But to
make an informed judgment, we must address the question in
its fullness and reality. That, I think, these technical requirements have not done. The common element is the amateur's
single-cause trust in mechanical solutions; and a corresponding
distrust of the human skill and judgment which are seamanship
—the same characteristics apparent in the efforts to control the
design and construction of the pipeline itself. It is predictable,
if past experience holds, that the bad weather of winter, when
the demand for imported oil is greatest, will continue to be
marked by serious tanker accidents in American waters—five
such in the harsh winter of 1976–77, each greeted by a chorus
of outrage and prophecies of environmental catastrophe. (The
fact that such predictions are *almost never* fulfilled does not,
I think, mean that they need not be taken seriously; we simply
do not know enough about the workings of the marine ecosystem
to say with confidence how it may be affected by massive in-
fusions of foreign substances, including oil.) When we examine
particular cases, however, it does not appear that mechanical
failure as such is a primary cause of tanker accidents. What we
find instead are ships where the captain may be the only licensed mariner on board; where engine room and bridge watch
at the time of the accident were manned by ordinary seamen
of limited experience; where the bridge is equipped with modern
navigation and communications instruments that are out of order
or giving erroneous information or simply turned off because
no one is competent to use them or keep them in repair; and so

on. This is the real issue: not simply "human error" but a sufficiency of minimal human competence measured by a recognized international standard, the master's license, the whole series of mates' and engineers' licenses by which seafarers are ranked and judged by their peers. To ignore this issue while pursuing a legally imposed mechanical cure-all is a serious confusion. And it is not an error for which reality shows much tolerance.

The whole question is unsettling: the righteous certainties of the lawyers on the one side, the steady demurrals of the American tanker men on the other; how is one to choose between these opposite self-interests? Seeking a neutral opinion, I tried it on a British brother-in-law who in thirty-five years from cadet to commander in the Royal Navy had put in a long tour as chief engineer of the North Atlantic Fleet. Eric's answer was a gruff sailor's judgment: in essence, trust in men and their seamanship, not machines; and don't tie their hands too tight with bloody regulations made in law courts and committee rooms. So in taking my leave of Commander Purdy I framed my stock question again—it stands for the whole array of lawyers' solutions—and what did *he* think of this idea of putting double bottoms on the tankers? His answer was a cheerful skepticism about cost and effectiveness; but of course if that was *the law*—. His is a dual role, divided by opposing loyalties: an officer speaking from within the Coast Guard's long maritime tradition, but also now an official of the Department of Transportation; another bureaucrat.

The entrance to the small-boat harbor is formed by a pair of breakwaters made of the universal black rock, new-built since my last time at Valdez, shielding a channel twenty-five yards wide by fifty long. On a steel tower at the base of one of them a radar antenna revolved in its pulsing heartbeat, fifty beats a minute, the source of the images Commander Purdy had shown me on the screen in his communications room. The other like an unfinished causeway pointed straight across the Port at the center of the terminal and its rows of storage tanks, a steel framework of some kind at the end, with a red warning light and a sign. Walking back to the Sheffield, I went out on the breakwater, setting my feet carefully on the shifting slabs of rock The sign was a crudely stenciled notice warning boaters out of a 200-yard security zone around the terminal and any transiting tankers, enforced by the Coast Guard—but not so far with any strictness; the several streams that fall out in and near the terminal are good spots for fishing and just now, with the Salmon Derby, were crowded with small boats and their hopeful anglers. The morning's cloud mass had lifted and broken up in feathery streamers caught on the mountain slopes, halfway up. The light pounded the water like a tropical rainstorm, so tangibly bright it was hard to see through. I got out field glasses, peered: one tanker in now at Berth 1 on the east end of Pine Point, riding high, not yet loading. Arco's dark-blue trapezoidal emblem on her stack—the *Prudhoe Bay*, a 70,000-tonner, the first built in the company's Alaskan tanker program, and at this distance, against the expanse of the terminal site and the cloud-wreathed mass of the mountains, with her deckhouse and bridge squeezed back against the stern, she looked dumpy as an elongated rowboat. I zipped up my parka and sat down to watch; it was as close as I could get. Even with the sun out it felt cool in the light breeze sliding off the mountains.

I spent a lot of my time in the next two days out at the end of the breakwater, watching, or at night, trying to read in the bare hotel room but drawn to the window, leaning on the sill to stare across the Port at the blue and pink lights scattered across the terminal slope. The *Prudhoe Bay* topped off and moved out, another still smaller tanker came and went; the heavy cloud returned and settled, stayed, dripping with fine rain. In between were bursts of activity: a morning going the rounds of the terminal with George James in his Alyeska Blazer, followed by a prolonged, moody lunch back at the Sheffield while we watched the boats and talked of how the job had gone and what he would be doing when it was finally over, and the redheaded California girl trotted chattily back and forth, pouring coffee. I put my name down for a rental car at the Avis agency and when one came through made a late-afternoon dash through Keystone Canyon and over Thompson Pass—out of the Valdez mists and into the brilliant fall sunlight of central Alaska—north 70 miles as far as the last pump station for a final look at the completed pipeline: not much to see—the rebuilding program on the

Richardson Highway, torn up most of the way (see Chapter 9); the buried pipe smoothed over with gray bedding and here and there sprouting grass and shoots of alder; and through the spindly Alaskan pines the shiny new insulation on the above-ground pipe gleaming in the twilight like a beer can in the grass of an empty lot—but that is how we wanted it, brothers, and that is what we got. And all the way, pushing the little mud-caked car past the glaciers and down the narrowing, north-sloping valleys, racing a light plane zigzagging back up the pass, a refrain pounded in my mind: there are no second chances, *there are no second chances;* take the experience as it comes or not at all. Returning to Valdez I stopped off at Sheep Creek to hunt another surveillance man, Bill Thompson, who had put me up there with my son the summer before, and make sure the camp was indeed working. There were two entrances now, one for the union men, the other for the nonunion maintenance group hired through a Native contractor, with pickets at both, sitting on folding chairs, grilling hotdogs on little fires; the strike had formally begun that afternoon and for them, however it came out, the job was over. The Wackenhut guard at the gate was busy—someone had tried to burn a truck inside the fence—but phoned around. Contrary to custom, Bill was not in his office or his room; for him too the job was about over.

On Thursday morning with the rising tide a small black Mobil tanker came into the Port and tied up at Berth 4; then gently maneuvered between two tankers, like a once *grande dame* in her decrepitude leaning on a pair of sticks and supported by retainers, a very big tanker with her bridge amidships in the old style, riding high in ballast, twenty feet of red-lead hull showing like petticoats below her black load line. The *Manhattan* was arriving, as Fisken's man had promised.

I telephoned Captain Fisken again and was breezily assured that a pass would be waiting at the terminal gate whenever I was ready to come aboard. The tanker was scheduled to deliver most of its cargo to a Socal refinery at El Segundo on the north side of Los Angeles, where an offshore pipe enabled it to tie up and unload in the deep water of Santa Monica Bay (I remembered the place—a failed housing development, cement streets and curbs and empty lots, glimpsed through a bus window on a ride into Los Angeles from the airport); thus lightened, the ship would be able to get over the bar in the ship channel through the Golden Gate and make its way to the Exxon refinery at Benicia in the inner recesses of San Francisco Bay. There was an immediate catch: no way of getting me the two miles or so from the gate to the ship; more seriously, no means of communicating with the *Manhattan*'s master to make sure I was indeed expected —late Friday morning by now, and if I wasn't, not likely with the four-hour time difference that I would catch Bram Mookhoek in Houston to straighten things out. The Captain suggested I get hold of Exxon's ship's agent, Dennis Clark, with an office in the airport building.

Joseph Conrad was as much as I knew of ships' agents, but this one was a young man with a full black beard, neatly trimmed, who had gone from sailoring to ship chartering in San Francisco, now settling in as manager of the new firm called into being to serve the oil ships; not a Lord Jim. Having free run of the terminal by virtue of his job, he would hunt out the *Manhattan*'s captain and find out where I stood; and the message came back—all was well, someone on board had a letter about me. Dennis would be going over after lunch with supplies, cases of bread and lettuce picked up from a local grocery store, and offered a lift.

As Dennis eased the Toyota out to the end of the Berth 5 dock, a short figure was stiffly descending the steep gangplank from the *Manhattan*'s main deck: the Old Man (the name seems not to have died out for all its use), Captain Leon Jean, in zippered blue windbreaker, and, except for the ramrod bearing that looked like old-time regular Navy and a tight-fitting beret on his close-cropped head, indistinguishable from the rest of the crew. I got out, introduced myself—the supercargo I thought Exxon had written him about. "Never heard of you!" the Captain shot back. Then, having gotten his effect: "But come on board anyway, sure, plenty of room—as long as you don't *eat* too goddamn much," with a small, hard, explosive laugh. He led the way back up the gangplank and, shouldering my pack, I followed, then up three flights of steel-pipe outside stairs to the Captain's deck, next below the bridge in the midships house. He pointed to a door on the port side of the ship—"You can sleep in there"—and went into his room with the agent. My room had been designed to house four Merchant Marine cadets and was so identified by a steel plaque on the doorframe, something to do with government approval for the ship's original financing, but had probably never been used as such: a small sitting room with a couch, a couple of chairs, coffee table, end tables with lamps, all bolted to the deck, 18th-century nautical prints screwed to the pale-green walls; down a short passage a stateroom with lockers, a dresser, double-deck bunks on either side, about as in tourist class on one of the old passenger ships. A crib, folded flat, stuck out from one upper bunk—somewhere in the course of an around-the-world voyage that started in Houston in April the suite had been home to a mate and his wife, a couple of small kids—but someone had left fresh sheets and a pillowcase on a lower bunk; I was not, apparently, altogether unexpected. Big windows moved by hand cranks looked out on a companionway up to the bridge wing, covered by Venetian blinds, lightproof blue drapes. I got them open to look out, struggled with a window, but it was long since rusted tight in its track. I stowed my gear in a locker and went to find the Captain.

Dennis and the Captain were sitting at a long, curved desk in the Captain's office across the landing from my cadet suite, working a calculator and checking the bills for the few stores the agent had brought out, figuring fuel on board, supplies to be delivered at El Segundo or Benicia; estimating loading time and the run to California. The Exxon charter specified loading rate in barrels per hour and a twenty-four-hour turnaround, but the ship had already been in more than twenty-four hours, the Old Man grumbled, and the loading arms were not yet connected; Bill Fisken and the Alyeska tanker men—and the Coast Guard, George James on the state's behalf and his own, with the naval architect's training he'd mentioned when I first met him—had taken up time inspecting the ship before they would allow it to load; it was a long time since the *Manhattan* had carried oil. She had come from two weeks' refitting in Osaka, among other things the spare radar system installed on the bridge to satisfy the new Coast Guard regulations and Alaskan law; a Japanese work crew had come over with her, rebuilding the intricate piping by which the oil tanks were filled and pumped out, painting it maritime gray. They were finished with their accounts for now. Captain Jean gestured at the piles of forms to be filled out—on his desk, on a shelf at one end of the room, goddamn paper work!—and offered a bottle of Heineken's from the refrigerator beside his desk. I sat down on a couch along one wall, foam-rubber cushions covered in indestructible gray-blue fabric and bolted to the deck; probably it had been there since the ship was finished. The style was period, 1950s movie-palace modern or the last refurbishing of the *Queen Elizabeth,* but cut down and economized—light-gray bare linoleum tiles on the floor; the room was spacious, perhaps twenty by forty feet, with a row of square roll-down windows looking forward and out the starboard side, the blank bulkhead of the Captain's bedroom to port. Over the beer he was offering Conradian civilities, propitiating the agent—whiskey, or maybe a bottle of Russian champagne, good stuff, sweet, or some honest-to-God Cuban cigars picked up in Russia. They had been there two months anchored off Odessa, too much draft to go into the port, unloading wheat into barges. The Russians provided girls, according to rank, his a twenty-three-year-old good at English, taught it in school, she said, but working for the KGB or at least reporting to them, keeping track when they went ashore. He went into the bedroom and brought out a framed, easel-mounted picture of the girl, in color, another of his young wife and the two current kids, at home in Tampa. Pretty good for an old guy, pushing sixty for Christ's sake, he thought, joking about it but pleased too: a young wife and a Russian girl friend!

Dinnertime. Captain Jean led the way down the stairways and back along the catwalk, raised above the oil pipes, to the afterhouse. It was Friday, red snapper on the menu—*rubber snapper,* the Captain called it, a round-the-world joke. The bearded messboy, in blue jeans and grimy mess jacket, brought the plates in silence and placed them before us on the gravy-stained white tablecloth. I thought the fish good and politely, unsure of myself, said it tasted fresh, had they *caught* it?—forgetting it is Gulf of Mexico, unknown up here; in the boredom of the week's wait at anchorage, someone had said, a few men had been fishing over the side of the ship. No, with a snort, rubber *snapper,* for Christ's sake; it is conventional, on a tanker as in the Army, to complain of the food and cooking, something harmless to talk about, and the Old Man played the game with the rest.

A stranger's first impression of the *Manhattan,* coming aboard that August afternoon, was rust: palm-size plates of rust lying loose on the decks, scars and scabs of rust blotching the white deckhouses, the new-painted oil lines already peeling to the iron-red corrosion underneath; there had in fact been efforts at chipping and spotting on the voyage from Osaka—such work by current sea rules is overtime and voluntary and not many offered, they had enough—but in the few days of the voyage the underlying rust had eaten through as if fresh paint had never been applied. Under the rust and the look of casual, penurious maintenance, the ship's lines were still beautiful, however, and the rebuilding that took her through the Northwest Passage had only refined the elegance of the original design. One's second impression—a great lady fallen on bad times, a famous actress working as a hat-check girl and pinching drinks on the side—was therefore not romantic or sentimental. Unlike the tankers and merchant ships that make up the world's fleets, built in multiples to standard patterns, the *Manhattan* was never one of a class. From the beginning, she was unique, the culmination of a line of development as old as the oil trade, as old, perhaps, as seafaring itself.

The ship was built in 1962, at the Bethlehem yard in Quincy, Massachusetts, for Stavros Niarchos, one of the golden Greeks of the tanker business. At the time, she combined several superlatives: biggest tanker in the world, probably the biggest merchant ship, certainly the biggest ship of any kind ever to fly the American flag; and as such, briefly, earned a place in the *Guinness Book of World Records,* first published the same year (and is still there—as the world's "largest converted icebreaker").

As originally built, the *Manhattan* was 940 feet in over-all length, 132.5 feet abeam, a little under 68 feet in depth from main deck to keel, with a loaded draft of close to 53 feet. These large dimensions, comparatively lean in their proportions and suggesting a hull that should be fast in the water, gave her a load capacity of around 115,000 dwt, good for nearly 795,000 bbl of Prudhoe crude. Since hauling crude was not an attractive proposition for a big American tanker, however, the premise of her design was that she would earn her way with higher priced, more profitable petroleum products—gasolines, lubricating and fuel oils—lighter in weight than crude of the same volume. To do this, her cargo tanks were built in 15 ranks, 45 altogether (the paired center tanks were undivided) and could in theory transport as many different products. As with other branches of engineering, the only test of a ship design that counts is the one that comes after commissioning, through years of service in its intended work; naval architecture in consequence evolves slowly —the penalty for unforeseen error is too great, total loss of ship, cargo, men. Because in size the *Manhattan* was at or beyond the limits of actual experience, she was conservative in her engineering: deck and hull plates of steel one to two inches thick, braced by a multiplicity of heavy steel beams. The large number of bulkheads for her many tanks were also structural, adding to her strength. This was one of the senses in which the *Manhattan* marked the end of a line of ship development: a long, fast hull that would be comparatively rigid in all dimensions (hence moving well in the water), built to withstand extremes of ocean weather even when sailing high in ballast, carrying about one-seventh of her dead-weight tonnage, with most of her tanks empty; and long-lived—by one estimate a ship rusts away perhaps 2 per cent of its plates a year, and if so, even looking bad from shoddy maintenance, the *Manhattan* could go many years before aging to the point of serious risk. Her power plant was in proportion: two steam-turbine systems picked up from smaller ships that had been ordered and canceled—hence also two drive shafts and propellers, two rudders; 43,000 forward horsepower, 17,000 astern. On the face of it, the dual system would seem to make the ship both highly maneuverable for her size and secure against breakdowns, but in practice the advantage, if any, is slight.

The same characteristics that made the *Manhattan* a good bet for Exxon's Arctic experiment seven years later also made her unattractive to a Greek shipowner interested in profit. At $28 million she had been expensive to build (today a ship of comparable capacity and simpler design might cost $40 million). It was also an expensive ship to operate: half a ton of fuel per mile to drive the big turbines; more seriously, a crew of about fifty, divided between engine room and deck—in that preautomation era, the large number of cargo tanks meant as many connecting pipes and hand-operated valves, with their pumps, and men on hand in proportion for loading and unloading; in between, an equally laborious job of washing the tanks. When the anticipated trade in products did not materialize to make good these costs, Niarchos dropped the *Manhattan* and within months of her commissioning sold her off to a bustling American shipping company, Seatrain Lines, Inc. There are interesting connections here. Seatrain has its headquarters in the Chase Manhattan building in New York—and what appears to be a long and intimate connection with Exxon and the Rockefellers as well as their bank. (Although the company has ridden the tanker trade's financial roller coaster, in the wake of the *Manhattan*'s Northwest Passage it could afford to borrow its way to a take-over of the Brooklyn Navy Yard and has kept it operating, though not largely or very profitably; and took a strong hand—20 per cent—in the Dallas combine that built the 150,000 b/d refinery at North Pole, near Fairbanks, premised on Alaska's state royalty share in the Prudhoe Bay oil.) Ownership of the *Manhattan* was originally vested in a Seatrain subsidiary, Hudson Waterways Corporation, more recently, for the same reasons of taxes and general liability, in yet another subsidiary, Manhattan Tankers, Inc., a one-ship corporation created for the purpose—typewritten notices strategically posted on the bulkheads announce this fact and the holder of the ship's first mortgage, the Prudential Insurance Company, synonymous as the Chase Manhattan itself with the center of American financial power. Seatrain's chairman and vice-chairman (and principal stockholders) are two long-time tanker operators and partners, Joseph Kahn and Howard M. Pack. It is their initials, the P reversed and imposed on the K, that provide the identifying cypher painted in red on the *Manhattan*'s stack and engines.

If the *Manhattan* was a final stage in the evolution of tanker design, she was also at least the precursor of the new kind of ship we call supertankers. By that standard she no longer seems big, or not in tonnage, not on paper. With the second closing of

the Suez Canal in 1967, it became obvious that Europe would
have to get its oil by a more secure route, around the Cape of
Good Hope; and for that greatly lengthened distance to be economic, much bigger tankers were essential. There are several
economies of scale. In any class of boat from a canoe on up, a
long hull moves better in the water—faster, with less effort—
than a short one. Recent tankers of the same tonnage as the
Manhattan move their cargoes at the same speed with little more
than half the horsepower; her power plant today would be considered sufficient for a vessel of 500,000 dwt. Since cubic volume is the essential measure, comparatively small increases in dimensions yield disproportionate bonuses in dead-weight tonnage.
Arco's new 120,000-tonners are 90 feet *shorter* than the *Manhattan* as first built; the grotesquely named *Globtik London*,
launched in 1973, is 1,242 feet long but is rated at 483,939 dwt.
Other costs decline with size: in terms of construction cost per
dead-weight ton of capacity, a tanker of 200,000 dwt can be built
(in a U.S. shipyard) for *half* the cost of one of 25,000 dwt;
thanks to fewer cargo tanks and automated loading, unloading,
and tank washing, the 120,000 dwt *Arco Fairbanks* runs with
about half the crew of the *Manhattan,* while the *Globtik London*'s complement is 39. The philosophy of design has changed,
with effects on costs. Where the *Manhattan* and her ancestors
were conceived as armored and braced against the outward pressure of the oil and the inward pressures of the sea, the new
tankers are more like great balloons: oil inside and water outside
are almost in equilibrium; the difference between these two
forces is balanced by a delicate skin of steel.

For all these reasons, most of the tankers built since 1967 have
been big. Today, substantially more than half the world's tanker
tonnage is made up of ships of 200,000 dwt or more. The vague
term "supertanker" will no longer do. A ship of around 200,000
dwt is known as a VLCC, "very large crude carrier." The *Globtik
London* required a new description, ULCC, "ultralarge crude
carrier," though it is already clear that it is not really an ultimate; tankers of a million tons have at least been planned,
though not yet named as a class. Although there are other kinds
of limits (not least, the depths of the ports and approach channels these deep ships must ply), I am not sure there are any

limits of engineering or design on tanker size; or not so long as oil, like love, is the resource that makes the world go round.

The meaning of these differences in tanker size and capacity is, of course, what it costs to haul a barrel of oil from source to market. For a tanker of 25,000 dwt, typical of the small ships that command the U.S. coastal trade, the current average tariff quoted by U.S. ship brokers is about 59¢ per barrel from the Texas Gulf ports to New York—something over 2,000 miles and roughly the same distance as from Valdez to Los Angeles. For a VLCC, on the other hand, reckoned as a ship of 160,000 dwt and up, the charge is much less than half, about 23¢ per barrel, even allowing for the fact that no tanker of this size can get in or out of any of these ports and must be lightered at both ends: and the rate for a tanker the size of the *Manhattan* is about 26¢, only a little higher and still attractive. (If we recall that the tariff for moving crude the 800 miles through the Trans-Alaska Pipeline will be anywhere from $4.50 to $6.50 per barrel, depending on whose arguments prevail, we can understand, if we have not already, why to the oil men the advantage of a short pipeline and a long tanker leg were so evident as hardly to require comment.) By contrast, the *Manhattan*'s charter cost to Exxon would be $29,000 per day, and with the cost of fuel on top of that—around three barrels per mile at $14 per barrel—the price of moving a full load from Valdez to El Segundo comes out to 39¢ per barrel; but assuming that Exxon bears the full cost of the empty return trip, the real tariff is more like 78¢. These are appealing numbers for the ship's owners, whose wages even for the *Manhattan*'s large and well-paid crew come to no more than $2,500 per day, but not for Exxon. Unless compelled by the ship market and other necessities, the *Manhattan* will *never* be hired to haul oil if there is any reasonable alternative.

These are current tanker rates, but the *Manhattan* must always have been at a disadvantage, a white elephant in terms of costs and profits.

She was like a natural leader born out of time to a dull generation; or held in reserve, destined by strength and size and power for the Northwest Passage. The Exxon plan involved a radical rebuilding, at considerably more than the original cost: $40 million was the announced figure for actual construction and, as we have seen, the cost of the whole two-year program was so large that Exxon has never allowed it to be made public. The ship was extended outward about twelve feet on either side by a sealed, heavily braced "ice belt" plated with tempered steel. The old bow was cut away and replaced by a new one 69 feet long of the same tough, low-temperature steel, rising from the keel at a rakish forward angle of 30 degrees—as in a conventional ice-breaker, the bow was designed to ride up over the pack ice with a surge of the ship's forward power, settle under the ship's weight, cutting through the ice with its sharp stem. Considerable additional bracing was built into the hull; new propellers of stainless steel (the old ones, of bronze, were stored on deck and are still there); a helicopter landing pad 125 feet in diameter, raised on steel posts behind the afterhouse; and, near the meat lockers below, a mortuary, used once, when a sailor committed suicide.

When the project finished and the ship went back to work—occasional voyages to Asia or Eastern Europe, hauling grain, not oil—these improvements were a hindrance. The hard steel of the ice belt smashed a dock at Gdynia, made the ship unwelcome, and was eventually removed, reducing her capacity to its present 113,919 dwt. (When I went aboard the eight-foot tires hung over the sides in the meantime as bumpers were still lashed down on deck, along with stacks of disused pipe and hose, other rem-nants; it was like the attic of a house where a family has lived too long and died, leaving a lifetime's accumulation of belong-ings.) The icebreaker bow remains. It is what ship designers call *fine:* unlike the blunt bows of most current tankers, it cuts the water like a racing scull and does not ride up over the waves.

The ship had been laid up off Hoboken for 18 months when Charlie Martin, who would be first engineer, came on to begin refitting for the charter that would take her to Alaska by way of the Black Sea, Suez, and Japan. A foot of water in the cabins of the afterhouse, for a start—the decks above leaked; patched now with thick plates of steel but still when it rained the drops came through in his own cabin and he set pails to catch them. A big, near-sighted American in an old style who since the age of twelve had set his hand to many kinds of work—digging tunnels for the Pennsylvania Turnpike, mining coal in Colorado, tending bar in New Jersey, manning oar-boat engines on the Great Lakes —and in 1945 had been one of those who greeted the Russians at the Elbe. The owners had put $5 million into that refurbishing, more in Japan. It was a start. The engine room came first and was in good shape, he thought, but an engineer does everything from replacing light bulbs and unclogging plumbing to rebuilding worn-out pumps and boilers. The job list hung over the desk in his room was 70 items long and growing.

The loading arms had finally been hooked up while we were aft eating dinner: four 14-inch steel pipes, called "checksans," mounted at the end of the dock. Counterweighted and joined in the middle, they arched against the sky like fishing poles, then bent to attach to the ship's intake pipes amidships, held by air-powered clamps. Hearing the steady clack of footsteps on the deck aft of my cabin, I went out to watch. It was the Captain, in beret and windbreaker, pacing back and forth—the Captain's Deck, it is called—and I fell in beside him. Nothing for him to do now, he said, it was the chief mate's job; this was how he got his after-dinner exercise, walking, till bedtime, nine o'clock—he liked to be up early, four o'clock, up on the bridge for the end of the night watch and a look at the weather. Before dinner in his office, there had been talk with the agent and Stan Barber, the Alyeska man who would be in charge, another BP tanker master with a North of England accent, as to how fast the ship could load. The Exxon charter agreement specified a rate of 50,000 barrels per hour, 16 hours to fill the tanks. The Captain called in the chief mate, Harry Feullerton, a youngish intense man, curly-haired and cigar-smoking, from Brooklyn by way of Fort Schuyler, the New York state maritime academy. With the refitting of the loading pipes on the way over from Japan, Harry insisted, he could take the oil as fast as they could put it through. Just now, from the gauges on the intake pipes, it looked as if the oil was pouring into the tanks at 80,000 barrels per hour or better, meaning only about twelve hours to load. Controlled by valves it flowed by gravity from the storage tanks up the terminal slope, with a throbbing rumble in the pipes like an express train coming through a tunnel.

I walked in silence beside the Captain, keeping pace, thinking of questions that would get him talking. What, I ventured, did he think of the pilots' training program that had run that spring? For two months masters scheduled for Alaska had been rehearsing at Valdez and through Prince William Sound, receiving, at the end of the week's course, a pilot's certificate exempting them from the hired local pilot through the Narrows otherwise required by Chancy Croft's new state law. So far, 99 masters had earned it, on tankers volunteered by the oil companies—first Arco's *Fairbanks,* later the smaller *Overseas Chicago,* chartered to Sohio; Captain Jean, of course, had been on his way to Russia while the program was on and in any case would be starting his six-months' leave when we reached California. We had reached the starboard railing of the deck. He stopped, turned, looked up at me: a flat, harsh face made ruddy by weather, a gamecock of a man, New England French by his name but with a dese-and-dose Boston accent; his master's license, posted on the bulkhead at the door to the bridge, carried pilot's endorsements for Boston Harbor and the Cape Cod Canal, where he had started out, forty years ago. Then, explosively: "A pilot's license for Prince William Sound! They might as well teach you to pilot the god-damned Atlantic Ocean, for Christ's sake!" He resumed his pacing. "And the Narrows!"—he gestured toward the bow, pointing west, the direction of the entrance to Port Valdez. "*Big deal.* The goddamn channel's nine hundred yards wide."

The hazards of Prince William Sound and the Narrows in particular had been argued since the pipeline was first proposed. Finally, in the fall of 1976, in an effort to get real answers, the state of Alaska had commissioned a simulation study at an onshore tanker training center, the Netherlands Ship Model Basin. The system used there is much like that for on-the-ground flight training: a set of ship's controls, taped data on conditions—wind, tide, current—fed through its instruments, a film showing the port studied as it would look through the windshield of the ship's bridge. The results were alarming. Of 160 test runs, about half were failures—the tanker, a moderately big one postulated at 165,000 dwt for the purpose, ran aground more or less violently, either on Middle Rock or avoiding it, on the steep, rocky shore opposite. Hence at meetings that winter in Juneau Chuck Champion, the State Pipeline Coordinator, in his characteristic aggressive style bullied the Alyeska marine committee and a reluctant Coast Guard into conducting the training program, probably his last big fight before he quit. So far as the safety of Valdez shipping is concerned, it was an exercise of questionable value. As was apparent afterward, the data used in the simulation represented conditions that do not occur or, if they did, would cause the Coast Guard to order the Narrows closed. For example, there is evidence for extraordinary winds in the mountains around Valdez—with the violence of the williwaws U.S. fliers encountered in the Aleutians during World War II, cyclonic winds that seem to blow straight up—but it is not at all clear that they occur at sea-level, in the comparatively confined waters of Valdez Arm or the Narrows; indeed, the scarcity of data from the Coast Guard station at one end of the Narrows suggests that the weather there is nothing special so far as mariners are concerned. Because of such uncertainties, the designer of the Dutch study added a fundamental qualification: that *no* conclusions were to be drawn from it. Nevertheless, the oil companies went along and treated the training program as a public relations gesture, part of the cost of having peace with the state: around $1.5 million for running the two ships and paying the masters while they trained, $160 a day plus expenses and travel.

"I mean, look," the Captain said after another round. "It's a straight, wide-open run. What's there to teach? Like giving a license to pilot the North Atlantic"—with a snort. We walked a little longer in silence, then: "You can go anywhere you like on the ship, make yourself at home, you know? The ship is yours," with an expansive gesture, and he said good night and went inside; a dismissal, keeping his captain's distance or simply not wanting to be bothered, which later he turned into a sardonic running joke —not going to teach me any more, answer any more questions, unless he got paid, by God. I offered to send a copy of my book if he'd give me his address but he was unimpressed.

It was dark by now. I leaned on the railing, looking down. The new gray paint of the loading arms gleamed damply under floodlights on the dock, others set up on masts along the maindeck rail. Taking the Captain at his word, I went down the ladder and made a slow circuit of the deserted main deck, climbed to the bridge, then up to the flying bridge above it, getting acquainted; and finally, bored and solitary, back to my room to try to read. The Captain had shut his door and gone to bed.

Toward midnight I went out on deck for a final look. The afternoon's rain had started again, wrapping the ship in thick layers of cloud, swallowing the town across the Port. A smaller tanker that had been loading in the afternoon at the dock behind us had finished now and was pulling away, moving slowly between two tugs. While I watched, the ship's outline blurred and faded in the fog, then its running lights, and within a hundred yards it was gone. Middle Rock: I wondered how it would show on their radar screens. Fog or no, at least we would be sailing in daylight.

By morning the rain had stopped, and the fog, changed back to cloud, had climbed halfway up the mountains and stuck; the terminal was visible again, Valdez still hidden. Under the weight of oil the ship had settled till yesterday's steep gangplank was level with the dock, more than thirty feet, and beside it, hanging from the railing, the leaveboard was chalked with the sailing time: noon, with the tide. At the berth behind us a new tanker was already tied up, taking the place of the one that had left in the night. When the Alyeska crew came on to disconnect the pipes, several gallons of mud-brown crude ran back from one loading arm and caught in a trough where it would drain into the tanks, and a little spilled on the deck, but harmless too—the overboard drains from the scuppers had been stopped with cement by port rule; a man ran with a roll of white cotton waste to sop it up. Word was around of a writer on board and Stan Barber was embarrassed: a spill, first time it ever happened; bad luck, eh? Promptly at seven-thirty the Captain left his cabin, descended, and stepped smartly along the catwalk to the afterhouse and breakfast. I prowled the deck: at intervals, heaps of garbage overflowed the containers, accumulated while the ship lay at anchor, waiting till we were back at sea to be dumped; another rule of the port.

A little before noon the pilot came on board, one of the natty young men I had noticed at the Sheffield, with an assistant and a third man in a new, dark-blue jumpsuit who turned out to be an academic psychologist from New York; commissioned by the Department of Transportation, it seemed, to study pilot and crew and make sure there was no letdown once the Valdez maneuvers became routine. The pilot stationed himself on the port wing of the bridge and with a walkie-talkie gave the orders for casting off. Already, with a puff of black exhaust from her stack, the ship was slipping imperceptibly away from the berth on her own power, with no feeling of motion—inching, literally; the berthing speeds are specified in Alyeska's Port Manual in inches per second. Still at dead-slow maneuvering speed, moving straight and silent as a canoe through the water, the ship gathered momentum and pointed down the Port toward the Narrows, attended on either side by a tug. The pilot had come into the bridge and taken a position in the center, behind him the helmsman at the wheel, the mate on watch, and Oscar Sewell, eyes fitted to the light shield of a radar screen, an amiable North Carolina man who had served as mate on the *Manhattan* ten years back, in the grain-hauling days—India, Poland—and now was on board to relearn the ship, take over as master when we reached San Francisco; to one side the folding chart table, set up, to the other the Captain leaning on the shelf in front of the windshield with a pair of binoculars, saying nothing—the ship was the pilot's till we were through the Narrows—but watchful. (Just now, casting off, a deck hand securing a boom had managed to drop it with a thunderous clang on the deck. The Captain, keeping an eye out from the bridge, had shouted, once, and gone inside, but made a note to deal with him at the end of the voyage; which he did, with a formal complaint to a Coast Guard board—still drunk on duty, he put it down, from his night in town.) On either side the indented shores rolled past in constantly changing perspective, as you see them from the water, like a film projected on a screen. I ran up to the flying bridge to look back, but Valdez and the terminal were already folded into the contours of the land and the long reach of the Port was empty and flat. Only the mountains, swathed like dancers in their veils of cloud, were big enough to see.

The ship made a wide swing to port and set her course south through the Narrows. Ahead two long low points of land, tree-grown, seemed to come together like fingers pointing in the damp light, widened beyond into Valdez Arm and twenty miles off opened out in the immense reaches of Prince William Sound, like a broad valley seen through a tunnel. I peered for Middle Rock but could see nothing, only the flat, pale-green water, gently lapping and waveless as a lake; then a hundred yards to starboard as we came abreast a single white warning buoy and behind it one of the tugs that had gone ahead, standing to, marking the hazard with its bow. The rock itself was under the tide but at either side I thought I could just make out a faint tracery of stirring water that to my canoeist's eye would mean shoal. At one point the Corps of Engineers, ever on the watch for large projects, had put forward a plan for dynamiting the entire obstacle out of the channel to a depth of 75 feet, at a cost of $18 million, and may yet bring it off if it can find someone to pay, but after all the years of anxiety and rhetoric it did not seem a very great hazard for a functioning ship, properly manned, and the Captain's scoff was not misplaced. With a turning diameter of something less than three times its length, it would probably be possible for a ship the size of the *Manhattan* to turn clear around between rock and shore—but pointless, given the nearly unlimited room for maneuver at either end of the Narrows. I went back to the bridge, then on down to the deck for another circuit.

The pilot boat came out, the word painted in spindly orange letters on the front of its deckhouse—PILOT—and the pilot and his companions went over the Jacob's ladder and stood waving up from the deck and smiling as the boat sheered off, heading for another tanker on its way in. To the west now the land opened and far in the distance a line of white lumps was picketed across the water like the lip of a rapids where the stream bed drops off, seen from upstream: growlers, someone said, the Alaskan word, tips of the icebergs guarding the mouth of Columbia Bay, six miles off by the chart, behind them the broad slope of the ice field narrowing into the mountains. It was another ingenious and much argued anxiety of the pipeline, now by its looks remote; it would be possible for a master to run his ship in there and destroy it, but only with the same hard deliberation he gives to holding his course and governing his men.

The Sound was opening out: to the east, a succession of deep bays defined by long points of land, to the west, bare islands—

Glacier, Naked, Smith, Knight—rising in steep brown rock from the water, recognizable by their shapes and the named points schematically emphasized on the chart. Ahead now the dark clouds are lifting and breaking up, through them in the distance, to the south, there are shreds of watery blue sky—we are leaving the mountains behind, leaving Alaska: the Hinchinbrook Entrance, to the west the sixty-mile bulk of Montague Island, to the east on the horizon white water breaking on what from the chart must be Seal Rocks; aligned with the ship, outside its wake—we are picking up sea speed, held down so far by Exxon orders to 13 knots—something swims with the humpbacked leap and dive of a porpoise, a walrus by its size and color, the latitude, swimming for the Rocks to lounge and sleep. There is nothing ahead: the infinite sea; the deep. So men go forth, for bravery, wages, profit, trusting their skill and the strength of their ship to the deep: that they do it now for oil does not make it less. Much has been written of the death of the old skills, the tradition of the sea. I think that mere sentiment, but if it is true of newer, bigger tankers, with their automated engine rooms and elaborately instrumented bridges, on this ship it is not; the tradition of sail lives, embedded in the language itself. The room where a seaman sleeps and keeps his gear is single and air conditioned but is still his fo'c'sle. The room in the midships house where at sea he can buy small necessities—cigarettes and pop, work gloves and underwear—is the slop chest. Down in the depths of the stern, behind the great cube of the engine room beneath the afterhouse, is a storeroom for cables and mooring lines known as the lazaret, a word as old in English sailing as Shakespeare and Drake. There is no carpenter's shop, no Chips—apart from some sticks of furniture in the officers' cabins, the only wood on board is the teak railings of the bridge wings, weathered black and in want of sanding and varnishing—but to one side of the engine room there is a spacious machine shop, with drill press, lathe, welder's tools, pipe-threading machines, and the capability in a pinch of making almost any broken part the ship may require. The radio operator (Tom Nichols, a gray-haired black man with gentle manners, looking to retire to a house on Long Island when this voyage is over) is known as Sparks, unself-consciously, as Captain Jean is the Old Man and Harry, the chief mate, is Chief, placed by his calling to one side of the two hierarchies of officers, mates and engineers. From master to bosun to the lowliest wiper in the engine room—"professional ordinaries" they are called, without condescension, ordinary seamen growing old in the job, with no will or ambition to rise to AB, able-bodied seamen, or the ranks above—these are men called to solitude and the ordered, impersonal intimacies of the life of ships, uncomfortable now with any other life, they have tried, most of them, breaking away to some job ashore, and have come back. The sea is their calling.

Dark now. I climb to the bridge again, feeling my way. Hooded light from the binnacle, from the instruments mounted above the windshield that give course bearing and engine speed, from green blips moving on the radar screen. Eyes adjust: the helmsman at the wheel, silent, a shadow; the mate staring forward through the darkness, moving restlessly to the wings and back, seeking the negative which means safety, but there are no lights ahead, no shapes of land or other ships; only the sea. Darkness and the sea, two solitudes, and overhead now the sky has cleared, its blackness pricked with stars, and you feel the lift and surge of the ship as she meets the ocean-combing wash of the eight-foot deep-sea swell, sailing the deep as earth sails the star-flecked deeps of the universe. You do not venture either voyage without gravely gathering in all that you have of courage and skill and strength, but venture you must. So we are called, by necessity, even when that necessity is only a load of crude that will fire a nation's furnaces and power its machines no more than an hour. The human freedom is in the response.

It is about 2,000 miles by sea from Valdez to the Socal refinery at El Segundo where most of the load would come off: 2,020 was the figure in the ship's log, measured from the charts where her daily position was marked, reckoned from a mate's shooting of the sun each noon with a sextant—the ancient system still, not trusting to Loran-C and its radio beacons. Engine time, counting revolutions of the twin screws, made it longer, 2,298 miles: against the steady lapping of sea waves outward from Southeast Alaska and the roil and flow of the great North Pacific currents, the old ship wandered a little on her course, and briefly, near the start, half the dual steering system (one system, complete, for each screw and rudder, invariably operated in tandem), the wheel no longer answered to the helm, and the Captain was called up from sleep in his cabin on the deck below; but small matter—it was only necessary to switch from one to the other and control was restored, and lasted till we reached our port, only another item to go on Charlie Martin's ever-lengthening repair list, for such time as the ship was briefly at rest and he could find out the trouble and order the spare part through a ship's agent. That voyage south is a book to itself. It has the classic shape: a beginning; incident measured by the passage of sea miles and time, provided by men as they do their work, under the stress of work and each other; a purposed and predestined end, though not, such is the hazard of the sea, to be known with certainty until it has arrived; and presiding over all a sardonic and ambiguous force, a little envied, a little feared, but after all the long months at sea not much liked or wanting to be—the Captain, keeping his thoughts in a drawer in his desk and his emotions locked in a suitcase under his bunk, save for the rare moment when one explodes into speech and he covers it with the hard shell of a joke. Maybe, granted life enough and time, it is a book I will write one day.

There will be a lot more of the ship in it, which at this point I was only beginning to learn. You walk the deck slowly, seeing a little more with each round by the changing lights of daytime and night. You see that what reporters have written of the bigness of these big ships, comparing them with football fields laid end to end, is sentimental hyperbole, and the environmental fears and predictions of catastrophe are fairy tales—grains of truth strung together and rigged up in dragon shapes to frighten children. The ship is human-scaled: man is her measure and her master. This one with her bridge amidships you can never see whole but only half at a time, fore and aft; and the sea is always close—15 feet of freeboard, the Chief says, with her load, but against memories of other ships it is like seeing the water close over the gunwales of a canoe. Most of the way in mid-sea birds follow the wake, narrow wings taut to the currents of air stirred by the ship's passage: albatross, sometimes half a dozen at a time, not the pale omens of Coleridge's Mariner but a dark species, body and wings; sometimes they settle and rest, then lift and catch up, still effortlessly gliding. Inquisitive, you climb down grease-blackened ladders to the oil pumps in the darkness of the pump rooms at the bottom of the ship, where an oily bilge washes, side to side, below the lowest catwalk, from a leaky valve not yet pumped out, and you feel the bulge of oil against the bulkheads of the tanks on either side—with a ruler, to measure the thickness of the supporting girders. Or opening a door a few steps forward of the mess you look down on a new world, the engine room, fresh-painted and oiled, four decks down, you had not imagined the ship contained such a volume of space and such depth, and you descend and wander, through blasts of coolness from the vents and heat from the fire boxes, ears numbed by the aortic throb of water, steam, and oil, pumping through metal pipes: the ship's heart, huge. A bad sailor, everyone tells you, with a frightening roll in heavy seas—she was built with stabilizer fins at bow and stern, removed for the ice voyage and never restored, but she has gone through a typhoon in the Indian Ocean on the way from Odessa, when the waves broke on the bridge, and weathered it. Now with your feet you feel the deck plates flex with the summer swells and wonder when the weather will come and if finally you will be sick, but it is a good hull, she takes the waves well enough; better flex than break.

Above all it will be a book of the men who man this ship. The ship's old design divides us, engineers and crew aft, deck officers amidships, each with their own quarters and lounges. Both the chief mate and the second (there is no first) have masters' papers and, though it does not affect their pay, so far as law is concerned are qualified to command the ship; it is the main thing meant by the superiority of American ships over foreign ones or anyway Panamanian and Liberian—more licensed men than there are jobs for in a shrinking trade. (Among the engineers there is one, a frail, white-haired elderly man, with a chief engineer's license, "taking a vacation," he told me, as a third engineer; that license was Charlie Martin's next goal, to be earned in his coming months ashore through a free course given by his union in preparation for the Coast Guard exam.) We do not come together at meals but straggle in and are seated apart: Captain and chief mate, chief and first engineers at one small table, Sparks and the three other mates at another, in the middle

six more engineers; to talk from one group to another it is necessary to turn one's chair around and shout, but there is not much
talk of any kind now at the end of this five months' round-the-
world voyage. The food is the same as in the crew mess through
the galley on the port side—between crew and officers there are
no differences of ceremony or dress on this democratic ship, only
the essential one of command. One night on the bridge I talked
with the helmsman about the tanker described in Noel Mostert's
Supership, a British one with a strong tradition of the Eastern
trade, where the officers were uniformed and gathered each evening in their lounge for cocktails, bathed and changed to mess
jackets, and at the ringing of a gong went in to a formal dinner,
the Captain presiding. The helmsman was a good-looking man of
thirty, well-spoken, who had tried shore life and found himself
stuck with the sea; overqualified, I thought, for an AB. Writing a
play, he said, though when I mentioned O'Neill and Conrad as
authors to read he had never heard of the names; the straining
darkness and the silence of the night watches will elicit poetry if
there is any in you. Yeah, the man said, still thinking of Mostert's
British ship: none of these guys have any class.

Seated apart for meals at a fourth table and having no regular
duties, Oscar Sewell, the new master, and I spent a lot of our
time together, lingering at table to talk, exploring the ship. He
had quit school after ninth grade, worked as a mess boy on a
Mississippi barge, joined the Navy in World War II and been
able afterward to collect a high school diploma, then finish college with an M.A. in education, and for years had taught school
in North Carolina before going back to sea for the money and a
growing family; and because of that interval ashore and, unlike
every other man on board, was both married and had stayed so,
husband of one wife, undivorced. (Hired for the ship in the
spring he was sent to Valdez for the pilots' course and, still
thinking of the money, seriously considered moving to Alaska
and trading his master's pay, $30,000 a year minimum for six
months at sea, for double that as a pilot, working regular hours
close to home.) A deeply inquisitive man, with a teacher's
eagerness to explain and be understood. The oil worried him and
one night he opened one of the tanks and fished up a coffee can
full, put it in the freezer of the pantry in the midships house

with a thermometer, to find out for himself what the pour point was. After all the years of analysis, the oil seemed to be rather different in its characteristics from what Alyeska and the oil companies had reported. It had come aboard about 80 degrees colder than predicted, not hot, at 120 degrees, but after the stop and go of start-up down to the air temperature of the Alaskan mountains through which it pumped, a little over 40. For him this meant that the wax content of the crude would precipitate in the tanks—hundreds of tons of it that would foul pumps and pipes and on this ship have to be shoveled into buckets after each voyage, drawn up to the deck and disposed of. My worry was different: if the oil was this cold at the end of August, with a winter shutdown it would gel in the pipe and the pipeline would be unstartable; on the other hand, if the pour point was in fact lower than expected, then perhaps most or all of the above-ground pipe, with its immense cost and complication, had been to no purpose, as the engineers insisted in the first place. (We had gotten the oil down to 20 in the zero temperature of the freezer compartment, still with no sign of solidifying, when one of the mates, perhaps even the Captain, tired of the experiment, took the can out and threw it overboard—Sewell's oil in the goddamn freezer, for Christ's sake! So the question, which is fundamental, was not answered, and so far has not been, but the oil still flows, and is cold.) Later, his curiosity undiminished, Oscar brought a soft-ball-sized lump of congealed wax up from a tank to study: black and sticky with oil but essentially paraffin—it is the same stuff sea-going environmentalists have reported as "oil" floating on Atlantic tanker routes, but it is wax, not oil, and harmless, I think, in any ecological sense. I wondered if it was dangerous, if it would burn, for instance. Sheltered under the midships house, we tried to light it with matches, then made a little fire of strips of cardboard in a bucket; the black lump of wax would melt all right, but slowly; it would not burn.

I divided the evenings between the two deckhouses, engineers aft, deck officers forward, helping drink up what seemed a nearly inexhaustible supply of beer—only one or two would be staying on after San Francisco, among them Charlie Martin (the beer still illegal, with all other alcohol, under archaic American law that makes a seaman something like a ward of the government).

For most of the voyage the two groups had come together several times a week for movies—a ripe selection of sex films and Westerns picked up in Houston—but by the time I came aboard the projector and the videotape machine were both ruined, and what remained was beer and talk. Among the more numerous engineers, mostly in Charlie's spacious cabin, the talk turned to the ship itself and how to keep its engines running; union matters and their different rules (three on this ship, each with a unique contract—another benefit, from the owners' standpoint, of incorporating her); the quality of the crew, most signed on simply for the round-the-world cruise, which they thought would improve when the ship fell into her regular Alaska run, by now sick of each other's company and venting displeasure by depositing mounds of feces in the common shower rooms or at the door of a disliked officer, though to me they seemed cheerful and competent enough; and often, in joking, tale-telling elaboration, the talk was of whores and marathon binges in ports from Seoul to Hamburg to Osaka, but told with an underlying savagery and sadness—the old talk of sailors, solitary men unfitted by their calling for a society ashore made normal by the presence of women. Among the deck officers gathered in Harry's cabin the talk ran more to money, what could be bought with it in Tokyo or Hong Kong, how best to preserve the big payoff at the end of a voyage—silver coins or gold, stocks, betting the Swiss money markets, land.

We made El Segundo after a week's steaming, in the midst of the Labor Day weekend, and then after Exxon's hurry-up order halfway down lay at anchor for a day, ten miles out in the Santa Barbara Channel, in sight of the planes taking off and landing at Los Angeles, while the refinery cleared enough tank capacity to take delivery. Then as the oil pumped ashore through the underwater pipe I noticed brown bubbles of crude rising and at slow intervals breaking on the surface, spreading in a sheen that surrounded the ship. The pumps were stopped but the bubbles kept coming. Commotion ensued. A natural seep from the sea floor, someone suggested, but the oil was sampled and seemed to be ours. A supply tug circled the ship, aerating and sinking the oil with its propeller. Ship's officers, refinery technicians, the Socal and Exxon environmental men who had come on board—gray-

haired men with master's papers, politely addressed as Captain
—hung over the railings, wondering what it was, where coming
from. After a long wait—the holiday weekend—a diver came out
in wet suit and flippers, astride the bow of his boat like a
figurehead: a balding, mustachioed, barrel-chested man who
might have been an Aegean sponge fisher from the old days or a
successful Sicilian *bandito,* accompanied by a young assistant
and, up on his boat's flying bridge with his pilot, a blond girl
friend in kerchief and Hollywood dark glasses. He plunged, surfaced to get his bearings, dived again, and came up to shout a
report up to the deck, treading water: could find no leak, only a
little oil, maybe from that leaky bilge, collected around an outlet
pipe; he had reached in and wiped it out and there seemed to be
no more. Pumping resumed. Later a Coast Guard team came out
to inspect and put the spill down in their report as two quarts;
nevertheless, it would go on Harry's record and they ordered
him to produce his license for the purpose—he was the man responsible.

(On the *Manhattan's* next run down, there was another spill at
El Segundo in what sounds like similar circumstances, though in
neither case could I find out exactly what happened. This time
the guess was 800 gallons, which is a minute fraction of the
ship's load—19 barrels—but makes a considerable slick; and Californians have separate and opposite sets of emotion for the oil
that runs their cars and the oil that dirties what's left of their
beaches.)

And then north by night to San Francisco through a shoal of
fishing boats converging on a reported catch, with a heavy, wet
autumnal fog all along the coast and the ship's horn blowing
from the stack, blasts a minute or two apart, and a watch posted
in the bow—Frenchy, a heavy-set Gloucesterman, fisherman by
trade, saving money for a boat—to clang the ship's bell when he
sees a light ahead. And finally, at dawn, a close glimpse of one
stone pylon and the bronze girders of the roadway high above,
the great bridge, as we lift through the Golden Gate—then Alca-
traz abandoned on its island and the city itself on its hills, with
the fog thinning and breaking up by now in hot sun; and the
long, slow morning following the windings of the bay inland to
Benicia.

There is a small cluster of people waiting on the dock as we tie up, the new crew, and among them several women, wives on hand for a brief visit with their men, Charlie's coming aboard for the next run or two to Alaska. A fragile old man, a new mate, coming up the gangplank in light shoes and flowered summer shirt, sweating with a suitcase on his shoulder and leading his wife to their cabin, stops, wonders if I am the chief mate, and asks about Valdez; he had served on a passenger ship when Fort Liscum still existed and that was where you docked, not back since—forty years it would be, most of a lifetime, at sea. We had been a crew for a while, disparate men joined in the ragged unity of common endeavor and authority. Now we are breaking up, returning to our private purposes and destinations. While the oil lines are attached and the last of the cargo is pumped ashore to the refinery hidden behind a barren California hill, you wait around the long table in the midships lounge where the new men are giving medical histories, producing vaccination certificates and signing the articles and two decks up in the Captain's office the paymaster, brought on board at El Segundo, is paying off, in combinations of cash and checks. Already as they sign on the new crew is down on deck, winching slingloads of supplies up from the dock and over the sides. Men change but the ship is constant, and its work; the sea. It is time: late afternoon and fifty miles of freeway and subway and buses to find your way through to the airport. You take up your duffel and with a few quick handshakes and goodbys are down the gangplank and persuading a ship's provisioner to give you a ride in his truck to the end of the BART line, where you begin. No looking back. Things end; this is over.

Leaving the Hinchinbrook Entrance and entering the oceanic darkness, I felt no elation. I remembered: crowds shrill on the New York piers and the ships' decks, champagne in the cabin, and I was bound for the limitless possibilities of Europe and the lifetime ahead, but always then as a passenger, living in the suspension of the Atlantic crossing to whatever was ahead. It had helped, being young; now there was chiefly work, to see, know, remember, find the words, and tell, a process made intent by the certainty that there would be no second chance. But there was more to the feeling than that. The great voyage is the one that began 225 million years ago in the mortal runoff of a continent to its Arctic rim: the voyage of the oil toward human use. By comparison the particular voyage of the *Manhattan* is only one immeasurably small link in an infinite chain—ah, but indispensable. In time, then, perhaps, I will have to write it too.

THE NEXT TIME

An Afterword

> It's the story of our nation in the last quarter of the twentieth century, all our weaknesses, bureaucratic and everything else, and in a way it's been a tragic story: while we may have won the war, we've lost every single battle.
> —Robert O. Anderson

So the pipeline was built, within the scheduled time laid down by its builders, and the oil flows. These are facts—but so qualified by frustrated intentions that no American can look back on the history of the Trans-Alaska Pipeline without a tremor of uneasiness for his nation's future. Nor is that feeling merely public and abstract. We are all of us implicated in the outcome and in the strategies by which it was achieved. More is at issue than a particular oil field and the means of developing it, more even than the collective source of social and national power we have come to call "energy." As it has developed, the project has taken the form of a gigantic experiment testing whether as a society we are capable of defining national goals rationally and by public means and whether, having done so, we can carry them to effective conclusion. In the process we have brought to bear most of the organs by which we conduct, regulate, and examine the public business—the several forms and levels of government, the courts, the press, the corporations, the private groups that claim to speak for conscience in the public interest—and none of these actors in the drama has come through with honor intact and competence untarnished. If that sounds sweeping, let us consider.

At the center of the pipeline's history is the more than four years' delay from first proposal to the congressional decision late in 1973 that made it possible for the work to begin. The oil companies have attributed these lost years to the irrational obstructionism of the environmental organizations, tolerated or encouraged by the courts and a complacent Congress; the environmentalists, on the other hand, have argued that the delay was due to the oil men's unreadiness to recognize the real difficulties inherent in Alaskan construction and development and to plan accordingly, their unwillingness to submit their plans to public scrutiny. Possibly the two arguments together make a truth, but neither, taken separately, is wholly true. It is not at all clear that in a national sense the delay procured any serious benefits—that the design of the system was more complete or final in 1974 than it had been in 1970, or that the engineering studies on which it was founded had in fact, in those years, more fully identified and defined the realities to be dealt with; or that, conventional public sentiment to the contrary, the system was any less likely to fail or better able to protect the environment in which it would be built.

The penalty for those years of national indecision has been enormous. If by 1973 TAPS had actually been in operation, as its owners first intended, or at least within sight of completion, it is most probable that the OAPEC embargo on oil would not have been attempted or, if it had been, would have had no effect on the United States and little on Japan and Western Europe—the oil producers are even more dependent on their exports than the manufacturing nations on oil supply—and the radical series of price increases would not have followed. As it has turned out, the effect goes beyond the arbitrary and disproportionate rise in the cost of one commodity, oil, and the general inflation of all costs that depend on it. If it is true that before 1973 the international oil trade had contrived to keep the price of oil unreasonably low and that the imposed shortage and the necessity for developing alternative sources of energy must sooner or later have come in any case, those are small benefits. What the United States and the world's other industrial democracies have lost is the control of their collective destiny. We react to events and decisions beyond our borders; we are less able to initiate. In the disruption of the international trade on which our national livelihood depends and the continued outpouring of our capital for unproductive uses—*im*balance of payments, it is costing us more to live than as a nation we can afford—we are discovering that the personal freedoms we take for granted are founded on national freedom of action, and that the foundation is less firm than we had supposed. It may even turn out that the democratic political forms which, however imperfectly, have flourished in the manufacturing nations of the West are not a culmination in the life of human societies, a stage that in time all must reach, but only a brief interval of relative freedom and prosperity in the long darkness of human tyranny. If so, then it is one more in the terrible transitions of that history that we are living through and witnessing. We have not so far been able to summon the collective urgency that this situation demands: it is the national survival that is at issue.

It would be some compensation if we could feel that the four years had been put to effective use; that the decision, when it came, was based on valid and pertinent information, and that the policy thus launched was commensurate with the need. Instead, the information generated by those years of apparent effort was seriously incomplete or misleading on most of the fundamental questions. The design of the pipeline system itself was at best unfinished and at key points depended on false assumptions and untested technology. The voluminous studies of its possible environmental effects were wishful, superficial, and intellectually infirm. The estimates of construction cost that figured in the decision bore only the sketchiest relation to reality, even when every allowance has been made for the effects of inflation and government regulation since then, and the project's real costs—and hence the pipeline tariffs and the general energy costs that would result—were never discussed. The arguments advanced in support of the TAPS system—the projections of West Coast supply and demand—were shown by critics at the time to be at least questionable and have since been so mocked by events as to cast doubt on the motives or the competence of those who made them: the Department of the Interior and its

Secretary; three of the leading pipeline partners, Sohio and BP, and Exxon. On the other hand, although the political and financial realities that made the alternatives to TAPS impossible were at least alluded to, they were put forward with so little force and substance as to be unpersuasive and were hooted down by opponents whose actual motives went unquestioned and unexpressed. The final phase of the pipeline debate was conducted in an atmosphere of crisis induced by artificial shortages that followed the OAPEC embargo, but the real meaning of these shortages—and hence the reasons for and means of dealing with them—has never been communicated to the public by our politicians or our journalists. We were told, that is, that the situation resulted on the one hand from the manipulations of unscrupulous international oil companies and, on the other, from wasteful consumption of fuel for private purposes, in private homes and cars; and could therefore be set right either by dismantling the companies or by a kind of moral revolution in personal habits —with the corollary assumption that since the American public is incapable of self-discipline, that change would have to be imposed. The only evidence against the oil men, however, is of failure to anticipate a complex of events beyond their control—a fault, certainly, but not worthy of hanging. The strictly private consumption of energy, on the other hand, is so small a part of the national budget that even stopping it altogether would hardly touch the real problem; but the people, unled, have at least restrained the national demand for oil in the years since the embargo, and they would, I think, do far more if offered something better than trivial reasons. The issue is not whether you and I drive our cars or ride a bus but whether our factories run and our farms are worked—and do so at costs in energy that will procure for us in the world's trade those goods which we cannot produce.

The apocalyptic view of energy constantly reiterated in these years is false; and in any case the panic induced by such a vision is the most dangerous frame of mind in which to face danger. Nevertheless, it does appear that in our lifetime the familiar forms of energy will become increasingly scarce and correspondingly expensive to find, develop, and use; and that at some time in the indefinite future, perhaps a century from now, if not literally exhausted, they will have become too rare and precious to consume as fuel—surely not by the inefficient means we now know. What we in the United States (and a few other nations) have, then, is a generation or two of grace in which to invent and bring to use replacements which have not so far been conceived. How much time we shall have depends, however, on the continuing availability of our present fuels, and oil in particular, the best of all in terms of abundance, ease of development, and variety of uses: on our ingenuity and determination in keeping these energies available while we hunt for better. The alternative is slow death, beginning with ourselves, accelerating with our children and theirs.

Our present energy predicament is not a static situation but one stage in a dynamic process of industrialization and development that has been evolving for more than a century and, so far as we can apprehend it from our own time-bound vantage point, will continue to unwind for as long again. The Prudhoe discovery and the strategy by which it would be put to social use—the pipeline system—were one incident in that process. In that perspective, it was one that demanded action based on accurate assessment of present facts as they are likely to shape the future and directed toward long-term goals; longer, probably, than as a nation we have ever before had consciously to consider. That is not what we got. The congressional investigations and the public debates were excessively narrow in scope, as if no more were at stake than the private interests, risk and profit, of a few oil companies. The legislative solution when it came—the Trans-Alaska Pipeline act—was a limited amendment of a regulatory law descended from what was literally the horse-and-wagon era of American industry and even in the time in which it was conceived was of at least doubtful validity, though over the years by administrative adjustment it had been made to serve. I am not sure, however, that any more effective response was possible within the system that we have. Long-range policy decisions can be no more meaningful than the annual budgets that at all levels of government give them substance—or than the legislators and administrators who make them and carry them out in a perennial turmoil of election and the narrowly bartered interests of the political market place. Our corporate institutions are not much better: a similar cycle of annual budgets in which the formation of capital for future goals is controlled by the necessities of taxation, government regulation, dividends and profits; and hardly a greater continuity in the managerial progression at the levels at which corporate policy is made. Yet it is doubtful that the alternative is tolerable, on either side: the further concentration of social power implied by programs measured in decades or generations rather than years; a further dilution of public accountability in the managerial continuity necessary to the carrying out of such programs.

Even the quite limited goals laid down for the three years of pipeline construction were very imperfectly realized: public review, verification, and approval of the original designs; enforcement of technical and environmental standards through Alyeska's quality-control and quality-assurance organizations and the federal and state surveillance offices; co-ordination of the numerous agencies of government endowed with legal and often conflicting interests in one or other aspect of the project. The practical consequences of these purposes make a short list of dubious benefits: the welding techniques and the tanker training program imposed by the State Pipeline Coordinator; the vulnerable miles of aboveground pipe which in the course of construction became a short-cut answer to most geophysical uncertainties, creating anomalies of design and engineering that may in the long run be more risky than those they were meant to avoid. At the same time, the Alaskan salmon and caribou populations that supplied much of the imagery and emotion surrounding the project and that were to be the immediate beneficiaries of these regulations have continued to decline, the two Arctic caribou herds disastrously—but for reasons that, though neither understood nor explained after the years of study and survey, seem to have nothing to do with oil development or pipeline work. Looking back on this particular turn of natural events, we might well conclude that the arrogance known as "wildlife management" had best be dropped from our mental baggage—in certain situations we can, after a fashion, master nature, sometimes we can learn to live with her, but we are not very good at managing—but nothing so obvious has so far happened.

Perhaps, then, if there is one thing to be learned from the history of TAPS, it is that such purposes are easier expressed in the generalizations of law and judicial opinion than carried out in the day-by-day contingencies of complex human activity. The law conferred on those charged with these tasks authority without responsibility, with the rigidity of law where what was needed was informed and flexible judgment responding to specific situations as they arose. For their surveillance to have been as specific and substantial as the law envisaged required an almost superhuman competence, technical versatility, and moral force; or a vastly enlarged surveillance staff, though still supported by the organization it was appointed to monitor. In effect, a different kind of government body created for the purpose would have had to assume the management of design and construction. The federal record in such situations—the Corps of Engineers and its dams, for instance—is not so flawless as to make that arrangement seem preferable to the workaday makeshift we actually got; but one way or other we must find some way of making responsibility match authority in the execution of large public works. TAPS was a beginning, a portent—an experiment, as I said. There will be others.

It was the newly found environmental viewpoint that provided most of the public fervor surrounding the pipeline debates and defined the issues and thereby the regulatory strategies that resulted. Although I dislike that viewpoint's ideological cast as much as I do most other forms of absolutism, that is where my natural sympathies lie; but emotion is a poor substitute for rational thought when matters of public policy must be decided. In its early stages the argument was clouded by hypothesis and serious misinformation—about the variable nature of permafrost, for instance. Throughout, excessive attention was paid to the pipeline itself, to the exclusion of the system's other components: that the pipe was the danger, would break in operation, pouring hot crude oil across the landscape. No one, however, anticipated the chronic spills that continued throughout the construction period, from tank trucks moving fuel in cold weather, from fuel lines heaved apart in the gravel pads on which the northern camps were built—thousands of incidents amounting to hundreds of thousands of gallons—but whether any permanent damage has in fact resulted is another question; the answer is at least far from certain. It was the same narrow focus, I think, that magnified the welding and radiography problems and led to thousands of repairs that, by laboratory tests and expert judgment, were of doubtful practical value but were nevertheless required by the law under which the pipeline was built. In the chronicle of errors that made up the testing and start-up programs, however, it was not the pipe or its welded joints that failed but the complex and self-canceling safety systems of the pump stations; aboveground pipe supports, technicians and their training, discipline. These present and continuing dangers—and there are others of the same order in the complexities of the terminal and its controls—were never touched on in the preliminaries to construction, but they threaten the system as a whole. The loss to pipeline capacity and Prudhoe production from the breakdown of Pump Station 8 is already immense. Law and ideology demanded absolute standards, but nothing human is absolute.

Behind these environmental confusions and misjudgments there were serious contradictions: denials of reality. Although they received only limited attention in the court rulings and the law, much of the four years of debate was eaten up by arguments about alternatives to TAPS: other routes, different technologies—gross impossibilities, as is obvious now from the TAPS experience and should have been when they were first advanced. And underlying these was another, mischievous or fanciful, but false: the alternative of leaving the oil in the ground, unproduced. We did not and do not have that choice, for two reasons. In the world we inhabit, *what is technically feasible will happen* —we can perhaps direct its course, but we can no more stop it than we can stop a steamroller with our bodies. And rationally: if the nation is to have leeway in which to develop new sources of energy, replacements for oil, and the social habits to go with them, then the twenty-year life of the Prudhoe field is an essential stopgap. No comparable opportunity existed before it was discovered, nor has there been any since.

Looking back on the manifold failures that distorted the planning and building of the Trans-Alaska Pipeline, we might conclude that oil, the common element in all, is the evil, fouling whatever it touches, coating the crystal of human intentions with sludge. That judgment would be simplistic even if true. We do not have the privilege of turning our backs and walking away with a shrug and a summarizing quip; we live here, and if the roof leaks and there are ants tunneling in the joists, the house is ours and all we have got, its problems ours to recognize and solve. Oil, moreover, is protean stuff, and sludge, if foul, is not its primary property but a by-product of use. So there is a truer image: misunderstood or carelessly used, oil is indeed a dangerous substance, but the nearly infinite range of products that can be extracted from it vastly outweighs the risks inherent in the process; for our time, it is indispensable. Much the same thing may be said of the public policies that concern oil and energy in general, and to that extent I am hopeful: these are rational problems, accessible to human will and reason.

In the field of energy, Prudhoe Bay and its pipeline have for eight years commanded most of the national attention. That is to say once more that the project has been a big one—big enough to touch and affect us all. Concurrent with it have been a number of other issues of energy policy no less vital to our survival and only a little smaller in scale. TAPS will have been more valuable even than the oil it carries if from its history we can derive more effectual principles of public policy and conduct and learn to apply them to the questions before us. We shall learn, or we shall perish; that is the natural order.

The accident of birth gave me a place in the audience for one of the long-running energy dramas of our time: the fading promise of nuclear power. The farther it has progressed, the clearer it has become that it is not likely to be a solution for the long run: with the technology we have, it is an absurdly inefficient use of a resource not at all abundant; and the storage of still-active atomic wastes—"disposal" is a misnomer—may prove to be the most seriously limiting factor of all, a problem still unsolved and perhaps not solvable in any final sense. I nevertheless believe that the legal obstructionism and civil disobedience that in this country have delayed nuclear power projects for as much as a decade are essentially hysterical—or if rational profoundly mischievous, serving no American national interest, though quite possibly the interest of others. It does not appear that in use, with our present technology (such as it is), nuclear fuel is either more or less dangerous than the gasoline that runs a car or truck, a risk to be taken consciously and with care because there is as yet no better alternative. Nuclear power, that is, represents another stopgap, like oil itself, whose time in man's history is likely to be short, but it will have served its purpose if it provides the interval in which to develop a more fundamental alternative. If that is to be based on some form of controlled nuclear fusion, then to judge from the snail's-pace progress of atomic energy since 1945—the life of my generation—and the present state of laboratory experiment, we will need all the time we can get. But it is essential to be clear on that point before we can turn our hands to the making of relevant and timely policies.

Several other energy issues have been argued during the pipeline years in the same forums and on the same terms, and only a little less vociferously than nuclear power and TAPS itself. All have to do with oil and U.S. reserves: how and where they should be developed—and whether. Thus, while oil prospecting since the 1950s has pushed ever farther north everywhere in the world, it has also moved out to sea wherever geological evidence has encouraged the hunt—with great success in the shallow inshore waters of Texas and Louisiana, rather less off California and in Alaska's Cook Inlet; and disastrously on the Gulf coast of Florida (from the standpoint of the oil companies that bet $1.5 billion on federal leases that produced no oil, but the loss in productive capital and potential reserves was national as well). Now we have reached what may well be the last great frontier in the world of oil, the deep waters (up to hundreds of feet) of our Atlantic, Gulf, and Pacific coasts and all the Alaskan seas, known in the industry's shorthand as the OCS—outer continental shelf. When or whether we shall venture across will apparently be decided in a succession of federal courts and perhaps finally in Congress, but the whole question has been clouded by legal obstructionism and misinformation. The issue, at any rate, is not one of technical feasibility—or not, ironically, so long as the value of oil holds at levels fixed by OPEC price-setting, high enough to sustain the considerable costs of OCS exploration, and production if it comes. While the United States has vacillated, it is American technology (provided, by and large, by American companies) in the service of Norway and Great Britain that has found and brought in the succession of new oil and gas fields in the deep and difficult waters of the North Sea.

In these same years both Western Europe and Japan—and, of course, the exporting nations from which their oil is shipped—have moved expeditiously to develop what the newspapers, for want of a better term, call superports: not ports at all in any traditional, land-bound sense but fixed mooring points in deep water far enough out for safe maneuvering by the current generation of giant tankers, where they can pump their cargo ashore (or take it on) through sea-floor pipes. As I write, the first such American installation, planned for the Louisiana Gulf, may at last be finding its way through the thicket of federal regulation and courtroom argument—to the point, at least, that the sponsoring consortium can begin to take up the serious questions of design and scheduling. As with the OCS question—and the tanker regulations laid down by the states of Alaska and Washington (the latter, if it stands, prohibiting tankers bigger than 125,000 dwt in Puget Sound)—a seriously complicating factor in this country has been one not present in Europe or Japan, touched on in my two final chapters: the balance of authority between state or local and federal sovereignty with respect to national resources.* The other points of conflict can be finally resolved through reason, valid and pertinent information, good will genuinely responsive to general need; this one, as I suggested earlier, is likely to remain in tension, be redefined, reapplied, case by case as the issues arise. The principle of sovereignty in this sense is central, I believe, to the theory and practice of American democracy and must not be lost—but neither can we permit it to work against the common good if we are to remain *united* states.

One other factor in the energy equation is worth noting here: the immense deposits of oil shale found in western Colorado and in Utah (and some other parts of the country), a form of hydro-

---

*The Puget Sound question has been answered since this was written. On March 6, 1978, the Supreme Court, deciding a suit brought by Arco, ruled that a state (Washington) could not constitutionally contradict federal law and regulation governing interstate commerce by prohibiting from its waters tankers of more than 125,000 dwt. The decision is probably of no immediate practical value to Arco, but it reaffirms an important principle.

carbon that can be thought of as dehydrated oil—somewhere in the course of its development the liquid has been lost, leaving a dark, hard, oil-like substance layered in sedimentary rock that is, incidentally, beautiful to eye and touch. Another irony: shale oil had been a likely candidate to replace scarce whale oil when in 1859 Colonel Drake's first Pennsylvania well came in, and it has run second ever since. The idea was revived during World War II (the Navy was still in charge of most oil shale reserves, along with its oil reserves), at intervals since, in demonstration plants sponsored by various oil companies experimenting with extraction techniques, but each time with the same result: the shale oil came out at a cost higher than whatever was current for conventional oil, and as one cost has increased, so has the other; but the experiments for now continue. It is probable that the oil shale reserves in the United States and elsewhere exceed all the world's oil reserves so far produced or discovered—if we had any economic means of getting the oil out. Because we do not, the possibility remains comparatively remote in the present context, like that of power from atomic fusion, a generation or more off, if ever; but sooner, perhaps, if growing scarcity of liquid oil and difficulty in obtaining it restrict its use to high-profit chemicals and plastics—profitable partly because the volume of the petroleum raw material that they require is comparatively small for the number and range of products derived. Even on their limited scale, however, these experiments have excited the same clamor as the other issues we have considered—but locally, in Colorado and Utah. This resistance has not been without reason. All the processes so far tested require huge quantities of water, a not plentiful commodity in those parts, and generate great volumes of spent shale, rubble, and dust to be somehow disposed of. Meanwhile, the existence of oil shale and the size of the reserves to be mined, when the time comes, should provide reassurance against desperation. The apocalypse of exhausted resources is an environmental fairy tale.

Our attempts at making sense of these serious alternatives have been accompanied by a drum fire of press agentry for others that are not: windmills; many solar generating schemes, for electricity or direct heating or cooling; ideas for producing power from tidal action or methane gas from garbage dumps;

and so on. There are limited and quite specific situations in which one or other of these proposals might be applicable, saving for the national budget small increments of energy that would otherwise be expended, but all have two characteristics in common. They are entirely incommensurate with the massed megawatts of power that an urban humanity actually requires for its survival; and if developed to the scale of city needs they would involve unthinkable expenditures of metals and other resources and a net deficit of energy (for mining, smelting, fabricating) that would almost certainly never be made good. Let us never dampen imagination, but these are preindustrial, machine-smashing fantasies fostered in the security of academic salaries and exurban spaciousness. They distract from the problems at hand.

The energy questions we have considered so far will be affected and perhaps determined by the public policy decisions we make today, but all are more or less remote from the present —years off, decades, generations. Two others are immediate and represent final stages in the planning and development that began with the discoveries at Prudhoe Bay: the destination of a substantial part of the present and future oil production; and the delivery of the natural gas to U.S. markets. In the course of our narrative we have touched on both matters several times in passing. It is time for a closer look.

It will be recalled (Chapter 3) that TAPS from the beginning faced a fundamental practical objection: it would inject more crude into PAD V, the West Coast market, than the market could absorb; that objection was argued at the time, though insufficiently, in my view, our minds being diverted to other aspects of the case, but on the other hand, as we have seen, the financial impossibility of the alternatives was never seriously considered. For a number of years, the demand for oil in PAD V has held steady at about 2 million b/d, about half of it locally produced, mainly in California but with some help from southern Alaska; the difference has been made up with imports of light, low-sulfur crude primarily from Indonesia and Alberta. In the aftermath of the OAPEC embargo, Congress determined that the Naval Petroleum Reserve at Elk Hills should be brought back into production and, with additional development, raised to 300,000 b/d (by late 1977, although behind schedule, the field was pumping at a rate of about 130,000 b/d); in addition, offshore exploration and development were cautiously reopened, chiefly in the Santa Barbara Channel. Thus, when Prudhoe Bay reached its first target of 1.2 million b/d sometime in 1978 (about equal to the rest of PAD V production), the West Coast would be receiving at least 400,000 b/d more crude than it could consume, even if all imports were excluded—swimming in oil, one might say, with no practical means of carrying the excess inland where it could be used. A few years later, when Prudhoe reached maximum production, perhaps 1.7 million b/d, the surplus might be as much as 900,000 b/d, barring a far more drastic increase in consumption than appears likely. The arithmetic is complicated by the fact that not all imports can in fact be excluded: a lot of the California crude, like that from Prudhoe Bay, is heavy and has a fairly high sulfur content—to be economically useful it must be blended with light, low-sulfur oil at a ratio that may be as great as three to one. In other words, while PAD V demand remains at 2 million b/d, as much as one fourth of the supply will necessarily be imported. What this means is that at a rate of 1.2 million b/d, at least half the Prudhoe production will be surplus and possibly a good deal more—anywhere from 600,000 to 900,000 b/d; and at 1.7 million b/d the surplus is likely to be well over a million barrels and might conceivably account for the entire daily production.*

The burden of this extraordinary predicament fell unequally on the several partners, as we know. For Arco and those with minor interests in the field and the pipeline, it was no burden at all—their refineries and retail outlets were sufficient to absorb their full share, with spare capacity for whatever oil Exxon could not funnel through its own limited system. Sohio and BP, on the other hand, had *no* means of disposing of the oil within PAD V

---

* This maximum surplus would, of course, be reduced if Elk Hills and the offshore wells are once more shut in and other California production declines; how much PAD V demand will rise in the meantime seems not very predictable. I think it certain that even without the development of other reservoirs on the Arctic Slope, hundreds of thousands of barrels a day of Prudhoe production will remain surplus for at least a decade, possibly for the entire life of the field.

on their own behalf, and although at each stage their more than 50 per cent share of production would about match the probable surplus, they had never considered entering the West Coast market and moved only slowly and late to develop means of delivering their oil to other American districts.

This lethargy (as it appears to one observing from the outside) is so puzzling that in the late spring of 1976 I traveled to Cleveland to spend most of a day arguing motives with Sohio's personable vice-president for public affairs, Richard M. Donaldson. By this time Sohio had apparently considered and rejected several schemes for pipelines to carry the oil to the Midwest (from the Puget Sound area or British Columbia, using in part lines that had formerly delivered Alberta crude to the United States); and was now pursuing a project that would utilize an abandoned natural gas pipeline to take the oil from southern California to Midland, Texas, where it could be sent on either to a Sohio refinery in Texas or else through the existing system to Cleveland—whence the project has been variously known as Caltex (California-Texas) or Latex (Los Angeles-Texas). On the other hand, this plan was developing so slowly—in part for reasons of California politics, which we shall come to presently—that on the most optimistic schedule it could not possibly be ready until at least a year after start-up and, more likely, a year or more beyond that; and the company had so managed its tanker deliveries that until about the same time it would not have sufficient capacity to deliver its oil to the West Coast, much less to make the longer run through the Panama Canal to the Gulf coast—or not, at any rate, in tankers that pass the Jones Act test (see Chapter 12).

Dick's explanation of this state of affairs was essentially that the company had made an honest mistake—or a series of mistakes. Their planning, he thought, had gone through three stages. When the Prudhoe field was first discovered, their projections of West Coast demand suggested a modest surplus by the time the pipeline was finished, 1971 or 1972, and they had at least looked at the possibilities for northern routes to the Midwest, running from Washington or British Columbia. These were dropped when TAPS was delayed by its legal troubles—it appeared that by the time it could be built, demand would have caught up with the supply thus provided. When West Coast demand leveled off in the aftermath of the OAPEC embargo, they looked again and settled on Caltex—but the timing was unfortunate.

To me that story made (and makes) no sense, and I said so. When one looks back at the actual oil production and consumption figures for the West Coast as they have developed since the early 60s, the projections and consequent indecision that Dick Donaldson described suggest not "honest mistakes" but errors so fundamental as to make one wonder how a company so managed could remain in business. What *does* seem to make sense of Sohio's strategy is export, and, politely, I suggested as much. Bound by John D. Rockefeller's original vision—control the refineries and you control the trade—Sohio for sixty-five years had been a buyer of crude, perennially short of its own supply; BP, more even than Exxon, has built its business on the international markets. Japan, intimately involved in TAPS from the beginning and Alaska's natural trading partner in every other area, was the obvious place for the oil to go, and if the pipeline act had seemed to preclude the export of Prudhoe oil, perhaps the real meaning of that clause was the opposite, a permission—the President had only to declare that export was necessary. So why not? *Oil has no nationality,* no more than money.

"What *amazes* me," Dick earnestly responded, developing his argument, "is people just like yourself saying, somehow you're going to take this stuff to Japan. The public gets confused, the public officials get aroused—. Let me ask you the question, what do you have to tell people to say that it isn't going to happen? Now if Congress decides and they do it on their own and they say, please send this oil to Japan and we'll take oil from Nigeria as an offset and the country's ahead—somebody can write a scenario on those economics but we're not wasting any time on doing that. And I guess my question to you is, what do we have to do to put this red herring aside?"

"What you have to do," I said slowly, "is to have had the west-east pipeline planned and built so that at least by January seventy-eight oil is going to go through it. And if you're not at that stage—if you really have a system that starts at Prudhoe

Bay and goes to Cleveland, then why in the hell isn't it designed and—"

"*O.K.,*" with resignation. "*O.K.*"

So far none of this has happened: the oil has not been exported, but neither has the Caltex pipeline advanced much beyond earnest discussion. After securing the approval of the dozens of federal, state, and local governments and agencies with a voice in the outcome, the plan was blocked by a California body known as the Air Resources Board, headed by a man named Tom Quinn, a close adviser of the governor, Jerry Brown. The ground of Mr. Quinn's argument was that the oil pipeline (and the new tanker terminal to feed it) would disastrously affect the quality of the atmosphere in the vicinity of Los Angeles. The argument, so far as I can determine, is a transparent falsehood,* but there the matter rests. In the meantime, Sohio was making do with tankers. This led to complicated arrangements. A few days before the *Manhattan* sailed on its first voyage from Valdez at the end of August, the Sohio-chartered *New York* loaded at the same berth—268,310 dwt, nearly 2 million barrels of oil, the greatest cargo ever taken on at an American port, but the record went almost unreported. From there the ship sailed south to a point 65 miles off Panama City where BP's still bigger *British Resolution* lay at anchor, there to deliver its load and return to Valdez. The oil was then pumped from the *Resolution*'s tanks into tankers small enough to negotiate the canal and thereby transported to Houston and the East Coast; three weeks later, the first Prudhoe crude reached the BP refinery at Marcus Hook on the Delaware River, on board the *Overseas Valdez,* chartered to Sohio. *None* of these ships met the Jones Act standard and exceptions were conveniently allowed, for six months. On the other hand, it would seem that any lawyer maintaining that the Jones Act does not apply to this circuitous routing would have an excellent case.

* And baffling as to motives, unless one supposes that Mr. Quinn meant to bargain Caltex for renewed shipments of natural gas; a second gas line is available for that purpose, but also in disuse and for the same reason—at the prevailing prices, there is not enough gas to fill it. *If* the gas could be supplied from some source and would drive a corresponding volume of oil out of the Los Angeles market, the exchange would have some beneficial effect on air quality.

All this is exceedingly puzzling, particularly in the light of Dick Donaldson's eloquent assurances, until one does the arithmetic. For profit, the unavoidable corporate standard, it's tankers all the way.

Sohio's preferred Caltex plan involved building a tanker terminal at Long Beach, the important port south of Los Angeles, and deepening the port's ship channel to accommodate moderately big tankers up to perhaps 100,000 dwt (by the standard of most other American ports Long Beach is not bad to begin with); buying the pipeline and right-of-way from its present owners (gas pipe is generally thicker walled than oil pipe, so the existing line was adaptable for the purpose); adding connecting links at both ends of the system; and supplying suitable pumps at appropriate points, presumably in about the same proportions as on the Trans-Alaska Pipeline. For two years or more Sohio's rule-of-thumb estimate for all this has been $500 million to bring the line to a capacity of 500,000 b/d, probably the minimum the company will have to dispose of when Prudhoe Bay is producing 1.2 million b/d of crude; on the same standard, it will take another $300 million to raise the system to 800,000 b/d—additional pumps, "loops" of pipe to double capacity and relieve pressure at key points along the route—which might possibly be enough to handle the Sohio share when Prudhoe is up to, say, 1.7 million b/d. (There has been talk that this would also leave room for carrying some of the excess from Elk Hills, if Congress allows it to continue producing, but this in my judgment is doubtful.) I suspect that these estimates are no more to be relied on than the early ones for TAPS, but let us take them at their face value and not quibble.

If we apply to the $500 million initial Caltex construction cost the same ICC financial standards as the oil companies used in their TAPS calculations (see Chapter 11), what we get, at a capacity of 500,000 b/d, is a tariff of around $2.64 per barrel. As a cost per barrel per mile (the Caltex distance is about a thousand miles), this is less than half what TAPS is likely to cost, but it is still a very stiff charge compared with the average tariff collected on all other existing U.S. pipelines. The calculation has several meanings. For one thing, it is one more of the essential cost factors that was never even considered in the original TAPS decision. Moreover, assuming that this tariff comes off the selling price of the oil and thereby proportionately reduces the wellhead price on which the Alaskan royalty and production tax are figured, it drastically reduces the income the state would receive on half the Prudhoe crude. Sohio, however, does not come out very well either. Allowing for its ICC profit on both TAPS and Caltex and the cost of tankers to Long Beach, I figure that its annual profit on 500,000 b/d of crude as it reaches Midland would be something less than $600 million before taxes—and the oil at that point is still a thousand miles or more from refining and its ultimate market.

Dick Donaldson's hint about Nigeria as a source of replacement for oil exported to Japan provides a superior alternative. For the East Coast or the Texas Gulf, Nigeria is the nearest producer not already largely committed to the U.S. market; its current production of around 2 million b/d—interestingly, about the same as that which should ultimately come out of Prudhoe Bay—was discovered and developed mainly by BP. At current VLCC tanker rates for more profitable foreign-flag ships, such an exchange is attractive from the Alaskan standpoint—only the tankerage to Tokyo, presumably, would come off the wellhead price and the state's revenue would be at least four times what it would be if the oil goes through Caltex, even if the Japanese pay no more than the present world market price. Sohio also would come out ahead: assuming that it would bear the entire tariff to Tokyo and from Lagos (neither assumption, I think, is at all certain), I figure its net income before taxes on the exchange, with the Nigerian crude conveniently landed at Houston or Philadelphia, would be on the rosy side of $900 million a year. Not at all bad.

We come finally to Sohio's present awkward makeshift, big tankers to the Gulf of Panama lightering into smaller ones for the balance of the voyage to Texas—and surprisingly this looks to be, on balance, the best deal of all. Averaging the probable tanker rates on long-term charter and allowing for the considerable fuel cost, it appears that the oil can be delivered from Valdez to Houston for about $2 per barrel, including the two empty return voyages. This does quite well by the Alaskan revenues—about three times what they would be with Caltex;

and for Sohio it should be altogether satisfactory, yielding an annual net income before taxes of at least $1.1 billion (and more if it owns the tankers outright instead of chartering and makes some profit on that leg of the system as well as its ICC 7 per cent from TAPS).

Contemplating the three sets of figures, you would surely, if you were a Sohio or BP manager, feel a warm glow at their renewed confirmation of the excellence of tankers and no great uneasiness at the prospect of selling some or all of your oil to Japan; but you would reckon that prospect politically remote, in Alaska or the nation, even if it would go far toward alleviating the national balance-of-payments problem and still further cement relations with the nation's chief Asian ally. On the other hand, coming to a rational conclusion, you would feel not the slightest urgency about refurbishing a thousand miles of abandoned gas line to move your oil, and you might prudently remember the Tom Quinns of this world in your prayers, petitioning on their behalf health, long life, and glorious bombast in perpetuity, world without end. Amen. There is, however, no reason to suppose that oil men are any more rational than politicians or the rest of us.

In September 1977, the Canadian and American governments signed an agreement in principle permitting Prudhoe Bay gas to be carried to the Lower 48 by pipeline through Canada. The agreement was a compromise within a compromise. It had seemed likely from the beginning that if one policy objective was to share the benefits of the construction bonanza between Alaskans and the Canadian cousins, then, eventually, a Canadian route for the gas line would be chosen to balance the Alaskan choice for oil. The Canadians in the meantime had shown their displeasure at TAPS in a number of ways: by withdrawing, for instance, the Alberta crude that formerly supplied the Northwest; by obstructing a deep-water tanker port planned for northern Maine on the ground that part of the approach route would lie in Canadian waters, a position not at all obvious as a matter of international law. For most of the eight years it took to reach the pipeline decision, the choice had seemed to be between two possible routes: one paralleling TAPS and ending at Cordova south of Valdez, where the gas, chilled to a liquid (LNG—liquefied natural gas), would be carried south in specially built insulated tankers; the other due east from Prudhoe through the Arctic National Wildlife Range, thence south along the Mackenzie. The route finally chosen borrowed from both: south to Fairbanks, then southeast along the Alaska Highway through Canada, for which it is known as the Alcan route; a compromise, as I said, within a compromise.

Besides the agreement itself, there were a couple of other modest accomplishments during this time. On behalf of the Canadian government, Thomas R. Berger, a justice of the Supreme Court of British Columbia, had spent four years making the rounds of the Northwest Territories and the Yukon, collecting the views of Natives and others on a pipeline through their land —and had concluded with a recommendation that any such project be postponed indefinitely; an opinion not, however, binding on anyone. The two governments had, in addition, worked out another treaty in which, essentially, Canada promised that its provincial governments would not make an American-owned pipeline inoperable through discriminatory taxation; though when we consider the dispersed powers of the Canadian constitution and the aggressive cage-rattling from Quebec and elsewhere, we may well wonder whether Canada's confederated government is in fact capable of enforcing such a guarantee, legally or by any other means. So far as the serious business of planning, designing, and scheduling the system is concerned, the project has hardly begun. The sequence of events provides one more answer, if any is needed, to those who throughout the debates on the oil pipeline insisted that one or other Canadian alternative could be prosecuted as expeditiously as TAPS. Had we heeded their advice, we would today be no farther advanced with Prudhoe's oil than with its gas; which was, of course, the purpose.

We are, that is, back where we were around 1970. It would be reassuring to think we had learned something in these intervening years, but from the way the gas project has started out it does not look as if that is the case.

Throughout their arguments, the three proponents have quoted construction costs ranging from $8 billion for the all-

Alaska route to $10 billion for the Alcan. None has made any reference to the costs of financing, which, as we have seen (Chapter 11), will assuredly be substantially greater and have a more serious effect on the tariff for moving gas through the system. The estimates are also incomplete in the same ways as all the TAPS estimates: they apparently make no allowance for the additional pipelines that will be needed to carry the gas from the U.S.-Canadian border to markets on the West Coast and in the Midwest (some lines exist and can be converted, but the costs will still be substantial); nor for a projected link to the Mackenzie delta, to provide a free ride for the gas reserves there, in case Canada decides to develop them. What the estimates in effect ask us to believe is that the far longer Alcan gas line can be built for roughly the same cost as the 800-mile TAPS oil line—2,200 miles to the Montana border, then another 1,200 to San Francisco (or 1,600 to Los Angeles) and 1,350 to Chicago.

The two systems are not entirely comparable. The gas will be pumped cold and except for some stream crossings the pipe will be entirely buried, avoiding the huge costs and the technical problems of elevated pipeline; on the other hand the costs of logistics and construction camps in Alaska and Canada are likely to be proportionate to those incurred by Alyeska (for a decade Canada has been engaged in building a gravel road to connect northern Alberta with Inuvik, the administrative center in the Mackenzie delta, but it is not finished); and this part of the line will require thirty to forty compressor plants to move the gas, at costs comparable to those for the twelve TAPS pump stations. With these and other adjustments, it does not appear from the Alyeska experience that the Alcan line can be carried to the Montana border for less that $16 billion, while extending the system to San Francisco looks like about $5 billion and on the same reasoning the Chicago link will come to more like $6 billion. These estimates are for construction alone and make no allowance for interest; and since they are based on TAPS costs, they allow nothing either for inflation—they are what we might expect the Alcan line to have cost *if it were already built*. We do not, I think, know all the private understandings that surround the Alcan agreement, but among them appears to be the Canadian position, maintained from the beginning, that the line is to be Canadian-controlled and that, therefore, more than half its financing must be from Canadian sources. When we look back on the difficulties Alyeska's owners have had in raising the cash for TAPS within the huge U.S. money market, it seems improbable that these far greater sums can be assembled from private Canadian sources; nor does the Canadian Government itself seem either able or willing to provide its share of the financing.

Just what these costs will do to the price of Prudhoe's gas—if, on these conditions, the Alcan line can be built at all—is hard to predict. For some idea of the effect, however, we may consider the Trans-Canadian Pipeline (TCP) once so enthusiastically argued. If actually completed in the same years and to approximately the TAPS standard, it looks as if the TCP bill would have come to $20 billion. Translated according to the ICC pipeline formula, that huge investment yields a tariff of about $18 per barrel for oil delivered to Chicago—about $4 more than the present inflated price for the oil itself. In other words, what we would have got for our $20 billion would be a marvelous engineering artifact—but one that we could not afford to use. And the Alcan gas line appears to be headed in the same direction.

If correct (as I believe they are), even approximately, these are astonishing considerations; and depressing. Once more a decision on a matter of fundamental public policy has been made on the basis of information that is incomplete and erroneous—or so we are left to infer, since, although this is *public* policy of the most serious and far-reaching kind, its authors have been at no pains to inform the public of the facts and motives on which it is based. Perhaps that is because they are not very well informed themselves.

Clearly we must manage our affairs better than this if we are to survive, but the generalization is easier made than reduced to useful specifics. In August 1977, in an attempt at something of the sort, the Department of the Interior, which had had the primary governmental responsibility for TAPS, convened a two-day conference of several hundred of those who had played some part in the proceedings and others with an interest in the forthcoming gas line, its actual route at that point still undecided but leaning toward Canada: representatives of Alyeska, among them

Ed Patton, and the various surveillance organizations, spokesmen for Alaskan and national environmental groups, Canadian oil men and officials, engineers on the lookout for contracts on the new pipeline. The meeting was held at the Anchorage campus of the University of Alaska, in the almost completed Arts Center—massive raw poured concrete, placed in one corner of a low academic quadrangle built in a complementary style and set against the surrounding rim of mountains; steps, perhaps, in the evolution of an architecture suited to the country, of its people growing to their landscape. I was on hand; something to do while waiting for the pipeline to restart and the tankers to sort into some kind of schedule.

Although intended as a critique, the meeting was too amiable to be very useful. Those in a position to act with some authority —Ed Patton, Jack Turner, now finally in charge of the federal surveillance team—were too set in the habit of defending established positions to say much that one had not already heard many times over; most others, returning soon to whatever careers they had come from, cautiously kept silent and, though invited to unburden themselves afterward, in writing and if need be anonymously, did not apparently seize the opportunity. About the only idea of substance that came forward in the two days was obvious enough to all concerned and indeed had been intended by Congress in the first place, though not provided for: that, next time around, all the federal and state government bodies with a legal voice in the outcome be united in a single agency created to monitor the project—"one-stop shopping," briefly put, but legally and politically more difficult to provide than it sounds; maybe impossible.

Although lamely expressed by the bureaucrats gathered in Anchorage, the idea points to an issue that I believe has been central to our experience of the Trans-Alaska Pipeline and must be reckoned in whatever improved arrangements we devise for future projects. That issue has to do with the mutual communication between the corporations and government, starting with Congress, that must focus on the management of such projects; with the interaction between private and public interests. Throughout, on the public side, we have been burdened by

doubt as to whether we were in fact getting complete and reliable information on which to base decisions from those best able
to provide it; and, within the corporations, by no assurance that
if all facts were indeed freely given the political and judicial
processes would issue in decisions rationally grounded on them.
Contrary to the old images of oil implanted in our collective
memory, too often in this history the relationship of government
to corporation has been like that of master to servant. As it has
worked out, it has not been a healthy relationship. More to the
point, if we care about results, it has not been efficient; an unwilling servant, however hectored with commands, does poor
work.

Let me lay out a few assumptions:

That the productive work of our society is most effectively
done by private interests directed toward the general good;

That public decisions concerning long-term necessities such
as energy must be based on correspondingly long-term and
accurate projections, secured early in the process of decision;

That the costs resulting from such decisions, which will be
borne by the public, must be accurately assessed and publicly accountable.

The means to all these goals exists. A corporate director has access to the information on which policy is based and a voice in
setting it, and he bears responsibility for the results; although
elected by stockholders, his effective power depends on his investment in the corporation on whose board he sits. Turning
these thoughts to oil, it seems to me that we might take a leaf
from the British book (the government has invested substantially
in BP and has a corresponding voice on its board and in its
policies) and find ways of seating public directors on oil-company boards; and public investment in corporate stock is the obvious way, providing some of the capital that will be needed for
energy development and a public balance to the private interest.

An arrangement on this order (or another to the same purpose) would assist the process of public policy but could not
substitute for it. The corresponding public official (and the
agency built around such an office) does not now exist. To me

the TAPS experience suggests the need for a permanent regulatory body for energy, combining the monitoring functions the Alaska Pipeline Office exercised over TAPS with considerable (but not unlimited) authority for policy and for decisions on major projects. Three characteristics seem to be required: broad technical competence supported by practical experience in the field; a degree of insulation from day-to-day congressional politics and therefore a comparatively long term of office; and a broader public responsibility and accountability than is now provided by presidential appointment subject to investigation and approval by the Senate,* the present system for choosing cabinet members and other high officials within the federal executive departments. The recently created Department of Energy has the size and, presumably, the technical capacity for early, detailed, and continuing review of the planning and design of future energy projects (*if* it is clear that that is what we want—the APO's effectiveness is reviewing the design of the Trans-Alaska Pipeline is pretty mixed) but seems to me insufficiently independent and accountable. In an ideal world, the body so charged would have the comparative independence and longevity of, say, the Federal Reserve Board but would be placed under the publicly determined responsibility of an elected chairman made accountable by a workable system of recall. There are no ideal worlds, of course, but these are functional characteristics and worth working toward.

One of the practical benefits we might hope for from these two kinds of officials—public directors on energy company boards, a permanent and comparatively independent agency monitoring projects concerned with energy—is control of costs. Realistic and complete estimates are an obvious essential. No less obviously, if estimates are to be translated into controlled construction costs, that will have to be done through fixed-cost contracts containing effective incentives. What this means, in turn, is fuller preliminary design than was developed for TAPS directly keyed to whatever system of public review and approval is finally established; and clear public standards consistently applied and expressed in terms general enough to allow reasonable

---

* When and if construction of the Alcan gas line becomes a reality, it appears that the official monitoring the American end of the project will be selected in this fashion.

latitude for the judgment of the monitoring authority, consistent with public policy, changing technology, and a decent respect for project costs. It does not seem to me that public and private goals of engineering effectiveness and economy need conflict.

The corollary of these administrative goals seems to me the necessity to limit (but not eliminate) the role of our judicial system, including the legislative processes on which it is based, in deciding essentially technical questions. There is value in the discipline required to express the technical foundations of public policy in terms intelligible to a court of law, to the extent that is possible, and the same thing may be said on behalf of the general public: the exercise exposes untenable assumptions to the light of common sense. But there are limits to what can be so clarified, and as our judicial and legislative systems now work, the process introduces distortions and absurdities—the spectacle of experts testifying to opposite conclusions and putting their knowledge to the service of unstated bias. It is no more reasonable to allow the nine gentlemen of the Supreme Court final discretion in deciding, say, the engineering effects of permafrost than to suppose they can arrive at a medically valid diagnosis of small pox or cancer.

To exercise any kind of meaningful role in national policy, the public is as dependent as its elected representatives on continuing and reliable information; an obvious but fundamental truth. As we have seen (Chapter 10), the efforts of the press to do this necessary job on the Trans-Alaska Pipeline were only partly successful and in several areas were, I believe, seriously misleading —in part because the numbers of reporters with at least a superficial interest in the project were greater than Alyeska's facilities could accommodate. The policy of restricted access that resulted was not entirely unreasonable, but the effect was to make it nearly impossible to cover the enterprise with the continuity, depth, and awareness of technical relationships that it warranted. There is an obvious solution: the kind of pool arrangement often used for other major stories for which unrestricted access is impractical, whereby a few reporters share their copy with all. Considering how vital such public information is in all matters of energy policy, perhaps we need to find a means of taking that idea a step further: something, for instance, on the order of publicly funded fellowships to enable a pool of newsmen to follow a major project long enough for full and responsible familiarity, rotated to avoid the risk of their going stale or becoming overly identified with official views—six months, say, at a stretch.

In 1974 six environmental groups, among them the Sierra Club, banded together in the Arctic Environmental Council for the purpose of periodic observation of Alaskan pipeline construction, with the co-operation of Alyeska and the Alaska Pipeline Office; perhaps because of that association the Wilderness Society and the Friends of the Earth, which had been prominent in the pipeline suit, disdained to participate. Instead, they have argued since for a public body with power to act as well as observe, provided by themselves: a fourth layer of environmental surveillance on top of those constituted by the press, the government monitors, the corporations, to represent the public. Although the idea found little favor at the Anchorage critique, it reappeared anyway as a recommendation in one of the reports. Volunteer public service is an honorable and distinctive American tradition, perhaps fundamental, but that is not the function I see in FOE, the Wilderness Society, and their allies; only more Washington pressure groups serving interests and ideologies as narrow as the rest, not the public at large. I speak from within that unorganized but not inarticulate mass. To grant further power to any such body not in some measure responsible and accountable to me is abhorrent. We have already surrendered too much.

Time passes. The pipeline is finished, as mixed in its motives and its carrying out as every human enterprise, but the oil moves south, is refined, transformed, consumed, and for now the immemorial millennia of its life are merged in the life of a nation. Meanings? Mixed also: human necessity and human good are not readily distinguished; but certain images remain.

A few weeks before start-up, sorting meanings, I flew to Los Angeles to see what could be learned from Arco's chairman, Robert Anderson, the only one of the oil men with a hand on the direction of Prudhoe Bay from start to finish—whatever consistency of purpose the project has had appears to be his. The meeting, months in negotiation, was to be on the top floor of the gray Arco tower built by, and in anticipation of, Prudhoe profit, and I arrived early, identified myself, was let into an ante-chamber through the motor-powered doors of inch-thick bul-letproof glass—and was told Mr. Anderson was not expected. Strategies of survival in the present age, if you are visible above the crowd: keep your movements to yourself and in appearance unplanned. I sat down to wait. The room was an immense dou-ble cube: pale marble, architectural chairs in steel and dark leather, rare plants, labeled, grown to the scale of an equatorial rain forest, sheets of glass looking down on the convolutions of the city: the kingdoms of the world in a moment of time, and the glory of them. All this, as Dr. Johnson observed of another great house, excludes only one evil, poverty. To have grown up among fortunes won, lost, remembered, won again is to be inoculated forever against that particular disease; one is not easily taken. At the appointed time a slight, gray man entered in unmatched pants and jacket and white Stetson, indistinguishable on a Los Angeles sidewalk except for the presidential cuff links at his shirt cuffs, emblems, though of an era already dead; security again. I followed Mr. Anderson into his office, another double cube; he made an explanatory allusion to the casual dress—had been rid-ing in the coastal hills, slept out in the rain that greeted me at the airport—and washouts, floods, tornados—the first in months of drought; and settled to talk.

For an hour we sparred through history, motives, strategies, practicing the conversational art. But, finally, what did it come to, how could we manage better—courts, their advocates and ad-vocacies, congressional hearings—the next time? The answer I got was a negative, or perhaps another question. "The adversary approach to any problem is the worst. It's the most costly, most expensive, least productive." Then, pursuing: "We have a society that is highly legalistic and built on the adversary concept of law. When we had time as a society to pursue things in that more leisurely way, I guess it worked out, but today the critical commodity in the world is time. We have run out of time, we are continuing to waste it, and before it's over it's going to be the one dimension that will cause more trouble than anything else." And finally, with purpose: "We've got a fantastic system—you know employment hit ninety-five million the other day? It's just incredible that any system could provide forty or fifty million jobs in a fairly short period of years, thirty or forty years. But you can't assume that will happen automatically forevermore."

Time passes and is not retrieved, one element and sense of loss, to opposite meanings. Brock Evans has a story of his meaning of the pipeline, from one of those tours on behalf of the Sierra Club and the Arctic Environmental Council, retold, practiced, with the Welsh eloquence of his name. It was thus when I heard it:

"We were down at Dietrich Camp, still in the Brooks Range. It was October but it was winter already down there, and I went off by myself, away from the others, down by the bank of the Koyukuk River, following wolf tracks in the snow, through a little patch of forest, and the wind sort of blowing in the dry grasses. And I followed these wolf tracks down to the bank of the river and looked far to the south, you could see the dim mountain ranges way off in the distance, and the ice was forming on the river. The whole thing was just so much the North, it was the whole north country the way Jack London and all the writers wrote about it: a spell came over me as it must have thousands before me, or millions. And then I turned around to go back, and all at once this awful cacophony started up, the bulldozers started in again—it was after lunch—planes started landing and trucks started roaring and grinding, all the machinery. And all at once I realized that was what we lost—we lost the whole sense and feeling of the North the way Jack London and the others wrote about it. That will never be the same again; it was just a narrow strip in that whole vastness . . . but it's broken the spine of that place, it's changed it, transformed it—so now you know it's not the wilderness from one end to the other of the continent. . . . We lost in that sense: the country lost. . . ."

Yes. Give me the freedom of the earth and I'd as soon be there, crouched to the wolf tracks by the side of the Koyukuk, and the smoke of my cook fire pressing upward against the weight of the cold; hearing the silence; hugging the warmth. And if a man must make camp in company, then as well with

Brock Evans, with his muscular forearms and his certainties. But we walk the earth in multitudes. They hedge our freedoms.

The bulldozers are still, the men who drove them mostly departed. The machines will return, though, the men, fall silent again, resume—man's life, the life of his race, uncoils in waves of action and quiescence, but truly nothing is ever the same again. So we know loss, losses—but in the natural and necessary order of all life, and therefore not to be mourned. The leaf feeds the flower, the flower begets the fruit, which falls, rots, and in the order of life spills seed, bears new life. Innocence is not in itself to be prized, it is a condition of incompleteness and a portent of death. The ripe body of a woman who has borne children, the breasts from which infants have sucked life, possess a completed beauty that her transient virginity cannot attain. We draw such images from the minute dimensions of man's life, but against the vast life of the earth they turn transparent and dissolve: *virgin* lands, *virgin* wilderness—in a world bred, made fecund, and harvested by a billion of earth's revolutions round the sun and thereby ceaselessly renewed. Nor is the human absence truer or more to be valued: in the ages of man's life on earth there can be no spot of forest, desert, or mountain range where he has not, some time or other, stepped his foot and left his mark; the two lives, of man and his earth, are united, intermingled, and equally to be reverenced. That is the final truth: earth turns to the pulsing of the sun, and among the substances with which that commerce fills her veins for the nourishing of man, there is oil; and now, for a time, there will be a little more.

SOME NOTES ON SOURCES

For the reader who may wish either to verify facts or to examine my sources for himself and draw his own conclusions, I have listed on the following pages the principal materials on which I have drawn. And since many of these are what scholars call primary sources, there is some value in assembling in one place the records of what I believe to have been a critical episode in the nation's history. This listing is not, however, and is not intended to be an exhaustive bibliography of energy, the oil business, or even of Alyeska and the Trans-Alaska Pipeline. Rather, it represents the materials I found useful in some way in preparing my narrative and that in the course of writing it I have either referred to or directly cited.

For the reader's convenience I have organized this listing by chapters. Such an arrangement is not entirely satisfactory: many ideas and themes recur and develop from chapter to chapter throughout the book; but I have tried, at least, to place each item under the heading to which it seems most relevant. Throughout this work I have drawn on the statistical compilations and informational publications of the two oil-trade organizations, the American Petroleum Institute and the Independent Petroleum Association of America; and in some places have used for general statistics such annuals as *The World Almanac* and *The Statistical Abstract of the United States*. The following have been particularly useful for oil production and consumption figures and for energy in general:

Annual Statistical Review. Washington: American Petroleum Institute.

BP Statistical Review of the World Oil Industry. London: The British Petroleum Company, Ltd., annual. The source most often cited for national production and consumption figures, with the virtue of consistency and comparative neutrality; each issue conveniently summarizes the preceding ten years as well as the current year and includes data on the movement of oil in the international trade, tanker fleets, and production and consumption of other important energy sources.

International Petroleum Encyclopedia. Tulsa: Petroleum Publishing Company, 1975. Statistics and general articles, organized by producing nations.

Monthly Energy Review. (Available from the National Technical Information Service, Springfield, Va.) Published during the period of pipeline construction by the former National Energy Administration; statistics on oil, gas, and coal production and consumption, other power sources, general articles.

The Oil Producing Industry in Your State. Washington: Independent Petroleum Association of America, periodically revised.

Platt's Oil Price Handbook and Oilmanac. New York: McGraw-Hill, annual. Crude and product prices, tanker tariffs, compiled from the daily *Platt's Oilgram.*

With a few exceptions, I have not cited specific articles in periodicals. I have, however, followed systematically the reportage of the New York *Times,* the *Wall Street Journal,* the Anchorage *Times,* the Anchorage *Daily News,* the Fairbanks *News-Miner,* and the Los Angeles *Times* (the latter with the assistance of my mother's friend Lucille Dreyer, who has assiduously collected relevant clippings); and sporadically the *All Alaska Weekly, Alaska* magazine, the *National Journal,* and the *Tundra Times,* published in Fairbanks by the late Howard Rock, a Native leader of the last generation; and, although handicapped by its lack of indexing, the *Oil and Gas Journal.* (Its material is often picked up in daily papers, particularly the Anchorage *Times.*) In establishing the chronology of events in the discovery and development of the Prudhoe Bay field and the evolution of the pipeline project, I have relied on the indexes of the New York *Times* and *Wall Street Journal,* and also on two chronologies kept by Anne Banville, who has handled Alyeska public relations in Washington since early 1970, and on some general materials originating with Alyeska. Among Alaskan journalists, Sally Jones and Richard Fineberg (formerly of the *Daily News* and *News-Miner,* respectively) and Tom Snapp, publisher of the *All Alaska Weekly,* have been helpful on a number of specific points; Drew Middleton of the New York *Times* offered advice on some of the strategic implications of my subject and introduced me to officials of the CIA from whom I obtained estimates of Soviet oil and gas production and consumption and some suggestions about Soviet international oil strategy.

In the course of my research I conducted about a hundred interviews, most of them tape-recorded. Many of those who provided information in this manner are quoted directly or otherwise referred to (and in several cases provided further elucidation in response to letters), and they are so identified in the index. The assistance of the following, also interviewed but not quoted or referred to by name, is likewise gratefully acknowledged (several have since moved to other jobs, but I have identified them by the positions they held when I met them):

Robert Arnold, Alaska Native Foundation, Anchorage

George Barril, executive director, Alaska Native Commission on Alcoholism and Drug Abuse, Anchorage

William S. Bicknell, manager, land operations, Atlantic Richfield Company, Dallas

Gene Dickason, manager, Alaska Operations Office, Environmental Protection Agency, Anchorage

Fred Ehrlich, Teamsters safety representative, Anchorage

Commander Loren D. Gordon, U. S. Coast Guard, Anchorage

William J. Hall, economist, National Bank of Alaska, Anchorage

Will Jennings, manager, Atlantic Richfield refinery, Philadelphia

B. Gil Johnson, attorney, Anchorage (senior partner in the firm that handles the Teamsters' legal trust)

Charles T. Keffer, North Slope project manager, Atlantic Richfield Company, Anchorage

William D. Leake, manager, Alaska coordination, Atlantic Richfield Company, Dallas

John C. Miller, petroleum engineer, Division of Oil and Gas, Alaska Department of Natural Resources, Anchorage

Quinn O'Connell, attorney, Washington (representing Alyeska from early 1970)

Edwin M. Piper, plant manager, Development Engineering, Inc. (oil shale experiment), Anvil Points, Colorado

John Real, Teamsters general council, Anchorage

Douglas Reger, archaeologist, Alaska Department of Natural Resources, Anchorage

Don Ryder, director, Alaska Labor & Management Employees Affairs, Inc., Anchorage

Glenn Simpson, manager, North Alaska District, Atlantic Richfield Company, Anchorage

Howard A. Slack, vice-president and resident manager, Atlantic Richfield Company, Anchorage

Armand C. Spielman, senior landman, Atlantic Richfield Company, Anchorage

G. E. Uthlaut, assistant to the president, Exxon Company, U.S.A., Houston

John Vasek, administrative assistant, Alaska Federation of Natives, Anchorage

William Workman, professor of archaeology, University of Alaska, Anchorage

Joseph E. Young, community relations officer, Anchorage Police Department

As our society now functions, the PR man, by whatever title, is generally the first source an inquiring outsider must turn to when he wants information about a corporation or a government agency. The Alyeska public affairs department appears directly in my narrative. The following members of the profession have all assisted at various points in my investigations (again, some have gone elsewhere since I dealt with them):

Anne Banville, Anne Banville Associates, Washington (Alyeska public relations since 1970)

Mim Dixon, Fairbanks North Star Borough Impact Information Center (and her successor Sue Fison)

Richard M. Donaldson, Standard Oil Company (Ohio), Cleveland

Robert Gastrock, Alaska Pipeline Office, Anchorage

Harry Hardy, American Petroleum Institute, Washington

Paul Hassler, Atlantic Richfield Company, Philadelphia

Michael C. Holland, El Paso Alaska Company, Anchorage

Claborn M. Hooper, Jr., Exxon Company, U.S.A., Anchorage

R. J. Howe, Exxon Company, U.S.A., Houston

Lieutenant Colonel George L. Jones, III, USAF, U.S. Readiness Command, MacDill Air Force Base

Arthur R. Kennedy, Bureau of Land Management, Anchorage

Charles E. Orr, Alaska Department of Community and Regional Affairs, Anchorage

Mickey Parr, Atlantic Richfield Company, Los Angeles

Les Rogers, Exxon Company, U.S.A., Houston

Stephen Stamas, Exxon Company, U.S.A., New York

Bert Tarrant, Alaska Arctic Natural Gas, Anchorage
Charles Towill, BP Alaska, Anchorage
Al Troche, Exxon Company, U.S.A., Houston
Mike Webb, Atlantic Richfield Company, Anchorage
Alane G. Williamson, Exxon Chemical Company, U.S.A., Houston

While I have been at work on my own book about the Trans-Alaska
Pipeline, a number of others have been published, treating the pipeline and the present state of Alaska (among them at least two novels)
and energy problems and alternatives in general. I have tried to keep
track of these books but have not looked closely at any of them, for
several reasons: in the time available—my purpose is history, but the
subject is timely—I have had all I could do to collect, verify, and
order the facts of the project without diluting my efforts with secondary sources; and I have not wanted to be influenced by other writers
on the project, in style, viewpoint, selection of evidence, or, most im-
portantly, interpretation. The efforts required simply to ascertain the
facts in any enterprise connected with oil are enormous, but it is their
interpretation, finally, that matters.

Nevertheless, given that purpose, it seems proper to list here those
related books of which I am aware, even though I have not referred
to them in my research or cited them in my text. In their various
ways, they form part of the record of the events I have endeavored
to set down:

Andrasko, Kenneth, and Halevi, Marcus. *Alaska Crude; Visions of the
Last Frontier.* Boston: Little, Brown, 1977.

Borgese, Elizabeth Mann. *The Drama of the Oceans.* New York:
Abrams, 1976.

DiCerto, Joseph J. *The Electric Wishing Well.* New York: Macmillan,
1976.

Gardey, Jon. *Alaska; The Sophisticated Wilderness.* New York: Stein
and Day, 1976.

Halacy, D. S. *Earth, Wind, Sun and Water; Our Energy Alternatives.*
New York: Harper, 1977.

Hanrahan, John D., and Gruenstein, Peter. *Lost Frontier; The Marketing of Alaska.* New York: Norton, 1977.

Hayes, Denis. *Rays of Hope; The Transition to a Post-Petroleum
World.* New York: Norton, 1977.

McGrath, Ed. *Inside the Alaska Pipeline.* Los Angeles: Celestial Arts,
1977.

McPhee, John. *Coming into the Country.* New York: Farrar, Straus & Giroux, 1977.

Novick, Sheldon. *The Electric War.* San Francisco: Sierra Club, 1976.

Roscow, James P. *800 Miles to Valdez; The Building of the Alaska Pipeline.* Englewood Cliffs: Prentice-Hall, 1977.

Solberg, Carl. *Oil Power; The Rise and Imminent Fall of an American Empire.* New York: Mason/Charter, 1976.

Chapter 1. The Land of Burning Snow

My account of the geology of the Arctic Slope reflects general information (such as encyclopedias); a number of specialized pamphlets published by the U. S. Geological Survey, not cited individually; information provided by James A. Savage, manager of North American exploration for the Atlantic Richfield Company and by several other Arco geologists; and personal observation. More general materials on Alaskan geography, history, and geology, which also figure here, are listed under Chapter 2.

Here and elsewhere, I have listed government publications, congressional reports, legal cases, and similar documents by title rather than by source. This seems to be contrary to academic practice but has the advantage of treating the varied materials consistently and placing the emphasis on the subject matter; and, with the accompanying notes on source and content, will, I hope, be useful to the reader interested in locating these documents in a suitable library.

Cooper, Bryan. *Alaska; The Last Frontier.* New York: Morrow, 1972. The Prudhoe Bay discovery by a Canadian-British writer, emphasizing the role of BP.

Herndon, Booton. *The Great Land.* New York: Weybright & Talley, 1971. A gossipy account of the Prudhoe discovery and its legislative aftermath, with some feeling for Alaskan urban life as it was in the late 1960s.

Jones, Charles S. *From the Rio Grande to the Arctic; The Story of the Richfield Oil Corporation.* Norman: University of Oklahoma Press, 1972. Memoirs of the company's last and long-time head before the merger with Atlantic Refining.

Lease Sale Reports. Alaska Department of Natural Resources. Summaries listing tracts offered and leased, successful bidders, bonus bid per acre for each tract, and the total realized by the state; I have collected and referred to those covering the four sales in which the state's Arctic Slope lands were leased, held in 1964, 1965, 1967, 1969.

Primer of Oil and Gas Production, 3rd ed. Washington: American Petroleum Institute, 1973. A textbook of drilling and other production techniques.

A Primer of Oilwell Drilling, 3rd ed. Austin: University of Texas, 1970.

Profile, December 1974. Magazine for employees of Exxon Company, U.S.A.; special issue on the process of oil exploration in Alaska. Similar material in *Exxon USA,* a quarterly for stockholders (dated First Quarter 1975).

Sobel, Lester A., ed. *Energy Crisis, 1969–73.* New York: Facts on File, 1974. Sequential compendium compiled from periodicals and various government sources.

Soviet Economy in a New Perspective. Washington: U. S. Government Printing Office, 1976. A collection of papers compiled on behalf of the Joint Economic Committee of Congress, including analyses of Soviet oil production and consumption prepared by oil economists of the CIA.

Springborn, Harold W. *The Story of Natural Gas Energy.* Arlington: Educational Services, American Gas Association, 1972.

Chapter 2. Great Land

Relevant maps, which in their way are as important as books as sources of information, are listed here in order, by title. As with Chapter 1, I have consulted a number of USGS pamphlets (on permafrost, for instance) which are not specifically identified here. Several books concerning Indian and Eskimo ethnography and archaeology are arbitrarily grouped together under Chapter 4, although they have a bearing on this chapter as well. Volney Richmond, Jr., in response to a letter, supplied a few details of the history of the Seattle-based Northern Commercial Company, of which he is chairman; a trading organization descended ultimately, under a succession of names, from Baranov's Russian American Company, whose principal business in Alaska during the pipeline years was leasing or selling Caterpillar construction equipment to Alyeska.

Alaska. Rand McNally, 1975. General topographic and political map, including the approximate pipeline route.

Alaska Statehood Act. United States Statutes at Large, Vol. 72, Part I, 1958. Public Law 85-508 (July 7, 1958). Conflicts of statehood with the body of existing law dealing with Alaska were reconciled in the *Alaska Omnibus Act* (Public Law 86–70, June 25, 1959,

United States Statutes at Large, Vol. 73, 1959)—including the contradictory wording of Section 4 with regard to Native rights.

Bancroft, Hubert Howe. *History of Alaska, 1730–1885.* San Francisco: Bancroft, 1886 (reprint, Antiquarian Press, 1959). Partly fanciful but still the best source for the Russian era, by virtue of access to Russian archival material by the author's Russian-Alaskan collaborator Ivan Petrof. In general, collection of the primary sources for serious Alaskan history has barely begun, and it is only recently that some material from the Russian archives has become available to Western scholars.

Barrow, John, ed. *Captain Cook's Voyages of Discovery.* Dutton: Everyman's Library. Edited from the logs of the three circumnavigations, including the third, with the Alaskan experiences of Cook and the lieutenants who succeeded to leadership after his death. A cheap and readily available edition, but a better version is that of J. C. Beaglehole, *The Journals of Captain Cook: The Voyage of the* Resolution *and* Discovery, *1776–1780* (Cambridge University Press for the Hakluyt Society, 1967).

The Blue Book. Greater Anchorage Area Borough, 1975. A substantial paperback describing available public services in Anchorage and telling how and where to obtain them.

Chevigny, Hector. *Russian America.* New York: Viking, 1965. A readable popular history by the Russian-speaking author of several other books on the Russian period.

Coal Fields of the United States: Alaska. United States Geological Survey, 1961. Map showing locations and characteristics of fields then known.

Directory of State Officials. Juneau: Legislative Affairs Agency, Alaska Department of Administration. Names, addresses, and telephone numbers of the state's legislators and leading officials, including judges of the three classes of courts and the heads of the twelve branches of the state college system.

Extent of Glaciations in Alaska. United States Geological Survey, 1965. Map I-415.

Garfield, Brian. *The Thousand-Mile War; World War II in Alaska and the Aleutians.* New York: Doubleday, 1969.

Gruening, Ernest. *The State of Alaska.* New York: Random House, 1968. A history, heavy on legislation and politics, by Alaska's last appointive and first elective governor.

Hulley, Clarence C. *Alaska: Past and Present,* 3rd ed. Portland: Binfords & Mort, 1970. A textbook of Alaskan history; weak on the economic and commercial motives of the Territorial period.

Hunt, William R. *Arctic Passage*. New York: Scribner's, 1975. There are innumerable books of Alaskan anecdotes but not much history. This, a history by intention, collects many of the personalities and their tales.

Major Ecosystems of Alaska. Joint Federal-State Land Use Planning Commission for Alaska, July 1973. Map.

Metal Provinces of Alaska. United States Geological Survey, 1974. Map I-834.

The Milepost. Anchorage: Alaska Northwest Publishing Co., annual. Guide to the Alaska Highway and the connecting roads in Canada and Alaska, with historical and tourist information about most of the places en route.

Morgan, Lael. *And the Land Provides; Alaska Natives in a Year of Transition*. New York: Doubleday, 1974. Personal impressions of a Fairbanks writer.

Muir, John. *Travels in Alaska*. Boston: Houghton Mifflin, 1915 (AMS Press reprint, 1971). Muir's accounts of his three journeys of exploration in Southeast Alaska, in 1879, 1880, 1890.

Nichols, Jeannette Paddock. *Alaska; A History of Its Administration, Exploitation, and Industrial Development During Its First Half Century Under the Rule of the United States*. Cleveland: Arthur H. Clark, 1924 (Russell & Russell reprint, 1963). An important but incomplete economic history.

Nienhueser, Helen. *55 Ways to the Wilderness in Southcentral Alaska*, 2nd ed. Seattle: The Mountaineers, 1975. Guide to hiking trails plus a few canoe kayak routes.

Organized Boroughs and Cities. Alaska Department of Community and Regional Affairs, January 1, 1975. Map showing boundaries of the twelve boroughs, with populations and dates of incorporation for the boroughs and the three classes of cities.

Permafrost Map of Alaska. United States Geological Survey, 1974. Map I-445.

Phillips, James W. *Alaska-Yukon Place Names*. Seattle: University of Washington Press, 1973. Sources and derivations for a fair number of names, with brief historical notes.

Roberts, Marjorie. *Alaska Earthquake*. Anchorage: Alaska Publications, 1964. Picture book with captions.

Selkregg, Lidia L., ed. *Alaska Regional Profiles*. Juneau: State of Alaska, 1974– . A series of six huge and lavishly produced volumes (when completed), combining the work of many contributors from the University of Alaska and the Joint Federal-State Land Use Planning Commission for Alaska, and describing the state's physical and human geography in the fullest sense (though not infallibly in matters of fact), with an abundance of maps, charts, tables, photographs. I have availed myself particularly of two volumes: *Arctic Region* and *Southcentral Region*.

Sherwood, Morgan, ed. *The Cook Inlet Collection; Two Hundred Years of Selected Alaskan History.* Anchorage: Alaska Northwest Publishing Company, 1974. Brief extracts from documents concerning the Cook Inlet region, from Cook's time to the present.

Surficial Geology of Alaska. United States Geological Survey, 1964. Map.

Van Deusen, Glyndon G. *William Henry Seward.* New York: Oxford University Press, 1967. An unexciting academic biography. Frederick Bancroft's (1900), probably less reliable on facts and documents, is still worth consulting.

Chapter 3. Lifting the Crude

In addition to the Alyeska maps listed below, for details of the pipeline route and its environs I have relied on the 1:250,000 maps (about 4 miles to the inch) in the Alaska Topographic series, issued in the middle 1950s by the United States Geological Survey, most of them revised ten to fifteen years later.

Arco, November–December 1966. A company periodical, this issue devoted to the history (with extracts from documents) of the Atlantic Refining Company.

Blair, John M. *The Control of Oil.* New York: Pantheon, 1976. A conspiracy interpretation of the oil industry, much praised by academic economists, intellectually superior to Sampson's *Seven Sisters* (below) and useful in detail, but, in my view, seriously distorted in its general argument.

Engler, Robert. *The Brotherhood of Oil.* Chicago: University of Chicago Press, 1977. Like the Blair and Sampson books in approach and viewpoint, but shorter.

Facts About Oil. Washington: American Petroleum Institute, undated. Overview of the history, functioning, and organization of the oil industry.

The Humble Way, First Quarter, 1971. Special issue of an employees' magazine devoted to the *Manhattan*'s Northwest Passage (Humble: former name of Exxon, U.S.A.).

Keating, Bern. *The Northwest Passage; From the* Mathew *to the* Manhattan, *1497 to 1969.* Skokie: Rand McNally, 1970. Despite the pretentious title, a picture book with accompanying text.

Petrochemical Flow Charts. Houston: Exxon Chemical Company, U.S.A., undated. Charts in booklet form that attempt to make visual sense of the bewildering numbers of chemicals and synthetic products derived from oil and natural gas.

Sampson, Anthony. *The Seven Sisters; The Great Oil Companies and the World They Made.* New York: Viking, 1975. A British journalist's conspiracy theory of the history of the major international oil companies, culminating in OPEC. Not bad as a brief introduction to the early period but seriously misleading on current oil company economics.

Smith, William. *Northwest Passage; The Historic Voyage of the* S.S. Manhattan. New York: American Heritage (New York *Times*), 1970. Journalism, but more informative than Keating's book on the same subject (above).

Tussing, Arlon R., Rogers, George W., and Fischer, Victor. *Alaska Pipeline Report.* Fairbanks: Institute of Social, Economic and Government Research, 1971. Economic analysis of anticipated effects of TAPS on Alaska.

Chapter 4. Land, Law, and Right

Since the discovery at Prudhoe Bay, various environmentally oriented periodicals have reported continuously on Alaska and the pipeline project. Although they are not specifically cited here, I have followed particularly the reportage and opinion of the magazines of the Wilderness and Audubon societies and on investigation have found much of what they have published on the technical aspects of the pipeline's design and construction and the tanker connection to be seriously misleading; some of the *Audubon* material on the public interest lands, on the other hand, although similarly biased, is more informative and at least useful in defining the issues in the continuing public debate, and one *Audubon* piece, on sealing in the Pribilofs (George Reiger, "Song of the Seal," September 1975), is original and of real historical interest. During the same period the *National Geographic* has produced at least two extended treatments of Alaska and the pipeline that are less stultifying than its usual standard.

An Act To Amend Section 28 of the Mineral Leasing Act of 1920. The Trans-Alaska Pipeline Authorization Act, Public Law 93-153, signed November 16, 1973.

Alaska. Bureau of Land Management, Department of the Interior, March 1974. Map showing the status of all Alaskan lands: state selections under the Alaska Statehood Act, withdrawals for Native villages and regional corporations under the Alaska Native Claims Settlement Act, and the various categories of proposed public interest ("D-2") lands.

Alaska Federation of Natives. *Annual Reports.* I also collected for the pipeline period the annual reports and other materials produced by seven of the twelve existing regional Native corporations: Arctic Slope Regional Corporation, NANA Regional Corporation, DOYON, Ltd., Calista Corporation, Bristol Bay Native Corporation, Cook Inlet Region, Inc., AHTNA, Inc., and relied on periodical sources (among them *Fortune, Alaska Industry,* and Alaskan and other newspapers) for all the corporations as a group.

Alaska Native Claims Settlement Act. United States Statutes at Large, Vol. 85, 1971. Public Law 92-203 (December 18, 1971).

Alaska Native Medical Center, Anchorage. *Annual Reports.*

Alaska Native Selections. Code of Federal Regulations, Part 2650, 38 F.R. 14218, May 30, 1973 (effective July 2, 1973). Administrative procedures published by the Bureau of Land Management, Department of the Interior, by which the provisions of the Alaska Native Claims Settlement Act were put into effect.

Alaska: Withdrawal of Unreserved Lands. Public Land Order 4582, *Federal Register,* Vol. 34, No. 15, January 23, 1969. The text of the final "land freeze" issued by Stewart Udall immediately before leaving office as Secretary of the Interior.

Berry, Mary Clay. *The Alaska Pipeline; The Politics of Oil and the Native Land Claims.* Bloomington: Indiana University Press, 1975. The intricacies of Native organization and congressional maneuver that produced the land claims settlement; journalism, not history, but useful; copiously documented.

Cicchetti, Charles J. *Alaskan Oil; Alternative Routes and Markets.* Washington: Resources for the Future (distributed by Johns Hopkins University Press), 1972. An often abstruse economic analysis that nevertheless raised several of the important questions.

Driver, Harold E. *Indians of North America.* Chicago: University of Chicago Press, 1961. The best general account I have found of the Native cultures of North America, including Alaska and the Canadian Northwest.

Federal Programs and Alaska Natives. Washington: Department of the Interior, 1975. A survey (four volumes) of demography and applicable federal programs, lavishly produced but of very moderate value.

Final Environmental Impact Statement—Proposed Trans-Alaska Pipeline. Washington: Department of the Interior, 1972 (available from the National Technical Information Service, Springfield, Va.). Five volumes plus a sixth of appendexes; a separate and collateral series, *An Analysis of the Economic and Security Aspects of the Trans-Alaska Pipeline*, was published simultaneously, prepared by the Department's Office of Economic Analysis (three volumes) and is available from the same source.

Giddings, J. Louis. *Ancient Men of the Arctic*. New York: Knopf, 1967. A popular account of Eskimo origins in northwest Alaska, based on the author's earlier scholarly work and incorporating his own important archaeological investigations.

Hearing Testimony. Alyeska Pipeline Service Company, undated. Statements by E. L. Patton and others associated with the planning of the pipeline system, extracted from hearings by the Department of the Interior held in Washington and Anchorage in February 1971.

Hopkins, David M., ed. *The Bering Land Bridge*. Palo Alto: Stanford University Press, 1967. Technical papers on the difficult question of Eskimo origins.

Ingstad, Helge. *Nunamiut; Among Alaska's Inland Eskimos*. New York: Norton, 1954. A personal narrative of the small group of caribou hunters around Anaktuvuk Pass, whose main village and migratory principal food source lay directly in the path of the original pipeline route (and the Walter J. Hickel Highway).

Klein, Robert M., et al. *Energy & Mineral Resources of Alaska & the Impact of Federal Land Policies on Their Availability*. Anchorage: Alaska Department of Natural Resources, Division of Geological and Geophysical Surveys, June 1974 (Petroleum Publications, Inc., Anchorage). Text and elaborate overlay maps showing "estimated speculative recoverable" reserves in relation to Native and proposed public interest lands. ("Discoverable reserves," the commoner term, expresses the same idea: statistical probability.)

Manning, Harvey, et al. *Cry Crisis! Rehearsal in Alaska*. San Francisco: Friends of the Earth, 1974. An apocalyptic tract intended to influence the content of the Trans-Alaska Pipeline Authorization Act.

Native Peoples and Languages of Alaska. Alaska Native Language Center, University of Alaska, Fairbanks, undated. Linguistic map of Alaska and the Canadian Northwest, including estimates of numbers of speakers of each language.

Natural Gas Regulation and the Trans-Alaska Pipeline. Washington: U. S. Government Printing Office, 1972. Transcript of the hearings of the 92nd Congress, 2nd Session, held before the Joint Economic Committee of Congress in June 1972.

North Slope Alaska: Man and the Wilderness. Anchorage: BP Alaska, Inc., undated. Public relations pamphlet on the company's environmental efforts in developing its half of the Prudhoe Bay field.

Oil and Natural Gas Pipeline Rights of Way. Washington: U. S. Government Printing Office, 1973. Transcripts (Parts I–III, three volumes) of the hearings of the 93rd Congress, 1st Session, held before the Subcommittee on Public Lands of the House Committee on Interior and Insular Affairs, April, May, and June 1973; includes "prepared testimony" and other documents submitted to the subcommittee.

Oversight of Oil Development Activities in Alaska. Washington: U. S. Government Printing Office, 1969. Transcript of hearings in August 1969 by the Special Subcommittee on Legislative Oversight of the Senate Committee on Interior and Insular Affairs, 91st Congress, 1st Session, chaired by Senator Henry Jackson.

Preparation of Environmental Impact Statements: Guidelines. Council on Environmental Quality. *Federal Register,* Vol. 38, No. 147, August 1, 1973. The Council was created by Executive Order 11514, signed by President Nixon March 5, 1970, and published in the *Federal Register,* Vol. 35, No. 46, March 7, 1970.

Review of Federal Actions Impacting the Environment. Environmental Protection Agency, November 30, 1972. The manual advising federal agencies on procedures for preparing environmental impact statements, produced prior to the administrative regulations on the same subject published in August 1973 by the Council on Environmental Quality (see above).

Sierra Club Archives, Seattle Office. Papers of Brock Evans and his colleagues covering the years of the pipeline debates, in the ms. collection of the University of Washington Library, Seattle (with, however, when I examined them in the summer of 1976, several important items still in the files of the club's Seattle office).

The Status of the Proposed Trans-Alaska Pipeline. Washington: U. S. Government Printing Office, 1969. Parts I and II (two volumes), transcripts of hearings in September and October 1969 by the Senate Committee on Interior and Insular Affairs, 91st Congress, 1st Session, under Senator Jackson; includes the first published draft of the Stipulations prepared by the Department of the Interior under Secretary Walter Hickel. Further material generated by the same committee in November 1969 was published as a separate volume with the title *Trans-Alaska Pipeline Hearings.* The full text of the answers to Russell Train's 79 questions, incorporated in the foregoing, is a letter of June 19, 1969, from R. G. Dulaney, then chairman of the TAPS management committee (president of the Arco Pipeline Company), which I found in the *Sierra Club Archives* (see above).

Vanstone, James W. *Athapaskan Adaptations; Hunters and Fishermen of the Subarctic Forests*. Chicago: Aldine, 1974. Introduction to the anthropology of the main Indian group of the North, more pertinent, despite the sweeping title, to the Slave Lake area than to Alaska.

Webster, Donald H., and Zibell, Wilfried. *Iñupiat Eskimo Dictionary*. Fairbanks: Summer Institute of Linguistics, 1970. The language as spoken in northwest Alaska, with puzzling contradictions to what little I know of Inuit speech elsewhere in the Arctic, which I have been unable to reconcile. Besides the dictionary, the Institute has produced instructional materials for adults and children and translations of the Bible (allied with the Wycliffe Bible Translators) in Inupiat, Yupik, Tlingit, and several Athabascan languages.

Wilderness Society et al. vs. *Secretary of the Interior*. I have studied particularly the following documents in the pipeline case: Judge Hart's *Preliminary Injunction* of April 23, 1970, filed in the United States District Court for the District of Columbia (Civil Action No. 928–70); *Plaintiffs' Brief on National Environmental Policy Act Issues* filed July 17, 1972 (same court, same civil action); *Reply Brief of the Plaintiffs-Appellants* filed September 13, 1972, in the United States Court of Appeals for the District of Columbia (No. 72–1797); the opinion and ruling of the Court of Appeals in the same case as published in 479 *Federal Reporter*, 2nd Series (pp. 842–912). In the separate question of the Bill of Costs arising from this case, I have studied particularly the Bill itself and its accompanying briefs as submitted to the United States Court of Appeals February 23, 1973; the *Ruling and Dissents of the Court of Appeals* dated April 4, 1974, as printed by the plaintiffs for purpose of appeal; and the *Ruling, Opinion, and Dissents* by the Supreme Court dated May 12, 1975, reversing the Court of Appeals ruling. Copies of most of these documents were made available to me by Dennis M. Flannery, Esquire, the plaintiffs' principal attorney, whose assistance in following the case's legal ramifications is gratefully acknowledged.

Chapter 5. The Build-up

Here and in the following chapters centered on the design and construction of the pipeline system, I necessarily drew on the substantial body of materials produced by the Alyeska public affairs department. Among these, in addition to the items listed separately here and elsewhere, were press kits put together annually or seasonally; press releases distributed every week or two, detailing construction progress and, occasionally, responding to criticism; *Alyeska Reports*, a glossy quarterly published from January 1975 to October 1977, by definition favorable to Alyeska and its owners but professionally written, illustrated, and produced, and often informative on particular aspects of the project; *The Camp Follower*, a weekly distributed to pipeline workers; miscellaneous maps and other materials intended for press use. Although from early fall of 1975 on I was supposedly on an Alyeska mailing list to receive everything the company put out, distribution was in fact quite selective, as I discovered each time I visited the Alyeska offices in Anchorage and Fairbanks; among the important items supplied through the unofficial generosity of one of the Alyeska PR men was a series of dated "data sheets" reproducing the engineering drawings for the main design features of the system.

Cameron, Raymond D., and Whitten, Kenneth P. *First Interim Report of the Effects of the Trans-Alaska Pipeline on Caribou Movements*. Anchorage: Joint State/Federal Fish & Wildlife Advisory Team, 1976.

Gavin, Angus. *Wildlife of the North Slope; A Five Year Study, 1969–1973*. Atlantic Richfield Company, undated. A lavishly illustrated report of surveys made by a respectable Canadian ecologist employed as an Arco consultant.

Summary Project Description of the Trans Alaska Pipeline System, September 1975. Alyeska Pipeline Service Company. Text and diagrams describing the basic designs for buried and elevated pipe, pump stations, terminal.

Trans-Alaska Report. Reprinted from *Oil & Gas Journal*, March 18, 1974. Technical description of the pipeline, terminal, and Prudhoe production facilities as then designed.

Van Ballenberghe, Victor. *First Interim Report of the Moose-Pipeline Technical Evaluation Study*. Anchorage: Joint State/Federal Fish & Wildlife Advisory Team, 1976.

Chapter 6. On the Line

Pipeline Department Control Manual. Anchorage: Alyeska Pipeline Service Company. General procedures required of Alyeska contractors, the final revision of which was issued in March 1976.

Trans-Alaska Oil Pipeline—Progress of Construction Through Novem-

ber 1975. A report to Congress by the Comptroller General, dated
February 17, 1976.

Trans Alaska Pipeline Atlas. Anchorage: Department of the Interior,
February 1972. Twenty-five maps, based on existing USGS maps,
showing the pipeline and its features as then planned, with com-
ments on engineering and environmental difficulties by the Department of the Interior. The maps were actually prepared by an
Alyeska contractor, Michael Baker, Jr., and the atlas was produced
by Alyeska.

Trans Alaska Pipeline Route Maps. Alyeska Pipeline Service Company, undated. Large-scale but schematic maps showing the main
pipelines features (pump stations, camps) and important topographic detail.

Chapter 7. Signing On

In addition to the sources listed, I have drawn on the series on the
Teamsters published in the Anchorage *Daily News* in December 1975
and January 1976, signed by Howard Weaver, Bob Porterfield, and
Jim Babb. This was the series for which the newspaper received a
Pulitzer prize, and it was afterward reprinted in tabloid form under
the heading "Empire; The Alaska Teamsters Story"; that is the form in
which I have it.

*Agreement Between Alaska Chapter Associated General Contractors
of America, Inc., and International Brotherhood of Teamsters,
Chauffeurs, Warehousemen and Helpers of America, Local 959.* The
labor contract, signed June 30, 1974, and effective till June 30, 1977,
which determined Teamsters wage scales for the duration of the
pipeline project (and elsewhere in Alaska), as well as work rules
not expressly modified by the *Project Agreement* (below).

Hazard Control Manual: Pipeline Operations. Morrison-Knudsen
Company. Workers' pocket pamphlet; similar materials were produced by the other TAPS execution contractors, apparently as a
contractual requirement of Alyeska's safety program.

*Rules of Conduct & Rules and Regulations for the Construction
Camp.* Alyeska Pipeline Service Company. Pocket pamphlet distributed to new hires.

Trans-Alaska Pipeline System Project Agreement. The labor contract
between Alyeska and most of the international and local unions involved in the pipeline, effective April 29, 1974 (though not signed
by all participants until at least a week later).

Chapter 8. The Watch on the Line

Agreement and Grant of Right-of-Way for Trans-Alaska Pipeline. The
printed version of the contract between the TAPS owners and the
Department of the Interior, which served as a manual for all surveillance officers; includes the final Stipulations and the agreement
between the state of Alaska and the Department of the Interior.
Furnished to me by the APO, but since it contains *no* identification
I am unable to cite a source.

The Alaska Historic Preservation Act. Alaska Statutes 41.35, published in booklet form by the Division of Parks of the Alaska Department of Natural Resources, along with the corresponding section
(11.16) of the *Alaska Administrative Code;* the legal basis for the
archaeological surveillance incorporated in the Stipulations.

Alaska Pipeline Commission. *Annual Report,* 1975. The APC also produced (1976) a public relations booklet, *The Alaska Pipeline Commission,* describing its activities.

Alaska Pipeline Commission Act. Alaska Statutes, 42.06. The 1972
state law creating the APC.

Campbell, John M. *Archeological Studies Along the Proposed Trans-
Alaska Oil Pipeline Route.* Washington: Arctic Institute of North
America, 1973. Work continued along the right-of-way through
1975, ahead of construction: by archaeologists of the state Department of Natural Resources, Doug Reger and Chuck Rutterman,
among others; and by two professors of the University of Alaska
funded by Alyeska, William Workman (Anchorage) and John
Cook (Fairbanks). I interviewed all but the latter, but the Alyeska-
sponsored reports were either unfinished or unobtainable.

Campbell, John M., and Cummings, Ellen M. *A Field Guide to the
Recognition of Archaeological Materials Along the Trans-Alaska
Pipeline.* Anchorage: Department of the Interior, 1975. Commissioned by a contractor of the Alaska Pipeline Office to satisfy
requirements expressed in the Right-of-Way Agreement and Stipulations (and paid for, presumably, by Alyeska) but not to my
knowledge much used.

Champion, Charles A. State of Alaska, Office of the Pipeline Coordinator: *Annual Report.* Mr. Champion prepared three reports covering the activities of his office in the calendar years 1974, 1975, and
1976. In addition, in January 1977, he produced an 18-page summary of recommendations for future state oversight, *The Trans
Alaska Pipeline: Government Monitoring and the Public Interest.*

Federal Surveillance Report, Trans-Alaska Pipeline System. Anchorage: Alaska Pipeline Office, January 1976. An uninformative public

relations booklet produced by the APO's contractor Mechanics Research, Inc.; an experiment not repeated.

Chapter 9. Impacts

Alaska Labor & Management Employee Affairs, Inc. *Activities Reports*. Summaries by this alcoholism counseling service covering intervals of from three to six months during the period of pipeline construction.

Alcohol and Health: New Knowledge. Washington: U. S. Government Printing Office, 1974. A report to Congress by the Department of Health, Education and Welfare; includes estimates by states of apparent per capita consumption in the form of equivalents in pure alcohol of distilled spirits, wine, and beer; the per capita figures are based on the population aged fifteen and over, realistically, perhaps, but not therefore comparable with other studies based on the legal drinking age (eighteen, twenty, or twenty-one).

Allocation of Adult Alcoholics in Alaska by Geographic, Sex & Racial Indices. Juneau: Office of Alcoholism, Alaska Department of Health and Social Services, July 1973. A comprehensive study of alcoholism, applying the Jellinek formula and making possible many kinds of comparisons of incidence.

Efron, Vera, Keller, Mark, and Gurioli, Carol. *Statistics on Consumption of Alcohol and Alcoholism*. New Brunswick: Publications Division, Rutgers Center for Alcohol Studies, annual. Generally about four years out of date; includes estimates of apparent consumption and of alcoholics, broken down by states, but from other sources the estimates for Alaska appear to be entirely too low. The Bureau of Alcohol, Tobacco and Firearms (U. S. Treasury Department) publishes annual *Summary Statistics* that are of some help in estimating apparent consumption nationally and by states; *Newsweek* magazine, as a service to advertisers, has since the late 1960s produced annual summaries of apparent consumption of distilled spirits by states—convenient, complete, up to date, but omitting the equivalent consumption of wine and beer.

Fairbanks North Star Borough Impact Information Center. *Reports*. A series of 34 reports prepared between July 1974 and February 1977, together with occasional special reports on aspects of the pipeline and its social effects, and a group of 13 summarizing final reports produced in the spring and summer of 1977.

Forecast of Alaska's Economy. Anchorage: National Bank of Alaska, annual. Review of the past year, projection of the next, prepared by the bank's economists, supplemented by an occasional *Midyear Re-*

view. I have studied these documents for the pipeline years, along with the bank's own annual reports.

Occupational Injuries in the United States, by Industry. Washington: Bureau of Labor Statistics, U. S. Department of Labor, annual. Statistical summary, with breakdowns by states as well as industries, generally about two years behind the years reported. For the pipeline I have also drawn on reports compiled by Robert Larson, manager of Alyeska's safety department (including voluminous written answers to specific questions) and summaries of work-related injuries and deaths from the Workmen's Compensation Division of the Alaska Department of Labor, based on reports from Alyeska; and two independent compilations of project deaths as well as my own estimates from reported deaths.

Performance Report on the Alaskan Economy. Juneau: Alaska Department of Commerce and Economic Development, annual and semiannual. Generally includes more detail on state revenues and expenditures than the economic studies of the National Bank of Alaska (see above); some variants in titles.

Peterson, W. Jack. *Attitudes Toward Alcohol, Drinking and Alcohol Abuse in Alaska.* Anchorage: The Center for Alcohol Addiction and Drug Abuse, University of Alaska, January 1975. Report on an elaborate questionnaire completed by a sampling of inhabitants of representative communities throughout Alaska.

Pipeline Yearly Report. Alaska State Troopers, annual during the pipeline period. I have cited the reports for 1975 and 1976, listing all police activities connected with the pipeline or along its route and including reports on thefts submitted by the two Alyeska security contractors. My general crime statistics are those compiled by the state's Department of Public Safety, supplemented by the records of the Anchorage and Fairbanks police departments.

Statewide Drug Report, November 1974–December 1975. Alaska State Troopers. Actually includes arrests and case dispositions, by localities, for 1973–75, which I supplemented with earlier and later data for Anchorage provided by the Metro Division, a co-operative body of the city and state police.

Vital Statistics of the United States. Health Resources Administration, Department of Health, Education, and Welfare, annual. The tables from Vol. II, Section I, *General Mortality,* available separately, give causes of death by states and regions, with numbers of deaths from each cause and death rates per 100,000 of population.

Chapter 10. Murphy's Law

Some further material bearing on the management of the project as well as its cost is listed under Chapter 11.

Alyeska Pipeline—Preliminary Report. September 8, 1976. Prepared for chairman John Dingell of the House Subcommittee on Energy and Power (Committee on Interstate and Foreign Commerce) by the subcommittee staff; I have amplified a few points from conversations with the two investigators who wrote it.

Alyeska Pipeline Service Company's Welding/Radiography Problems and Solutions. Anchorage: Alaska Pipeline Office, June 1977. Three volumes; I have referred particularly to the summary Volume III.

Oil Spill Contingency Response Plan. Anchorage: Alyeska Pipeline Service Company, 1976. Five volumes: General Provisions plus separate treatment for the Prudhoe Bay area, Valdez, and Prince William Sound, prepared by Woodward-Clyde Consultants; an important component of the *Operating Manual* (next item).

Operating Manual for the Trans-Alaska Pipeline System. Anchorage: Alyeska Pipeline Service Company, 1976. Ten volumes but several are subdivided: includes the *Port Information Manual for Valdez*, eleven manuals of surveillance procedures in two volumes, the *Oil Spill Contingency Response Plan* (see above, seven manuals in five volumes), and general procedures.

Start-Up Presentation/T.A.P.S. Two-volume summary of doubtful cases with documents showing their resolution according to the Stipulations, a joint effort (apparently) by Alyeska and the Alaska Pipeline Office in the spring of 1977.

Trans-Alaska Pipeline—Special Study; An Overview Study with Respect to Effectiveness of the Stipulations During the Construction Phase. Anchorage: Alaska Pipeline Office, Commissioned by the APO to its subcontractor Mechanics Research, Inc., and actually prepared by a Toronto research organization in June and August 1977. Particular attention to Alyeska's managerial organization and effectiveness and the relation of its organization to that of the APO.

Welding on the Trans Alaska Pipeline. Alyeska Pipeline Service Company, September 1976. A press kit, including a chronology of the problem and definitions of most technical terms; similar material in an issue of *Alyeska Reports* published at the same time with the title *Special Report: Welding/Radiography*.

"*The White Book*" (untitled). Alyeska Pipeline Service Company.

The confidential report on welding and radiography produced in the spring of 1976.

Chapter 11. The Price of the Pipe

Besides the materials having to do with costs, I made some use of the press kit provided for the January 1976 Arctic training exercise organized by USREDCOM, in which I participated as an observer, Operation Jack Frost (mentioned in passing in this chapter); its theme was the defense of the pipeline.

Alaska Administrative Code. The general regulations governing the development of Prudhoe Bay (and other oil and gas fields) are given in Title 11, Part 3, Chapter 22, "Oil and Gas Conservation," deriving its authority from AS 31.05; the regulations are administered by the Division of Oil and Gas of the Department of Natural Resources.

Alaska Statutes. The oil tax laws cited are: *Gross Production Tax,* AS 43.55, effective January 1, 1974; *Conservation Tax,* AS 43.57, same date; *Property Taxes,* AS 43.56, same date; *Ad Valorem Tax,* AS 43.58, effective January 1, 1976–December 31, 1977.

Atlantic Richfield Company. *Annual Reports* and *Supplements,* 1974–76. Also, *Annual Report to the Securities and Exchange Commission for 1974,* required prior to the company's offer of stock and debentures for early financing of the pipeline; includes, besides financial information since 1970, some material on the company's oil reserves at Prudhoe Bay and elsewhere, outstanding debts, and a list of subsidiaries.

Chase Manhattan Bank, Energy Economics Division. Annual analyses, in booklet form, of the international petroleum industry, capital investment and financial performance of selected companies as a group. Titles vary; among recent issues I have found helpful are *Capital Investments of the World Petroleum Industry* and *Financial Analysis of a Group of Petroleum Companies.*

Dismembering the Oil Companies. American Petroleum Institute, undated (1976). Pamphlet opposing divestiture.

Domestic Petroleum Industry Capital Needs and Availability 1975–1985. Cleveland: Standard Oil Company (Ohio), April 14, 1975. Prepared by the company's economic research and public information departments.

Exxon Corporation. *Annual Reports,* 1974–76. Also, *This Is Exxon* (July 1975), a public relations booklet that includes useful maps showing the company's centers of operations throughout the world.

The Impact on U.S. Energy Self-Reliance of Various Crude Oil Price Proposals. Cleveland: Standard Oil Company (Ohio), undated (1976).

Interstate Commerce Commission, Docket No. 36611 1977– . The record of evidence submitted by all parties to the TAPS tariff proceedings, including complaints, replies, ICC orders, and the transcript of the hearing on June 27, 1977, which led to the ruling of July 11, 1977. The material when I examined it in late July comprised three large volumes (Volume II consisting of the transcript), of which I have made particular note of the following: regulations and proposed tariffs of the eight TAPS owners; briefs supporting the proposed tariffs submitted at the request of the ICC by BP Pipelines Inc. (June 15, 1977), Exxon Pipeline Company (June 22, 1977), Phillips Alaska Pipeline Corp. (July 20, 1977), Sohio Pipe Line Company (June 13, 1977), Union Alaska Pipeline Company (June 14, 1977); and Order No. 36611 of the Interstate Commerce Commission, dated July 11, 1977, including its appendices summarizing the owners' financial evidence and the Commission's analysis of the effects of its proposed tariffs. Some of the briefs missing from the docket when I examined it I was able to obtain from the companies, but not all.

Johnson, William, and Messick, Richard E. *Competition in the Oil Industry*. Unpublished economic study arguing against divestiture produced within the Energy Policy Research Project of George Washington University; summarized in the authors' testimony in January 1976 before the Subcommittee on Antitrust and Monopoly of the Senate Committee on the Judiciary.

Lenzner, Terry F. *Preliminary Statement*. This report on Alyeska's management and costs, dated June 15, 1977, was prepared for the Alaska Pipeline Commission and submitted, in turn, as a brief to the Interstate Commerce Commission in its consideration of the pipeline tariffs.

––––. *The Management, Planning and Construction of the Trans-Alaska Pipeline System; Report to the Alaska Pipeline Commission.* The complete version of the preceding, dated August 1, 1977.

Oil Profits: How Much? Where Do They Go? Cleveland: Standard Oil Company (Ohio), August 18, 1975.

Operations Center. BP Alaska, undated. An illustrated booklet describing the BP base camp at Prudhoe Bay.

Prudhoe Bay. Alaska Map Service AS-11 D. Large-scale map (1:96,000, 1 inch=8,000 feet) showing lease ownerships, drill sites and wells, other features of development, including the boundaries of the Prudhoe Bay Unit. Periodically revised; I have studied and compared those dated August 7, 1974, and August 1, 1977.

Standard Oil Company (Ohio). *Annual Reports, 1974–76.* Also, the *Prospectus of the Sohio Pipe Line Company* produced for its April 27, 1976, offering of debentures (about $250 million), to finance the pipeline; includes indebtedness and other financial details to that point, among them the complicated relationship with BP.

Tanzer, Michael. *Alaska's Prudhoe Bay Oil: Profitability and Taxation Potential.* The report commissioned by the state legislature, dated January 9, 1976. Among the legislative testimony elicited by this document and the tax proposals that followed, I have had reference, in addition to news reports, to transcripts of the statements of H. H. Goerner, senior vice-president, Exxon, U.S.A.; Atlantic Richfield Company (unattributed, but prepared under the direction of William Leake, manager, Alaska co-ordination); Richard M. Donaldson, vice-president, Standard Oil Company (Ohio); Charles E. Spahr, Chairman, Standard Oil Company (Ohio).

Unit Agreement, Prudhoe Bay Unit, State of Alaska. The agreement dated April 1, 1977, between the state and the unit participants; a supplemental volume of exhibits includes an exact list of tract owners and maps showing the tracts included in the unit and the approximate boundaries of the oil and gas reservoirs.

Unit Operating Agreement, Prudhoe Bay. The agreement among the unit participants alone (same date as preceding); since it is regarded as "proprietary," I have been able to obtain only fragments of the complete document.

Chapter 12. Across the Wide Pacific

Among the important questions treated here (to which I return briefly in my Afterword) is that of the markets to which Prudhoe's oil was destined, a question complicated by the related ones of West Coast production and demand estimates, refinery capacity, tanker availability, and the characteristics of Prudhoe Bay oil. All these subsidiary aspects of the main question were treated early and accurately by Sally Jones in a series of six articles in the Anchorage *Daily News* in November 1975 (with two follow-ups in March 1976). Although the series was overshadowed by the Pulitzer prize series on the Teamsters in the *Daily News* at about the same time, it is work, in my view, of the same caliber.

An Act Relating to Oil Terminal Facilities and the Marine Transportation of Crude Oil. Alaska Statutes 30.20. The bill introduced by Senators Croft, Kerttula, and Poland, effective July 1, 1977, governing the design of tankers operating in Alaskan waters. I have also examined the testimony of the oil company tanker men on this bill, among them Frank Mosier of Sohio and A. D. Mookhoek of Exxon.

Environmental Defense Fund et al. vs. United States Coast Guard. Suit brought by the Center for Law and Social Policy to require that double bottoms be specified in the pending tanker regulations; I have studied particularly the Center's briefs dated August 19, 1974, and November 12, 1975, copies of which were kindly furnished me by Eldon Greenberg, who prepared them.

Final Environmental Impact Statement—Regulations for Tank Vessels Engaged in the Carriage of Oil in Domestic Trade. Washington: U. S. Coast Guard, 1976.

Large Tankers—Our Energy Lifelines. Washington: American Petroleum Institute, undated (1975?). Pamphlet with some useful diagrams and statistics.

Merchant Marine Act, 1920. United States Code, Chapter 24, Section 46:883. The Jones Act.

Mostert, Noel. *Supership.* New York: Knopf, 1975. A provocative introduction to VLCCs, built on a narrative of a voyage on a British tanker; but to be taken cautiously on many matters of fact concerning this technical subject.

Oil Transportation by Tankers: An Analysis of Marine Pollution and Safety Measures. Washington: U. S. Government Printing Office, 1975. A study by the Office of Technology Assessment of the U. S. Congress.

Operating Manual, Vessel Traffic Service, Prince William Sound. Juneau: U. S. Coast Guard, July 1977.

Prince William Sound. National Oceanic and Atmospheric Administration, 16700. The current (1977) marine chart for the area.

Rules and Regulations for Tank Vessels. Washington: U. S. Coast Guard, January 1, 1973. The latest version in effect (early 1978) but in process of revision.

Tanker Double Bottoms: Yes or No? Washington: American Institute of Merchant Shipping, 1974. Pamphlet.

Very Large Crude Carriers (VLCCs). New York: Exxon Corporation, 1975. Pamphlet in the Exxon Background Series.

The Next Time: An Afterword

Although energy alternatives are subordinate to my main subject, their public discussion has formed part of the context in which the

TAPS project was carried out, and I have tried at least to keep track. Besides the materials on the gas-line proposals described below, the reader is referred to the more general works listed at the outset of these Notes.

Alaskan Arctic Gas Pipeline Co. *Press Kits.* A series of elaborate presentations by this U.S.-Canadian consortium of its plan for a gas line routed east from Prudhoe Bay through the Arctic National Wildlife Range, then south through the Mackenzie valley to the Montana border; it was to have carried gas from the Mackenzie delta as well as the larger Prudhoe supply.

El Paso Alaska Company. *Press Kit.* Presentation of the pipeline plan that competed with that of Arctic Gas: the "all-American" route south from Prudhoe Bay to Cordova, where the gas was to be chilled to liquid form (LNG) and carried to U.S. markets by LNG tankers.

Energy Outlook 1975–1990. Exxon Company, U.S.A., periodically revised. Pamphlet projecting U.S. demand by sectors and supply by sources.

Oil Shale. Grand Junction, Colorado: *The Daily Sentinel,* 1974. A review of Colorado oil shale projects.

Report on the Third On-Site Visit to the Trans Alaska Pipeline System, July 1976. Washington: The Arctic Environmental Council (distributed by the Arctic Institute of North America). The Council also produced reports of its members' visits in 1974 and 1975, which I have not seen.

Summary of the Trans-Alaska Oil Pipeline System Critique Session Held August 18–19, 1977, University of Alaska, Anchorage. Washington: Department of the Interior, October 27, 1977.

INDEX

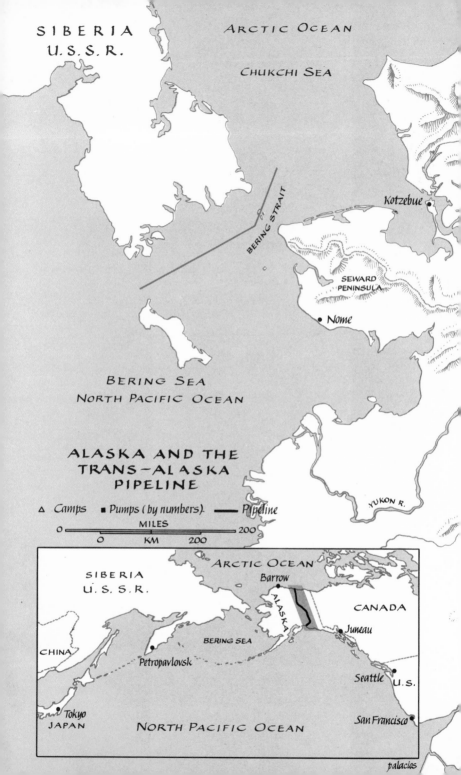

SIBERIA
U.S.S.R.

ARCTIC OCEAN

CHUKCHI SEA

Kotzebue

BERING STRAIT

SEWARD
PENINSULA

Nome

BERING SEA
NORTH PACIFIC OCEAN

ALASKA AND THE
TRANS–ALASKA
PIPELINE

△ Camps   ▪ Pumps (by numbers).   —— Pipeline

0   MILES   200

0   KM   200

SIBERIA
U. S. S. R.

ARCTIC OCEAN

Barrow

ALASKA

CANADA

Juneau

CHINA

BERING SEA

Petropavlovsk

Seattle   U.S.

Tokyo
JAPAN

San Francisco

NORTH PACIFIC OCEAN

palacios